Strings, Branes and Dualities

NATO ASI Series

Advanced Science Institute Series

A Series presenting the results of activities sponsored by the NATO Science Committee, which aims at the dissemination of advanced scientific and technological knowledge, with a view to strengthening links between scientific communities.

The Series is published by an international board of publishers in conjunction with the NATO Scientific Affairs Division

A Life Sciences **B Physics**	Plenum Publishing Corporation London and New York
C Mathematical and Physical Sciences **D Behavioural and Social Sciences** **E Applied Sciences**	Kluwer Academic Publishers Dordrecht, Boston and London
F Computer and Systems Sciences **G Ecological Sciences** **H Cell Biology** **I Global Environment Change**	Springer-Verlag Berlin, Heidelberg, New York, London, Paris and Tokyo

PARTNERSHIP SUB-SERIES

1. Disarmament Technologies	Kluwer Academic Publishers
2. Environment	Springer-Verlag / Kluwer Academic Publishers
3. High Technology	Kluwer Academic Publishers
4. Science and Technology Policy	Kluwer Academic Publishers
5. Computer Networking	Kluwer Academic Publishers

The Partnership Sub-Series incorporates activities undertaken in collaboration with NATO's Cooperation Partners, the countries of the CIS and Central and Eastern Europe, in Priority Areas of concern to those countries.

NATO-PCO-DATA BASE

The electronic index to the NATO ASI Series provides full bibliographical references (with keywords and/or abstracts) to about 50,000 contributions from international scientists published in all sections of the NATO ASI Series. Access to the NATO-PCO-DATA BASE is possible via a CD-ROM "NATO Science and Technology Disk" with user-friendly retrieval software in English, French, and German (©WTV GmbH and DATAWARE Technologies, Inc. 1989). The CD-ROM contains the AGARD Aerospace Database.

The CD-ROM can be ordered through any member of the Board of Publishers or through NATO-PCO, Overijse, Belgium.

Series C: Mathematical and Physical Sciences – Vol. 520

Strings, Branes and Dualities

edited by

Laurent Baulieu
CNRS and Université Pierre et Marie Curie (Paris VI),
Paris, France

Philippe Di Francesco
University of North Carolina at Chapel Hill,
Chapel Hill, NC, U.S.A.

Michael Douglas
Rutgers University,
Piscataway, NJ, U.S.A.
and Institut des Hautes Études Scientifiques,
Bures-sur-Yvette, France

Vladimir Kazakov
Université Pierre et Marie Curie (Paris VI),
and École Normale Supérieure,
Paris, France

Marco Picco
CNRS and Université Pierre et Marie Curie (Paris VI),
Paris, France

and

Paul Windey
Université Pierre et Marie Curie (Paris VI),
Paris, France

Kluwer Academic Publishers

Dordrecht / Boston / London

Published in cooperation with NATO Scientific Affairs Division

Proceedings of the NATO Advanced Study Institute on
Strings, Branes and Dualities
Cargèse, France
May 26–June 14, 1997

A C.I.P. Catalogue record for this book is available from the Library of Congress.

ISBN 0-7923-5344-7

Published by Kluwer Academic Publishers,
P.O. Box 17, 3300 AA Dordrecht, The Netherlands.

Sold and distributed in North, Central and South America
by Kluwer Academic Publishers,
101 Philip Drive, Norwell, MA 02061, U.S.A.

In all other countries, sold and distributed
by Kluwer Academic Publishers,
P.O. Box 322, 3300 AH Dordrecht, The Netherlands.

Printed on acid-free paper

Table of Contents

Preface

The Cargèse Summer School "Strings, Branes and Dualities" was held from May 26th till June 14th 1997.

This session focused on the recent progress in supersymmetric quantum field theory and in string theory. As the developments of the last few years amply demonstrate, these two areas of research are intimately related and often advances in one shed light on the other. The organizing ideas of most of these advances have been the notion of duality and the physics of higher dimensional objects or p-branes.

The school featured a large set of detailed reviews, topical seminars and a very successful "Gong Show" were participants made short presentation of their current work. This volume offers in Part I a very complete written account of the material presented during the lectures, often in greatly extended versions. Part II is devoted to the work presented during the seminars, most of them developing some of the themes presented in the lecture while Part III contains very short accounts of the "Gong Show" presentations.

The topics covered include duality in field theory, in particular in supersymmetric field theory and in supergravity, and in string theory. The theory of Seiberg-Witten and its recent developments are also covered in detail. A large fraction of this volume is devoted to the current state of the art in M-theory, in particular its underlying superalgebra as well as its connection with superstrings and $N = 2$ strings. The physics of D-branes and its essential rôle in the beautiful computation of the black hole entropy is also carefully covered. Finally the last two sets of lectures are devoted to the exciting matrix approach to non-perturbative string theory.

Some contributions constitute a very thorough introduction to their subject matter and should be helpful to advanced students in the field while others are geared to the specialist.

We are greatly indebted to the NATO Division for Scientific Affairs for funding and for their constant attention to our meeting, to the European Commission for a Summer School Grant ERBFMMACT970188, to the *Bureau de la formation permanente du Centre national de la recherche scientifique* for financial support and to the *Ministère des affaires étrangères*. It is also a pleasure to thank all the people who contributed to the successful organization of the school and in particular the staff of the Cargèse Institute and its director Mme. Dubois-Violette.

Finally, we would like to thank all the participants for creating an excellent working atmosphere and especially the contributors of this volume who wrote wonderful sets of lectures.

Laurent Baulieu
Philippe Di Francesco
Michael Douglas
Vladimir Kazakov
Marco Picco
Paul Windey

Part I

Lectures

EXACT ELECTROMAGNETIC DUALITY

Introductory Lectures

DAVID I OLIVE
Dept of Physics
University of Wales Swansea
Swansea SA2 8PP, UK

1. Introduction

Electromagnetic duality is an idea with a long pedigree that addresses a number of old questions, for example:-

- Why does space-time possess four dimensions?
- Why is electric charge quantised?
- What is the origin of mass?
- What is the internal structure of the elementary particles?
- How are quarks confined?

During the last forty years our understanding of quantum field theory has advanced dramatically on a number of fronts that, at first sight, appeared unrelated:-

- The theory of solitons
- The idea of integrable quantum field theories as deformations of conformally invariant QFTs.
- Unified gauge theories with Higgs
- Supersymmetry
- Instanton theory

and so on. As I shall explain, the old idea of electromagnetic duality can be considerably enhanced in the light of these developments. The bonus is a compelling framework of ideas within which these apparently disparate developments become much more synthesised. This new vantage point on elementary particle theory then provides a springboard for extensions of the basic ideas to more ambitious theories such as supergravity and the various versions of string theory. This is what the other lecturers will present.

Let me begin by elaborating three of the themes that will ultimately be woven together.

Theme I - Quantum excitations and solitons

Because the principles of special relativity and quantum mechanics are so soundly based in the physics of the present century, any theory of particle physics must combine the two in a seamless way. The most promising approach involves

L. Baulieu et al. (eds.), Strings, Branes and Dualities, 3–31.

relativistic quantum field theory, even though, after more than sixty years of development it has to be admitted that it remains not totally satisfactory. Traditionally particles have appeared as the quantum excitations of the fields entering the equations of motion. However there is another, conceptually distinct notion of how a particle can appear in a field theory. This is as a soliton, originally a concept in hydrodynamic theory. This is a classical solution to the field equations possible when they exhibit a particular sort of intense nonlinearity. When stationary, its energy concentrates in a small region of space and does not dissipate as time evolves. It can be boosted to move at any speed slower than light. Its stability is often associated with the fact that it carries some knottedness property that cannot be unravelled. All this enables it to be interpreted as a particle with structure.

Is it possible that the distinction between these two apparently different notions of particle phenomena dissolve when the theory is fully quantised? Skyrme [1961] was probably the first to ask this question and he found that in a certain model, in a space-time of two dimensions, where the question was tractable, that the distinction did indeed disappear. He and Coleman [1975] showed that the soliton solution of the sine-Gordon equation describing a single scalar field could be considered equivalently as the quantum fluctuation of what is known as the massive Thirring field. Indeed the two actions, sine-Gordon and massive Thirring, are quantum equivalent. This was established by a version of what is now known as the vertex operator construction (Mandelstam [1975]).

Theme II - Unified Theories and Conformal Symmetry

With the increasing flow of data from high energy particle accelerators, so has grown the urge to construct unified theories that codify the observed patterns of particle behaviour in a simple mathematical structure. Happily the most successful theories have been based on principles of an aesthetic and geometrical nature. Examples are the Yang-Mills gauge theories generalising Maxwell's equations in a non-linear way, and Einstein's theory of general relativity, relating gravitation to the Riemannian geometry of space-time. These theories have two distinctive features, non-linearity and conformal symmetry. Unifying them are the string theories in their various guises and these also display a conformal symmetry, this time on the world sheet. Conformal transformations in space-time are those that preserve angles but not necessarily distances. Hence a conformally invariant field theory will exclude particle masses. This is perfectly acceptable for the photon, the quantum of the Maxwell field, but not for the more general Yang-Mills equations as the intermediate bosons W^+ and W^- are known to be massive. Therefore a mechanism must be found that destroys the conformal symmetry without corrupting too much of the aesthetic attraction of the theory.

Two possibilities are known, one now more than thirty years old, the Higgs mechanism whereby it is the vacuum, rather than the equations of motion, that fail to respect all the gauge symmetry. As a consequence certain scalar fields, called Higgs fields, may fail to vanish there, and as a consequence produce masses for those gauge particles associated with generators of the gauge group not annihilating the vacuum. A second type of mechanism, due to Zamololodchikov [1989] came to light ten years ago. This insight came from the study of second order phase transitions in two dimensions in terms of models that possessed certain integrability properties. It was realised that the integrability was a relic of the conformal symmetry that held at the critical temperature. This can happen because the integrable theory is a special kind of deformation of the conformal theory. The sinh-Gordon theory illustrates this phenomenon. Having broken the sinh term into its two constituent exponentials we can multiply one of these pieces by a real positive coefficient which can be absorbed by a redefinition of the scalar field and the mass parameter. However, when this coefficient vanishes, the equation changes radically, collapsing to the Liouville equation, so named because Liouville was the first to solve it, exploiting its conformal symmetry. Reversing this process, we can say that the sinh-Gordon equation is an integrable deformation of the Liouville equation. Making the coupling in sinh-Gordon imaginary produces the sine-Gordon theory already mentioned.

Does this procedure which works so well in two dimensions have an analogue in a space-time of four dimensions? Can it then be applied to unified gauge theories? If so, what is its relation to the Higgs mechanism?

Theme III - Duality symmetry of Maxwell theory

The basic idea is surely very old; simply that there is a tantalizing similarity between the electric and magnetic fields $\underline{E}$ and $\underline{B}$. This resemblance was made more precise once Maxwell's equations governing their behaviour were established. In vacuo these equations display several symmetries of physical importance, Poincaré rather than Galilean space-time symmetry, and beyond that, conformal symmetry (with respect to transformations preserving angles and not just lengths). Even more sensitive to the precise nature of the Minkowski space-time metric is the electromagnetic duality rotation symmetry of these equations

$$\underline{E}+i\underline{B} \to e^{i\phi}(\underline{E}+i\underline{B}). \tag{1.1}$$

It is the extension of this fascinating symmetry that is the guiding theme in what follows. Notice that we can form two real, quadratic expressions invariant with respect to (1.1);

$$\frac{1}{2}|\underline{E}+i\underline{B}|^2 = \frac{1}{2}(E^2+B^2), \tag{1.2a}$$

$$\frac{1}{2i}(\underline{E}+i\underline{B})^* \wedge (\underline{E}+i\underline{B}) = \underline{E}\wedge\underline{B}. \tag{1.2b}$$

These are both physically observable quantities, respectively recognised as the energy and momentum densities of the electromagnetic field.

It is natural to try to extend the duality rotation symmetry (1.1) to include matter carrying charges and, beyond that, to the non-abelian gauge theories of the type already mentioned as being thought to unify the fundamental particle interactions.

The first step is easy: point particles carrying electric and magnetic charges q and g can satisfy Newton's equations of motion with the driving force specified by a natural generalisation of the Lorentz force law. The complete system of equations is indeed duality rotation invariant if (1.1) is augmented by

$$q+ig \to e^{i\phi}(q+ig). \tag{1.3}$$

The price to be paid for this is the necessity of magnetic charge g never observed in isolation. A second difficulty is that the symmetry only works classically and not when quantum mechanics is introduced. The reason is that electrically charged particles must then be described by complex wave functions whose equations of motion require the introduction of the gauge potentials ϕ and $\underline{A}$ forming a four-vector A_μ.

The possibility of a gauge four potential A_μ such that

$$\mathbf{F}_{\mu\nu} = \partial_\mu A_\nu - \partial_\nu A_\mu. \tag{1.4}$$

was already implied by Maxwell's equations for the field strengths $\mathbf{F}_{\mu\nu}$. These latter are observable physical quantities and so exist globally in any sort of region of space-time, whatever its topology. The gauge potentials are not so directly physical and so need not exist globally. From a strict mathematical point of view they can only be constructed locally, that is, in regions of space time that are topologically trivial in the sense that they can be contracted continuously to a point.

This mathematical nicety will fit in with another feature of (1.4), that the there are ambiguities in the potentials giving rise to a given field strength. Thus for any reasonable function $\chi(x)$ on space time, two gauge potentials A_μ and A'_μ give rise to the same field strength if they are related to

$$A'_\mu = A_\mu + \partial_\mu \chi. \tag{1.5}$$

So perhaps these "gauge transformations" should be used to relate gauge potentials valid in different patches of space-time in a region where they overlap.

After the discovery of quantum mechanics, in which electrically charged particles are described by complex wave functions $\psi(x)$ it was realised by Weyl and by Fock that gauge covariance of the overall equations of motion could only be achieved if wave functions corresponding to different choices of gauge potential were related by

$$\psi'(x) = e^{iq\chi(x)/\hbar}\psi(x). \tag{1.6}$$

where q is the electric charge carried by the particle with the wave function in question and the gauge function χ is the same as in (1.5).

Now consider a spherical surface S_2 fixed in three dimensional space surrounding a centre which is an excluded point. This is an example of a space-time region with non trivial topology since it cannot be contracted continuously to a point without a tear whilst avoiding the origin. This sphere can be covered by two patches which we can think of as northern and southern hemispheres overlapping on the equator. Theses hemispheres are topologically trivial as they can be retracted to their respective poles. If the gauge potential A_μ were to exist globally on S_2, the flux of magnetic field out of the sphere would vanish by Stokes' theorem since $\underline{B} = \nabla \wedge \underline{A}$ and

$$\int_{S_2} \underline{B}.\underline{dS} = \int_{S_2} \nabla \wedge \underline{A}.\underline{dS} = 0$$

This means that the net magnetic charge inside the sphere would vanish. On the other hand, if we allow two different gauge potentials, A_μ and A'_μ in the northern and southern hemispheres respectively, Stokes' theorem can only be applied to the hemispheres and yields for the magnetic flux the following integral around the equator, S_1,

$$\int_{S_2} \underline{B}.d\underline{S} = \int_{S_1} \left(\underline{A}(x) - \underline{A}'(x)\right).d\underline{x}$$

By (1.5) this is the line integral of the gradient of the gauge function χ around the equator. This integrates to give

$$\chi(\phi = 2\pi) - \chi(\phi = 0)$$

where ϕ is the longitude (azimuthal angle). On the other hand, for the wave functions ψ and ψ' to be single valued in their respective hemispheres, equation (1.6) implies that this jump in χ around the equator must equal an integral multiple of $2\pi\hbar/q$. But the flux evaluated equals the magnetic charge g inside the sphere. The final conclusion is not that this vanishes, but rather that it satisfies the condition

$$qg = 2\pi\hbar\mathbb{Z}$$

This is the celebrated Dirac [1931] quantisation condition, derived according to an argument of Wu and Yang [1975].

This condition was extended to greater generality by Zwanziger [1968] and Schwinger [1969] and restricted the charges on any pair of isolated dyons to satisfy

$$q_1 g_2 - q_2 g_1 = 2\pi n\hbar, \qquad n = 0, \pm1, \pm2, \pm3 \ldots \tag{1.7}$$

Notice that this condition respects the duality rotation (1.3) applied to dyons 1 and 2 simultaneously. Dyon is the epithet coined by Schwinger [1969] for a particle carrying both electric and magnetic charges.

In 1979 Witten realised that it is possible to solve these quantisation conditions. The most general family of solutions is given by

$$q + ig = q_0(m\tau + n), \qquad m, n \in \mathbb{Z} \tag{1.8}$$

where the the complex parameter τ takes the form

$$\tau \equiv \frac{\theta}{2\pi} + \frac{2\pi i n_0 \hbar}{q_0^2}. \tag{1.9}$$

Notice that apart from an integer n_0, the imaginary part of τ is the inverse of the fine structure constant and hence fundamentally positive. The real part is given by a new angular parameter θ that Witten [1979] recognised as being what is called the vacuum angle already familiar in certain specific theories.

Equation (1.8) means that the allowed dyon charges constitute points of a discrete lattice to be called the "charge lattice". Notice that this lattice structure "spontaneously breaks" the duality rotation symmetry (1.3). Associated with this is the attractive consequence that electric and magnetic charges are quantised. To

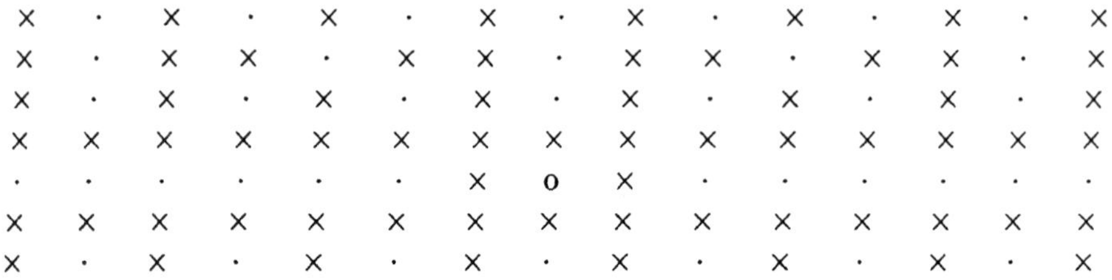

Figure 1. The charge lattice of Eq. (1.8)

make this clearer it is helpful to look at the charge lattice (1.8) in the $q+ig$ plane depicted in Fig.1. The origin, denoted by zero, represents particles carrying no charge, such as the photon. The points of the lattice with coordinate m equal to zero lie on the horizontal line through the origin and correspond to states carrying only electric charge q and no magnetic charge. They form a one dimensional lattice and this means that for magnetically neutral states electric charge is quantised, always occurring as an integer multiple of q_0, the charge of the state corresponding to the $\times$ to the right of the origin. This can be thought of as the electron with the positron occurring to the left of the origin. Of course the quantisation of electric charge has long been one of the most visible features of the spectrum of observed elementary particles. Dirac [1931] was very impressed that this fact could be explained by the existence of a single magnetic charge somewhere in the universe, not otherwise observed. Since alternative explanations turn out to be elusive, the theoretical study of magnetic monopoles has provided a source of fascination ever since this insight.

The reader may wonder why there is a distinction in the diagram between the points of the charge lattice outside the origin, some being denoted by dots and some by crosses. The crosses correspond to primitive vectors of the lattice, namely vectors that can be joined to the origin by a straight line avoiding any other points of the lattice. Equivalently the integer coordinates m, n in (1.8) are coprime. The physical significance of this will become clearer later when specific models are considered.

Although the lattice has spontaneously broken the duality rotation symmetry (1.3) of the quantisation condition (1.7) which it solves, it is still conceivable that the symmetry survives in a formula for the particle masses just as it applies to the energy density (1.2a). Could particle mass be a function of $|q+ig|$?

2. Unified Gauge Theories

Themes I and III of the previous chapter date back to the previous century whereas the ideas of unified gauge theories are relatively new. Because they allow the electromagnetic gauge group $U(1)$ to be unified with the other gauge groups they offer the possibility of an alternative explanation of the quantisation of electric charge that avoids introducing magnetic charge. To see this in more detail consider the lagrangian density describing a non-abelian gauge theory, with a Higgs field in the adjoint representation,

$$\mathcal{L} = -\frac{1}{4}Tr(\mathbf{F}_{\mu\nu}\mathbf{F}^{\mu\nu}) + \frac{1}{2}Tr(\mathcal{D}_\mu\mathbf{\Phi}\mathcal{D}^\mu\mathbf{\Phi}) - V(\mathbf{\Phi}), \tag{2.1}$$

where the covariant derivative of a field in the representation D of the gauge group G (assumed simple and compact) satisfies the standard properties

$$\mathcal{D}_\mu \equiv \partial_\mu + ieD(\mathbf{W}_\mu), \tag{2.2a}$$

$$[\mathcal{D}_\mu, \mathcal{D}_\nu] = ieD(\mathbf{F}_{\mu\nu}) = ieD(\partial_\mu\mathbf{W}_\nu - \partial_\nu\mathbf{W}_\mu + ie[\mathbf{W}_\mu, \mathbf{W}_\nu]). \tag{2.2b}$$

The scalar field $\mathbf{\Phi}$ is the (Lie algebra valued) Higgs field, and so is assumed not to vanish in the vacuum. Because it lies in the adjoint representation of the gauge group G, that is the same representation as the gauge potential $\mathbf{W}_\mu$, then automatically G is broken to the subgroup whose adjoint action on

$$Q = e\hbar\mathbf{\Phi}/a = e\hbar\phi.\mathbf{T}/a \tag{2.3}$$

preserves (2.3). a denotes the magnitude of the Higgs field in vacuo, $a^2 = \phi^2$ and is a constant as the covariant derivative of the Higgs field $\mathcal{D}_\mu\mathbf{\Phi}$ vanishes there. This subgroup, which is associated with long range gauge fields has the structure

$$(U(1)_Q \times K)/Z \tag{2.4}$$

where Z is a discrete diagonal subgroup that need not concern us. In particular if G is $SU(2)$, the Higgs field is an isovector of length a in vacuo. The only rotations leaving this axis invariant are precisely those about the axis itself as given by (2.4) since in this case K is trivial. For more detailed reviews see Goddard and Olive [1978], Olive [1982].

Q given by (2.3) can be interpreted as the electric charge operator as it generates an invariant $U(1)$ subgroup of the exact gauge group (2.4) just as the Maxwell $U(1)$ should. Furthermore it is correctly normalised. The eigenvalues q of the matrix $D(Q)$ then specify the electric charges of the particles created by a field in the representation D. These eigenvalues are indeed independent of the space-time point at which the Higgs field is evaluated since the matrix varies by gauge conjugation in vacuo.

According to the Higgs mechanism, (Englert and Brout [1964], Higgs [1966], Kibble [1967]), the formula for the mass of a gauge particle with electric charge q in this situation is simply

$$\text{MASS(Gauge particle)} = a|q|, \tag{2.5}$$

whatever the gauge group G and whatever the direction the adjoint Higgs field chooses in the vacuum.

Putting these results together we see that when G is $SU(2)$ the three gauge particles have masses

$$\text{MASS}(W^\pm) = a|e\hbar|, \qquad \text{MASS}(\gamma) = 0, \tag{2.6}$$

where γ denotes the photon.

Notice that in this case the electric charge (2.3) is proportional to an angular momentum matrix and hence has quantised eigenvalues. Thus unification has indeed led to the quantisation of electric charge without overt magnetic monopoles.

This $SU(2)$ model is known as the Georgi-Glashow model but was superseded by the Salam-Weinberg model as a description of electroweak gauge interactions because of its lack of a weak neutral current. Despite this we shall continue studying it because of the interesting properties that emerge and the possibilities of different sorts of generalisation.

Amongst these properties is the possibility of knotted solutions of a type first realised by 't Hooft [1974] and Polyakov [1974]. The occurrence of these can be inferred by considering the energy E corresponding to the above Lagrangian (2.1) for a coupled adjoint Higgs-gauge system:

$$2E = \int d^3x \left(Tr(\mathbf{E}_i^2 + \mathbf{B}_i^2) + (\mathcal{D}_i\mathbf{\Phi})^2 + (\mathcal{D}_0\mathbf{\Phi})^2 + 2V(\mathbf{\Phi})\right).$$

It is expressed as a sum of positive terms. In order for any classical field configuration to have finite energy, all the terms must vanish faster than the inverse of the radius of a very large two-sphere in three-space, when evaluated on that sphere. In particular the space components of the covariant derivative of the Higgs field must vanish there, so

$$\mathcal{D}_i\hat{\underline{\phi}} \equiv \nabla_i\hat{\underline{\phi}} + e\underline{W}_i \wedge \hat{\underline{\phi}} = 0$$

where the hat indicates that the Higgs field has been divided by a so as to be a unit vector in the vacuum. We are now considering the gauge group to be $SU(2)$ and use appropriate vector notation. Taking the vector product of this with the Higgs field yields the following behaviour for the gauge potential in terms of the Higgs field

$$\underline{W}_i = A_i\hat{\underline{\phi}} - \frac{1}{e}\hat{\underline{\phi}} \wedge \nabla_i\hat{\underline{\phi}},$$

where the $SU(2)$ invariant potential A_i is undetermined and will eventually drop out of the argument. Notice that the undetermined term is parallel to the Higgs field while the behaviour perpendicular to it is fully determined. Since the commutator of two covariant derivatives gives the field strength we have

$$0 = [\mathcal{D}_i, \mathcal{D}_j]\hat{\underline{\phi}} = e\underline{F}_{ij} \wedge \hat{\underline{\phi}}.$$

Hence, at larger distances, the magnetic field is parallel to the Higgs field in $SU(2)$ group space. In fact explicit calculation yields

$$\underline{F}_{ij} = \hat{\underline{\phi}}\left(\nabla_i A_j - \nabla_j A_i - \frac{1}{e}\hat{\underline{\phi}}.(\nabla_i\hat{\underline{\phi}} \wedge \nabla_j\hat{\underline{\phi}})\right) = F_{ij}\hat{\underline{\phi}}.$$

The Maxwell magnetic charge of the field configuration of finite energy can be found my calculating the magnetic flux out of the large sphere surrounding it. Since the above expression for the magnetic field holds on this sphere, it yields

$$g = \frac{1}{2}\int dS_i\epsilon_{ijk}F_{jk} = \frac{-1}{2e}\int dS_i\epsilon_{ijk}\hat{\underline{\phi}}.(\nabla_j\hat{\underline{\phi}} \wedge \nabla_k\hat{\underline{\phi}})$$

Notice that the undetermined potential A_μ has integrated out by Stokes' theorem leaving an integrand that has a very nice geometrical interpretation. It is the jacobian of the map provided by the Higgs field mapping from the two-sphere at large distances to the two-sphere that is the Higgs vacuum in the $SU(2)$ theory ($\underline{\hat{\phi}}^2 = 1$). Hence the integral is the winding number m, say, of this map between two two-spheres times the area, 4π, of the unit sphere. Thus

$$g = -\frac{4\pi}{e}m = -\frac{4\pi\hbar}{q(W)}m. \tag{2.7}$$

Thus the winding number m counts the number of units of magnetic charge and so can be identified precisely with the coordinate in equation (1.8). If fields carrying half integral $SU(2)$ spin are introduced the smallest possible electric charge in the theory, q_0, equals $q(W)/2$. Otherwise it is $q(W)$, the charge carried by the W^+ gauge particle. Hence, taking the imaginary part of (1.8) and comparing with (2.7), we see that the Dirac-Schwinger-Zwanziger condition is satisfied, with n_0 equal to 1 or 2 in the respective situations.

There are actually solutions to the equations of motion for all integer values of m and the winding number 1 solution (corresponding to the identification map between the spheres) will describe a magnetic monopole soliton and -1 its antiparticle, the antimonopole, also a soliton.

We learn that the idea that the quantisation of electric charge could be explained by unification without the occurrence of magnetic charge is illusory. Dirac's original argument has been vindicated.

Given a stationary solution, it should be possible to insert it in the energy integral following from (2.1) and hence evaluate its mass. In fact, by neatly completing a square in the energy expression above, Bogomolny [1976] found the lower bound

$$\text{MASS(Magnetic monopole)} \geq a|g|. \tag{2.8}$$

Because of the remarkable similarity to the Higgs formula (2.5) it is natural to enquire whether this "Bogomolny bound" could be saturated. Indeed, the energy integral shows that this can be done if two conditions are satisfied. Firstly the self-interaction of the Higgs field specified by $V(\mathbf{\Phi})$ in (2.1) should vanish identically. This is known as the Prasad-Sommerfield limit [1975]. Secondly the following first order field equations, known as the Bogomolny equations [1976], should hold:

$$\mathcal{D}_0\mathbf{\Phi} = 0, \qquad \underline{\mathbf{E}} = 0, \tag{2.9a}$$

$$\mathcal{D}_i\mathbf{\Phi} = \pm\mathbf{B}_i. \tag{2.9b}$$

It is easy to check that any solution to (2.9) automatically solves the full second order equations of motion in the Prasad-Sommerfield limit. The innocuous facade that the Bogomolny equations present is deceptive. It will emerge that they exhibit a remarkably rich mathematical structure which will have profound physical implications in what follows.

As a first step note that, if a fifth, space-like dimension x^5 is introduced in which

$$\mathbf{W}^5 = \mathbf{\Phi}, \tag{2.10}$$

then equations (2.9b) can be regarded as the self-dual Yang-Mills equations in four euclidean dimensions (1,2,3 and 5). Here we have exploited the fact that the Higgs field $\mathbf{\Phi}$ lies in the same adjoint representation of the gauge group as the gauge potentials $\mathbf{W}^{\mu}$. Normally the solutions to such equations describe instantons but here the solutions are independent of both x^5 and time and will have a different character describing what are called BPS monopoles. To justify this we have to determine the nature of the moduli spaces $\mathcal{M}(m)$ of solutions to the Bogomolny equations (2.9b) labelled by the different values of the winding number m which also specifies the magnetic charge.

Suppose we seek small deviations about a solution that still solve the equations. These will satisfy linear differential equations in the background of the original solution. After discounting small gauge transforms of the original equation as irrelevant this linear equation can be recast as a Dirac equation for adjoint spinors in the Bogomolny background. Its solutions form a vector space whose dimension is finite and given by a version of the Atiyah-Singer index theorem as being

$$\mathrm{Dim}\mathcal{M}(m) = 4|m|. \tag{2.11}$$

Furthermore $\mathcal{M}(m)$ is a manifold. Thus it has no singular points but is not compact, for good physical reasons that we shall see. These results were due to E. Weinberg [1979].

The physical consequences are remarkable. (2.11) can be restated as saying that a solution with magnetic quantum number m has $4m$ bosonic zero modes. In other words there is a $4m$ dimensional space of continuous and physically significant alterations that can be made without changing the energy. When $m = 1$ three of the four possibilities correspond to moving the monopole soliton sideways. Thus three of the coordinates on the moduli space $\mathcal{M}(1)$ are the spatial coordinates of the monopole. The fourth is more subtle: it describes a coordinate conjugate to the electric charge, just as the space coordinate is conjugate to momentum. Momentum is excited when the space coordinate moves in time. Likewise electric charge can be excited when its conjugate coordinate moves in time. Thus when we consider motions on the moduli space of BPS solutions we can endow monopoles with momentum and with electric charge (so that they become dyons). However this is jumping ahead and we must just accept this charge coordinate for the time being.

Since $\mathrm{Dim}\mathcal{M}(m) = m\mathrm{Dim}\mathcal{M}(1)$, we conclude that $\mathcal{M}(m)$ describes m magnetic monopoles, each with three space coordinates and its own charge coordinate.

So each point of $\mathcal{M}(m)$ describes a configuration of m like monopoles. Since the solution is independent of time, this configuration is stationary. By Newton's equations of motion it follows that the monopoles, rather surprisingly, exert no forces on each other when they are at relative rest. In other words, intermonopole forces vanish at relative rest and so must be velocity dependent. Notice that this differs from the situation with sine-Gordon solitons. There the analogue of the Bogomolny first order differential equation is satisfied by the soliton and by the antisoliton individually but not by any multisoliton configuration as, in this case, solitons do exert static forces on each other.

The Bogomolny equations do not describe mixed configurations of monopoles and antimonopoles which must be described by solutions of the full second order equations and so be time dependent. Thus monopole antimonopole pairs must exert static forces on each other.

Finally notice that there is no zero mode corresponding to the angular orientation of any monopole. This means that they cannot acquire intrinsic angular momentum by classical motion.

We can now summarise the particle spectrum of the $SU(2)$ gauge theory as far as it is understood so far. Four particles arise as quantum excitations of the field in (2.1), the photon, the two heavy gauge particles $W^{\pm}$ and the Higgs which is neutral and massless, because its mass term vanishes in the Prasad-Sommerfield limit. The remaining spectrum consists of the solitons with winding number ± 1. These can be monopoles as already discussed or, possibly, dyons (Julia and Zee [1975]). A Bogomolny style argument [Coleman et al 1977] shows that their mass is given by

$$\mathrm{MASS}(q, g) = a\sqrt{q^2 + g^2} = a|q + ig|. \tag{2.12}$$

In fact all the particles we have mentioned satisfy this single universal mass formula which unifies Higgs and Bogomolny mechanisms, quantum excitations and classical topological solitons. Furthermore it does respect the duality rotation symmetry (1.3), in the simplest possible way. It is quite remarkable that this theory should display such properties and it is natural to seek a deeper understanding of it.

It is interesting that the vacuum expectation value parameter is explicitly responsible for all forms of mass as it alone breaks the conformal symmetry of the theory. In other words, this theory seems to provide a version in four space-time dimensions of Zamolodchikov's mechanism [1989] for the deformation of a conformally invariant field theory (cf Theme II above). Maybe it is indeed close to being integrable.

3. Electromagnetic Duality Conjectures

We have just seen that an $SU(2)$ gauge theory with triplet Higgs has remarkable and unexpected properties. Some of the particle states, those with magnetic charge, arise as solitons and so have an extended classical structure, quite unlike the quantum excitations of the original fields which are magnetically neutral and point-like. Yet, despite this, the mass formula displays an unusual symmetry with respect to rotations in the complex charge space, $q + ig$, between these states. In view of Skyrme's result that sine-Gordon solitons can be viewed as being created by massive Thirring fields in a reformulation of the theory, it is tempting to investigate the possibility of a second formulation of the gauge theory with quantum fields ascribed to the monopole solitons rather than the original gauge particles.

As the monopole solitons carry magnetic charge they emanate a Coulomb magnetic field which is presumably described by a gauge theory whose potential couples to the magnetic charge. This means that there is a "magnetic" gauge group with coupling strength inversely related to the original "electric coupling" by

$$q_0 \to g_0 = \frac{4\pi\hbar}{q_0}, \tag{3.1}$$

because of the relation (2.7) respecting the Dirac-Schwinger-Zwanziger quantisation condition (1.4). Notice the characteristic feature that this transformation exchanges strong and weak coupling regimes.

Thinking along these lines, two specific conjectures were proposed in 1977. First, considering a more general context, with a less restrictive exact symmetry group H than (2.4), Goddard, Nuyts and Olive [1977] established a generalised version of the Dirac quantisation condition and used it to propose conjectures for the precise global structure of the magnetic, or dual gauge group H^{V}. However it is only the situation leading to (2.4) that yields a specific mass formula. Montonen and Olive [1977] therefore considered the setup as described above, restricted to $G = SU(2)$, even though the mass formula described holds for any gauge group G with adjoint Higgs.

The possible quantum states of the theory correspond to points of the lattice considered in the first chapter. Ignoring the dyons, the single particle states correspond to five points of this lattice. The photon and Higgs particle correspond to the origin, the heavy gauge particles $W^{\pm}$ to the points $(\pm q_0, 0)$. Thus the particles created by the fundamental fields in the original "electric" formulation of the action all lie on the real, electric axis. Supposing they have no electric charge, the magnetic monopole solitons $M^{\pm}$ lie on the imaginary axis at $(0, \pm g_0)$ while any dyon solitons would lie on the horizontal lines through these points, but we shall temporarily ignore them.

Now, if we follow the transformation (3.1) by a rotation through a right angle in the $q+ig$ plane, the four points just described away from the origin are rearranged. This suggests that the "dual" or magnetic formulation in which the $M^{\pm}$ monopoles are created by fields present in the action will likewise be a spontaneously broken gauge theory, but with coupling constants changed according to (3.1). In this alternative formulation it is the $W^{\pm}$ states that will occur as solitons.

This is the electromagnetic duality conjecture of Montonen and myself as originally formulated [1977]. In principle, it could be proven by finding a generalisation of Skyrme's vertex operator construction, but, this still seems beyond reach despite the intervening advances in mathematical knowledge. Notice that the sine-Gordon quantum field theory was described by two quite dissimilar actions whereas in the present case the two hypothetical actions have a very similar structure but refer to electric and magnetic formulations with inversely related coupling strengths.

Physical predictions should coincide whichever action is chosen as the starting point. The conjecture will immediately pass at least two tests of this kind, showing that it is not obviously inconsistent. First, the mass formula (2.12) satisfies this criterion, precisely because of the universal property that has already been emphasised, and indeed motivated the idea of two dual formulations.

The fact that the forces between an M^{+} pair are velocity dependent thereby vanishing at relative rest, as discussed in the previous chapter, had just been realised by Manton [1977]. It was therefore a valid test of the conjecture to establish whether this force vanishes in the magnetic formulation. This is equivalent to checking the absence of the static $W^+ - W^+$ force in the electric formulation and this is simply a question of Feynman rules in the Born approximation. Then two diagrams contribute, photon exchange and Higgs exchange. Photon exchange produces the expected q^2/r^2 Coulombic repulsion which is precisely counterbalanced by a q^2/r^2 Coulombic attraction due to the Higgs exchange. This is possible because the adoption of the Prasad-Sommerfield limit forces the Higgs to be massless. To check this in more detail we have to extract the couplings from the original Lagrangian density (2.1).

The first step is to choose a gauge in which the Higgs field is aligned in a standard direction, the 3 axis in $SU(2)$ space, say:

$$\phi_i(x) = \delta_{i3}(a + \sigma(x)). \tag{3.2}$$

This is possible because we are considering fluctuations about the vacuum rather than a monopole background. The field $\sigma(x)$ is the neutral Higgs left after its charged partners have been absorbed by $W^\pm$.

Inserting (3.2) into the term in (2.1) describing the Higgs kinetic energy we obtain

$$\frac{(\mathcal{D}_\mu\phi)^2}{2} = \frac{(\partial_\mu\sigma)^2}{2} + \frac{e^2a^2}{2}W_\mu^+W_\mu^- + e^2a\sigma W_\mu^+W_\mu^- + \dots \tag{3.3}$$

As well as confirming that σ is massless and that $W^\pm$ hass a mass $M = ea\hbar$, this expression shows that σ has a Yukawa coupling to $W^\pm$ of strength $e^2a = eM/\hbar$. Thus the Feynman rule for the corresponding vertex is $\epsilon_1.\epsilon_2 Me/\hbar$, where ϵ_1 and ϵ_2 are the polarisation vectors of the two W^+ lines incident at the vertex. This is to be compared with the corresponding rule for the photon coupling $\epsilon_1.\epsilon_2(p_1+p_2)_\mu e/2\hbar$.

Now it is easy to calculate the contributions to $W^+ - W^+$ scattering due to Higgs and photon exchange respectively. These are

$$\epsilon_1.\epsilon_2\epsilon_3.\epsilon_4\frac{M^2e^2}{\hbar^2k^2}$$

and

$$-\epsilon_1.\epsilon_2\epsilon_3.\epsilon_4\frac{e^2(p_1+p_2).(p_3+p_4)}{4\hbar^2k^2}$$

where k^μ is the momentum transfer, that is the momentum carried by the virtual particles exchanged. In the static limit the momentum transfer is very small and the momenta p_1, p_2, p_3 and p_4 are equal. Then the two contributions above are equal and opposite and hence cancel, yielding the claimed result. On the other hand, if $W^+ - W^-$ scattering were considered $p_1, p_2, -p_3$ and $-p_4$ would be equal in the static limit leading to the equality of the two contributions and hence the doubling of the expected Coulomb attraction.

The electromagnetic duality conjecture immediately provokes the following questions

(1) How can the magnetic monopole solitons possess the unit spin characteristic of heavy gauge particles?

(2) Surely the effect of quantum corrections will vitiate the universal mass formula?

(3) How would the proper inclusion of dyon states affect the picture?

We have already noted that the monopoles are spinless on the classical level as the moduli space of Bogomolny solutions has no degree of freedom to accommodate their rotation. Hence any spin must originate in the quantisation procedure applied to the theory.

In fact it was soon realised that there were several options in quantising the theory. It is possible to add fermionic and other fields to the original theory which play no role in the solutions discussed so far since putting them to zero is consistent with the new equations of motion. Furthermore, this can be done in a way that enhances the symmetry of the theory so that it achieves supersymmetry (D'Adda, Horsley and Di Vecchia [1978]). The quantum fluctuations of the added fields do have an important effect. They can cancel the existing fluctuations so as to ensure that the mass formula is quantum exact and they can create spin. Before understanding how supersymmetry achieves these feats we shall now explain in more detail what supersymmetry is and the relevant features of its algebraic structure.

4. Algebraic Structure of Supersymmetry

Many of the key features of supersymmetry are very well illustrated by the supersymmetric harmonic oscillator, probably the simplest toy model for it. Consider two annihilation-creation oscillators a and b which are bosonic and fermionic respectively:

$$[a, a^\dagger] = 1 = \{b, b^\dagger\}, \qquad b^2 = 0 \tag{4.1}$$

With these, form the supercharges

$$Q = a^\dagger b, \qquad Q^\dagger = b^\dagger a \tag{4.2}$$

whose anticommutator is the Hamiltonian H

$$\{Q, Q^\dagger\} = H = a^\dagger a + b^\dagger b = (a^\dagger a + \frac{1}{2}) + (b^\dagger b - \frac{1}{2}) \tag{4.3a}$$

The two bracketed terms are the Hamiltonians for the individual bosonic and fermionic oscillators respectively with their zero point energies. Notice that these are equal and opposite and so cancel from the total Hamiltonian. Furthermore

$$Q^2 = Q^{\dagger 2} = 0 = [Q, H] \tag{4.3b}$$

Equations (4.3) constitute the supersymmetry algebra satisfied by the supercharges and Hamiltonian. The Heisenberg equation of motion for a reads:

$$\frac{da}{dt} = i[a, H] = ia$$

Hence, repeating, a indeed satisfies the harmonic oscillator equation with unit frequency, as does b by a similar calculation.

One of the important general features of a supersymmetry algebra is that

$$H \geq 0 \tag{4.4}$$

as can be seen from (4.3a). In fact, if $|\Omega>$ is the ground state, that is the state for which H has its smallest possible eigenvalue, zero, then it is the state annihilated by both a and b, the oscillator vacuum. Thus the vanishing of the zero point energy is an intrinsic feature of a supersymmetric system and this is the reason why quantum corrections tend to cancel to zero. Linear fluctuations in a more complex supersymmetric system can be decomposed into independent supersymmetric oscillators, each of whose zero point energy vanishes.

It is easy to calculate all the energy eigenstates of the supersymmetric oscillator and an interesting pattern emerges. For energy eigenvalue n, $(n \geq 1)$ there are precisely two eigenstates, $a^{\dagger n}|\Omega>$ and $a^{\dagger n-1}b^{\dagger}|\Omega>$. Thus the dimension of the energy eigenspace depends on the whether the energy eigenvalue is zero or not, being one when it is zero and two otherwise. Corresponding to this we talk of "short" and "long" representations of supersymmetry.

Now let us see how these features extend to supersymmetry algebras in Minkowski space time, that is with signature (1,3). There the (Dirac) gamma matrices have some special properties that we want to exploit. The Dirac equation reads

$$(i\gamma.\partial + m)\psi = 0,$$

provided the γ^μ satisfy the Clifford algebra

$$\{\gamma^\mu, \gamma^\nu\} = 2g^{\mu\nu} = 2diag(1,-1,-1,-1).$$

Anticommuting with these is

$$\gamma^5 = \gamma^0\gamma^1\gamma^2\gamma^3, \qquad (\gamma^5)^2 = -1.$$

It will be very important that γ^5 behaves like the imaginary unit i, as does the Hodge star operator, implicit in the duality transformation (1.1). The Clifford algebra has only one irreducible representation and that has dimension 4. It is possible to choose a "Majorana" basis in which the matrices are pure imaginary. In it $i\gamma^\mu$ are real and given by

$$\begin{pmatrix} 0 & 0 & 0 & 1 \\ 0 & 0 & -1 & 0 \\ 0 & 1 & 0 & 0 \\ -1 & 0 & 0 & 0 \end{pmatrix}, \begin{pmatrix} 0 & 0 & 0 & 1 \\ 0 & 0 & 1 & 0 \\ 0 & 1 & 0 & 0 \\ 1 & 0 & 0 & 0 \end{pmatrix}, \begin{pmatrix} 0 & 0 & 1 & 0 \\ 0 & 0 & 0 & -1 \\ 1 & 0 & 0 & 0 \\ 0 & -1 & 0 & 0 \end{pmatrix}, \begin{pmatrix} 1 & 0 & 0 & 0 \\ 0 & 1 & 0 & 0 \\ 0 & 0 & -1 & 0 \\ 0 & 0 & 0 & -1 \end{pmatrix}.$$

This means that it is perfectly consistent for the solution ψ to the Dirac equation to be real and this will be exploited. Notice that the last three matrices are symmetric while the first, $i\gamma^0$, is real antisymmetric, as is

$$\gamma^5 = \begin{pmatrix} 0 & 0 & -1 & 0 \\ 0 & 0 & 0 & -1 \\ 1 & 0 & 0 & 0 \\ 0 & 1 & 0 & 0 \end{pmatrix}$$

Now we can introduce the supercharges Q^i_α carrying a Dirac spinor index α running from 1 to 4 and an internal symmetry index i running from 1 to N. Because of the previous comment these can conveniently be taken to be hermitian. Because of their spinor nature they satisfy the anticommutation relations of type N supersymmetry:

$$\{Q^i_\alpha, Q^j_\beta\} = 2(\gamma^\mu P_\mu \gamma^0)_{\alpha\beta}\delta^{ij} \tag{4.5a}$$

Note that everything in this equation is real and that the right hand side of the equation displays symmetry under interchange of $\alpha \leftrightarrow \beta$ and $i \leftrightarrow j$ separately, even though the left hand side only requires the joint symmetry. There are further commutation relations expressing translation invariance,

$$[P^\mu, Q^i_\alpha] = 0, \tag{4.5b}$$

$$[P^\mu, P^\nu] = 0, \tag{4.5c}$$

as well as the statement of the Lorentz transformation properties. We can solve (4.5a) for the Hamiltonian by taking the trace in the spinor indices α and β and choosing i equal to j without summing :

$$H = P^0 = \frac{1}{4}\sum_{\alpha=1}^{4}(Q^i_\alpha)^2, \qquad i = 1, 2 \ldots N. \tag{4.6}$$

Hence $H \geq 0$ because the supercharges are hermitian. Again, as in the toy oscillator, positivity of energy is guaranteed by the algebraic structure.

Furthermore the vacuum, the state with zero energy, is annihilated by all the supercharges and hence totally supersymmetric. It automatically has zero momentum by (4.5a).

Equations (4.5b) and (4.5c) mean that the superalgebra must be represented on the set of states with any given momentum. Let us consider massive particle states, choosing a rest frame in which $P^\mu = (M, 0, 0, 0)$ so that (4.5a) reads

$$\{Q^i_\alpha, Q^j_\beta\} = 2M\delta_{\alpha\beta}\delta^{ij} \tag{4.7}$$

This is recognisable as a Clifford algebra in $4N$ Euclidean dimensions [Nahm 1978]. Its unique irreducible representation therefore has dimension 2^{2N}.

This is to be contrasted with a massless particle state whose momentum can, at best, be chosen as $P^\mu = (E, E, 0, 0)$ so that (4.5a) reads

$$\{Q^i_\alpha, Q^j_\beta\} = 2E(1 - \gamma^1\gamma^0)_{\alpha\beta}\delta^{ij} \tag{4.8}$$

Because $(1 - \gamma^1\gamma^0)$ is 4×4 matrix with two eigenvalues equal to 0 and the other two equal to 2 we see that half the supercharges must annihilate the massless state and hence effectively vanish while the other half generate a Clifford algebra in $2N$ Euclidean dimensions with a unique irreducible representation of dimension 2^N (the square root of the dimension in the massive case).

In practice N, the number of spinor supercharges, will not exceed 4 and we can tabulate the number of dimensions required to represent the Clifford algebra on the different types of state:

	$N=1$	$N=2$	$N=4$
$P^2>0$	4	16	256
$P^2=0$	2	4	16
$P_\mu=0$	1	1	1

The massless case is said to form the "short" representation.

To understand the spin content of the massless multiplets better, consider the helicity, the component of angular momentum in the direction of motion,

$$J_1=\frac{\underline{p}.\underline{J}}{|\underline{p}|}=\frac{i\gamma^2\gamma^3}{2}=\frac{-i\gamma^5\gamma^1\gamma^0}{2}$$

using the definitions. But according to (4.8) the non vanishing supercharges satisfy $\gamma^1\gamma^0 Q=-Q$ Hence

$$J_1 Q=\frac{i\gamma^5}{2}Q. \tag{4.9}$$

As a consequence, of the $4N$ spinor supercharges, $2N$ vanish leaving N with helicity $1/2$ and N with helicity $-1/2$. These can be thought of as N fermionic annihilation-creation operators destroying and creating helicity in units of $1/2$. Acting on a ground state they form a Fock space of dimension 2^N upon which the range of helicities is

$$\Delta J_1=\frac{N}{2} \tag{4.10}$$

(allowing for the ground state to have a helicity value). This range equals $1/2, 1$ or 2 as $N=1,2$ or 4.

In contrast the corresponding result for massive states is more complicated to prove but yields

$$\Delta J_1=N \tag{4.11}$$

which is twice as large.

Returning to the spontaneously broken gauge theory of chapters 2 and 3 we recall that, in the first instance, the two fields, the Higgs and gauge field, are massless and carry helicities ±1 and 0. Since they transform similarly under the gauge group it is tempting to fit them into a single irreducible representation of the Clifford algebra on massless states. This requires $\Delta(\text{helicity})=1$, at least, so that $N\geq 2$. Consider first $N=2$. The representation of the supercharge algebra on massless states is four dimensional with a helicity span of one. Letting the helicity axis be vertical this yields a supermultiplet as follows

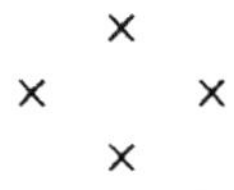

If we want to describe helicity one and no higher we must include helicity -1 and this is done with two copies of the above pattern

$$\begin{array}{ccc} & \times & \\ \times & & \times \\ \times & & \otimes \\ \otimes & & \otimes \\ & \otimes & \end{array} \tag{4.12}$$

Here the helicity runs from one at the top down to minus one at the bottom. Thus, in order to construct an $N = 2$ massless supermultiplet containing the gauge and Higgs field, we have to add a second scalar field as well as the two adjoint spin half fields. We shall call this the gauge multiplet. There is a second sort of massless $N = 2$ multiplet possible which has no helicity ± 1 components, called the hypermultiplet, also made of two copies of the fundamental unit:

$$\begin{array}{cccccc} & \times & & & \otimes & \\ \times & & \times & \otimes & & \otimes \\ & \times & & & \otimes & \end{array} \tag{4.13}$$

The second copy is needed to realise the internal symmetry associated with the supersymmetry, known as $\mathcal{R}$-symmetry. The fundamental anticommutation relation (4.5a) has a manifest $O(N)$ symmetry according to which the supercharges transform as an N. This $O(N)$ can be enlarged to $U(N)$ by a complexification in which the role of the imaginary unit acting on the supercharges is simply γ_5 as suggested above. In the $N = 2$ massless gauge multiplet the fermions form an $SU(2)_{\mathcal{R}}$ doublet and the two spinless particles form two singlets. On the other hand in the hypermultiplet the fermions are singlets while the spinless particles form a doublet. Since the doublet is a complex (pseudoreal) representation and γ^5 does not act on spinless particles there have to be four real scalars.

Any other possible massless $N = 2$ multiplets must possess helicities exceeding one in magnitude and hence are not considered.

It turns out that $N = 3$ gives the same structure as $N = 4$ so we shall move on to the latter. Since the helicity span given by (4.9) is 2, there is only one possible massless $N = 4$ supermultiplet with helicity not exceeding one. Its helicities run from 1 down to -1:

$$\begin{array}{ccccccc} & & & \times & & & \\ & \times & \times & & \times & \times & \\ \times & \times & \times & & \times & \times & \times \\ & \times & \times & & \times & \times & \\ & & & \times & & & \end{array} \tag{4.14}$$

In this case the fermions form a $4_L \oplus \bar{4}_R$ of $SU(4)_{\mathcal{R}} \equiv SO(6)_{\mathcal{R}}$ and the spinless particles a 6 (which is real). Notice that this supermultiplet consists of a single irreducible representation of the Clifford algebra (4.7).

Higher values of N imply massless supermultiplets with helicities larger than one in magnitude and that is why those values are excluded. The problems worsen if massive supermultiplets are considered. This leads to a paradox that we now discuss.

5. Supersymmetry to the Rescue

We have seen how the massless states of the spontaneously broken gauge theory can be fitted into a single (gauge) supermultiplet of either the $N = 2$ or the $N = 4$ theory but it still has to be determined which of these degrees of supersymmetry is appropriate. Before addressing this there is a more urgent problem. It was seen earlier that mass was acquired from the Higgs mechanism and it is a familiar feature of this mechanism that the total number of helicity states does not change. Yet, according to the Clifford algebra analysis just discussed, the dimensionality of the representation changes radically with mass thereby begetting a paradox. (This is a problem peculiar to $N = 2$ and $N = 4$ and not $N = 1$ supersymmetry because in $N = 1$ the massless gauge and Higgs particles have to be in different supermultiplets). The resolution of this paradox is that the supersymmetry algebra (4.5a) as it stands is incomplete as extra terms may occur on the right hand side (Haag, Lopuszanski and Sohnius [1975]). For example if $N = 2$ we can have

$$\{Q^i_\alpha, Q^j_\beta\} = 2(\gamma^\mu P_\mu \gamma^0)_{\alpha\beta}\delta^{ij} + 2i\epsilon^{ij}\left((Z_1 + \gamma^5 Z_2)\gamma^0\right)_{\alpha\beta}, \tag{5.1}$$

where ϵ^{ij} is the antisymmetric matrix with two rows and columns. Z_1 and Z_2 are real quantities, called central charges because they commute with the supercharges and all the other generators of the algebra. Notice that the right hand side still reproduces the joint symmetry under interchange of $\alpha \leftrightarrow \beta$ and $i \leftrightarrow j$ of the left hand side. Furthermore the structure is consistent with all the generalised Jacobi identities.

The reason that the superalgebra realised on massless states, (4.8), unlike the superalgebra on massive states (4.7) had a shorter representation was that the matrix on the right hand side was singular. This possibility is now attainable for massive states with the extra terms on the right hand side of (5.1). To see this consider the algebra acting on a massive particle at rest carrying charges Z_1 and Z_2. Then

$$\{Q^i_\alpha, Q^j_\beta\} = 2M\delta_{\alpha\beta}\delta^{ij} + 2i\epsilon^{ij}\left((Z_1 + \gamma^5 Z_2)\gamma^0\right)_{\alpha\beta}. \tag{5.2}$$

As $((Z_1 + \gamma^5 Z_2)\gamma^0)^2 = (Z_1 + \gamma^5 Z_2)(Z_1 - \gamma^5 Z_2) = Z_1^2 + Z_2^2$, the right hand side of (5.2) has eigenvalues $2(M \pm \sqrt{Z_1^2 + Z_2^2})$, each with fourfold multiplicity. Because of the hermiticity of the supercharges, the left hand side of (5.2) is a positive definite matrix and hence the mass M is bounded below

$$M \geq \sqrt{Z_1^2 + Z_2^2}. \tag{5.3}$$

When this bound is saturated, that is, an equality, half the eigenvalues of the right hand side of (5.2) vanish and the other half equal $4M$. As a consequence the superalgebra has a short representation exactly as in the massless case (Witten and Olive [1978]). This means that mass can be acquired without changing the number of states, provided the bound (5.3) is saturated.

Because the supercharges are conserved so are the central charges. As the only conserved charges available are the electric and magnetic ones, Q and G, we expect the central charges to be functions of these. A simple and traditional way of checking a superalgebra is to check the algebra of supertransformations on the field of the theory and this results in

$$Z_1 = aQ, \tag{5.4a}$$

while Z_2 apparently vanishes. This method has two limitations: (1) As it involves transformations of classical fields it is immune to possible quantum corrections. (2) As none of the fields in the action carry magnetic charge its possible appearance is invisible to this argument.

Nevertheless we already see something interesting: the Higgs mass formula is a consequence of the supersymmetry algebra, combined with the notion that mass requires no extra states.

The calculation can be refined by expressing the supercharges as integrals over space, thereby demonstrating that

$$Z_2 = aG, \tag{5.4b}$$

at least semiclassically (Witten and Olive 1978]). Thus the saturation of the bound (5.3) is tantamount to the universal mass formulae (2.12) that originally provoked the electromagnetic duality conjecture of Chapter 3.

In the $N = 4$ theory the superalgebra can also be modified by the addition of extra charges on the right hand side. In this case $\epsilon^{ij} Z_p$ on the right hand side of (5.1) is replaced by $Z^{ij}_{(p)}$ where $p = 1, 2$ and the matrices $Z_{(1)}$ and $Z_{(2)}$ are real antisymmetric matrices with four rows and columns each transforming as the real 6 of $SU(4)_R$. These matrices carry information about the electric and magnetic charges and again lead to the validity of the mass formula (2.12) being equivalent to the shortness of the relevant supermultiplet.

The argument that the states of a given momentum must represent the supersymmetry algebra is perfectly democratic: it applies equally to any sort of particle whether it is a quantum excitation of an original field, whether it is a soliton or whether it arises by some other mechanism (as indeed will be the case). In particular it means that monopole dyon solitons do, willy nilly, possess spin. If the maximum helicity does not exceed one in magnitude there are two possibilities in $N = 2$, the gauge multiplet (4.12) or the hypermultiplet (4.13). In $N = 4$ there is only one possibility, the gauge multiplet (4.14). As these are short representations the mass formula has to hold.

In chapter 2 we mentioned that within the moduli space of Bogomolny solutions describing configurations of like monopole solitons there was no bosonic zero mode that could be interpreted as an angular coordinate for any monopole and hence capable of excitation to give angular momentum. What then is the origin of the spin required by supersymmetry? The answer has to to with quantum fluctuations of zero modes of the fermion fields needed for supersymmetry. In a Bogomolny background, the equations of motion for the fermion fields take the form of a massless Dirac equation coupled to the Euclidean self dual gauge configuration described in chapter 2. The number of fermionic zero modes is then determined by the version of the Atiyah-Singer index theorem appropriate to the monopole background, as elucidated by Callias [1978] and E Weinberg [1979]. It is proportional to the magnetic charge and depends on the transformation properties of the fermions under the $SU(2)$ gauge group as well as the supersymmetry algebra.

When quantised canonically, these zero modes generate a Clifford algebra whose representation space yields a space of states all with the same energy (because they are zero modes). However the states carry a variety of quantum numbers of the type carried by the fermion fields in question, such as spin. This phenomenon first became familiar in the context of the Ramond fermionic string in the early 1970's.

So far there is only one type of fermion field, that in the gauge multiplet and so transforming under the adjoint representation of the gauge group. The effects on spin of its zero modes is to create a hypermultiplet of monopole states in the $N = 2$ theory and a gauge multiplet in the $N = 4$ theory. Thus it is only in the $N = 4$ theory that monopoles can be gauge particles and hence this is the only case in which the electromagnetic duality conjecture of Montonen and myself can be valid as first realised by Osborn [1979]. The idea cannot be valid as it stands in the $N = 2$ theory but nevertheless a variety of interesting phenomena can occur.

It is possible to introduce extra hypermultiplets in the $N = 2$ theory that transform as doublets under the gauge $SU(2)$. If there are N_F of these, all massless in the first instance, this introduces an $U(N_F)$ flavour symmetry accidentally augmented to $SO(2N_F)$. These can be thought of as quarks. Their zero modes do not affect the monopole spin but they do contribute $SO(2N_F)$ spinor flavour quantum numbers to them. This is the phenomenon sometimes known as charge fractionalisation. The results mean that in the $N = 2$ theory it is possible to contemplate a duality symmetry which exchanges monopoles with quarks, rather than gauge particles, as in the $N = 4$ theory. This is the idea that has been developed and refined by Seiberg and Witten.

The internal $U(N)_{\mathcal{R}}$ symmetry plays an important role in both the $N = 2$ and $N = 4$ theories. Because the imaginary unit i acting on the fermions is realised by γ_5, the invariant subgroup $U(1)_{\mathcal{R}}$ is a chiral symmetry and hence vulnerable to quantum anomalies of the type related to the Atiyah-Singer index theorem. In the $N = 2$ supersymmetric gauge theory the divergence of the $U(1)_{\mathcal{R}}$ current is proportional to the instanton number density with a coefficient which is one loop exact and possesses a nice group theoretic factor expressing the gauge group transformation properties of the field content of the theory,

$$\partial_\mu j_5^\mu = \frac{e^3\hbar}{(4\pi)^2}(\tilde{h}(G) - x(H))Tr(*\mathbf{F}_{\mu\nu}\mathbf{F}^{\mu\nu}). \tag{5.5}$$

Here $x(H)$ is the Dynkin index for the (possibly reducible) representation of the gauge group G carried by the hypermultiplet H. Since the gauge multiplet is automatically in the adjoint representation of G its Dynkin index is given by the dual Coxeter number of G, $\tilde{h}(G)$. Such quantities are familiar from work on conformal field theory (see Goddard and Olive [1986]). Recall that the definition of the Dynkin index $x(D)$ for a representation D of a group G is given in terms of an orthonormal basis of Lie algebra generators T^i by

$$Tr(D(T^i)D(T^j)) = x(D)\psi^2\delta^{ij}.$$

The reason that the difference of these two integers (or half-integers) occurs as a factor on the right hand side of (5.5) is special to $N = 2$ supersymmetry. It is that the fermions of given helicity possess equal, and opposite $U(1)_{\mathcal{R}}$ charges in the gauge and hypermultiplets. This is because the supercharges possess $U(1)_{\mathcal{R}}$ charge equal to twice their helicity and the gauge particles and the hypermultiplet scalars possess zero charge.

The result has several consequences. First the $U(1)_{\mathcal{R}}$ symmetry is broken by the presence of instantons to a cyclic subgroup, usually $Z(4(\tilde{h}(G) - x(H)))$, but there may be special effects causing an alteration of the order by a factor of 2. The nonvanishing of the vacuum expectation value of the Higgs (gauge multiplet scalars) will break this cyclic subgroup further. All this is important in the Seiberg-Witten analysis of $N = 2$ and its generalisations.

A second feature is that $N = 2$ supersymmetry relates this breaking of $U(1)_{\mathcal{R}}$ symmetry to the breaking of conformal symmetry. Hence the Callan-Symanzik beta function for the running of the gauge coupling constant is proportional to the same coefficient as appears on the right hand side of (5.5). This means that the necessary and sufficient condition for asymptotic freedom is that

$$x(H) \leq \tilde{h}(G)$$

with equality corresponding to exact conformal symmetry.

An obvious way of achieving the latter is to consider the hypermultiplet to be in the adjoint representation of the gauge group, just like the gauge multiplet. Then the eight states in each multiplet add to sixteen, the same number and helicity content as in the $N = 4$ gauge multiplet. Indeed this yields the $N = 4$ supersymmetric gauge theory which is thus seen to have a Callan-Symanzik beta function which vanishes identically, at least in perturbation theory (Sohnius and West [1981], Brink, Lindgren and Nilsson [1983], Mandelstam [1983]).

This ensures several features favourable to the validity of the electromagnetic duality conjecture. It means that the couplings, electric or magnetic, do not run, eliminating the quandary as to whether it is the renormalised or unrenormalised couplings which should satisfy the Dirac-Schwinger-Zwanziger quantisation condition (Rossi [1981]). The $N = 4$ theory was the first quantum field theory with exact quantum conformal symmetry to be identified in four dimensions. It seems that in some sense, yet to be fully understood, it is the quantum field theory closest to its classical version because of the vanishing of quantum effects such as the beta function. It is certainly the most symmetrical renormalisable quantum field theory in four dimensions and is the most inviting candidate for exact construction, presumably by mathematical technology not yet available.

The masses occur because of the scale introduced by the non-vanishing Higgs vacuum expectation values which supply the deformation from exact conformal symmetry in accordance with the insight of Zamolodchikov in two dimensions.

It is easy to spot other ways for the beta function to vanish. $\tilde{h}(SU(N_c)) = N_c$ while each flavour of hypermultiplet quark in the defining representation contributes 1/2 to the Dynkin index. Hence $2N_c$ copies of these yield conformal symmetry. This has been understood in detail by Seiberg and Witten [1994b] when $N_c = 2$.

We now have two strong reasons for thinking that $N = 4$ supersymmetry is the correct arena for the exact electromagnetic duality conjecture of Montonen and myself, the fact that only then do the monopole-dyon solitons form a gauge multiplet and the conformal symmetry. The third and decisive reason was provided by Sen and concerns the dyons of higher magnetic charge as we now discuss.

6. Electromagnetic Duality and the Modular Group

Now it is established that the $N = 4$ supersymmetric $SU(2)$ gauge theory is the most promising candidate framework for the exact electromagnetic duality conjecture of chapter 3, it is instructive to reexamine the charge lattice introduced at the end of chapter 1 in this new light. We have an explicit theory whose particle spectrum is partly known. In particular it is clear that the only magnetically neutral particles are the three gauge $N = 4$ supermultiplets corresponding to the photon and the $W^{\pm}$. These have electrical charges 0 and $\pm q_0$, that is, coordinates $m = 0$ and $n = 0, \pm 1$. The monopole dyon solitons have $m = \pm 1$ with any value of n and all likewise form gauge supermultiplets according to the analysis of fermionic zero modes. Thus the single particle states with $m = 0, \pm 1$ are all known and, for these values of m correspond precisely to the primitive vectors of the charge lattice, that is, the points denoted by a cross in the figure at the end of chapter 1. This contrasts with the $N = 2$ supersymmetric $SU(2)$ gauge theory including $SU(2)$ doublet flavours. There the points $m = 0$, $n = 0, \pm 1, \pm 2$ all correspond to particles created by fields in the defining lagrangian.

Another specific feature of the $N = 4$ supersymmetric gauge theory concerns the angle θ occurring in the real part of the complex variable τ, (1.9) which was defined by the ratio of periods of the charge lattice. As Witten explained [1979], this variable can be identified with the "vacuum angle" which is an extra coupling parameter occurring in the $N = 4$ supersymmetric gauge lagrangian. It multiplies an appropriate multiple of the instanton number density. Because this density is a total derivative it affects neither the classical equations of motion nor the energy momentum density. But, as Witten showed by a Noether argument, it does affect the identification of electric charge in the way indicated by equation (1.8). Sometimes this parameter can be defined away by changing the phase in the elementary fermion fields, but not when there is $N = 4$ supersymmetry which lacks $U(1)_{\mathcal{R}}$ chiral symmetry.

Because of the absence of quantum corrections, the universal mass formula (2.12) is believed to be quantum exact in the $N = 4$ supersymmetric gauge theory. This means that mass is given by Euclidean distance from the origin in the charge lattice as far as concerns the short supermultiplets (with sixteen components) corresponding to single particle states. An attractive consequence of this mass formula is that any state satisfying the mass formula and given by a primitive vector is absolutely stable with respect to any decays conserving the two charges, electric and magnetic. This is because the straight lines defined by the original state and its two decay products form the sides of a non-degenerate triangle. Hence, by the triangle inequality, the decay products are heavier. This argument breaks down when the decay of a non primitive vector state is considered. Its coordinate vector is automatically an integer multiple, k, say, of a primitive vector and its mass therefore k times the mass of the primitive vector state and so on the verge of instability.

Now let us recall the electromagnetic duality conjecture of Montonen and myself. According to that, a new action can be chosen with monopoles created by elementary fields. This is defined by a choice of primitive vector (and its negative). A second primitive vector is needed to define the θ angle. Thus the choice of action is defined by a pair of primitive vectors in the charge lattice, that is, a basis for the lattice. As remarked earlier, with this new action the roles of quantum excitations and solitons change, and we now expect to find a new set of single particle states. These must still correspond to primitive vectors of the charge lattice as that concept is basis independent. This lead Ashoke Sen [1994] to conjecture that in the $N = 4$ supersymmetric $SU(2)$ gauge theory there is a precise correspondence between all the single particle states and all the primitive vectors of the charge lattice. This would imply a quite concrete statement as to which of the states $(m = \pm 2, n)$ with double magnetic charge could be single particle states, namely only those in which the coordinate n is an odd integer, as depicted in the lattice diagram at the end of chapter 1.

It would therefore be a test of the electromagnetic duality conjecture to find the dyon states with double magnetic charge and verify that n is odd and this is what Sen did in 1994. Since from our previous work we know that these missing states can be neither quantum excitations of the original fields nor classical soliton solutions, a new mechanism is needed for single particle states. Sen realised that a pair of like $m = 1$ monopoles could form quantum bound states in a very precise sense.

To do this he considered the moduli manifold $\mathcal{M}(2)$ of solutions to the Bogomolny equations (2.9) with $m = 2$. This is a manifold whose physical significance implies certain nice mathematical properties that we need to describe. The coordinates on the manifold are the bosonic zero modes which are interpreted as the collective coordinates of the two monopole system as already discussed. It is possible to set these coordinates in motion and find an approximate form of the action expressed in terms of these coordinates and their time derivatives. For slow motions it is quadratic in these velocities and the corresponding quadratic form for the kinetic energy automatically defines a Riemannian metric for the manifold (Manton [1982]). In principle this works for any $\mathcal{M}(m)$, as Manton first realised, but it is only when $m = 2$ that it has been possible to find a closed analytic expression for this metric. This construction, due to Atiyah and Hitchin [1985], involves elliptic functions and was discovered by indirect mathematical arguments. This metric exhibits the special feature that there are three anticommuting complex structures on it, and is an example of a hyper-Kähler manifold (as are all the $\mathcal{M}(m)$).

It follows that the slow motion of a configuration of monopoles is determined by the Euler-Lagrange equations of motion of this action and hence follow geodesics on the monopole moduli spaces. This is sufficient to determine the classical scattering of two monopoles at low relative velocity and yields surprisingly complex behaviour (Gibbons and Manton [1986]), including a type of incipient breathing motion perpendicular to the scattering plane, visible on a video prepared by IBM. Despite these beautiful results, there is no idea of how to describe relative motion of monopole solitons with unlike charge. The duality conjecture does predict pair annihilation, in contrast with the behaviour of conventional solitons in two dimensions.

When the fermionic zero modes are included the action describes a supersymmetric mechanics on $\mathcal{M}$, made possible because of the hyper-Kähler metric.

Sen realised that, when this is quantised so that the wave function of the theory satisfies a Schrödinger equation involving the Laplacian for the Riemannian manifold, then zero energy eigenvalues, that is solutions to Laplace's equation, yield bound states satisfying the mass formula. Duality requires that this bound state be unique for odd n and absent for even n. This means that the "wave function" would have to be a two-form (when the overall motion is factored out leaving the relative motion between the two monopoles described by a four-manifold). This is exactly what Sen [1994] found, namely two-monopole quantum bound states occurring if and only if n is odd. Further, these fell into the 16 dimensional short $N = 4$ gauge supermultiplet.

This leaves the question of establishing a similar result for primitive vectors of the charge lattice at higher values of $|m|$. Sen's argument cannot be repeated without an explicit form of the metric but it is possible to make encouraging confirmatory progress with indirect arguments using Hodge's theorem relating zero modes of the Laplacian to the Betti numbers of $\mathcal{M}(m)$ (Segal and Selby [1996]). Topological information about this manifold is available through its characterisation in terms of rational maps between two Riemann spheres.

Armed with the new insight that the spectrum of single particle states correspond to the primitive vectors of the charge lattice, augmented by the origin, rather than the five points previously considered, we can see that the original Montonen-Olive conjecture was too modest. Instead of possessing two equivalent choices of action, the $N = 4$ supersymmetric gauge theory apparently possesses an infinite number of them, all with an isomorphic structure, but with different values of the parameters.

Roughly speaking, the reason is that physical reality is described by the charge lattice endowed with an orientation given by the sign of the imaginary part of τ, (1.9), or, since $n_0 = 2$ in the $N = 4$ supersymmetric theory,

$$\tau = \frac{\theta}{2\pi} + \frac{4\pi i\hbar}{q_0^2}. \tag{6.1}$$

Choices of action correspond to choices of basis in the oriented lattice, that is an ordered pair of non collinear primitive vectors (or, periods). As the charge lattice is two dimensional, these choices are related by the action of the modular group, an infinite discrete group containing the previous transformation (3.1).

Let us choose a primitive vector in the charge lattice, represented by a complex number, q_0', say. Then we may ascribe short $N = 4$ supermultiplets of quantum fields to each of the three points $\pm q_0'$ and 0. The particles corresponding to the origin are massless and neutral whereas the particles corresponding to $\pm q_0'$ possess complex charge $\pm q_0'$ and mass $a|q_0'|$. We may form an $N = 4$ supersymmetric action with these fields. It is unique, given the coupling $|q_0'|$, apart from the vacuum angle whose specification requires a second primitive vector, $q_0'\tau'$, say, non-collinear with q_0'. The remaining single particle states are expected to arise as monopole solitons or as quantum bound states of them as discussed above.

Since the two non-collinear primitive vectors q_0' and $q_0'\tau'$ form an alternative basis for the charge lattice, they can be expressed as integer linear combinations of the original basis, q_0 and $q_0\tau$:

$$q_0'\tau' = aq_o\tau + bq_0, \tag{6.2a}$$

$$q_0' = cq_0\tau + dq_0, \tag{6.2b}$$

where

$$a, b, c, d \in \mathbb{Z}. \tag{6.2c}$$

Equally, $q_0\tau$ and q_0 can be expressed as integer linear combinations of $q_0'\tau'$ and q_0'. This requires that the matrix of coefficients in (6.2a) and (6.2b) has determinant equal to ± 1,

$$ad - bc = \pm 1. \tag{6.2d}$$

Only the plus sign preserves the sign of the imaginary part of τ, (6.1), and hence the sign of the fine structure constant, which is the inverse of this. Such matrices

$$\begin{pmatrix} a & b \\ c & d \end{pmatrix}$$

form a group, $SL(2, \mathbb{Z})$, whose quotient by its centre, $(\pm I)$, is called the "modular group". This is an infinite discrete group, of course. Equation (6.2a) divided by (6.2b) yields

$$\tau' = \frac{a\tau + b}{c\tau + d}.$$

These are the transformations which form the modular group and preserve the sign of the imaginary part of τ. This gives the relation between the values of the dimensionless parameters in the two choices of action corresponding to the two choices of basis. It is customary to think of the modular group as being generated by elements T and S where

$$T : \tau \to \tau + 1 \qquad S : \tau \to -\frac{1}{\tau}.$$

According to (6.1), T increases the vacuum angle by 2π. This is a trivial symmetry of the quantum theory. If the vacuum angle vanishes, S yields the transformation (3.1) above, as considered by Montonen and myself.

Notice that the modular group can be used to map between any pair of primitive vectors of the charge lattice. Hence the infinite number of single particle massive states, each forming a sixteen dimensional gauge supermultiplet of $N = 4$ supersymmetry and corresponding to a primitive vector, can be thought of as an infinite multiplet with respect to the modular group. Beware that the masses of these states can be indefinitely large, even for $m = 1$.

Proof of the quantum equivalence of all the actions associated with each choice of basis in the charge lattice would presumably require a generalised vertex operator transformation relating the corresponding quantum fields, and still provides a formidable challenge.

What we have described so far are the results that are believed to be exact. This faith is based on the remarkable confluence of a large number of quite disparate arguments, some of them of quite a sophisticated mathematical nature. For example, the Atiyah-Singer index theorem has played an ubitiquous role. Nevertheless, final proof presumably awaits further mathematical breakthroughs.

Despite this, the last three years has seen progress burgeon on a number of fronts. Admittedly the arguments are even more speculative but there really has been a quite decisive advance in the conceptual understanding of string theory as a truly unified theory incorporating gravity in a convincing way.

One of the hopes underlying the $N = 4$ supersymmetric theory is that it is the most symmetrical quantum field theory possible in flat space time with four dimensions. This means that gravity is ignored but it would be undeniably attractive to reinstate it. According to current wisdom, quantum consistency requires consideration of a superstring theory in ten dimensional space-time. Now it is a historical fact that the $N = 4$ supersymmetric gauge theory was first discovered by dimensionally reducing a rather simple supersymmetric gauge theory in ten dimensions down to four (Gliozzi, Scherk and Olive [1977]). Furthermore that supersymmetric gauge theory was constructed as what was then called the zero slope limit of the type I superstring. This is the superstring with just one real chiral supercharge in ten dimensions. Because of this pedigree it is not unreasonable to imagine that superstrings possess duality properties too and indeed this is what has been learnt, leading to a revitalisation of the subject.

The way is now open for the other speakers to describe these and the other more recent developments.

Acknowledgements

I am grateful to the organisers of this NATO Advanced Study Institute for making it possible for me to deliver these lectures in such pleasant surroundings. Most of these notes were written during my stay at the Newton Institute for Mathematical Sciences in Cambridge and I wish to thank it for support. I would like to thank many of my colleagues for numerous helpful discussions.

References

Atiyah MF and Hitchin NJ 1985 *Phys Lett* **107A** 21-25, "Low energy scattering of non-abelian monopoles"

Bogomolny EB 1976 *Sov J Nucl Phys* **24** 449-454, "The stability of classical solutions"

Brink L, Lindgren O and Nilsson BEW 1983 *Phys Lett* **123B** 323-328 "The ultraviolet finiteness of the $N = 4$ Yang-Mills theory"

Callias C 1978 *Commun Math Phys* **62** 213-234, "Axial anomalies and index theorems on open spaces"

Coleman S 1975 *Phys Rev* **D11** 2088-2097, "Quantum sine-Gordon equation as the massive Thirring model"

Coleman S, Parke S, Neveu A, and Sommerfield CM 1977 *Phys Rev* **D15** (77) 544-545, "Can one dent a dyon?"

D'Adda A, Horsley R and Di Vecchia P 1978 *Phys Lett* **76B** 298-302 "Supersymmetric monopoles and dyons"

Dirac PAM 1931 *Proc Roy Soc* **A33** 60-72, "Quantised singularities in the electromagnetic field"

Englert F and Brout R 1964 *Phys Rev Lett* **13** 321-323, "Broken symmetry and the mass of gauge vector bosons"

Gibbons G and Manton N 1986 *Nucl Phys* **B274** 183-224, "Classical and quantum dynamics of monopoles"

Gliozzi F, Scherk J and Olive DI 1977 *Nucl Phys* **B122** 253-290, "Supersymmetry, supergravity theories and the dual spinor model"

Goddard P, Nuyts, J and Olive DI 1977 *Nucl Phys* **B125** 1-28, "Gauge theories and magnetic charge"

Goddard P and Olive DI 1978 *Reports on Prog in Phys* **41** 1357-1437, "Magnetic monopoles in gauge field theories"

Goddard P and Olive DI 1986 *Int J Mod Phys* **A1** 303-414, "Kac-Moody and Virasoro algebras in relation to quantum physics"

Haag R, Lopuszanski JT and Sohnius M 1975 *Nucl Phys* **B88** 257-274, "All possible generators of supersymmetry of the S-matrix"

Higgs P 1966 *Phys Rev* **145** 1156-1163, "Spontaneous symmetry breakdown without massless bosons"

't Hooft G 1974 *Nucl Phys* **B79** 276-284, "Magnetic monopoles in unified gauge theories"

Julia B and A Zee A 1975 *Phys Rev* **D11** 2227-2232, "Poles with both magnetic and electric charges in non-Abelian gauge theory"

Kibble TWB 1967 *Phys Rev* **155** 1554-1561, "Symmetry breaking in non-abelian gauge theories"

Mandelstam S 1975 *Phys Rev* **D11** 3026-3030, "Soliton operators for the quantized sine-Gordon equation"

Mandelstam S 1983 *Nucl Phys* **B213** 149-168, "Light-cone superspace and the ultraviolet finiteness of the $N = 4$ model"

Manton N 1977 *Nucl Phys* **B126** (77) 525-541, "The force between 't Hooft-Polyakov monopoles"

Manton N 1982 *Phys Lett* **110B** 54-56, "A remark on the scattering of BPS monopoles"
Montonen C and D Olive DI 1977 *Phys Lett* **72B** 117-120, "Magnetic monopoles as gauge particles?"
Nahm W 1978 *Nucl Phys* **B135** 149-166, "Supersymmetries and their representations"
Olive DI 1982 "Magnetic monopoles and electromagnetic duality conjectures", 157-191 in "Monopoles in Quantum Field theory", edited by NS Craigie, P Goddard and W Nahm (World Scientific)
Osborn H 1979 *Phys Lett* **83B** 321-326, "Topological charges for $N = 4$ supersymmetric gauge theories and monopoles of spin 1"
Polyakov AM 1974 *JETP Lett* **20** 194-195, "Particle spectrum in quantum field theory"
Prasad MK and Sommerfield CM 1975 *Phys Rev Lett* **35** (75) 760-762, "Exact classical solution for the 't Hooft monopole and the Julia-Zee dyon"
Rossi P 1981 *Phys Lett* **99B** 229-231, "$N = 4$ supersymmetric monopoles and the vanishing of the β function"
Schwinger J 1969 *Science* **165** 757-761, "A magnetic model of matter"
Segal G and Selby A 1996 *Commun Math Phys* **177** 775-787, "The cohomology of the space of magnetic monopoles"
Seiberg N and Witten E 1994a *Nucl Phys* **B426** 19-52, *Erratum* **B430** 485-486, "Electromagnetic duality, monopole condensation, and confinement in $N = 2$ supersymmetric Yang-Mills theory".
Seiberg N and Witten E 1994b *Nucl Phys* **B431** 484-550, "Monopoles, duality and chiral symmetry breaking in $N = 2$ supersymmetric QCD"
Sen A 1994 *Phys Lett* **329B** 217-221, "Dyon-monopole bound states, self-dual harmonic forms on the multi-monopole moduli space, and $SL(2, \mathbb{Z})$ invariance in string theory"
Skyrme THR 1961 *Proc Roy Soc* **A262** 237-245, "Particle states of a quantized meson field"
Sohnius MF and West PC 1981 *Phys Lett* **100B** 245-250, "Conformal invariance in $N = 4$ Yang-Mills theory"
Weinberg E 1979 *Phys Rev* **D20** 936-944, "Parameter counting for multimonopole solutions"
Witten E 1979 *Phys Lett* **86B** 283-287, "Dyons of charge $e\theta/2\pi$"
Witten E and Olive DI 1978 *Phys Lett* **78B** 97-101, "Supersymmetry algebras that include topological charges"
Wu TT and Yang CN 1975 *Phys Rev* **D12** 3845-3857, "Concept of non-integrable phase factors and global formulation of gauge fields"
Zamolodchikov AB 1989 *Advanced Studies in Pure Mathematics* **19** 642-674, "Integrable Field Theory from Conformal field Theory"
Zwanziger D 1968 *Phys Rev* **176** 1489-1495, "Quantum field theory of particles with both electric and magnetic charges"

SUPERSYMMETRY AND DUALITIES IN VARIOUS DIMENSIONS

BERNARD DE WIT
Institute for Theoretical Physics, Utrecht University
Princetonplein 5, 3508 TA Utrecht, The Netherlands

AND

JAN LOUIS
Martin–Luther–Universität Halle–Wittenberg,
FB Physik, D-06099 Halle, Germany

1. Introduction

Since the seventies string theory has been discussed as a possible candidate for a theory which unifies all known particle interactions including gravity. Until recently, however, string theory has only been known in its perturbative regime. That is, the (particle) excitations of a string theory are computed in the free theory ($g_s = 0$), while their scattering processes are evaluated in a perturbative series for $g_s \ll 1$. The string coupling constant g_s is a free parameter of string theory but for $g_s = \mathcal{O}(1)$ no method of computing the spectrum or the interactions had been known. This situation dramatically changed during the past three years. For the first time it became possible to go beyond the purely perturbative regime and to reliably compute some of the nonperturbative properties of string theory. The central point of these developments rests on the idea that the strong-coupling limit of a given string theory can be described in terms of another, weakly coupled, 'dual theory'. This dual theory can take the form of either a different string theory, or the same string theory with a different set of perturbative excitations, or a new theory termed M-theory.

The precise nature of the strong-coupling limit sensitively depends on the number of (Minkowskian) spacetime dimensions and the amount of supersymmetry. Supersymmetry has played a major role in the recent developments in two respects. First of all, it is difficult (and it has not been satisfactorily accomplished) to rigourously prove a string duality, since it necessitates a full nonperturbative formulation, which is not yet available. Nevertheless it has been possible to perform nontrivial checks of the conjectured dualities for quantities or couplings whose quantum corrections are under (some) control. It is a generic property of supersymmetry that it protects a subset of the couplings and implies a set of non-

L. Baulieu et al. (eds.), Strings, Branes and Dualities, 33–101.

renormalization theorems. The recent developments heavily rely on the fact that the mass (or tension) of BPS-multiplets is protected and that holomorphic couplings obey a nonrenormalization theorem. Thus, they can be computed in the perturbative regime of string theory and, under the assumption of unbroken supersymmetry, reliably extrapolated into the nonperturbative region. It is precisely for these BPS-states and holomorphic couplings that the conjectured dualities have been successfully verified.

Second of all, for a given spacetime dimension D and a given representation of supersymmetry there can exist perturbatively different string theories. For example, the heterotic SO(32) string in $D = 10$ and the type-I string in $D = 10$ share the same supersymmetry, but their interactions are different in perturbation theory. However, once nonperturbative corrections are taken into account, it is believed that the two theories are identical and merely different perturbative limits of the same underlying quantum theory. A similar phenomenon is encountered with other string theories in different dimensions and the moduli space of string theory is much smaller than was previously known. In the course of these lectures we will see that many nontrivial relations exist among perturbatively distinct string vacua and furthermore, that, what were thought to be disconnected components of the moduli space, can in fact be different perturbative regions of one and the same component. Thus, in determining the properties of the underlying quantum theory, supersymmetry seems to play a much more prominent role than had previously been appreciated.

The purpose of these lectures is to review some of these recent developments with particular emphasis on the role played by supersymmetry.[1] In section 2 we collect the representations of supersymmetry in spacetime dimensions $3 \leq D \leq 11$ from a common point of view. Many features are only displayed in appropriate tables but we present slightly more detail in the dimensions $D = 11, 10$ and 6 as representative cases. We explain a number of features of the dimensional reduction of supergravity, such as the emergence of hidden symmetries, the low-energy action in different frames and other aspects relevant in the string context. In section 3 we first recall the different perturbative string theories, their (Calabi-Yau) compactifications and the dualities which already exist at the perturbative level. Then we discuss the various types of possible strong-coupling limits (S-duality, U-duality, M-theory, F-theory) and the corresponding string vacua. This leads to a 'web' of interrelations which we attempt to visualize at the end of section 3. Finally, in appendix A we review some basic properties of the free-field representation for states of different spin, while in appendix B we present a more detailed discussion of the relation between the parameters of string theory and those of the corresponding low-energy effective field theory.

[1]This set of lectures notes is an expanded version of [1].

D	$Q_{\rm irr}$	$H_{\rm R}$	type
1, 3, 9, 11, mod 8	$2^{(D-1)/2}$	SO(N)	M
5, 7, mod 8	$2^{(D+1)/2}$	USp(2N)	D
4, 8, mod 8	$2^{D/2}$	U(N)	M
6, mod 8	$2^{D/2}$	USp($2N_+$)×USp($2N_-$)	W
2, 10, mod 8	$2^{D/2-1}$	SO(N_+)×SO(N_-)	MW

TABLE 1. The supercharges in various spacetime dimensions D. In the second column, $Q_{\rm irr}$ specifies the real dimension of an irreducible spinor in a D-dimensional Minkowski spacetime. The third column specifies the group $H_{\rm R}$ for N-extended supersymmetry, defined in the text, acting on N-fold reducible spinor charges. The fourth column denotes the type of spinors: Majorana (M), Dirac (D), Weyl (W) and Majorana-Weyl (MW).

2. Supersymmetry in various dimensions

2.1. THE SUPERSYMMETRY ALGEBRA

An enormous amount of information about supersymmetric theories is contained in the structure of the underlying representations of the supersymmetry algebra [2]. Here we distinguish the supermultiplet of the fields, which transforms irreducibly under the supersymmetry transformations, and the supermultiplet of states described by the theory. The latter will depend on the dynamics encoded in a supersymmetric action or Hamiltonian. The generators of the super-Poincaré algebra comprise the supercharges, transforming as spinors under the Lorentz group, the energy and momentum operators, the generators of the Lorentz group, and possibly additional generators that commute with the supercharges. For the moment we ignore these additional charges, often called *central* charges[2]. There are other relevant superalgebras, such as the supersymmetric extension of the anti-de Sitter algebra. These will not be considered here, but they are unavoidable when considering supersymmetry in theories with a cosmological term.

Ignoring the central extensions for the moment, the most important anticommutation relation is the one between the supercharges,

$$\{Q_\alpha, \bar{Q}_\beta\} = -2iP^\mu\,(\Gamma_\mu)_{\alpha\beta}\,. \tag{2.1}$$

Here Γ_μ are the gamma matrices that generate the Clifford algebra with Minkowskian signature $(-,+,+,\cdots)$.

The size of a supermultiplet depends sensitively on the number of independent supercharge components Q. The first step is therefore to determine Q for any

[2]The terminology adopted in the literature is not always very precise. Usually, all charges that commute with the supercharges, but not necessarily with all the generators of the Poincaré algebra, are called 'central charges'. We will adhere to this nomenclature.

given number of spacetime dimensions D. The result is summarized in Table 1. As shown, there exist five different sequences of spinors, corresponding to spacetimes of particular dimensions. When this dimension is odd, it is possible in certain cases to have Majorana spinors. These cases constitute the first sequence. The second one corresponds to those odd dimensions where Majorana spinors do not exist. The spinors are then Dirac spinors. In even dimension one may distinguish three sequences. In the first one, where the number of dimensions is a multiple of 4, charge conjugation relates positive- with negative-chirality spinors. All spinors in this sequence can be restricted to Majorana spinors. For the remaining two sequences, charge conjugation preserves the chirality of the spinor. Now there are again two possibilities, depending on whether Majorana spinors can exist or not. The cases where we cannot have Majorana spinors, whenever $D = 6 \bmod 8$, comprise the fourth sequence. For the last sequence Majorana spinors exist and we restrict the charges to Majorana-Weyl spinors.

One can consider *extended supersymmetry*, where the spinor charges transform reducibly under the Lorentz group and comprise N irreducible spinors. For Weyl charges, one can consider combinations of N_+ positive- and N_- negative-chirality spinors. In all these cases there exists a group $H_{\rm R}$ that rotates the irreducible components such that the supersymmetry algebra is left invariant. This group acts exclusively on the spinor charges and commutes with the Lorentz transformations. The group $H_{\rm R}$ is thus the part of the automorphism group of the supersymmetry algebra that commutes with the Lorentz group. This group is often realized as a manifest invariance group of a supersymmetric field theory.

Another way to present the various cases is shown in Table 2. Here we list the real dimension of an irreducible spinor charge and its corresponding spacetime dimension. In addition we have included the number of states of the shortest[3] supermultiplet of massless states, written as a sum of bosonic and fermionic states.

2.2. MASSLESS REPRESENTATIONS

Because the momentum operators P^μ commute with the supercharges, we may consider the states at arbitrary but fixed momentum P^μ, which, for massless representations, satisfies $P^2 = 0$. The matrix $P^\mu\Gamma_\mu$ on the right-hand side of (2.1) has therefore zero eigenvalues. In a positive-definite Hilbert space some (linear combinations) of the supercharges must therefore vanish. To exhibit this more explicitly, let us rewrite (2.1) as

$$\{Q_\alpha, Q^\dagger_\beta\} = 2\,(\not{P}\Gamma^0)_{\alpha\beta}\,. \tag{2.2}$$

For light-like $P^\mu = (P^0, \vec{P})$ the right-hand side is proportional to a projection operator $(\mathbf{1}+\Gamma_{\parallel}\Gamma^0)/2$. Here $\Gamma_{\parallel}$ is the gamma matrix along the spatial momentum $\vec{P}$ of the states. The supersymmetry anti-commutator can then be written as

$$\{Q_\alpha, Q^\dagger_\beta\} = 2\,P^0\Big(\mathbf{1} + \tilde\Gamma_D\tilde\Gamma_\perp\Big)_{\alpha\beta}\,. \tag{2.3}$$

[3]By the *shortest* multiplet, we mean the multiplet with the helicities of the states as low as possible. This is usually (one of) the smallest possible supermultiplet(s).

$Q_{\rm irr}$	D	shortest supermultiplet
32	$D = 11$	$128 + 128$
16	$D = 10, 9, 8, 7$	$8 + 8$
8	$D = 6, 5$	$4 + 4$
4	$D = 4$	$2 + 2$
2	$D = 3$	$1 + 1$

TABLE 2. Simple supersymmetry in various dimensions. We present the dimension of the irreducible spinor charge with $2 \leq Q_{\rm irr} \leq 32$ and the corresponding spacetime dimensions D. The third column represents the number of bosonic + fermionic massless states for the shortest supermultiplet.

Here $\tilde{\Gamma}_D$ consists of the product of all D independent gamma matrices, and $\tilde{\Gamma}_\perp$ of the product of all $D-2$ gamma matrices in the transverse directions (i.e., perpendicular to $\vec{P}$), with phase factors such that

$$(\tilde{\Gamma}_D)^2 = (\tilde{\Gamma}_\perp)^2 = \mathbf{1}\,, \qquad [\tilde{\Gamma}_D, \tilde{\Gamma}_\perp] = 0\,. \tag{2.4}$$

This shows that the right-hand side of (2.3) is proportional to a projection operator, which projects out half of the spinor space. Consequently, half the spinors must vanish on physical states, whereas the other ones generate a Clifford algebra. Denoting the real dimension of the supercharges by Q, the representation space of the charges decomposes into the two chiral spinor representations of SO$(Q/2)$. When confronting these results with the last column in Table 2, it turns out that the dimension of the shortest supermultiplet is not just equal to $2^{Q_{\rm irr}/4}$, as one might naively expect. For $D = 6$, this is so because the representation is complex. For $D = 3, 4$ the representation is twice as big because it must also accommodate fermion number (or, alternatively, because it must satisfy the correct CPT properties). The derivation for $D = 4$ is presented in many places. For $D = 3$ we refer to [3].

The two chiral spinor spaces correspond to the bosonic and fermionic states, respectively. For the massless multiplets, the dimensions are shown in Table 2. Bigger supermultiplets can be obtained by combining various irreducible multiplets in a nontrivial way. We will demonstrate this below in three relevant cases, corresponding to $D = 11, 10$ and 6 spacetime dimensions. For the convenience of the reader we present Fig. 1, which lists the pure supergravity theories in dimensions $4 \leq D \leq 11$ with $Q = 32, 16, 8, 4$.[4] Some of these theories will be discussed later

[4]In $D = 4$ there exist additional theories with $Q = 12, 20$ and 24; in $D = 5$ there exists a theory with $Q = 24$ [4] and most likely there also exits a $Q = 24$ supergravity in $D = 6$. So far these supergravities have played no role in string theory and hence we omit them from our discussion here.

D \ Q	32	16	8	4
11	○			
10	○ ○ IIB IIA	○		
9	○	○		
8	○	○		
7	○	○		
6	○ (2,2)	○ ○ (2,0) (1,1)	○ (1,0)	
5	○	○	○	
4	○ N=8	○ N=4	○ N=2	○ N=1

Figure 1. Pure supergravity theories in dimensions $4 \leq D \leq 11$ with the number of independent supercharges equal to $Q = 32$, 16, 8 and 4. In 3 spacetime dimensions, pure supergravity does not describe propagating degrees of freedom and is a topological theory.

in more detail (in particular supergravity in $D = 11$ and 10 dimensions).

2.2.1. $D = 11$

In 11 dimensions we are dealing with 32 independent real supercharges. In odd-dimensional spacetimes irreducible spinors are subject to the eigenvalue condition $\tilde{\Gamma}_D = \pm\mathbf{1}$. Therefore (2.3) simplifies and shows that the 16 nonvanishing spinor charges transform according to the chiral spinor representation of the helicity group SO(9).

On the other hand, when regarding the 16 spinor charges as gamma matrices, it follows that the representation space constitutes the spinor representation of SO(16), which decomposes into two chiral subspaces, one corresponding to the bosons and the other one to the fermions. To determine the helicity content of the bosonic and fermionic states, one considers the embedding of the SO(9) spinor representation in the SO(16) vector transformation. It then turns out that one of the **128** representations branches into helicity representations according to $\mathbf{128} \to \mathbf{44} + \mathbf{84}$, while the second one transforms irreducibly according to the **128** representation of the helicity group.

The above states comprise precisely the massless states corresponding to $D = 11$ supergravity [5]. The graviton states transform in the **44**, the antisymmetric tensor states in the **84** and the gravitini states in the **128** representations of SO(9). Rather than showing all this in detail, we continue with other cases, where the representations are smaller and the group theory is more transparent. The helicity representations of the graviton, gravitino and tensor gauge fields are discussed in the appendix. Bigger supermultiplets consist of multiples of 256 states. For instance, without central charges, the smallest massive supermultiplet comprises $32768 + 32768$ states. These multiplets will not be considered here.

2.2.2. $D = 10$

In 10 dimensions the supercharges are both Majorana and Weyl spinors. The latter means that they are eigenspinors of $\tilde{\Gamma}_D$. According to (2.3), when we have simple (i.e., nonextended) supersymmetry with 16 charges, the nonvanishing charges transform in a chiral spinor representation of the SO(8) helicity group. With 8 nonvanishing supercharges we are dealing with an 8-dimensional Clifford algebra, whose irreducible representation space corresponds to the bosonic and fermionic states, each transforming according to a chiral spinor representation. Hence we are dealing with three 8-dimensional representations of SO(8), which are inequivalent. One is the representation to which we assign the supercharges, which we will denote by $\mathbf{8}_s$; to the other two, denoted as the $\mathbf{8}_v$ and $\mathbf{8}_c$ representations, we assign the bosonic and fermionic states, respectively. The fact that SO(8) representations appear in a three-fold variety is known as *triality*, which is a characteristic property of the group SO(8). With the exception of certain representations, such as the adjoint and the singlet representation, the three types of representation are inequivalent. They are traditionally distinguished by labels s, v and c (see, for instance, [6]).

The smallest massless supermultiplet has now been constructed with 8 bosonic and 8 fermionic states and corresponds to the vector multiplet of supersymmetric Yang-Mills theory in 10 dimensions [7]. Before constructing the supermultiplets that are relevant for $D = 10$ supergravity, let us first discuss some other properties of SO(8) representations. One way to distinguish the inequivalent representations, is to investigate how they decompose into representations of an SO(7) subgroup. Each of the 8-dimensional representations leaves a different SO(7) subgroup of SO(8) invariant. Therefore there is an SO(7) subgroup under which the $\mathbf{8}_v$ representation branches into

$$\mathbf{8}_v \longrightarrow \mathbf{7} + \mathbf{1}.$$

Under this SO(7) the other two 8-dimensional representations branch into

$$\mathbf{8}_s \longrightarrow \mathbf{8}\,, \qquad \mathbf{8}_c \longrightarrow \mathbf{8}\,,$$

supermultiplet	bosons	fermions
vector multiplet	$\mathbf{8}_v$	$\mathbf{8}_c$
graviton multiplet	$\mathbf{1}+\mathbf{28}+\mathbf{35}_v$	$\mathbf{8}_s+\mathbf{56}_s$
gravitino multiplet	$\mathbf{1}+\mathbf{28}+\mathbf{35}_c$	$\mathbf{8}_s+\mathbf{56}_s$
gravitino multiplet	$\mathbf{8}_v+\mathbf{56}_v$	$\mathbf{8}_c+\mathbf{56}_c$

TABLE 3. Massless $N=1$ supermultiplets in $D=10$ spacetime dimensions containing 8 + 8 or 64 + 64 bosonic and fermionic degrees of freedom.

where **8** is the spinor representation of SO(7). Corresponding branching rules for the 28-, 35- and 56-dimensional representations are

$$\begin{aligned} \mathbf{28} &\longrightarrow \mathbf{7}+\mathbf{21}\,, & & \\ \mathbf{35}_v &\longrightarrow \mathbf{1}+\mathbf{7}+\mathbf{27}\,, & \mathbf{56}_v &\longrightarrow \mathbf{21}+\mathbf{35}\,, \\ \mathbf{35}_{c,s} &\longrightarrow \mathbf{35}\,, & \mathbf{56}_{c,s} &\longrightarrow \mathbf{8}+\mathbf{48}\,. \end{aligned} \tag{2.5}$$

In order to obtain the supersymmetry representations relevant for supergravity we consider tensor products of the smallest supermultiplet consisting of $\mathbf{8}_v+\mathbf{8}_c$, with one of the 8-dimensional representations. There are thus three different possibilities, each leading to a 128-dimensional supermultiplet. Using the multiplication rules for SO(8) representations,

$$\begin{aligned} \mathbf{8}_v\times\mathbf{8}_v &= \mathbf{1}+\mathbf{28}+\mathbf{35}_v\,, & \mathbf{8}_v\times\mathbf{8}_s &= \mathbf{8}_c+\mathbf{56}_c\,, \\ \mathbf{8}_s\times\mathbf{8}_s &= \mathbf{1}+\mathbf{28}+\mathbf{35}_s\,, & \mathbf{8}_s\times\mathbf{8}_c &= \mathbf{8}_v+\mathbf{56}_v\,, \\ \mathbf{8}_c\times\mathbf{8}_c &= \mathbf{1}+\mathbf{28}+\mathbf{35}_c\,, & \mathbf{8}_c\times\mathbf{8}_v &= \mathbf{8}_s+\mathbf{56}_s\,, \end{aligned} \tag{2.6}$$

it is straightforward to obtain these new multiplets. Multiplying $\mathbf{8}_v$ with $\mathbf{8}_v+\mathbf{8}_c$ yields $\mathbf{8}_v\times\mathbf{8}_v$ bosonic and $\mathbf{8}_v\times\mathbf{8}_c$ fermionic states, and leads to the second supermultiplet shown in Table 3. This supermultiplet contains the representation $\mathbf{35}_v$, which can be associated with the states of the graviton in $D=10$ dimensions (the field-theoretic identification of the various states will be clarified in the appendix). Therefore this supermultiplet will be called the *graviton multiplet.* Multiplication with $\mathbf{8}_c$ or $\mathbf{8}_s$ goes in the same fashion, except that we will associate the $\mathbf{8}_c$ and $\mathbf{8}_s$ representations with fermionic quantities (note that these are the representations to which the fermion states of the Yang-Mills multiplet and the supersymmetry charges are assigned). Consequently, we interchange the boson and fermion assignments in these products. Multiplication with $\mathbf{8}_c$ then leads to $\mathbf{8}_c\times\mathbf{8}_c$ bosonic and $\mathbf{8}_c\times\mathbf{8}_v$ fermionic states, whereas multiplication with $\mathbf{8}_s$ gives $\mathbf{8}_s\times\mathbf{8}_c$ bosonic and $\mathbf{8}_s\times\mathbf{8}_v$ fermionic states. These supermultiplets contain fermions transforming according to the $\mathbf{56}_s$ and $\mathbf{56}_c$ representations, respectively, which can be associated with gravitino states (see the appendix for the helicity assignment of gravitino states), but no graviton states as those transform in the $\mathbf{35}_v$ representation. Therefore these two supermultiplets are called *gravitino multiplets.* We have thus established the existence of two inequivalent gravitino multiplets. The explicit SO(8) decompositions of the vector, graviton and gravitino supermultiplets are shown in Table 3.

By combining a graviton and a gravitino multiplet it is possible to construct an $N = 2$ supermultiplet of 128 + 128 bosonic and fermionic states. However, since there are two inequivalent gravitino multiplets, there will also be two inequivalent $N = 2$ supermultiplets containing the states corresponding to a graviton and two gravitini. According to the construction presented above, one $N = 2$ supermultiplet may be be viewed as the tensor product of two identical supermultiplets (namely $\mathbf{8}_v + \mathbf{8}_c$). Such a multiplet follows if one starts from a supersymmetry algebra based on *two* Majorana-Weyl spinor charges Q with the *same* chirality. The states of this multiplet decompose as follows:

Chiral $N = 2$ supermultiplet (IIB)

$$(\mathbf{8}_v + \mathbf{8}_c) \times (\mathbf{8}_v + \mathbf{8}_c) \implies \begin{cases} \textbf{bosons}: \\ \mathbf{1} + \mathbf{1} + \mathbf{28} + \mathbf{28} + \mathbf{35}_v + \mathbf{35}_c \\ \\ \textbf{fermions}: \\ \mathbf{8}_s + \mathbf{8}_s + \mathbf{56}_s + \mathbf{56}_s \end{cases} \tag{2.7}$$

This is the multiplet corresponding to IIB supergravity [8]. Because the supercharges have the same chirality, one can perform rotations between these spinor charges which leave the supersymmetry algebra unaffected. Hence the automorphism group H_{R} is equal to SO(2). This feature reflects itself in the multiplet decomposition, where the $\mathbf{1}$, $\mathbf{8}_s$, $\mathbf{28}$ and $\mathbf{56}_s$ representations are degenerate and constitute doublets under this SO(2) group.

A second supermultiplet may be viewed as the tensor product of a $(\mathbf{8}_v + \mathbf{8}_s)$ supermultiplet with a second supermultiplet $(\mathbf{8}_v + \mathbf{8}_c)$. In this case the supercharges constitute two Majorana-Weyl spinors of opposite chirality. Now the supermultiplet decomposes as follows:

Nonchiral $N = 2$ supermultiplet (IIA)

$$(\mathbf{8}_v + \mathbf{8}_s) \times (\mathbf{8}_v + \mathbf{8}_c) \implies \begin{cases} \textbf{bosons}: \\ \mathbf{1} + \mathbf{8}_v + \mathbf{28} + \mathbf{35}_v + \mathbf{56}_v \\ \\ \textbf{fermions}: \\ \mathbf{8}_s + \mathbf{8}_c + \mathbf{56}_s + \mathbf{56}_c \end{cases} \tag{2.8}$$

This is the multiplet corresponding to IIA supergravity [9]. It can be obtained by a straightforward reduction of $D = 11$ supergravity. The latter follows from the fact that two $D = 10$ Majorana-Weyl spinors with opposite chirality can be combined into a single $D = 11$ Majorana spinor. The formula below summarizes the massless states of IIA supergravity from an 11-dimensional perspective. The massless states of 11-dimensional supergravity transform according to the $\mathbf{44}$, $\mathbf{84}$ and $\mathbf{128}$ representation of the helicity group SO(9). They correspond to the degrees of freedom described by the metric, a 3-rank antisymmetric gauge field and the gravitino field, respectively. We also show how the 10-dimensional states can subsequently be branched into 9-dimensional states, characterized in terms of representations of the helicity group SO(7):

$$
\begin{array}{lcl}
\mathbf{44} & \Longrightarrow & \left\{\begin{array}{lcl} \mathbf{1} & \longrightarrow & \mathbf{1} \\ \mathbf{8}_v & \longrightarrow & \mathbf{1}+\mathbf{7} \\ \mathbf{35}_v & \longrightarrow & \mathbf{1}+\mathbf{7}+\mathbf{27} \end{array}\right. \\
\mathbf{84} & \Longrightarrow & \left\{\begin{array}{lcl} \mathbf{28} & \longrightarrow & \mathbf{7}+\mathbf{21} \\ \mathbf{56}_v & \longrightarrow & \mathbf{21}+\mathbf{35} \end{array}\right. \\
\mathbf{128} & \Longrightarrow & \left\{\begin{array}{lcl} \mathbf{8}_s & \longrightarrow & \mathbf{8} \\ \mathbf{8}_c & \longrightarrow & \mathbf{8} \\ \mathbf{56}_s & \longrightarrow & \mathbf{8}+\mathbf{48} \\ \mathbf{56}_c & \longrightarrow & \mathbf{8}+\mathbf{48} \end{array}\right.
\end{array}
\qquad (2.9)
$$

Clearly, in $D=9$ we have a higher degeneracy of states, related to the automorphism group SO(2). We note the presence of graviton and gravitino states, transforming in the **27** and **48** representations.

One could also take the states of the IIB supergravity and decompose them into $D=9$ massless states. This leads to precisely the same supermultiplet as the reduction of the states of IIA supergravity. Indeed, the reductions of IIA and IIB supergravity to 9 dimensions, yield the same theory [10, 11, 12]. To see this at the level of the Lagrangian requires certain duality transformations, which we discuss in section 3. Hence $Q=32$ supergravity is unique in all spacetime dimensions, except for $D=10$. Maximal supergravity will be discussed in subsection 2.4. The field-content of the maximal $Q=32$ supergravity theories for dimensions $3 \leq D \leq 11$ will be presented in two tables (cf. Table 6 and 7).

2.2.3. $D=6$

In 6 dimensions we have chiral spinors, which are not Majorana. The charge conjugated spinor has the same chirality, so that the chiral rotations of the spinors can be extended to the group USp($2N_+$), for N_+ chiral spinors. Likewise N_- negative-chirality spinors transform under USp($2N_-$). This is already incorporated in Table 1. In principle we have N_+ positive- and N_- negative-chirality charges, but almost all information follows from first considering the purely chiral case. In Table 4 we present the decomposition of the various helicity representations of the smallest supermultiplets based on $N_+ = 1,2,3$ or 4 supercharges. In $D=6$ dimensions the helicity group SO(4) decomposes into the product of two SU(2) groups: SO(4)$\cong$ (SU$_+$(2) $\times$ SU$_-$(2))$/Z_2$. When we have supercharges of only one chirality, the smallest supermultiplet will only transform under one SU(2) factor of the helicity group, as is shown in Table 4.[5]

Let us now consider specific supermultiplets. All these multiplets are summarized in Table 5. The helicity assignments of the states described by gravitons, gravitini, vector and tensor gauge fields, and spinor fields are presented in the appendix. The simplest case is $(N_+,N_-)=(1,0)$, where the smallest supermultiplet is the (1,0) *hypermultiplet*, consisting of a complex doublet of spinless states and

[5]The content of this table also specifies the smallest *massive* supermultiplets in four dimensions. The SU(2) group is then associated with spin in three space dimensions.

$SU_+(2)$	$N_+ = 1$	$N_+ = 2$	$N_+ = 3$	$N_+ = 4$
5				1
4			1	8
3		1	6	27
2	1	4	14	48
1	2	5	14	42
	$(2+2)_\mathbf{C}$	$(8+8)_\mathbf{R}$	$(32+32)_\mathbf{C}$	$(128+128)_\mathbf{R}$

TABLE 4. Shortest massless supermultiplets of $D = 6$ N_+-extended chiral supersymmetry. The states transform both in the $SU_+(2)$ helicity group and under a $USp(2N_+)$ group. For odd values of N_+ the representations are complex, for even N_+ they can be chosen real. Of course, an identical table can be given for negative-chirality spinors.

a chiral spinor. Taking the tensor product of the smallest supermultiplet with the $(2,1)$ helicity representation gives the (1,0) *tensor multiplet*, with a selfdual tensor, a spinless state and a doublet of chiral spinors. The tensor product with the $(1,2)$ helicity representation yields the (1,0) *vector multiplet*, with a vector state, a doublet of chiral spinors and a scalar. Multiplying the latter with the (2,3) helicity representation, one obtains the states of (1,0) *supergravity*. Observe that the selfdual tensor fields in the tensor and supergravity supermultiplet are of opposite selfduality phase.

Next consider $(N_+, N_-) = (2,0)$ supersymmetry. The smallest multiplet, shown in Table 4, then corresponds to the (2,0) *tensor multiplet*, with the bosonic states decomposing into a selfdual tensor, and a five-plet of spinless states, and a four-plet of chiral fermions. Multiplication with the (1,3) helicity representation yields the (2,0) supergravity multiplet, consisting of the graviton, four chiral gravitini and five selfdual tensors [13]. Again, the selfdual tensors of the tensor and of the supergravity supermultiplet are of opposite selfduality phase.

Of course, there exists also a nonchiral version with 16 supercharges, namely the one corresponding to $(N_+, N_-) = (1,1)$. The smallest multiplet is now given by the tensor product of the supermultiplets with (1,0) and (0,1) supersymmetry. This yields the vector multiplet, with the vector state and four scalars, the latter transforming with respect to the (2,2) representation of $USp(2) \times USp(2)$. There are two doublets of chiral fermions with opposite chirality, each transforming as a doublet under the corresponding USp(2) group. Taking the tensor product of the vector multiplet with the (2,2) representation of the helicity group yields the states of the (1, 1) *supergravity* multiplet. It consists of 32 bosonic states, corresponding to a graviton, a tensor, a scalar and four vector states, where the latter transform under the (2,2) representation of $USp(2) \times USp(2)$. The 32 fermionic states comprise two doublets of chiral gravitini and two chiral spinor doublets, transforming

multiplet	#	bosons	fermions
(1,0) hyper	$4+4$	$(1,1;2,1)+\text{h.c.}$	$(2,1;1,1)$
(1,0) tensor	$4+4$	$(3,1;1,1)+(1,1;1,1)$	$(2,1;2,1)$
(1,0) vector	$4+4$	$(2,2;1,1)$	$(1,2;2,1)$
(1,0) supergravity	$12+12$	$(3,3;1,1)+(1,3;1,1)$	$(2,3;2,1)$
(2,0) tensor	$8+8$	$(3,1;1,1)+(1,1;5,1)$	$(2,1;4,1)$
(2,0) supergravity	$24+24$	$(3,3;1,1)+(1,3;5,1)$	$(2,3;4,1)$
(1,1) vector	$8+8$	$(2,2;1,1)+(1,1;2,2)$	$(2,1;1,2)+(1,2;2,1)$
(1,1) supergravity	$32+32$	$(3,3;1,1)$ $+(1,3;1,1)+(3,1;1,1)$ $+(1,1;1,1)+(2,2;2,2)$	$(3,2;1,2)+(2,3;2,1)$ $+(1,2;1,2)+(2,1;2,1)$
(2,2) supergravity	$128+128$	$(3,3;1,1)$ $+(3,1;1,5)+(1,3;5,1)$ $+(2,2;4,4)+(1,1;5,5)$	$(3,2;4,1)+(2,3;1,4)$ $+(2,1;4,5)+(1,2;5,4)$

TABLE 5. Some relevant $D=6$ supermultiplets with (N_+,N_-) supersymmetry. The states $(n,m;\tilde{n},\tilde{m})$ are assigned to (n,m) representations of the helicity group $\mathrm{SU}_+(2)\times\mathrm{SU}_-(2)$ and $(\tilde{n},\tilde{m})$ representations of $\mathrm{USp}(2N_+)\times\mathrm{USp}(2N_-)$. The second column lists the number of bosonic + fermionic states for each multiplet.

as doublets under the appropriate USp(2) group.

Finally, we turn to the case of $(N_+,N_-)=(2,2)$. The smallest supermultiplet is given by the tensor product of the smallest (2,0) and (0,2) supermultiplets. This yields the 128 + 128 states of the (2,2) *supergravity* multiplet. These states transform according to representations of $\mathrm{USp}(4)\times\mathrm{USp}(4)$.

In principle, one can continue and classify representations for other values of (N_+,N_-). As is obvious from the construction that we have presented, this will inevitably lead to states transforming in higher-helicity representations. As we will discuss in subsection 2.4, the higher-spin gauge fields associated with these representations can not be coupled to gravity. Although the representations exist and can be described by appropriate free-field theories, they have no future as nontrivial quantum field theories.

2.3. MAXIMAL SUPERSYMMETRY: $Q\leq 32$

In the above we have restricted ourselves to (massless) supermultiplets based on $Q\leq 32$ supercharge components. From the general analysis it is clear that increasing the number of supercharges leads to higher and higher helicity representations. Obviously some of these representations will also occur in lower-Q supermultiplets, by multiplying shorter multiplets by suitable helicity representations. It is not so easy to indicate in arbitrary dimension what we mean by a higher helicity representation, but we have in mind those representations that are described by gauge fields that are *symmetric* Lorentz tensors. Symmetric tensor gauge fields for arbitrary helicity states can be constructed (in four dimensions, see, for instance, [14]). However, it turns out that symmetric gauge fields cannot consistently couple to themselves or to other fields. An exception is the graviton field, which can interact with itself as well as to low-spin matter, but not to other higher-spin gauge fields. By consistent, we mean that their respective gauge invariances of the higher-spin fields (or appropriate deformations thereof) cannot be preserved at the interacting level.

There have been many efforts to circumvent this apparent no-go theorem. What seems clear, is that one needs a combination of the following ingredients in order to do this (for a recent review, see [15]): (i) an infinite tower of higher-spin gauge fields; (ii) interactions that are inversely proportional to the cosmological constant; (iii) extensions of the super-Poincaré or the super-de Sitter algebra with additional fermionic and bosonic charges.

Conventional supergravity theories are not of this kind. This is the reason why we have avoided (i.e. in Table 5) to list supermultiplets with states transforming in higher-helicity representations. The bound $Q \leq 32$ originates from the necessity of avoiding the higher-spin fields. It implies that supergravity does not exist for spacetime dimensions $D \geq 11$ (at least, if one assumes a single time coordinate), because Lorentz spinors have more than 32 components beyond $D = 11$ [16].

Most of the search for interacting higher-spin fields was performed in four spacetime dimensions [17]. When one increases the number of supercharges beyond $Q = 32$, then a supermultiplet will contain several massless states of spin-2 and at least spin-5/2 fermions. In higher spacetime dimensions, more than 32 supercharges are excluded (in the absence of higher-spin gauge fields), because, upon dimensional reduction, these theories would give rise to theories that are inconsistent in $D = 4$. There is also direct evidence in $D = 3$, where graviton and gravitini fields do not describe dynamic degrees of freedom. Hence, one can write down supergravity theories based on a graviton field and an arbitrary number of gravitino fields, which are topological. However, when coupling matter to this theory, described by scalars and spinors, the theory supports not more than 32 supercharges. Beyond $Q = 16$ there are four unique theories with $Q = 18, 20, 24$ and 32 [3].

2.4. MAXIMAL SUPERGRAVITIES

In this section we review the maximal supergravities in various dimensions. These theories have precisely $Q = 32$ supercharge components. We restrict our discussion to $3 \leq D \leq 11$.

The bosonic fields always comprise the metric tensor for the graviton field and a number of antisymmetric gauge fields. For the antisymmetric gauge fields, it is a priori unclear whether to choose a $(p+1)$-rank gauge field or its dual $(D-3-p)$-rank partner, but it turns out that the interactions often prefer the rank of the gauge field to be as small as possible. Therefore, in Table 6, we restrict ourselves to

D	$H_{\rm R}$	graviton	$p=-1$	$p=0$	$p=1$	$p=2$	$p=3$
11	1	1	0	0	0	1	0
10A	1	1	1	1	1	1	0
10B	SO(2)	1	2	0	2	0	1*
9	SO(2)	1	$2+1$	$2+1$	2	1	
8	U(2)	1	$5+1+\bar{1}$	$3+\bar{3}$	3	[1]	
7	USp(4)	1	14	10	5		
6	USp(4)×USp(4)	1	(5,5)	(4,4)	$(5,1)+(1,5)$		
5	USp(8)	1	42	27			
4	U(8)	1	$35+\overline{35}$	[28]			
3	SO(16)	1	128				

TABLE 6. Bosonic field content for maximal supergravities. The $p=3$ gauge field in $D=10$B has a self-dual field strength. The representations [1] and [28] (in $D=8,4$, respectively) are extended to U(1) and SU(8) representations through duality transformations on the field strengths. These transformations can not be represented on the vector potentials. In $D=3$ dimensions, the graviton does not describe propagating degrees of freedom.

$p\leq 3$, as in $D=11$ dimensions, $p=3$ and $p=4$ are each other's dual conjugates. This table presents all the field configurations for maximal supergravity in various dimensions. Obviously, the problematic higher-spin fields are avoided, because the only symmetric gauge field is the one describing the graviton. In Table 7 we also present the fermionic fields, always consisting of gravitini and simple spinors. All these fields are classified as representations of the automorphism group $H_{\rm R}$. In order to compare these tables to similar tables in the literature, one may need to use the (local) equivalences: USp(4)~SO(5), USp(2)~SU(2) and SU(4)~SO(6).

The supersymmetry algebra of the maximal supergravities comprises general coordinate transformations, local supersymmetry transformations and the gauge transformations associated with the antisymmetric gauge fields[6]. These gauge transformations usually appear in the anticommutator of two supercharges, and may be regarded as central charges. In perturbation theory, the theory does not contain charged fields, so these central charges simply vanish on physical states. However, at the nonperturbative level, there may be solitonic or other states that carry charges. An example are magnetic monopoles, dyons, or extremal black holes. On such states, some of the central charges may take finite values. Without further knowledge about the kind of states that may emerge at the nonperturbative level,

[6]There may be additional gauge transformations that are of interest to us. As we discuss in subsection 2.5.1, it is possible to have (part of the) the automorphism group $H_{\rm R}$ realized as a local invariance. However, the corresponding gauge fields are then composite and do not give rise to physical states (at least, not in perturbation theory).

D	$H_{\rm R}$	gravitini	spinors
11	1	1	0
10A	1	1+1	1+1
10B	SO(2)	2	2
9	SO(2)	2	$2+2$
8	U(2)	$2+\bar{2}$	$2+\bar{2}+4+\bar{4}$
7	USp(4)	4	16
6	USp(4)×USp(4)	$(4,1)+(1,4)$	$(4,5)+(5,4)$
5	USp(8)	8	42
4	U(8)	$8+\bar{8}$	$56+\overline{56}$
3	SO(16)	16	128

TABLE 7. Fermionic field content for maximal supergravities. For $D = 5,6,7$ the fermion fields are counted as symplectic Majorana spinors. For $D = 4,8$ we include both chiral and antichiral spinor components, which transform in conjugate representations of $H_{\rm R}$. In $D = 3$ dimensions the gravitino does not correspond to propagating degrees of freedom.

we can generally classify the possible central charges, by considering a decomposition of the anticommutator. This anticommutator carries at least two spinor indices and two indices associated with the group $H_{\rm R}$. Hence we may write

$$\{Q_\alpha, Q_\beta\} \propto \sum_r (\Gamma^{\mu_1\cdots\mu_r}C)_{\alpha\beta}\, Z_{\mu_1\cdots\mu_r}\,, \tag{2.10}$$

where $\Gamma^{\mu_1\cdots\mu_r}$ is the antisymmetrized product of r gamma matrices, C is the charge-conjugation matrix and $Z_{\mu_1\cdots\mu_r}$ is the central charge, which transforms as an antisymmetric r-rank Lorentz tensor and depends on possible additional $H_{\rm R}$ indices attached to the supercharges. The central charge must be symmetric or antisymmetric in these indices, depending on whether the product of the gamma matrices with C is symmetric or antisymmetric, so that the product is symmetric in the combined indices of the supercharges. For given spacetime dimension all possible central charges can be classified.[7] For the maximal supergravities in spacetime dimensions $3 \leq D \leq 11$ this classification is given in Table 8. Because we have 32 supercharge components, the sum of the independent momentum operators and the central charges must be equal to $(32\times 33)/2 = 528$.

[7]For a related discussion see for example [18, 19] and references therein.

D	$H_{\rm R}$	$r=0$	$r=1$	$r=2$	$r=3$	$r=4$	$r=5$
11	1			1			1
				[55]			[462]
10A	1	1	1	1		1	$1+1$
		[1]	[10]	[45]		[210]	[126]
10B	SO(2)		2		1		$1+2$
			[10]		[120]		[126]
9	SO(2)	$1+2$	2	1	1	$1+2$	
		[1]	[9]	[36]	[84]	[126]	
8	U(2)	$3+\bar{3}$	3	$1+\bar{1}$	$1+3$	$3+\bar{3}$	
		[1]	[8]	[28]	[56]	[35]	
7	USp(4)	10	5	$1+5$	10		
		[1]	[7]	[21]	[35]		
6	USp(4)×USp(4)	$(4,4)$	$(1,1)+(5,1)$	(4,4)	$(10,1)$		
			$+(1,5)$		$+(1,10)$		
		[1]	[6]	[15]	[10]		
5	USp(8)	$1+27$	27	36			
		[1]	[5]	[10]			
4	U(8)	$28+\overline{28}$	63	$36+\overline{36}$			
		[1]	[4]	[3]			
3	SO(16)	120	135				
		[1]	[3]				

TABLE 8. Decomposition of the central extension in the supersymmetry algebra with $Q=32$ supercharge components in terms of r-rank Lorentz tensors. The second row specifies the number of independent components for each r-rank tensor charge. The total number of central charges is equal to $528-D$, because we have not listed the D independent momentum operators

2.4.1. $D=11$

Supergravity in 11 spacetime dimensions is based on an "elfbein" field $E_M^{\;A}$, a Majorana gravitino field Ψ_M and a 3-rank antisymmetric gauge field C_{MNP}. With chiral (2,0) supergravity in 6 dimensions, it is the only $Q\geq 16$ supergravity theory without a scalar field. Its Lagrangian can be written as follows [5],

$$\begin{aligned}\mathcal{L}_{11} &= \frac{1}{\kappa_{11}^2}\Big[-\tfrac{1}{2}E\,R(E,\Omega)-\tfrac{1}{2}E\bar{\Psi}_M\Gamma^{MNP}D_N(\Omega)\Psi_P-\tfrac{1}{48}E(F_{MNPQ})^2\\ &\qquad -\tfrac{1}{3456}\sqrt{2}\,\varepsilon^{MNPQRSTUVWX}\,F_{MNPQ}\,F_{RSTU}\,C_{VWX}\end{aligned} \tag{2.11}$$

$$-\tfrac{1}{192}\sqrt{2}E\Big(\bar\Psi_R\Gamma^{MNPQRS}\Psi_S+12\,\bar\Psi^M\Gamma^{NP}\Psi^Q\Big)F_{MNPQ}+\cdots\Big]\,,$$

where the ellipses denote terms of order Ψ^4, $E=\det E_M^A$ and $\Omega_M{}^{AB}$ denotes the spin connection. The supersymmetry transformations are equal to

$$\begin{aligned}\delta E_M^A &= \tfrac12\bar\epsilon\Gamma^A\Psi_M\,,\\ \delta C_{MNP} &= -\tfrac18\sqrt2\,\bar\epsilon\,\Gamma_{[MN}\Psi_{P]}\,,\\ \delta\Psi_M &= D_M(\hat\Omega)\,\epsilon+\tfrac{1}{288}\sqrt2\left(\Gamma_M{}^{NPQR}-8\,\delta_M^N\,\Gamma^{PQR}\right)\epsilon\,\hat F_{NPQR}\,.\end{aligned}\tag{2.12}$$

Here the covariant derivative is covariant with respect to local Lorentz transformation

$$D_M(\Omega)\,\epsilon=\Big(\partial_M-\tfrac14\Omega_M{}^{AB}\Gamma_{AB}\Big)\epsilon\,,\tag{2.13}$$

and $\hat F_{MNPQ}$ is the supercovariant field strength

$$\hat F_{MNPQ}=24\,\partial_{[M}C_{NPQ]}+\tfrac32\sqrt2\,\bar\Psi_{[M}\Gamma_{NP}\Psi_{Q]}\,.\tag{2.14}$$

Note the presence in the Lagrangian of a Chern-Simons-like term $F\wedge F\wedge C$, so that the action is only invariant up to surface terms. We also wish to point out that the quartic-Ψ terms can be included into the Lagrangian (2.11) by replacing the spin-connection field Ω by $(\Omega+\hat\Omega)/2$ in the covariant derivative of the gravitino kinetic term and by replacing F_{MNPQ} in the last line by $(\hat F_{MNPQ}+F_{MNPQ})/2$. These substitutions ensure that the field equations corresponding to (2.11) are supercovariant. The Lagrangian is derived in the context of the so-called "1.5-order" formalism, in which the spin connection is defined as a dependent field determined by its (algebraic) equation of motion, whereas its supersymmetry variation in the action is treated as if it were an independent field [20]. The supercovariant spin connection is the solution of the following equation,

$$D_{[M}(\hat\Omega)\,E_{N]}^A-\tfrac14\bar\Psi_M\Gamma^A\Psi_N=0\,.\tag{2.15}$$

The left-hand side is the supercovariant torsion tensor.

We have the following bosonic field equations and Bianchi identities,

$$\begin{aligned}R_{MN} &= \tfrac{1}{72}g_{MN}\,F_{PQRS}F^{PQRS}-\tfrac16F_{MPQR}\,F_N{}^{PQR}\,,\\ \partial_M\Big(E\,F^{MNPQ}\Big) &= \tfrac{1}{1152}\sqrt2\,\varepsilon^{NPQRSTUVWXY}\,F_{RSTU}\,F_{VWXY}\,,\\ \partial_{[M}F_{NPQR]} &= 0\,,\end{aligned}\tag{2.16}$$

which no longer depend on the antisymmetric gauge field. An alternative form of the second equation is [21]

$$\partial_{[M}H_{NPQRSTU]}=0\,,\tag{2.17}$$

where $H_{MNPQRST}$ is the dual field strength,

$$H_{MNPQRST}=\frac{1}{7!}E\,\varepsilon_{MNPQRSTUVWX}F^{UVWX}-\tfrac12\sqrt2\,F_{[MNPQ}\,C_{RST]}\,.\tag{2.18}$$

One could imagine that the third equation of (2.16) and (2.17) receive contributions from charges that would give rise to source terms on the right-hand side of the equations. These charges are associated with the 'flux'-integral of $H_{MNPQRST}$ and F_{MNPQ} over the boundary of an 8- and a 5-dimensional spatial volume, respectively. This volume is transverse to a $p = 2$ and $p = 5$ brane configuration, and the corresponding charges are 2- and 5-rank Lorentz tensors. These are just the charges that can appear as central charges in the supersymmetry algebra, as one can verify in Table 8. Solutions of 11-dimensional supergravity that contribute to these charges were considered in [22, 23, 24, 18].

It is straightforward to evaluate the supersymmetry algebra on these fields. The commutator of two supersymmetry transformations yields a general-coordinate transformation, a local Lorentz transformation, a supersymmetry transformation and a gauge transformation associated with the tensor gauge field,

$$[\delta(\epsilon_1), \delta(\epsilon_2)] = \delta_{\rm gct}(\xi^M) + \delta_Q(\epsilon_3) + \delta_L(\lambda^{AB}) + \delta_A(\xi_{MN})\,. \tag{2.19}$$

The parameters of the transformations on the right-hand side are given by

$$\begin{aligned} \xi^M &= \tfrac{1}{2}\bar\epsilon_2\Gamma^M\epsilon_1\,, \\ \epsilon_3 &= -\xi^M\Psi_M\,, \\ \lambda^{AB} &= -\xi^M\hat\Omega_M^{AB} + \tfrac{1}{288}\sqrt{2}\,\bar\epsilon_2\Big[\Gamma^{ABCDEF}\hat F_{CDEF} + 24\,\Gamma_{CD}\hat F^{ABCD}\Big]\epsilon_1\,, \\ \xi_{MN} &= -\tfrac{1}{8}\sqrt{2}\,\bar\epsilon_2\Gamma_{MN}\epsilon_1\,. \end{aligned} \tag{2.20}$$

Note that the normalizations differ from the ones used in the supersymmetry algebra in previous subsections. The tensor gauge field transforms under gauge transformations as $\delta C_{MNP} = \partial_{[M}\xi_{NP]}$.

Finally, the constant $1/\kappa_{11}^2$ in front of the Lagrangian (2.11), which has the dimension $[\text{length}]^{-9} \sim [\text{mass}]^9$, is undetermined and depends on fixing some length scale. To see this consider a continuous rescaling of the fields,

$$E_M^{\,A} \to \mathrm{e}^{-\alpha}E_M^{\,A}\,,\qquad \Psi_M \to \mathrm{e}^{-\alpha/2}\Psi_M\,,\qquad C_{MNP} \to \mathrm{e}^{-3\alpha}C_{MNP}\,. \tag{2.21}$$

Under this rescaling the Lagrangian changes according to

$$\mathcal{L}_{11} \to \mathrm{e}^{-9\alpha}\mathcal{L}_{11}\,. \tag{2.22}$$

This change can then be absorbed into a redefinition of κ_{11},[8]

$$\kappa_{11}^2 \to \mathrm{e}^{-9\alpha}\kappa_{11}^2\,. \tag{2.23}$$

The indetermination of κ is not a special property of $D = 11$ but occurs in any spacetime dimension where the Einstein-Hilbert action displays a similar scaling property,

$$g_{\mu\nu}^D \to \mathrm{e}^{-2\alpha}g_{\mu\nu}^D\,,\qquad \mathcal{L}_D \to \mathrm{e}^{(2-D)\alpha}\mathcal{L}_D\,,\qquad \kappa_D^2 \to \mathrm{e}^{(2-D)\alpha}\kappa_D^2\,. \tag{2.24}$$

[8]Note that the rescalings also leave the supersymmetry transformation rules unchanged, provided the supersymmetry parameter ϵ is changed accordingly.

Newton's constant, $(\kappa_D^2)^{\rm physical}$, does not necessarily coincide with the parameter κ_D^2 but also depends on the precise value adopted for the (flat) metric in the ground state of the theory. Up to certain convention-dependent normalization factors one defines

$$(\kappa_D^2)^{\rm physical} := \kappa_D^2 \, \lambda^{(2-D)/2} \,, \tag{2.25}$$

where $g_{\mu\nu}^D$ is expanded about $\lambda\, \eta_{\mu\nu}^D$, with $\eta_{\mu\nu}^D$ equal to the Lorentz-invariant flat metric with diagonal elements equal to ± 1. Note that $(\kappa_D^2)^{\rm physical}$ is invariant under the scale transformations (2.24) and thus a physically meaningful scale.

When the Lagrangian contains additional terms, for instance, of higher order in the Riemann tensor, then the corresponding coupling constant will scale differently under (2.24). Its physical value will therefore depend in a different way on the parameter λ that parametrizes the (flat) metric in the ground state. An even simpler example is a scalar massive field, added to the Einstein-Hilbert Lagrangian. Its physical mass is equal to λ times the mass parameter in the Lagrangian. However, we should stress that the physics never depends *explicitly* on λ, provided one expresses all physical quantities in terms of physical parameters, all determined for the same value of λ. We return to the issue of frames and scales in section 2.5.1 and in appendix B.

2.5. DIMENSIONAL REDUCTION AND HIDDEN SYMMETRIES

The maximal supergravities in various dimensions are related by dimensional reduction. Here some of the spatial dimensions are compactified on a hyper-torus whose size is shrunk to zero. In this situation some of the gauge symmetries that are related to the compactified dimensions survive and take the form of internal symmetries. The aim of our discussion here is to elucidate a number of features related to these symmetries, mainly in the context of the reduction of $D = 11$ supergravity to $D = 10$ dimensions.

We denote the compactified coordinate by x^{11} which now parameterizes a circle of length L.[9] The fields are thus decomposed as periodic functions in x^{11} on the interval $0 \leq x^{11} \leq L$. This results in a spectrum of massless modes and an infinite tower of massive modes. The massless modes form the basis of the lower-dimensional supergravity theory. Because a toroidal background does not break supersymmetry, the resulting supergravity has the same number of supersymmetries as the original one. For compactifications on less trivial spaces than the hyper-torus (which we will discuss in section 3) this is not necessarily the case and the number of independent supersymmetries can be reduced. Actually, fully supersymmetric compactifications are rare. For instance, in 11-dimensional supergravity 7 coordinates can be compactified in precisely two ways such that all supersymmetries remain unaffected [25]. One is the compactification on a torus T^7, the other one the compactification of a sphere S^7. However, in the latter case the resulting 4-dimensional supergravity theory acquires a cosmological term. In

[9]Throughout these lectures we enumerate spacetime coordinates by $0, 1, \ldots, D-1$. Nevertheless, we denote the compactified coordinate by x^{11}, to indicate that it is the eleventh spacetime coordinate.

the context of these lectures, such compactifications are less relevant and will not be discussed.

In the formulation of the compactified theory, it is important to decompose the higher-dimensional fields in such a way that they transform covariantly under the lower-dimensional gauge symmetries, and in particular under diffeomorphisms of the lower-dimensional spacetime. This ensures that various complicated mixtures of massless modes with the tower of massive modes will be avoided. It is a key element in ensuring that solutions of the lower-dimensional theory remain solutions of the original higher-dimensional one. Another point of interest concerns the nature of the massive supermultiplets. Because these originate from supermultiplets that are massless in higher dimensions, these multiplets must be shortened by the presence of central charges. The central charge here originates from the momentum operator in the compactified dimension. We return to this issue shortly.

The emergence of new internal symmetries in theories that originate from a higher-dimensional setting, is a standard feature of Kaluza-Klein theories [26]. Following the discussion in [27] we distinguish between symmetries that have a direct explanation in terms of the symmetries in higher dimensions, and symmetries whose origin is obscure from a higher-dimensional viewpoint. Let us start with the symmetries associated with the metric tensor. The 11-dimensional metric can be decomposed according to

$$\mathrm{d}s^2 = g_{\mu\nu}\,\mathrm{d}x^\mu \mathrm{d}x^\nu + \mathrm{e}^{4\phi/3}(\mathrm{d}x^{11} + V_\mu \mathrm{d}x^\mu)(\mathrm{d}x^{11} + V_\nu \mathrm{d}x^\nu)\,, \tag{2.26}$$

where the indices μ, ν label the 10-dimensional coordinates and the factor multiplying ϕ is for convenience later. The massless modes correspond to the x^{11}-independent parts of the 10-dimensional metric $g_{\mu\nu}$, the vector field V_μ and the scalar ϕ. Here the x^{11}-independent component of V_μ acts as a gauge field associated with reparametrizations of the circle coordinate x^{11} with an arbitrary function $\xi(x)$ of the 10 remaining spacetime coordinates x^μ. Specifically, we have $x^{11} \to x^{11} - \xi(x)$ and $x^\mu \to x^\mu$, corresponding to

$$V_\mu(x) \to V_\mu(x) + \partial_\mu \xi(x)\,. \tag{2.27}$$

The massive modes, which correspond to the Fourier modes in terms of x^{11}, couple to this gauge field with a charge that is a multiple of

$$e_{\rm KK} = \frac{2\pi}{L}\,. \tag{2.28}$$

Another symmetry of the lower-dimensional theory is more subtle to identify.[10] In the previous subsection we identified certain scale transformations of the $D = 11$ fields, which did not leave the theory invariant but could be used to adjust the coupling constant κ_{11}. In the compactified situation we can also

[10]There are various discussions of this symmetry in the literature. Its existence in 10-dimensional supergravity was noted long ago (see, e.g. [4, 28]) and an extensive discussion can be found in [12]. Our derivation here was alluded to in [27], which deals with isometries in $N = 2$ supersymmetric Maxwell-Einstein theories in $D = 5, 4$ and 3 dimensions.

involve the compactification length into the dimensional scaling. The integration over x^{11} introduces an overall factor L in the action (we do not incorporate any L-dependent normalizations in the Fourier sums, so that the 10-dimensional and the 11-dimensional fields are directly proportional). Therefore, the coupling constant that emerges in the 10-dimensional theory equals

$$\frac{1}{\kappa_{10}^2} = \frac{L}{\kappa_{11}^2}\,, \tag{2.29}$$

and has the dimension [mass]^8. However, because of the invariance under diffeomorphisms, L itself has no intrinsic meaning. It simply expresses the length of the periodicity interval of x^{11}, which itself is a coordinate without an intrinsic meaning. Stated differently, we can reparameterize x^{11} by some diffeomorphism, as long as we change L accordingly. In particular, we may rescale L according to

$$L \to \mathrm{e}^{-9\alpha} L\,, \tag{2.30}$$

corresponding to a reparametrization of the 11-th coordinate,

$$x^{11} \to \mathrm{e}^{-9\alpha} x^{11}\,, \tag{2.31}$$

so that κ_{10} remains invariant. Consequently we are then dealing with a symmetry of the Lagrangian.

In the effective 10-dimensional theory, the scale transformations (2.21) are thus suitably combined with the diffeomorphism (2.31) to yield an invariance of the Lagrangian. For the fields corresponding to the 11-dimensional metric, these combined transformations are given by[11]

$$e_\mu^a \to \mathrm{e}^{-\alpha} e_\mu^a\,, \qquad \phi \to \phi + 12\alpha\,, \qquad V_\mu \to \mathrm{e}^{-9\alpha} V_\mu\,. \tag{2.32}$$

The tensor gauge field C_{MNP} decomposes into a 3- and a 2-rank tensor in 10 dimensions, which transform according to

$$C_{\mu\nu\rho} \to \mathrm{e}^{-3\alpha} C_{\mu\nu\rho}\,, \qquad C_{11\mu\nu} \to \mathrm{e}^{6\alpha} C_{11\mu\nu}\,. \tag{2.33}$$

The presence of the above scale symmetry is confirmed by the resulting 10-dimensional Lagrangian for the massless (i.e., x^{11}-independent) modes. Its purely bosonic terms read

$$\begin{aligned} \mathcal{L}_{10} \;=\; \frac{1}{\kappa_{10}^2}\Big[&-\tfrac{1}{2} e\, \mathrm{e}^{2\phi/3} R(e,\omega) - \tfrac{1}{8} e\, \mathrm{e}^{2\phi}(\partial_\mu V_\nu - \partial_\nu V_\mu)^2 \\ &-\tfrac{1}{48} e\, \mathrm{e}^{2\phi/3}(F_{\mu\nu\rho\sigma})^2 - \tfrac{3}{4} e\, \mathrm{e}^{-2\phi/3}(H_{\mu\nu\rho})^2 \\ &+\tfrac{1}{1152}\sqrt{2}\,\varepsilon^{\mu_1-\mu_{10}}\, C_{11\mu_1\mu_2}\, F_{\mu_3\mu_4\mu_5\mu_6}\, F_{\mu_7\mu_8\mu_9\mu_{10}} \Big]\,, \end{aligned} \tag{2.34}$$

[11]Note that this applies to all Fourier modes, as they depend on x^{11}/L, which is insensitive to the scale transformation.

where $H_{\mu\nu\rho} = 6\partial_{[\mu}C_{\nu\rho]11}$ is the field strength tensor belonging to the 2-rank tensor gauge field.

The above reduction allows us to discuss a number of characteristic features. First of all, the metric tensor produces an extra vector and a scalar, when dimensionally reducing the dimension by one unit. The scalar is invariant under certain shift symmetries, as shown above, which act multiplicatively on the other fields. Secondly, tensor fields generate tensor fields of a rank that is one unit lower. When this lower-rank field is a scalar field (or equivalent to it by a duality transformation), it will be subject to shifts by a constant parameter, which is simply associated with a gauge transformation that is linearly proportional to the extra higher-dimensional coordinate. Because of this, these shifts must leave the Lagrangian invariant. This last feature is still missing in the above discussion, as the rank-3 tensor decomposes into a rank-3 and a rank-2 tensor. But when descending to lower dimensions than 10, additional scalars will emerge and this phenomenon will be present.

When consecutively reducing the dimension, this pattern repeats itself.[12] In this way, for each scalar field that is generated by the dimensional reduction, there is also an extra symmetry. The dimension of the isometry group is thus (at least) equal to the dimension of the manifold. Furthermore it is easy to see that the symmetries indicated above act *transitively* on the manifold, so that this manifold is homogeneous. The corresponding algebra of these isometries is solvable and the rank of the algebra is equal to $\mathbf{r} = 11 - D$, where D is the spacetime dimension to which we reduce. This is because its Cartan subalgebra is precisely associated with the scale symmetries connected with the scalars that originate from the metric.

The above isometries not only leave the scalar manifold invariant but the whole supergravity Lagrangian. In $D = 4$, or 8, these symmetries do not leave the Lagrangian, but only the field equations invariant. The reason for this is that the isometries act by means of a duality transformation on the field strengths associated with the vector or 3-rank gauge field, respectively. They cannot be implemented directly on the gauge fields themselves. The presence of these duality invariances is a well-known feature of supergravity theories, which was first observed many years ago [29, 30, 31, 32, 27]. However, it is easy to see that the scalar manifold (as well as the rest of the theory) must possess additional symmetries, simply because the isometries corresponding to the solvable algebra do not yet contain the automorphism group $H_{\rm R}$ of the underlying supermultiplet. We expect that $H_{\rm R}$ is realized as a symmetry, because the maximal supergravity theories have no additional parameters, so there is nothing that can break this symmetry.

[12]For instance, in (2.34) one considers the scale transformations (for the bosonic fields),

$$\begin{aligned} e_\mu^{\ a} &\to e^{-\beta}\, e_\mu^{\ a}\,, & C_{\mu\nu 11} &\to e^{-2\beta}\, C_{\mu\nu 11}\,,\\ V_\mu &\to e^{-\beta}\, V_\mu\,, & C_{\mu\nu\rho} &\to e^{-3\beta}\, C_{\mu\nu\rho}\,, \end{aligned} \qquad (2.35)$$

while ϕ remains invariant. These transformations change the Lagrangian by an overall factor $\exp[-8\beta]$, which can be absorbed into $1/\kappa_{10}^2$. When compactifying one more dimension to a circle, these scale transformations yield another isometry in 9 spacetime dimensions, that commutes with the scale transformations (2.32,2.33).

D	G	H	dim [G] − dim [H]
11	1	1	$0-0=0$
10A	$SO(1,1)/Z_2$	1	$1-0=1$
10B	SL(2)	SO(2)	$3-1=2$
9	GL(2)	SO(2)	$4-1=3$
8	$E_{3(+3)} \sim SL(3)\times SL(2)$	U(2)	$11-4=7$
7	$E_{4(+4)} \sim SL(5)$	USp(4)	$24-10=14$
6	$E_{5(+5)} \sim SO(5,5)$	USp(4)×USp(4)	$45-20=25$
5	$E_{6(+6)}$	USp(8)	$78-36=42$
4	$E_{7(+7)}$	SU(8)	$133-63=70$
3	$E_{8(+8)}$	SO(16)	$248-120=128$

TABLE 9. Homogeneous scalar manifolds G/H for maximal supergravities in various dimensions. The type-IIB theory cannot be obtained from reduction of 11-dimensional supergravity and is included for completeness. The difference of the dimensions of G and H equals the number of scalar fields, listed in Table 6.

So we expect an homogeneous space with an isometry group whose algebra is the sum of the solvable algebra and the one corresponding to (part of) $H_{\rm R}$. A counting argument (of the type first used in [30]) then usually reveals what the structure of the homogeneous space is. In Table 9 we list the isometry and isotropy groups of these scalar manifolds for maximal supergravity in dimensions $3 \leq D \leq 11$. Earlier version of such tables can, for instance, be found in [4, 28]. A more recent discussion of these isometry groups can be found in, for example, [34, 33]. We return to this discussion and related issues in section 3.

We should add that it is generally possible to realize the group $H_{\rm R}$ as a *local* symmetry of the Lagrangian. The corresponding connections are then composite connections, governed by the Cartan-Maurer equations. In such a formulation most fields (in particular, the fermions) do not transform under the duality group, but only under the local $H_{\rm R}$ group. The scalars transform linearly under both the rigid duality group as well as under the local $H_{\rm R}$ group. After fixing a gauge, the isometries become nonlinearly realized. The fields which initially transform only under the local $H_{\rm R}$ group, will now transform under the duality group through field-dependent $H_{\rm R}$ transformations. This phenomenon is also realized for the central charges, which transform under the group $H_{\rm R}$ as we have shown in Table 8.

2.5.1. *Frames and scales*

The Lagrangian (2.34) does not contain the standard Einstein-Hilbert term for gravity, while a standard kinetic term for the scalar field ϕ is lacking. This does not pose a serious problem. In this form the gravitational field and the scalar field are entangled and one has to deal with the scalar-graviton system as a whole. To separate the scalar and gravitational degrees of freedom, one may apply a so-called Weyl rescaling of the metric $g_{\mu\nu}$ by an appropriate function of ϕ. In the case that we include the massive modes, this rescaling may depend on the extra coordinate x^{11}. In the context of Kaluza-Klein theory this factor is therefore known as the 'warp factor'. For these lectures two different Weyl rescalings are particularly relevant, which lead to the so-called Einstein and the string frame, respectively. They are defined by

$$e_\mu^a = \mathrm{e}^{-\phi/12}\, [e_\mu^a]^{\text{Einstein}}\,, \qquad e_\mu^a = \mathrm{e}^{-\phi/3}\, [e_\mu^a]^{\text{string}}\,. \tag{2.36}$$

After applying the first rescaling (2.36) to the Lagrangian (2.34) one obtains the Lagrangian in the Einstein frame. This frame is characterized by a standard Einstein-Hilbert term and by a graviton field that is invariant under the scale transformations (2.32,2.33). The corresponding Lagrangian reads[13]

$$\begin{aligned} \mathcal{L}_{10}^{\text{Einstein}} = \frac{1}{\kappa_{10}^2}\Big[& e\Big[-\tfrac{1}{2}R(e,\omega) - \tfrac{1}{4}(\partial_\mu\phi)^2\Big] - \tfrac{1}{8}e\,\mathrm{e}^{3\phi/2}(\partial_\mu V_\nu - \partial_\nu V_\mu)^2 \\ & -\tfrac{3}{4}e\,\mathrm{e}^{-\phi}(H_{\mu\nu\rho})^2 - \tfrac{1}{48}e\,\mathrm{e}^{\phi/2}(F_{\mu\nu\rho\sigma})^2 \\ & +\tfrac{1}{1152}\sqrt{2}\,\varepsilon^{\mu_1-\mu_{10}}\,C_{11\mu_1\mu_2}\,F_{\mu_3\mu_4\mu_5\mu_6}\,F_{\mu_7\mu_8\mu_9\mu_{10}}\Big]\,. \end{aligned} \tag{2.37}$$

Supergravity theories are usually formulated in this frame, where the isometries of the scalar fields do not act on the graviton.

The second rescaling (2.36) leads to the Lagrangian in the string frame,

$$\begin{aligned} \mathcal{L}_{10}^{\text{string}} = \frac{1}{\kappa_{10}^2}\Big[& e\,\mathrm{e}^{-2\phi}\Big[-\tfrac{1}{2}R(e,\omega) + 2(\partial_\mu\phi)^2 - \tfrac{3}{4}(H_{\mu\nu\rho})^2\Big] \\ & -\tfrac{1}{8}e\,(\partial_\mu V_\nu - \partial_\nu V_\mu)^2 - \tfrac{1}{48}e\,(F_{\mu\nu\rho\sigma})^2 \\ & +\tfrac{1}{1152}\sqrt{2}\,\varepsilon^{\mu_1-\mu_{10}}\,C_{11\mu_1\mu_2}\,F_{\mu_3\mu_4\mu_5\mu_6}\,F_{\mu_7\mu_8\mu_9\mu_{10}}\Big]\,. \end{aligned} \tag{2.38}$$

This frame is characterized by the fact that R and $(H_{\mu\nu\rho})^2$ have the same coupling to the scalar ϕ, or, equivalently, that $g_{\mu\nu}$ and $C_{11\mu\nu}$ transform with equal weights under the scale transformations (2.32,2.33). In string theory ϕ coincides with the dilaton field that couples to the topology of the worldsheet and whose vacuum-expectation value defines the string coupling constant according to $g_{\rm s} = \exp(\langle\phi\rangle)$. We shall return to this in section 3, but here we already indicate the significance of the dilaton factors in the Lagrangian above. The metric $g_{\mu\nu}$, the antisymmetric tensor $C_{\mu\nu 11}$ and the dilaton ϕ always arise in the Neveu-Schwarz sector and couple universally to $\mathrm{e}^{-2\phi}$. On the other hand the vector V_μ and the 3-form $C_{\mu\nu\rho}$ describe Ramond-Ramond (R-R) states and the specific form of their vertex operators

[13]Note that under a local scale transformation $e_\mu^a \to \mathrm{e}^{\Lambda}e_\mu^a$, the Ricci scalar in D dimensions changes according to

$$R \to \mathrm{e}^{-2\Lambda}\Big[R + 2(D-1)D^\mu\partial_\mu\Lambda + (D-1)(D-2)g^{\mu\nu}\,\partial_\mu\Lambda\,\partial_\nu\Lambda\Big]\,.$$

forbids any tree-level coupling to the dilaton [33, 12]. In particular the Kaluza-Klein gauge field V_μ corresponds in the string context to the R-R gauge field of type-II string theory. The infinite tower of massive Kaluza-Klein states carry a charge quantized in units of $e_{\rm KK}$, defined in (2.28). In the context of 10-dimensional supergravity, states with a R-R charge are solitonic. In string theory, the R-R charges are carried by the D-brane states.

As we already discussed in the previous section, Newton's constant is only defined after a choice of the metric in the ground state is made. Expanding the metric in the Einstein frame around $\lambda\,\eta_{\mu\nu}$ one obtains from (2.24,2.37)

$$(\kappa_{10}^2)^{\rm physical} = \kappa_{10}^2\,\lambda^{-4}\,, \tag{2.39}$$

while expanding the metric in the string frame around $\lambda\,\eta_{\mu\nu}$ leads to

$$(\kappa_{10}^2)^{\rm physical} = \kappa_{10}^2\,\lambda^{-4}\,{\rm e}^{2\langle\phi\rangle}\,. \tag{2.40}$$

Note that one cannot expand both metrics simultaneously around $\lambda\,\eta_{\mu\nu}$.

For later purposes let us note that the above discussion can be generalized to arbitrary spacetime dimensions. The Einstein frame in any dimension is defined by a gravitational action that is just the Einstein-Hilbert action, whereas in the string frame the Ricci scalar is multiplied by a dilaton term $\exp(-2\phi)$, as in (2.37) and (2.38), respectively. The Weyl rescaling which connects the two frames is given by,

$$[e_\mu^{\;a}]^{\rm string} = {\rm e}^{2\phi/(D-2)}\,[e_\mu^{\;a}]^{\rm Einstein}\,. \tag{2.41}$$

In arbitrary dimensions (2.39) and (2.40) read

$$\begin{aligned} \text{Einstein frame}: &\qquad (\kappa_D^2)^{\rm physical} = \kappa_D^2\,\lambda^{(2-D)/2}\,,\\ \text{string frame}: &\qquad (\kappa_D^2)^{\rm physical} = \kappa_D^2\,\lambda^{(2-D)/2}\,{\rm e}^{2\langle\phi\rangle}\,. \end{aligned} \tag{2.42}$$

This frame dependence does not only apply to κ_D^2 but to any dimensionful quantity. For example, a mass, when measured in the same flat metric but specified in the two frames, is related by

$$M^{\rm string} = {\rm e}^{-2\langle\phi\rangle/(D-2)}\,M^{\rm Einstein}\,. \tag{2.43}$$

Of course this is consistent with the relation (2.42). The physical masses in the above relation depend again on the value of λ. In the remainder of this subsection we choose $\lambda = 1$ for convenience.

Let us now return to 11-dimensional supergravity with the 11-th coordinate compactified to a circle so that $0 \leq x^{11} \leq L$. As we stressed already, L itself has no intrinsic meaning and it is better to consider the geodesic radius of the 11-th dimension, which reads

$$R_{11} = \frac{L}{2\pi}\,{\rm e}^{2\langle\phi\rangle/3}\,. \tag{2.44}$$

This result applies to the frame specified by the 11-dimensional theory[14]. In the string frame, the above result reads

$$(R_{11})^{\rm string} = \frac{L}{2\pi}\, {\rm e}^{\langle\phi\rangle}\,. \qquad (2.45)$$

It shows that a small 11-th dimension corresponds to small values of $\exp\langle\phi\rangle$ which in turn corresponds to a weakly coupled string theory. We come back to this crucial observation in section 3. Observe that L is fixed in terms of κ_{10} and κ_{11} (cf. (2.29)).

From the 11-dimensional expressions,

$$E_a^{\;M}\partial_M = e_a^{\;\mu}(\partial_\mu - V_\mu\,\partial_{11})\,, \qquad E_{11}^{\;M}\partial_M = {\rm e}^{-2\phi/3}\,\partial_{11}\,, \qquad (2.46)$$

where a and μ refer to the 10-dimensional Lorentz and world indices, we infer that, in the frame specified by the 11-dimensional theory, the Kaluza-Klein masses are multiples of

$$M^{\rm KK} = \frac{1}{R_{11}}\,. \qquad (2.47)$$

Hence Kaluza-Klein states have a mass and Kaluza-Klein charge (cf. (2.28)) related by

$$M^{\rm KK} = |e_{\rm KK}|\,{\rm e}^{-2\langle\phi\rangle/3}\,. \qquad (2.48)$$

In the string frame, this result becomes

$$(M^{\rm KK})^{\rm string} = |e_{\rm KK}|\,{\rm e}^{-\langle\phi\rangle}\,. \qquad (2.49)$$

Massive Kaluza-Klein states are always BPS states, meaning that they are contained in supermultiplets that are 'shorter' than the generic massive supermultiplets because of nontrivial central charges. The central charge here is just the 11-th component of the momentum, which is proportional to the Kaluza-Klein charge.

The surprising insight that emerged in recent years, is that the Kaluza-Klein features of 11-dimensional supergravity have a precise counterpart in string theory [34, 35, 33]. There one has nonperturbative (in the string coupling constant) states which carry R-R charges. We return to this phenomenon in section 3.

2.6. NONMAXIMAL SUPERSYMMETRY

In previous subsections we discussed a large number of supermultiplets. Furthermore, in Fig. 1 we presented an overview of all supergravity theories in spacetime dimensions $4 \leq D \leq 11$ with $Q = 32, 16, 8$ or 4 supercharges. In this section we summarize a number of results on nonmaximal supersymmetric theories with $Q = 16$ supercharges, which are now restricted to dimensions $D \leq 10$.

For $Q = 16$ the automorphism group $H_{\rm R}$ is smaller. Table 10 lists this group in various dimensions and shows the field representations for the vector multiplet in dimension $3 \leq D \leq 10$. This multiplet comprises $8+8$ physical degrees of freedom.

[14]This is the frame specified by the metric given in (2.26), which leads to the Lagrangian (2.34).

D	$H_{\rm R}$	A_μ	ϕ	χ
10	1	1	0	1
9	1	1	1	1
8	U(1)	1	$1+\bar{1}$	$1+\bar{1}$
7	USp(2)	1	3	2
6	USp(2)×USp(2)	1	(2,2)	(2,1)+(1,2)
5	USp(4)	1	5	4
4	U(4)	1	6*	$4+\bar{4}$
3	SO(8)		8	8

TABLE 10. Field content for maximal super-Maxwell theories in various dimensions. All supermultiplets contain a gauge field A_μ, scalars ϕ and spinors χ. In $D=3$ dimensions the vector field is dual to a scalar. The 6* representation of SU(4) is a selfdual 2-rank tensor.

We also consider the $Q=16$ supergravity theories. The Lagrangian can be obtained by truncation of (2.34). However, unlike in the case of maximal supergravity, we now have the option of introducing additional matter fields. For $Q=16$ the matter will be in the form of vector supermultiplets, possibly associated with some nonabelian gauge group. Table 11 summarizes $Q=16$ supergravity for dimensions $3\leq D\leq 10$. In $D=10$ dimensions the bosonic terms of the supergravity Lagrangian take the form [36],

$$\begin{aligned}\mathcal{L}_{10} &= \frac{1}{\kappa_{10}^2}\Big[-\tfrac{1}{2}e\,\mathrm{e}^{2\phi/3}R(e,\omega)\\ &\qquad -\tfrac{3}{4}e\,\mathrm{e}^{-2\phi/3}(H_{\mu\nu\rho})^2-\tfrac{1}{4}e(\partial_\mu A_\nu-\partial_\nu A_\mu)^2\Big]\,,\end{aligned}\tag{2.50}$$

where, for convenience, we have included a single vector gauge field, representing an abelian vector supermultiplet. A feature that deserves to be mentioned, is that the field strength $H_{\mu\nu\rho}$ associated with the 2-rank gauge field acquires a Chern-Simons term $A_{[\mu}\partial_\nu A_{\rho]}$. Chern-Simons terms play an important role in the anomaly cancellations of this theory. Note also that the kinetic term for the Kaluza-Klein vector field in (2.34), depends on ϕ, unlike the kinetic term for the matter vector field in the Lagrangian above. This reflects itself in the extension of the symmetry transformations noted in (2.32, 2.33),

$$e_\mu^a\to\mathrm{e}^{-\alpha}e_\mu^a\,,\quad \phi\to\phi+12\alpha\,,\quad C_{11\mu\nu}\to\mathrm{e}^{6\alpha}C_{11\mu\nu}\,,\quad A_\mu\to\mathrm{e}^{3\alpha}A_\mu\,.\tag{2.51}$$

where A_μ transforms differently from the Kaluza-Klein vector field V_μ.

In this case there are three different Weyl rescalings that are relevant, namely

$$e_\mu^a = \mathrm{e}^{-\phi/12}\,[e_\mu^a]^{\rm Einstein}\,,\qquad e_\mu^a=\mathrm{e}^{-\phi/3}\,[e_\mu^a]^{\rm string}\,,$$

D	$H_{\rm R}$	graviton	$p=-1$	$p=0$	$p=1$
10	1	1	1		1
9	1	1	1	1	1
8	U(1)	1	1	$1+\bar{1}$	1
7	USp(2)	1	1	3	1
6A	USp(2)×USp(2)	1	1	(2,2)	(1,1)
6B	USp(4)	1			5*
5	USp(4)	1	1	5	
4	U(4)	1	$1+\bar{1}$	[6]	
3	SO(8)	1	$8k$		

TABLE 11. Bosonic fields of nonmaximal supergravity with $Q=16$. In 6 dimensions type-A and type-B correspond to (1,1) and (2,0) supergravity. Note that, with the exception of the 6B and the 4-dimensional theory, all these theories contain precisely one scalar field. In $D=4$ dimensions, the SU(4) transformations cannot be implemented on the vector potentials, but act on the (abelian) field strengths by duality transformations. In $D=3$ dimensions supergravity is a topological theory and can be coupled to scalars and spinors. The scalars parameterize the coset space $\mathrm{SO}(8,k)/\mathrm{SO}(8)\times\mathrm{SO}(k)$, where k is an arbitrary integer.

$$e_\mu^a \;=\; \mathrm{e}^{\phi/6}\,[e_\mu^a]^{\rm string'}\,. \tag{2.52}$$

It is straightforward to obtain the corresponding Lagrangians. In the Einstein frame, the graviton is again invariant under the isometries of the scalar field. The bosonic terms read

$$\begin{aligned}\mathcal{L}_{10}^{\rm Einstein} \;=\; \frac{1}{\kappa_{10}^2}\Big[&-\tfrac{1}{2}e\,R(e,\omega)-\tfrac{1}{4}e(\partial_\mu\phi)^2\\ &-\tfrac{3}{4}e\,\mathrm{e}^{-\phi}(H_{\mu\nu\rho})^2-\tfrac{1}{4}e\,\mathrm{e}^{-\phi/2}(\partial_\mu A_\nu-\partial_\nu A_\mu)^2\Big]\,.\end{aligned} \tag{2.53}$$

The second Weyl rescaling leads to the following Lagrangian,

$$\begin{aligned}\mathcal{L}_{10}^{\rm string} \;=\; \frac{1}{\kappa_{10}^2}\mathrm{e}^{-2\phi}\Big[&-\tfrac{1}{2}e\,R(e,\omega)+2e(\partial_\mu\phi)^2\\ &-\tfrac{3}{4}e\,(H_{\mu\nu\rho})^2-\tfrac{1}{4}e(\partial_\mu A_\nu-\partial_\nu A_\mu)^2\Big]\,,\end{aligned} \tag{2.54}$$

which shows a uniform coupling with the dilaton. This is the low-energy effective Lagrangian relevant to the heterotic string. Eventually the matter gauge field has to be part of an nonabelian gauge theory based on the group SO(32) or $\mathrm{E}_8\times\mathrm{E}_8$, in order to be anomaly-free.

Finally, the third Weyl rescaling yields

$$\mathcal{L}_{10}^{\text{string}'} = \frac{1}{\kappa_{10}^2}\Big[e\,e^{2\phi}\Big[-\tfrac{1}{2}R(e,\omega)+2(\partial_\mu\phi)^2\Big] \\ -\tfrac{3}{4}e\,(H_{\mu\nu\rho})^2-\tfrac{1}{4}e\,e^{\phi}(\partial_\mu A_\nu-\partial_\nu A_\mu)^2\Big]. \tag{2.55}$$

Here the dilaton seems to appear with the wrong sign. As it turns out, this is the low-energy effective action of the type-I string, where the type-I dilaton must be associated with $-\phi$. This will be further elucidated in section 3.

3. String theories in various dimensions

3.1. PERTURBATIVE STRING THEORIES IN $D = 10$

The perturbative expansion. In string theory the fundamental objects are one-dimensional strings which, as they move in time, sweep out a 2-dimensional worldsheet Σ [37]. Strings can be open or closed and their worldsheet is embedded in some higher-dimensional target space which is identified with a Minkowskian spacetime. States in the target space appear as eigenmodes of the string and their scattering amplitudes are generalized by appropriate scattering amplitudes of strings. These scattering amplitudes are built from a fundamental vertex, which for closed strings is depicted in Fig. 2. It represents the splitting of a string or the joining of two strings and the strength of this interaction is governed by a dimensionless string coupling constant g_s. Out of the fundamental vertex one composes all possible closed string scattering amplitudes $\mathcal{A}$, for example the four-point amplitude shown in Fig. 3. The expansion in the topology of the Riemann surface

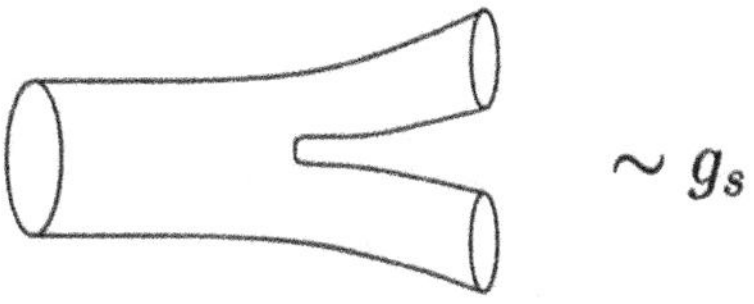

Figure 2. The fundamental closed string vertex.

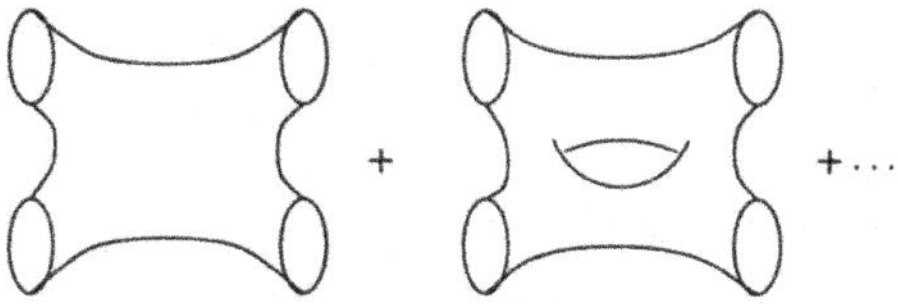

Figure 3. The perturbative expansion of string scattering amplitudes. The order of g_s is governed by the number of holes in the world sheet.

(i.e. the number of holes in the surface) coincides with a power series expansion in the string coupling constant formally written as

$$\mathcal{A} = \sum_{n=0}^{\infty} g_s^{-\chi} \mathcal{A}^{(n)} , \tag{3.1}$$

where $\mathcal{A}^{(n)}$ is the scattering amplitude on a Riemann surface of genus n and $\chi(\Sigma)$ is the Euler characteristic of the Riemann surface

$$\chi(\Sigma) = \frac{1}{4\pi} \int_{\Sigma} R^{(2)} = 2 - 2n - b . \tag{3.2}$$

$R^{(2)}$ is the curvature on Σ and b the number of boundaries of the Riemann surface (for the four-point amplitude of Fig. 3 one has $b = 4$).[15]

In all string theories there is a massless scalar field ϕ called the dilaton which couples to $\sqrt{h}R^{(2)}$ and therefore its vacuum-expectation value determines the size of the string coupling; one finds [38, 37]

$$g_s = e^{\langle\phi\rangle} . \tag{3.3}$$

g_s is a free parameter since ϕ is a flat direction (a modulus) of the effective potential. Thus, string perturbation theory is defined in that region of the parameter space (which is also called the moduli space) where $g_s < 1$ and the tree-level amplitude (genus-0) is the dominant contribution with higher-loop amplitudes suppressed by higher powers of g_s. Until three years ago this was the only regime accessible in string theory.

The spacetime spectrum of the string. The propagation of a free string ($g_s = 0$) is governed by the 2-dimensional action

$$S_{\text{free}} = -\frac{1}{4\pi\alpha'} \int_{\Sigma} \partial_i X^{\mu}(\sigma,\tau)\, \partial^i X^{\nu}(\sigma,\tau)\, \eta_{\mu\nu} , \tag{3.4}$$

where ∂_i denotes $\partial/\partial\sigma$ and $\partial/\partial\tau$. Here σ parameterizes the spatial direction on Σ while τ denotes the 2-dimensional time coordinate. The coordinates of the D-dimensional target spacetime in which the string moves, are represented by X^{μ}, with $\mu = 0, \ldots, D-1$; in terms of the 2-dimensional field theory they appear as D scalar fields. For S to be dimensionless α' has dimension [length]^2 $\sim$ [mass]$^{-2}$; it is the fundamental mass scale of string theory which is also denoted by M_s with the identification $\alpha' = M_s^{-2}$. The mass of all perturbative string states is a multiple of M_s. Just as the coupling constant κ in the supergravity Lagrangians in section 2, this scale has no intrinsic meaning and must be fixed by some independent criterion. Demanding that string theory contains Einstein gravity as its low-energy limit relates the characteristic scales of the two theories. By comparing

[15]For open strings different diagrams contribute at the same order of the string loop expansion. See [37] for further details.

for example physical graviton-graviton scattering amplitudes in both theories one finds in the following expression for Newton's constant in D dimensions [39],[16]

$$(\kappa_D^2)^{\text{physical}} = \alpha'^{(D-2)/2}\, e^{2\langle\phi\rangle}\,, \tag{3.5}$$

where we dropped convention-dependent numerical proportionality factors.

The equations of motion of the action (3.4) are given by

$$(\partial_\tau^2 - \partial_\sigma^2)X^\mu = 0\,, \tag{3.6}$$

with the solutions

$$X^\mu = X_L^\mu(\sigma+\tau) + X_R^\mu(\sigma-\tau)\,. \tag{3.7}$$

A closed string satisfies the boundary condition $X^\mu(\sigma) = X^\mu(\sigma+2\pi)$, which does not mix X_L^μ and X_R^μ and leaves them as independent solutions. This splitting into left (L) and right (R) moving fields has the consequence that upon quantizing the 2-dimensional field theory, also the Hilbert space splits into a direct product $\mathcal{H} = \mathcal{H}_L \otimes \mathcal{H}_R$ where $\mathcal{H}_L(\mathcal{H}_R)$ contains states built from oscillator modes of $X_L(X_R)$. These states also carry a representation of the D-dimensional target space Lorentz group and thus can be identified as perturbative states in spacetime of a given spin and mass.[17]

In open string theory one has a choice to impose at the end of the open string either Neumann (N) boundary conditions, $\partial_\sigma X^\mu = 0$, or Dirichlet (D) boundary conditions, $X^\mu =$ constant. The boundary conditions mix left- and right-movers and the product structure of the closed string is not maintained. As a consequence a perturbative spectrum of states is built from a single Hilbert space. Neumann boundary conditions leave the D-dimensional Lorentz invariance unaffected.

Dirichlet boundary conditions, on the other hand, lead to very different types of objects and a very different set of states (D-branes) in spacetime [40]. In this case the end of an open string is constrained to only move in a fixed spatial hyper-plane. This plane must be regarded as a dynamical object with degrees of freedom induced by the attached open string. A careful analysis shows that the corresponding states in spacetime are not part of the perturbative spectrum but rather correspond to nonperturbative solitonic type excitations[18]. It is precisely these states which dramatically affect the properties of string theory in its nonperturbative regime. These aspects will be subject of section 3.3.

So far we discussed the free string governed by the action (3.4); its interactions are incorporated by promoting S_{free} to a nonlinear 2-dimensional σ-model. The

[16]The relation (3.5) holds in arbitrary dimensions with $e^{\langle\phi\rangle}$ being the string coupling constant for a string moving in D spacetime dimensions. Later in these lectures we consider compactifications of string theory and then there is a volume-dependent relation between the string couplings defined in different dimensions. This relation is discussed in appendix B.

[17]These are perturbative states since the quantization procedure is a perturbation theory around the free string theory with $g_s = 0$.

[18]They are nonperturbative in that their mass (or rather their tension for higher-dimensional D-branes) goes to infinity in the weak coupling limit $g_s \to 0$.

closed string theories	worldsheet supersymmetry	$D_{\max}$
bosonic string	$(0,0)$	26
superstring	$(1,1)$	10
heterotic string	$(0,1)$	10

TABLE 12. The closed-string theories, their worldsheet supersymmetry and the maximal possible spacetime dimension.

amplitude $\mathcal{A}$ can be interpreted as a unitary scattering amplitude in the target space whenever this 2-dimensional field theory is conformally invariant. The action is found to be

$$\begin{aligned} S \;=\; & -\frac{1}{4\pi\alpha'}\int_\Sigma \partial_i X^\mu(\sigma,\tau)\,\partial^i X^\nu(\sigma,\tau)\; g_{\mu\nu}(X(\sigma,\tau)) \\ & -\frac{1}{4\pi\alpha'}\int_\Sigma \varepsilon^{ij}\,\partial_i X^\mu(\sigma,\tau)\,\partial_j X^\nu(\sigma,\tau)\; b_{\mu\nu}(X(\sigma,\tau)) \\ & +\frac{1}{4\pi}\int_\Sigma R^{(2)}\,\phi(X(\sigma,\tau)) + \cdots\,, \end{aligned} \tag{3.8}$$

where $g_{\mu\nu}(X)$ is the metric[19] of the target space, $b_{\mu\nu}(X)$ is the antisymmetric target-space tensor and $R^{(2)}$ is the curvature scalar of the 2-dimensional worldsheet Σ. The target-space field $\phi(X)$ represents a scalar coupling and corresponds to the dilaton, since the coefficient of its constant vacuum-expectation value $\langle\phi\rangle$ is the Euler number $\chi(\Sigma) = \frac{1}{4\pi}\int_\Sigma R^{(2)}$. The ellipses denote further terms depending on the type of string theory and the number of spacetime dimensions.

The spacetime properties of a string theory significantly change once one introduces supersymmetry on the worldsheet. In two dimensions the irreducible supercharges are Majorana-Weyl spinors (see Table 1). In addition there are independent left- and right-moving supercharges Q_L, Q_R, so that in general one can have p supercharges Q_L and q supercharges Q_R; this is also termed (p,q) supersymmetry. A supersymmetric version of the action (3.8) requires the presence of Majorana-Weyl worldsheet fermions χ^μ with appropriate couplings; for example a scalar supermultiplet of $(1,0)$ supersymmetry contains the fields $\big(X_L(\sigma+\tau),\chi_L(\sigma+\tau)\big)$. Depending on the amount of worldsheet supersymmetry one defines the different *closed* string theories: the bosonic string, the superstring and the heterotic string (see Table 12).

For *open* string theories the left- and right-moving worldsheet supercharges are not independent. One can either have a bosonic open string (with no worldsheet

[19] As we mentioned already in subsection 2.5, the metric $g_{\mu\nu}(X)$ in (3.8) is the metric in the string frame.

supersymmetry) or an open superstring with one supercharge which is a linear combination of Q_L and Q_R. The latter string theory is called type-I. It contains (unoriented) open and closed strings with SO(32) Chan-Paton factors coupling to the ends of the open string.

The bosonic string (open or closed) is tachyonic and cannot accommodate spacetime fermions; for these reasons we omit it from our subsequent discussion. The superstring, the heterotic string and the type-I string can all be tachyon-free and do have spacetime fermions in the massless spectrum. In addition, in most cases they are also spacetime supersymmetric and contain (at least) a massless gravitino. There are also tachyon-free non-supersymmetric string theories [41] but they have a dilaton tadpole at one-loop and thus do not seem to correspond to stable vacuum configurations.[20] For this reason we solely focus on supersymmetric string theories henceforth.

The worldsheet fermions χ^μ can have two distinct type of boundary conditions when transported around the closed string,

$$\chi^\mu(\sigma) = \begin{cases} +\chi^\mu(\sigma + 2\pi) & \text{Ramond (R)}, \\ -\chi^\mu(\sigma + 2\pi) & \text{Neveu-Schwarz (NS)}. \end{cases} \tag{3.9}$$

Consequently the states of the closed string Hilbert space can arise in four different sectors of fermion boundary conditions:

$$\left.\begin{array}{lcl} \mathrm{NS}_L & \otimes & \mathrm{NS}_R \\ \mathrm{R}_L & \otimes & \mathrm{R}_R \end{array}\right\} \quad \text{spacetime bosons}$$
$$\left.\begin{array}{lcl} \mathrm{NS}_L & \otimes & \mathrm{R}_R \\ \mathrm{R}_L & \otimes & \mathrm{NS}_R \end{array}\right\} \quad \text{spacetime fermions}\,.$$

The first two sectors contain the spacetime bosons, while the last two sectors generate spacetime fermions. The bosons from the R-R sector are built from bi-spinors and thus the representation theory of the Lorentz group constrains these bosons to always be antisymmetric Lorentz tensors of varying rank. Furthermore, in the effective action and in all scattering processes these tensors can only appear via their field strength and thus there are no states in perturbative string theory which carry any charge under the antisymmetric tensors in the R-R sector. However, it turns out that this is an artifact of perturbation theory and states carrying R-R charge do appear in the nonperturbative spectrum; they are precisely the states generated by appropriate D-brane configurations [40].

Conformal invariance on the worldsheet (or equivalently unitarity in spacetime) imposes a restriction on the maximal number of spacetime dimensions and the spacetime spectrum. All supersymmetric string theories necessarily have $D \leq 10$ and they are particularly simple in their maximal possible dimension $D = 10$.[21]

[20] For a recent discussion of non-supersymmetric string theories, see [42].

[21] For closed strings an additional constraint arises from the requirement of modular invariance of one-loop amplitudes which results in an anomaly-free spectrum of the corresponding low-energy effective theory [43]. For open strings anomaly cancellation is a consequence of the the absence of tadpole diagrams [37].

In $D = 10$ there are only five consistent spacetime supersymmetric string theories: type-IIA, type-IIB, heterotic $E_8 \times E_8$ (HE8), heterotic SO(32) (HSO) and the type-I SO(32) string. The first two have $Q = 32$ supercharges and thus there is a unique massless multiplet in each case with a field content given in Table 6. As we already indicated perturbative string theory distinguishes between states arising in the NS-NS sector from states of the R-R sector in that the coupling to the dilaton is different. In the type-IIA theory one finds the graviton $g_{\mu\nu}$, an antisymmetric tensor $b_{\mu\nu}$ and the dilaton ϕ in the NS-NS sector while an abelian vector V_μ and a 3-form $C_{\mu\nu\rho}$ appear in the R-R sector. The corresponding low-energy effective Lagrangian was already given in section 2.5.1. In type-IIB one has exactly the same states in the NS-NS sector, but in the R-R sector one has a 2-form $b'_{\mu\nu}$, an additional scalar ϕ' and a 4-form $c^*_{\mu\nu\rho\sigma}$ whose field strength is selfdual. Its field equations can be found in [8].

The other three string theories all have $Q = 16$ supercharges. In this case, the supersymmetric representation theory alone does not completely determine the spectrum. The gravitational multiplet is unique (see Table 11), but the gauge group representation of the vector multiplets (see Table 10) is only fixed if also anomaly cancellation is imposed. The low-energy effective Lagrangian for the two heterotic theories is displayed in (2.53) and (2.54), with the abelian vector appropriately promoted to vector fields of $E_8 \times E_8$ or SO(32), respectively. The type-I string has the same supersymmetry but $b_{\mu\nu}$ arises in the R-R sector and thus has different (perturbative) couplings to the dilaton. The corresponding low-energy effective Lagrangian is given by (2.55) with ϕ replaced by $-\phi$. In Table 13 we summarize the bosonic massless spectra for the five string theories, which is in direct correspondence with some of the material collected in the Tables 6, 10 and 11, presented in section 2.

3.2. CALABI–YAU COMPACTIFICATIONS AND PERTURBATIVE DUALITIES

So far we discussed the various string theories in 10 spacetime dimensions. Lower-dimensional theories can be obtained by compactifying the $D = 10$ theories on an internal, 'curled up' compact manifold Y.[22] Unitarity in spacetime requires Y to be a Calabi-Yau manifold [37].[23] Calabi-Yau manifolds are Ricci-flat Kähler manifolds of vanishing first Chern class ($c_1(Y) = 0$) with holonomy group SU(M) where M is the complex dimension of Y. A one dimensional (complex) Calabi-Yau manifold is topologically always a torus T^2 and toroidal compactifications leave all supercharges intact. For $M = 2$ all Calabi-Yau manifolds are topologically equivalent to the 4-dimensional $K3$ surface [44] and as a consequence of the nontrivial SU(2) holonomy half of the supercharges are broken. For $M = 3$ there

[22]There are also string vacua which cannot be viewed as a compactification of the 10-dimensional string theories. Their duality properties have been much less investigated and for a lack of space we neglect them in our discussion here and solely focus on string vacua with a geometrical interpretation.

[23]By a slight abuse of terminology we include in our discussion here also the circle T^1, which strictly speaking is not a Calabi-Yau manifold but does give unitary S-matrices.

type	Q		bosonic spectrum
IIA	32	NS-NS	$g_{\mu\nu}$, $b_{\mu\nu}$, ϕ
		R-R	V_μ, $C_{\mu\nu\rho}$
IIB	32	NS-NS	$g_{\mu\nu}$, $b_{\mu\nu}$, ϕ
		R-R	$c^*_{\mu\nu\rho\sigma}$, $b'_{\mu\nu}$, ϕ'
HE8	16		$g_{\mu\nu}$, $b_{\mu\nu}$, ϕ A_μ in adjoint of $E_8 \times E_8$
HSO	16		$g_{\mu\nu}$, $b_{\mu\nu}$, ϕ A_μ in adjoint of SO(32)
I	16	NS-NS	$g_{\mu\nu}$, ϕ
		R-R	$b_{\mu\nu}$
		open string	A_μ in adjoint of SO(32)

TABLE 13. Supersymmetric string theories in $D = 10$ and their fields describing the bosonic massless spectrum.

exist many topologically distinct Calabi–Yau threefolds Y_3 and they all break $\frac{3}{4}$ of the supercharges [45]. We summarize this situation in the following table:

$$\begin{array}{lll} M=1: & T^2 & \text{breaks no supercharges} \\ M=2: & K3 & \text{breaks 1/2 of the supercharges} \\ M=3: & Y_3 & \text{breaks 3/4 of the supercharges} \end{array}$$

The different theories obtained by compactifying on such Calabi–Yau manifolds are depicted in Fig. 4. Note that different compactifications can have the exact same representation of supersymmetry.

3.2.1. *Toroidal compactifications and T-duality.*
Compactifying the 10-dimensional string theories on an n-dimensional torus T^n leads to string theories with $D = 10 - n$. Supersymmetry remains unbroken and thus one moves down vertically in the D-Q plane of Fig. 4. The massless spectrum can be obtained by dimensional reduction of the appropriate 10-dimensional theories. For simplicity we start by considering closed string theories with one compact dimension which we take to be a circle T^1. In this case there are nine spacetime coordinates X^μ satisfying the boundary conditions[24]

$$X^\mu(\sigma = 2\pi, \tau) = X^\mu(\sigma = 0, \tau), \tag{3.10}$$

[24]Throughout this section the indices μ, ν always denote the uncompactified spacetime directions.

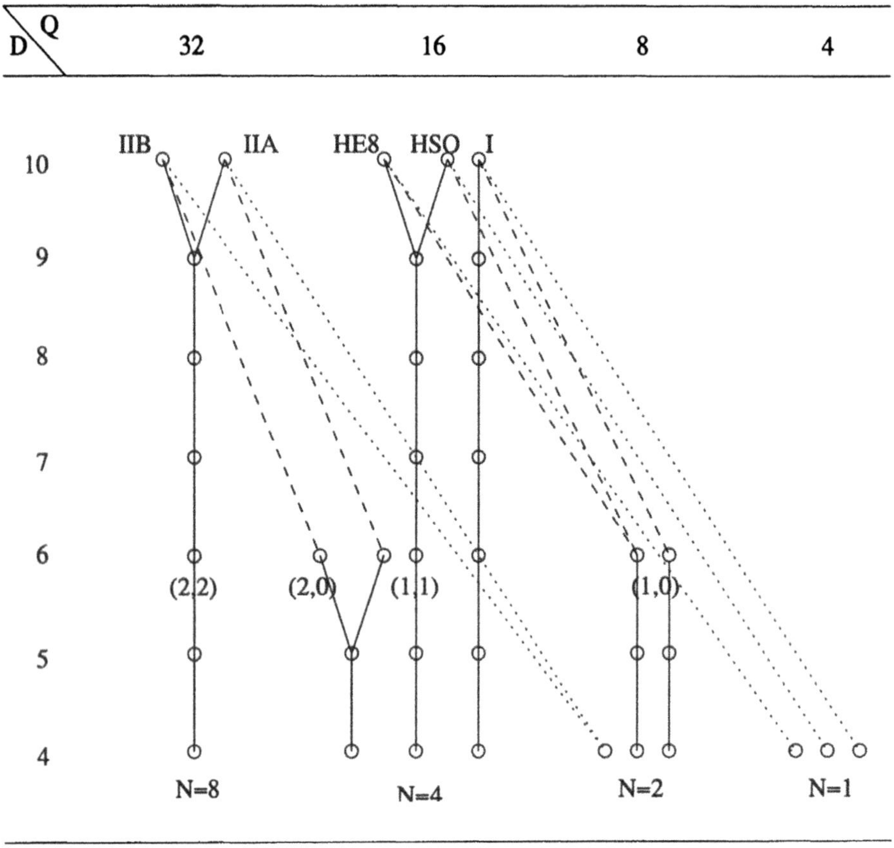

Figure 4. Calabi-Yau compactifications of the 10-dimensional string theories. The solid line (−) denotes toroidal compactification, the dashed line (−−) denotes $K3$ compactifications and the dotted line $(\cdots)$ denotes Y_3 compactifications. Whenever two compactifications (two lines) terminate in the same point, the two string theories are related by a perturbative duality. (A line crossing a circle is purely accidental and has no physical significance.)

and one internal coordinate X^{10}, which can wrap m times around the T^1 of radius R,

$$X^{10}(\sigma = 2\pi, \tau) = X^{10}(\sigma = 0, \tau) + 2\pi m R \,. \qquad (3.11)$$

The massless spectrum of the 9-dimensional theory includes the two abelian Kaluza-Klein gauge bosons $g_{\mu\,10}$ and $b_{\mu\,10}$ as well as a massless scalar field $g_{10\,10}$ which is a flat direction of the effective potential and whose vacuum-expectation value parameterizes the size of the internal T^1 (cf. section 2.5) The appearance of flat directions is a generic feature of string compactifications and such scalar fields are called moduli. For the case at hand the moduli space is one-dimensional and hence there is a one-parameter family of inequivalent string vacua.[25] The bound-

[25]The different solutions of a given string theory are often referred to as the vacuum states of that string theory or simply as the string vacua.

ary condition (3.11) leads to a quantization of the internal momentum component $p^{10} = k/R$ and a whole tower of massive Kaluza–Klein states labelled by the integer k. In addition there are also massive winding modes labelled by m and altogether one finds a mass spectrum

$$M^2 = \tfrac{1}{2}(P_L^2 + P_R^2) + \cdots , \qquad P_{L,R} = \frac{k}{R} \pm \frac{mR}{\alpha'} . \tag{3.12}$$

The ellipses stand for R-independent contributions of the oscillator modes. The mass spectrum is invariant under the exchange of $R \leftrightarrow \alpha'/R$ if simultaneously one also exchanges $m \leftrightarrow k$. This Z_2 invariance of the spectrum is the first example of a *T–duality* [46].

These considerations can be generalized to higher-dimensional toroidal compactifications on a torus T^n [47]. The boundary conditions on the n compact coordinates are

$$X^i\,(\sigma = 2\pi, \tau) = X^i\,(\sigma = 0, \tau) + 2\pi R^i \ , \tag{3.13}$$

where $R^i = \sum_I m_I e_I^i$ are vectors on an n-dimensional lattice with basis e_I^i.[26] The momenta p^i live on the dual lattice and one finds

$$M^2 = \tfrac{1}{2}(P_L^2 + P_R^2) + \cdots , \qquad P_{L,R}^i = p^i \pm \frac{R^i}{\alpha'} . \tag{3.14}$$

Modular invariance constrains the lattice to be an even, selfdual Lorentzian lattice. The inequivalent lattices of this type can be labelled by points in the coset space [47]

$$\mathcal{M} = \frac{\mathrm{SO}(n,n)}{\mathrm{SO(n)} \times \mathrm{SO}(n)} \Big/ \Gamma_T \ , \tag{3.15}$$

where

$$\Gamma_T = \mathrm{SO}\,(n, n, Z) \tag{3.16}$$

is the T–duality group identifying equivalent lattices or in other words equivalent toroidal compactifications. The mass formula (3.14) (as well as the entire partition function) shares the invariance under Γ_T.

Type-II compactified on T^n. Toroidal compactifications of type-II string theory all have the maximal number of supercharges $Q = 32$. The associated supergravities have been discussed in section 2.4 and for each of these cases there is a unique gravitational multiplet containing all massless fields.[27] However in perturbative

[26]The index i runs over the internal dimensions, i.e. $i = 1, \ldots, n = 10 - D$.

[27]We briefly mentioned in section 2.2.2 that below $D = 10$ the type-IIA and type-IIB supergravities are equivalent. A careful analysis in $D = 9$ reveals a flip of the chiralities of the space-time fermions in the limits $R \to 0$ versus $R \to \infty$ and thus the non-chiral type-IIA and the chiral type-IIB theory can be viewed as two distinct limits of one and the same type-II theory in $D = 9$ [10, 11]. One also often refers to this fact by stating that in $D = 9$ type-IIA and type-IIB are T-dual to each other in that type-IIA at a large compactification radius is equivalent to type-IIB at a small compactification radius and vice versa.

string theory there is a clear distinction between states arising from the NS-NS versus the R-R sector. In the NS-NS sector one finds the Kaluza–Klein gauge bosons $g_{\mu i}$ and $b_{\mu i}$ of a gauge group $G = U(1)^{2n}$, the dilaton ϕ and the moduli g_{ij}, b_{ij}; the latter are precisely the n^2 coordinates of the toroidal moduli space $\mathcal{M}$ given in (3.15). The dilaton and the scalars of the R-R sector are not part of this $\mathcal{M}$. However, all scalar fields – NS-NS and R-R – reside in the same gravitational multiplet and so the supergravity considerations discussed in section 2.5 suggest that they combine into a larger moduli space with nontrivial mixings. For a long time this state of affairs seemed incompatible with perturbative string theory since the specific form of the vertex operator of the R-R scalars implies that they can have no nontrivial couplings to the NS-NS scalars [40]. However, nonperturbative corrections alter this conclusion and by now it is believed that taking perturbative and nonperturbative contributions together exactly reproduces the geometrical structures suggested by supergravity. We will return to this point in more detail in section 3.3.

Heterotic string compactified on T^n. The heterotic string compactified on T^n has 16 supercharges and there are n additional scalars A_i^a transforming in the adjoint representation of $E_8 \times E_8$ or $SO(32)$. However, only the $16n$ scalars in the Cartan subalgebra are flat directions and their (generic) vacuum expectation values break the non-abelian gauge symmetry to $U(1)^{16}$. Together with the $2n$ Kaluza–Klein gauge bosons $g_{\mu i}$ and $b_{\mu i}$ they form[28] the total gauge group $G = U(1)^{2n+16}$. On special subspaces of the moduli space there can be nonabelian enhancement of the $U(1)^{16}$, at most up to the original $E_8 \times E_8$ or SO(32) (at least perturbatively).

The $16n$ scalars in the Cartan subalgebra parametrize together with the toroidal moduli g_{ij}, b_{ij} and the dilaton ϕ the $n(n+16)+1$ dimensional moduli space [47]

$$\mathcal{M} \;=\; R^+ \times \left. \frac{\mathrm{SO}(n,n+16)}{\mathrm{SO}(n)\times \mathrm{SO}(n+16)} \right/ \Gamma_T, \tag{3.17}$$

with the T–duality group

$$\Gamma_T = \mathrm{SO}\,(n, n+16, Z)\ . \tag{3.18}$$

The toroidal moduli all reside in $n+16$ (abelian) vector multiplets and the heterotic dilaton is the unique scalar in the gravitational multiplet (see Table 11). Supergravity implies that there can be no mixing between the dilaton and the other $n(n+16)$ scalars and thus the dilaton spans the R^+ component of $\mathcal{M}$.[29] Locally, the moduli space is uniquely determined by supersymmetry [32] and so already from this point of view the moduli space (3.17) is likely to be exact.

[28] $n+16$ of these vectors reside in vector multiplets while the remaining n vectors are part of the gravitational multiplet (see Tables 10 and 11).

[29] In $D=4$ the antisymmetric tensor in the gravitational multiplet is dual to a pseudoscalar and thus can be combined with the dilaton into one complex scalar field. Since this scalar still resides in the gravitational multiplet it does not mix with the other toroidal moduli and again spans a separate component of the moduli space which is found to be locally equivalent to the $SU(1,1)/U(1)$ coset space, which replaces R^+ in (3.17).

It has also been shown that below ten dimensions the heterotic $E_8 \times E_8$ theory and the heterotic SO(32) theory are continuously connected in the moduli space. That is, the two theories sit at different points of the same moduli space of one and the same heterotic string theory [48]. The continuous path which connects the two theories in $D = 9$ involves a transformation $R \to \alpha'/R$ and hence they are also called T–duals of each other.

Type-I compactified on T^n. Toroidal compactifications of the type-I theory are slightly more involved. Locally the moduli space is dictated by supersymmetry to be the coset $R^+ \times SO(n, n+16)/SO(n) \times SO(n+16)$ but perturbatively there is no T–duality symmetry and thus the global moduli space does not coincide with (3.17). However, once D-branes are included as possible open string configurations, type-I theories also have T–duality and the moduli space is given by (3.17) and (3.18) [11, 49, 40]. In fact, establishing T–duality was a guiding motive in the original discovery of D-branes [11]. From the open string point of view T–duality is not a perturbative symmetry since it necessarily involves the presence of solitonic-like excitations.

3.2.2. *K*3 *compactifications*

So far we discussed toroidal compactifications which leave all supercharges intact. Compactifications on a 4-dimensional $K3$ surface break half of the supercharges and hence one moves one column to the right and four rows down in the D-Q plane of Fig. 4; the resulting string theories therefore have $D = 6$ and either 16 or 8 supercharges. The massless modes of $K3$ compactifications arise from nontrivial deformations of the metric and from nontrivial harmonic forms on the $K3$ surface [44]. The moduli space of nontrivial metric deformations is known to be 58-dimensional and given by the coset space

$$\mathcal{M} = R^+ \times \frac{SO(3,19)}{SO(3) \times SO(19)}, \tag{3.19}$$

where the second factor is the Teichmüller space for Einstein metrics of unit volume and the first factor is associated with the size of the $K3$.

On any (complex) Kähler manifold, the differential forms can be decomposed into (p, q)-forms with p holomorphic and q antiholomorphic differentials. The harmonic (p, q)-forms form the cohomology groups $H^{p,q}$ of dimension $h^{p,q}$ and for $K3$ one has a Hodge diamond

$$\begin{matrix} & & h^{0,0} & & \\ & h^{1,0} & & h^{0,1} & \\ h^{2,0} & & h^{1,1} & & h^{0,2} \\ & h^{2,1} & & h^{1,2} & \\ & & h^{2,2} & & \end{matrix} \;=\; \begin{matrix} & & 1 & & \\ & 0 & & 0 & \\ 1 & & 20 & & 1 \\ & 0 & & 0 & \\ & & 1 & & \end{matrix}\,. \tag{3.20}$$

Thus there are 22 harmonic $p + q = 2$–forms which represent the nontrivial deformations of an antisymmetric tensor b_{ij}. On a 4-dimensional manifold an antisymmetric tensor can be constrained to a selfdual or an anti-selfdual tensor and on

$K3$ one finds that the 22 2-forms decompose into 3 selfdual and 19 anti-selfdual 2-forms [44]. For later reference we also record that the Euler number of $K3$ is found to be

$$\chi(K3) = \sum_{p,q} (-)^{p+q} h^{p,q} = 24. \tag{3.21}$$

Type-IIA compactified on K3 Compactifying the type-IIA string on $K3$ breaks half of the 32 (non-chiral) supercharges in $D = 10$ and thus results in a $(1,1)$ supergravity in $D = 6$ coupled to vector multiplets which we already discussed in section 2.2.3. The massless bosonic modes of such compactifications are given by $g_{\mu\nu}$, $b_{\mu\nu}$, ϕ, g_{ij}, b_{ij} all of which arise from the NS-NS sector. g_{ij} denotes the 58 zero modes of the metric on $K3$ and b_{ij} are the 22 harmonic 2-forms.[30] In the R-R sector one finds $V_\mu, C_{\mu\nu\rho}$ and 22 vectors $C_{\mu ij}$. In $D = 6$ a 3-form is dual to a vector field so that altogether there are 24 vector fields in the R-R sector and 81 scalars in the NS-NS sector. The multiplets of $(1,1)$ supergravity are discussed in section 2.2.3. and one infers that the bosonic states of the $K3$ compactification fill out one gravity multiplet and 20 vector multiplets.

All 81 scalars arise in the NS-NS sector. The deformations of the metric span the moduli space given in (3.19). Together with the 22 harmonic 2-forms they combine into the 81-dimensional moduli space [50, 51]

$$\mathcal{M} \;=\; R^+ \;\times\; \frac{\mathrm{SO}(4,20)}{\mathrm{SO}(4)\times\mathrm{SO}(20)}\Big/\,\Gamma_T\;, \tag{3.22}$$

where the factor R^+ is again parameterized by the single scalar in the gravitational multiplet which can be identified with the 6-dimensional dilaton. The second coset factor is spanned by the scalar fields of the vector multiplets. Similar to toroidal compactifications one finds perturbative identifications of the parameter space which are directly related to the properties of the underlying 2-dimensional conformal field theory. Such equivalences are also termed T-duality and for the case at hand the T–duality group is found to be [50, 51]

$$\Gamma_T = \mathrm{SO}(4,20,Z)\;. \tag{3.23}$$

Type-IIB compactified on K3. Compactifying the type-IIB string on $K3$ breaks half of the 32 chiral supercharges in $D = 10$ and thus results in a $(2,0)$ supergravity in $D = 6$ coupled to tensor multiplets which we already discussed in section 2.2.3. The massless bosonic modes of such compactifications are given by $g_{\mu\nu}, b_{\mu\nu}, \phi, g_{ij}, b_{ij}$ in the NS-NS sector and $b'_{\mu\nu}, \phi', b'_{ij}, C_{\mu\nu\rho\sigma}, C_{\mu\nu ij}$ in the R-R sector; both b_{ij} and b'_{ij} are 22 harmonic 2-forms on $K3$. In $D = 6$ a 4-form tensor $C_{\mu\nu\rho\sigma}$ describes only one physical degree of freedom and is dual to a real scalar. Furthermore, there are 22 spacetime tensors $C_{\mu\nu ij}$, proportional to the 22 harmonic forms on $K3$. Since C is chosen selfdual in $D = 10$, the tensor fields $C_{\mu\nu ij}$ are selfdual or anti-selfdual and their selfduality phase is correlated with the

[30]Contrary to toroidal compactifications there are no massless vectors $g_{\mu i}$ or $b_{\mu i}$ since there are no one-forms on $K3$ (cf. (3.20)).

(anti)selfduality of the corresponding $K3$ harmonic forms. Hence the $C_{\mu\nu ij}$ decompose into 3 selfdual and 19 anti-selfdual $D = 6$ antisymmetric tensors. Altogether there are thus 81 NS-NS and 24 R-R scalars and 5 selfdual and 21 anti-selfdual R-R antisymmetric tensors. The corresponding supermultiplets were already discussed in section 2.2.3.; we immediately infer that the massless modes arising from the $K3$ compactification combine into one gravitational and 21 tensor multiplets of $(2,0)$ supersymmetry. This theory being chiral is potentially anomalous; however, it was shown in [13] that precisely this combination of multiplets is anomaly free.

The 81 scalars of NS-NS sector span the same moduli space as in (3.22) and similar to toroidal compactifications the scalars from the R-R cannot have any nontrivial mixing with the NS-NS scalars at the perturbative level. However, the $(2,0)$ gravitational multiplet contains no scalar at all but rather all scalars appear in the 21 tensor multiplets. On the basis of supersymmetry it was conjectured [52] that all 105 scalars locally parameterize the moduli space

$$\mathcal{M} = \frac{\mathrm{SO}(5,21)}{\mathrm{SO}(5)\times\mathrm{SO}(21)}\,. \tag{3.24}$$

Indeed, once nonperturbative corrections of string theory are taken into account this moduli space is generated; we return to this point in section 3.3.4.

The heterotic string compactified on $K3$ The heterotic string compactified on $K3$ has 8 unbroken supercharges or $(1,0)$ supergravity in $D = 6$ coupled to vector-, tensor- and hypermultiplets. Contrary to the previously discussed type-II compactifications this heterotic string theory does not uniquely fix the massless spectrum but instead one finds distinct families of string vacua with different contents of massless states.[31] However, $(1,0)$ supersymmetry is chiral and thus gauge and gravitational anomaly cancellation do impose some constraints on the allowed spectra of supermultiplets. One finds the condition [53]

$$n_H - n_V + 29\,n_T - 273 = 0, \tag{3.25}$$

where n_H, n_V and n_T are the numbers of hyper, vector and tensor multiplets, respectively. (The specific field content of these multiplets is displayed in Table 5.) In the perturbative spectrum of the heterotic string there is only one tensor multiplet which contains the selfdual part of $b_{\mu\nu}$ (the anti-selfdual piece resides in the gravitational multiplet) and the dilaton and hence anomaly cancellation in any perturbative heterotic string vacua demands $n_H - n_V = 244$. In addition the Green–Schwarz mechanism requires a modified field strength H for the antisymmetric tensor $H = db + \omega^L - \sum_a v_a\, \omega_a^{YM}$ where ω^L is a Lorentz–Chern–Simons term and ω_a^{YM} are the Yang–Mills Chern–Simons terms [53]. The index a labels the factors G_a of the gauge group $G = \otimes_a G_a$ and v_a are some computable constants which depend on the specific massless spectrum. In order to ensure a globally

[31] Of course, it is a general property of supersymmetry and string theory that fewer unbroken supercharges lead to a much richer variety of low-energy spectra.

well-defined H on the compact $K3$ the integral $\int_{K3} \mathrm{d}H$ has to vanish. This implies

$$\sum_a n_a \equiv \sum_a \int_{K3} \left(\mathrm{tr}\, F^2\right)_a = \int_{K3} \mathrm{tr}\, R^2 = 24, \qquad (3.26)$$

where the last equation used the fact that 24 is the Euler number of $K3$. From (3.26) we learn that in any compactification of the heterotic string on $K3$ the original 10-dimensional gauge symmetry ($E_8 \times E_8$ or SO(32)) is necessarily broken since consistency requires a non-vanishing instanton number n_a.

As before we can also ask for perturbative equivalences on the space of heterotic $K3$ compactifications. It has been shown that the $K3$ compactifications of the 10-dimensional heterotic string with gauge group SO(32) lie in the same moduli space as (particular) $K3$ compactifications of the 10-dimensional heterotic string with gauge group $E_8 \times E_8$ [54, 55]. More precisely, the gauge group is really $\mathrm{Spin}(32)/Z_2$ and one has to distinguish two different types of instantons which are characterized by the second Stieffel–Whitney class [55]. The corresponding distinct compactifications of the $\mathrm{Spin}(32)/Z_2$ heterotic string are also called compactifications with and without vector structure. It has been shown that compactifications with vector structure have a common moduli space with $E_8 \times E_8$ compactifications of instanton numbers $n_1 = 8, n_2 = 16$ [54] while compactifications without vector structure have a common moduli space with $E_8 \times E_8$ compactifications of instanton numbers $n_1 = n_2 = 12$ [55]. The continuous path which connects the two pairs of theories involves a transformation $R \to \alpha'/R$ and hence they are also called T–dual. Furthermore, the $E_8 \times E_8$ compactifications with instanton numbers $n_1 = n_2 = 12$ are part of the same moduli space as the compactifications with instanton numbers $n_1 = 10, n_2 = 14$ [56, 54].

Type-I compactified on $K3$. Also this compactification leads to $(1, 0)$ supersymmetry and thus anomaly cancellation imposes the same constraint (3.25) on the massless spectrum. However, in type-I compactification there can be more than one tensor multiplet and as a consequence also a generalized Green-Schwarz mechanism can be employed [57]. The resulting spectra are much less investigated and we refer the reader to the literature for further details [57, 58].

3.2.3. *Calabi-Yau threefolds compactifications*

Compactifications on a 6-dimensional Calabi-Yau threefold Y_3 break 3/4 of the supercharges present in $D = 10$ and hence one moves two columns to the right and six rows down in the *D-Q* plane of Fig. 4; the resulting string theories therefore have $D = 4$ and either 8 or 4 supercharges. The massless modes of such compactification arise from the nontrivial harmonic forms on Y_3 which again are most conveniently summarized by the Hodge diamond

$$\begin{array}{ccccccc} & & & 1 & & & \\ & & 0 & & 0 & & \\ & 0 & & h^{1,1} & & 0 & \\ 1 & & h^{1,2} & & h^{1,2} & & 1 \\ & 0 & & h^{1,1} & & 0 & \\ & & 0 & & 0 & & \\ & & & 1 & & & \end{array} , \tag{3.27}$$

where $h^{1,1}$ and $h^{1,2}$ are arbitrary integers [37, 45]. The corresponding $(1,1)$ and $(1,2)$ forms are the deformations of the Calabi-Yau metric and the complex structure, respectively. Using the definition (3.21) one finds the Euler number of a threefold to be $\chi(Y_3) = 2(h^{1,1} - h^{1,2})$. It is believed that most (if not all) Calabi–Yau threefolds Y_3 have an associated mirror manifold $\tilde{Y}_3$ with the property that its Hodge numbers are exactly reversed, i.e. $h^{1,1}(\tilde{Y}_3) = h^{1,2}(Y_3)$ and $h^{1,2}(\tilde{Y}_3) = h^{1,1}(Y_3)$, so that $\chi(Y_3) = -\chi(\tilde{Y}_3)$ [59].

The moduli space of Calabi–Yau threefolds is locally a direct product space

$$\mathcal{M}_{Y_3} = \mathcal{M}_{(1,1)} \otimes \mathcal{M}_{(1,2)} , \tag{3.28}$$

where $\mathcal{M}_{(1,1)}$ ($\mathcal{M}_{(1,2)}$) is the moduli space parameterized by the $(1,1)$-forms ($(1,2)$-forms). Both factors are constrained to be special Kähler manifolds [31, 60, 61, 62]. The metric $G_{i\bar{j}}$ of a Kähler manifold is determined by a real scalar function, the Kähler potential K

$$G_{i\bar{j}} = \frac{\partial}{\partial z^i}\frac{\partial}{\partial \bar{z}^{\bar{j}}} K(z,\bar{z}), \tag{3.29}$$

where z^i are the (complex) coordinates on the moduli space. For a special Kähler manifold the Kähler potential satisfies the additional constraint

$$K = -\log\Big[2(\mathcal{F}(z) + \bar{\mathcal{F}}(\bar{z})) - (z^i - \bar{z}^{\bar{i}})(\mathcal{F}_i(z) - \bar{\mathcal{F}}_{\bar{i}}(\bar{z}))\Big] , \quad \mathcal{F}_i \equiv \frac{\partial \mathcal{F}}{\partial z^i} . \tag{3.30}$$

That is, K is determined by a single holomorphic function $\mathcal{F}(z)$.

Type-II compactified on Y_3. Such compactifications have 8 unbroken supercharges which is also called $N = 2$ supersymmetry in $D = 4$. The multiplets are the gravitational multiplet, the vector multiplet and the hypermultiplet.

Compactifications of the type-IIA string results in the massless modes $g_{\mu\nu}$, $b_{\mu\nu}$, ϕ, g_{ij}, b_{ij} from the NS-NS sector and $A_\mu, C_{\mu ij}, C_{ijk}$ from the R-R sector. From g_{ij}, b_{ij} one obtains $h^{1,1} + h^{1,2}$ complex massless scalar fields in the NS-NS sector. $C_{\mu ij}$ leads to $h^{1,1}$ abelian vectors while $h^{1,2}$ complex scalars arise from C_{ijk} (all in the R-R sector) [50, 63]. These states (together with their fermionic partners) combine into $h^{1,1}$ vector multiplets and $h^{1,2}$ hypermultiplets. Furthermore, in $D = 4$ an antisymmetric tensor is dual to a scalar and hence $\phi, b_{\mu\nu}$ and two R-R scalars from C_{ijk} form an additional hypermultiplet. Thus the total number of vector multiplets is $n_V = h^{1,1}$ while the number of hypermultiplets is given by $n_H = h^{2,1} + 1$.

	IIA	IIB
n_H	$h^{1,2}+1$	$h^{1,1}+1$
n_V	$h^{1,1}$	$h^{1,2}$
G	$\mathrm{U}(1)^{h^{1,1}+1}$	$\mathrm{U}(1)^{h^{1,2}+1}$

TABLE 14. Massless spectra of type-II vacua.

For type-IIB vacua one also has $h^{1,1}+h^{1,2}$ complex massless scalar fields in the NS-NS sector but now $h^{1,2}$ abelian vectors together with $h^{1,1}$ complex scalars in the R-R sector (the universal hypermultiplet containing the dilaton is again present) [50, 63]. Hence, $n_V = h^{2,1}$ and $n_H = h^{1,1}+1$ holds for the type-IIB theory.

The gauge group is always abelian and given by $(h^{1,1}+1)$ U(1) factors in type-IIA and $(h^{1,2}+1)$ $U(1)$ factors in type-IIB (the extra U(1) is the graviphoton of the gravitational multiplet). We summarize the spectrum of type-II vacua in Table 14.

As we see the role of $h^{1,1}$ and $h^{1,2}$ is exactly interchanged between type-IIA and type-IIB. Therefore, compactification of type-IIA on a Calabi–Yau threefold Y_3 is equivalent to compactification of type-IIB on the mirror Calabi–Yau $\tilde{Y}_3$. This is another example of a perturbative equivalence of two entire classes of string vacua.

Heterotic string compactified on Y_3. Such compactification have 4 unbroken supercharges which corresponds to $N=1$ supersymmetry in $D=4$. Now the $(1,1)$ and $(1,2)$ forms both correspond to massless chiral multiplets. Generically, there are many distinct families of string vacua with varying gauge groups and matter content and relatively little can be said in general about their properties.

As in $K3$ compactifications of the heterotic string the field strength H of the antisymmetric tensor has to be modified by appropriate Lorentz– and Yang–Mills Chern–Simons terms. However, the requirement for a globally defined H on Y_3 is slightly more involved since the compact manifold is now 6-dimensional and the integral over $\mathrm{d}H$ is no longer a topological invariant.

A special class of consistent compactifications is obtained by embedding the spin connection in the gauge connection of $\mathrm{E}_8\times\mathrm{E}_8$ or SO(32) [64]. In the first case one obtains a gauge group $\mathrm{E}_8\times\mathrm{E}_6$ with $h^{1,1}$ chiral multiplets in the $\mathbf{27}$ and $h^{1,2}$ chiral multiplets in the $\overline{\mathbf{27}}$ representation of E_6. In addition there are also $h^{1,1}+h^{1,2}$ gauge neutral moduli multiplets.

Similarly, compactifications of the SO(32) heterotic string lead to a gauge group $\mathrm{SO}(26)\times\mathrm{U}(1)$ with $h^{1,1}$ chiral multiplets in the $\mathbf{26}_1\oplus\mathbf{1}_{-2}$ and $h^{1,2}$ chiral multiplets in the $\mathbf{26}_{-1}\oplus\mathbf{1}_2$ representation of $\mathrm{SO}(26)\times\mathrm{U}(1)$. In addition, there are $h^{1,1}+h^{1,2}$ gauge neutral moduli multiplets.

Again we encounter a perturbative equivalence: A heterotic string compactified on a given Y_3 leads to the exact same string vacuum as a compactification on the mirror manifold $\tilde{Y}_3$. This can be seen from the above assignment of the massless spectrum (it is only convention what is called **27** versus $\overline{\mathbf{27}}$) but also has been shown more generally for the full string theory [59].

There are many compactifications with different embeddings than the one discussed above and they can lead to very different spectra. The space of heterotic Y_3 compactifications displays a much bigger variety of spectra than any of the compactifications discussed so far and many of the properties can only be discussed on a case-by-case basis [65].

Type-I string compactified on Y_3. This class of string compactification also has 4 supercharges ($N = 1$) but has been much less investigated as the heterotic string. There are also many distinct families of string vacua with varying spectra and couplings [66].

3.3. DUALITY IN STRING THEORY

The concept of duality is very common in physics. Generically it means that there are two (or more) different descriptions of the same physical system. Frequently the different descriptions are only valid in specific domains of the parameter space and only together they can be used to cover the entire parameter space of the physical system. The past few years have shown [67, 18, 40] that also the various string theories of Fig. 4 are interrelated by a complicated 'web' of duality relations; they are not at all independent but instead are different regions of a common parameter space. In fact, it seems that a given representation of the supersymmetry algebra (with a given number of supersymmetries and spacetime dimensions) lead only to one distinct quantum theory with a parameter space that can encompass various perturbatively distinct string theories. In Fig. 4 we plotted the different perturbative compactifications, some of which share the same representation of supersymmetry. As we will see in this section they all turn out to be different regions in a common parameter space.

One distinguishes perturbative and nonperturbative dualities. Perturbative dualities already hold at weak string coupling and the map which identifies the perturbative theories does not involve the dilaton. On the other hand nonperturbative dualities identify regions of the parameter space which are not simultaneously at weak coupling and the duality map involves the dilaton in a nontrivial way. Such nonperturbative dualities are of utmost importance since they map the strong-coupling region of a given (string) theory to the weak-coupling region of a dual theory where perturbative methods are applicable and hence the strong-coupling limit gets (at least partially) under quantitative control.

The perturbative dualities are well established and we have already seen them in the previous section. The nonperturbative dualities are more difficult to deal with and they cannot be proven at present. Rather their validity has only been checked for quantities or couplings which do not receive quantum corrections. Such quantities or couplings exist in supersymmetric (string) theories; they are the BPS states of the theory as well as the holomorphic couplings (such as the prepotential $\mathcal{F}(z)$ of $N = 2$ supergravity in $D = 4$) of the effective action. It is precisely for this reason that supersymmetry has played such an important (technical) role in establishing nonperturbative dualities.

Let us first briefly discuss the perturbative dualities from a common point of view. Then we focus on the nonperturbative dualities and discuss the various relations in turn.

3.3.1. *T–dualities*

All perturbative dualities are now called T–dualities but one can divide them into two classes. In toroidal compactifications (which we discussed in section 3.2.1.) different points in the parameter space of the compactification correspond to equivalent theories with the exact same spectrum and interactions. As a consequence there is a discrete symmetry Γ_T acting on the space of toroidal compactifications. The same situation is encountered in $K3$ compactifications of type-IIA string theories where also a discrete symmetry Γ_T has been identified (cf. (3.23)).

A different situation occurs in T^1 or $K3$ compactifications of the heterotic string. What was thought are two distinct perturbative string theories – the $E_8 \times E_8$ heterotic string compactified on T^1 or $K3$ and the SO(32) string compactified on T^1 or $K3$ – turn out to be merely different regions of a common parameter space. In other words there is a continuous path which connects the two theories and thus also their respective parameter spaces are continuously connected. Finally, the equivalence of type-IIA compactified on Y_3 with type-IIB compactified on the mirror $\tilde{Y}_3$ identifies compactifications on geometrically distinct manifolds as identical and hence maps the parameter space of type-IIA compactifications onto the parameter space of type-IIB compactifications. The common feature of all of these examples is a perturbative equivalence between string compactifications. Let us now turn to nonperturbative equivalences.

3.3.2. *S–dualities*

Let A and B be two perturbatively distinct string theories each with its own string coupling g_A and g_B, respectively. However, it is possible that once all quantum corrections (including the nonperturbative corrections) are taken into account A and B are equivalent as quantum theories and one has

$$A \equiv B\,. \tag{3.31}$$

This situation can occur in two different ways:

(a) The strong-coupling limit of A is mapped to the weak coupling limit of B

$$\lim_{g_A \to \infty} A \to \lim_{g_B \to 0} B\,, \tag{3.32}$$

or in other words $g_A \sim g_B^{-1}$. Using (3.3) one finds in terms of the corresponding dilatons the identification

$$\phi_A = -\phi_B\,. \tag{3.33}$$

Along with this strong-weak coupling relation goes a map of the elementary excitations of theory A to the nonperturbative, solitonic excitations of theory B and vice versa. The theories A and B are called S–dual and one also refers to this situation as a 'string–string duality'.
Examples of S–dual string theories are:

- The heterotic SO(32) string and the type-I string are S–dual in $D = 10$. The evidence for this duality is the agreement of the low-energy effective actions[32] once one identifies $\phi_{\rm HSO} = -\phi_{\rm I}$ [33, 68, 69] and the fact that the perturbative heterotic SO(32) string has been identified as the D–string of the type-I theory [70]. In the limit of strong coupling in the type-I theory ($g_{\rm I} \to \infty$) the heterotic SO(32) string becomes the 'lightest' and thus perturbative object.
- The type-IIA string compactified on $K3$ and the heterotic string compactified on T^4 are S–dual [71, 34, 33, 72, 73]. Both theories have $(1,1)$ supersymmetry in $D = 6$ with exactly the same massless spectrum. Furthermore, from (3.17, 3.22) one learns that also the moduli spaces of the two string compactifications coincide. The effective actions of the two perturbative theories agree if one identifies [33]

$$\begin{aligned} \phi_{\rm H} &= -\phi_{\rm IIA} \,, \\ H_{\rm H} &= e^{-2\phi_{\rm IIA}}\, {}^*H_{\rm IIA} \,, \\ (g_{\mu\nu})_{\rm H} &= e^{-2\phi_{\rm IIA}} (g_{\mu\nu})_{\rm IIA} \,, \end{aligned} \tag{3.34}$$

where $H = db$ is the field strength of the antisymmetric tensor and ${}^*H$ is its Poincare dual. The first equation in (3.34) again implies a strong-weak coupling relation while the second is the equivalent of an electric-magnetic duality. Further evidence for this S-duality arises from the observation that the zero modes in a solitonic string background of the type-IIA theory compactified on $K3$ have the same structure as the Kaluza–Klein modes of the heterotic string compactified on T^4 [72, 73].

(b) There is a variant of the above situation where the dilaton of theory A is not mapped to the dilaton of theory B as in (3.33), but rather to any of the other perturbative moduli R_B of theory B. In this case one has the identifications

$$\phi_A \sim R_B \,, \qquad \phi_B \sim R_A \,, \tag{3.35}$$

or in other words the strong-coupling limit of A is independent of g_B

$$\lim_{g_A \to \infty} A \;=\; \text{independent of } g_B \,. \tag{3.36}$$

As in case (a) also here the strong-coupling limit of A is controlled by the perturbative regime of theory B and thus accessible in perturbation theory (at least in principle).
This situation is found in the following examples:

[32] In (2.53)–(2.55) we displayed the heterotic Lagrangian in different frames. The last frame (2.55) shows the equivalence with type-I.

- The type-II string compactified on Y_3 and the heterotic string compactified on $K3\times T^2$ are S–dual in the sense just defined [74, 75]. The heterotic dilaton is a member of a vector multiplet and mapped to one of the geometrical moduli of the Calabi-Yau threefold Y_3.[33] Conversely, the type-II dilaton is part of a hypermultiplet and mapped to one of the geometrical moduli of the $K3$. The validity of this duality has been checked in a variety of ways for quite a number of dual string vacua [76, 77, 78, 79]. In particular it has been shown that the prepotential $\mathcal{F}(z)$ appearing in (3.30) agrees for dual pairs of string vacua.
- In the same sense the heterotic string compactified on $K3$ and the type-I string compactified on $K3$ are S–dual [55, 80].

Let us summarize the known S–dualities in the following table

$$
\begin{array}{llcl}
D=10: & \mathrm{HSO} & \stackrel{S}{\longleftrightarrow} & \mathrm{I} \\
D=6\;: & \mathrm{IIA}/K3 & \stackrel{S}{\longleftrightarrow} & \mathrm{H}/T^4 \\
 & \mathrm{H}/K3 & \stackrel{S}{\longleftrightarrow} & \mathrm{I}/K3 \\
D=4\;: & \mathrm{II}/Y_3 & \stackrel{S}{\longleftrightarrow} & \mathrm{H}/K3\times T^2
\end{array}
$$

In Fig. 5 these S-dualities are denoted by a horizontal bar (—).

3.3.3. *Self-duality and U-duality*

Another situation is encountered when the strong-coupling limit of a theory A is controlled not by a distinct theory B, but rather by a different perturbative region of the same theory A. That is, the strong-coupling regime of A has an alternative weakly-coupled description within the same theory A but in terms of a different set of elementary degrees of freedom. The new perturbation theory is often called the magnetic theory and its perturbative degrees of freedom are called magnetic degrees of freedom. This stems from the fact that the first duality (in $D=4$) put forward by Montonen and Olive [81] suggested that an electric U(1) gauge theory is dual to a magnetic U(1) gauge theory with a magnetic photon and magnetic monopoles as perturbative degrees of freedom. This situation is more general and can appear also for extended objects. However, for such a self-duality to hold the theory A has to have a nontrivial (discrete) symmetry which maps the strong-coupling region to a region of weak coupling and simultaneously the different elementary excitations onto each other. One has to make a further subdivision of this case:

(a) The symmetry group is $\Gamma_S = \mathrm{SL}(2,Z)$ which acts on a single complex scalar field containing the dilaton as its real (or imaginary) part. Unfortunately this situation is also called S–duality and the associated symmetry group Γ_S is called the S–duality group. (We prefer to call it a special case of a U–duality.) Examples of this case are:

[33] More precisely, Y_3 has to be a $K3$-fibration and the heterotic dilaton is mapped to the modulus parameterizing the size of the $\mathbf{P}^1$ base of the fibration [76, 77, 78].

- The type-IIB string in $D = 10$ is conjectured to have $\Gamma_S = \mathrm{SL}(2, Z)$ [34, 33, 69]. The corresponding supergravity theory has a $\mathrm{SL}(2, R)$ as a symmetry group [8] (see also Table 9) but quantum corrections break this continuous symmetry to its discrete version $\mathrm{SL}(2, Z)$. This exact symmetry predicts an infinite number of equivalent weakly coupled type-IIB strings which carry R-R charge; such strings have indeed been identified as appropriate D-strings [82, 83].
- A second example is the heterotic string compactified on T^6 which has $D = 4$ and also $\Gamma_S = \mathrm{SL}(2, Z)$. In toroidal compactifications of the heterotic string the dilaton is the unique scalar in the gravitational multiplet and parameterizes the R^+ component of the heterotic moduli space (3.17). However, in $D = 4$ an antisymmetric tensor of rank 2 is on-shell equivalent to a pseudoscalar and thus can be combined together with the dilaton into one complex scalar of the gravitational multiplet spanning the SU(1,1)/U(1) component of the moduli space. It is this complex scalar on which Γ_S acts, leaving all other toroidal moduli invariant.[34] The vacuum-expectation value of the pseudoscalar plays the role of the θ-angle and so this duality is nothing but the string theoretical version of the original electric–magnetic Montonen–Olive duality which in many respects started the subject of string dualities [81, 84, 85, 86].

In Fig. 5 we mark the theories with $\Gamma_S = \mathrm{SL}(2, Z)$ by an 'S' next to it. There also is a nontrivial generalization of this case:

(b) The product of Γ_S and the T-duality group Γ_T is contained as a maximal subgroup in a bigger group Γ_U, called the U–duality group [34]. This situation is encountered in toroidal T^n compactifications of the type-II string. Γ_S is 'inherited' from the type-IIB string in $D = 10$ and Γ_T already exists at the perturbative level (cf. (3.16)). As we already discussed extensively in section 2.5, the corresponding supergravities do have a much larger continuous symmetry group which transforms all scalar fields into each other irrespective of their NS-NS or R-R origin (cf. Table 9). This is a consequence of the fact that the supergravities have a unique gravitational multiplet, which contains all scalar fields on an equal footing. Furthermore, they are constructed as toroidal compactifications of the 11-dimensional supergravity while the string vacua arise as compactifications of 10-dimensional string theories. Within the perturbative regime there can never be a symmetry which mixes NS-NS scalars with their R-R 'colleagues' due to their rather different dilaton couplings. However, nonperturbatively, when also D-brane configurations are taken into account, such a symmetry is no longer forbidden and evidence in its favour has been accumulated [34, 33]. The necessary states carrying R-R charge have been identified and the nonperturbative BPS-spectrum assembles in representations of Γ_U. The U–duality groups in arbitrary dimensions are summarized in

[34]Note that this is rather different than the previous case where Γ_S acted on the two scalars of the IIB string.

D	Γ_T	Γ_U
10A	1	1
10B	1	SL(2,Z)
9	Z_2	$\mathrm{SL}(2,Z)\times Z_2$
8	$\mathrm{SO}(2,2,Z)$	$\mathrm{SL}(3,Z)\times\mathrm{SL}(2,Z)$
7	$\mathrm{SO}(3,3,Z)$	$\mathrm{SL}(5,Z)$
6	$\mathrm{SO}(4,4,Z)$	$\mathrm{SO}(5,5,Z)$
5	$\mathrm{SO}(5,5,Z)$	$\mathrm{E}_{6(+6)}(Z)$
4	$\mathrm{SO}(6,6,Z)$	$\mathrm{E}_{7(+7)}(Z)$

TABLE 15. U–duality groups for type-II strings.

Table 15 [34]; they are just the discrete version of the group G in Table 9. In Fig. 5 these theories are marked with a 'U'.

3.3.4. *M-theory*

The various dualities discussed so far relate different perturbative string theories. In these cases the strong-coupling limit of a given string theory is controlled by another (or the same) perturbative string theory. However, not all strong-coupling limits are of this type. Instead it is possible that the strong-coupling limit of a given theory is something entirely new, not any of the other string theories [33]. Only limited amount of information is so far known about this new theory which is called M-theory. Examples of this situation are:

- The strong-coupling limit of the type-IIA theory in $D = 10$. The low-energy effective action was discussed in section 2.5 where we also indicated how it can be constructed as a T^1 compactification of 11-dimensional supergravity. This implied a relation between the radius R_{11} of the 11-th dimension and the string coupling constant $g_s = e^{\langle\phi\rangle}$ [33] (cf. (2.44))

$$R_{11} = \frac{L}{2\pi}\, g_s^{\frac{2}{3}} \,, \tag{3.37}$$

where L is the length of the 11-th dimension introduced in section 2.5. Moreover, the Kaluza-Klein spectrum of this theory obeys (in the string frame)

$$M^{\mathrm{KK}} = \frac{2\pi|n|}{g_s L} \,, \tag{3.38}$$

where n is an arbitrary integer (cf. (2.49)). These KK-states are not part of the perturbative type-IIA spectrum since they become heavy in the weak-coupling limit $g_s \to 0$. However, in the strong-coupling limit $g_s \to \infty$ they become light and can no longer be neglected in the effective theory. This infinite number of light states (which can be identified with D-particles of type-IIA string theory, or extremal black holes of IIA supergravity) signals that the theory effectively decompactifies, which can also be seen from (3.37). Supersymmetry is unbroken in this limit and hence the KK-states assemble in supermultiplets of the 11-dimensional supergravity. Since there is no string theory which has 11-dimensional supergravity as the low-energy limit, the strong-coupling limit of type-IIA string theory has to be a new theory, called M-theory, which cannot be a theory of (only) strings. M-theory is supposed to capture all degrees of freedom of all known string theories, both at the perturbative and the nonperturbative level [35, 33, 87, 18]. There exists a conjecture according to which the degrees of freedom of M-theory are captured in U(N) supersymmetric matrix models in the $N \to \infty$ limit [88]. These matrix models have been known for some time [89] and were also known to describe supermembranes [90] in the lightcone gauge [91]. The same quantum-mechanical models describe the short-distance dynamics of N D-particles, caused by the exchange of open strings [83, 40]. A review of these developments is beyond the scope of these lectures and we refer the reader to the literature [92].

- A second and maybe even more surprising result shows that also the strong-coupling limit of the heterotic $E_8 \times E_8$ string is captured by M-theory. In this case, 11-dimensional supergravity is not compactified on a circle but rather on a Z_2 orbifold of the circle [87]. The space coordinate x^{11} is odd under the action of Z_2 and hence the three-form $C_{\mu\nu\rho}$ as well as $g_{\mu 11}$ are also odd. The Z_2-invariant spectrum in $D = 10$ consists of the metric $g_{\mu\nu}$, the antisymmetric tensor $C_{\mu\nu 11}$ and the scalar $g_{11\,11}$. Up to the gauge degrees of freedom this is precisely the massless spectrum of the 10-dimensional heterotic string. The $E_8 \times E_8$ Yang-Mills fields have to arise in the twisted sector of the orbifold. One way to see this is to note that the Z_2 truncation of 11-dimensional supergravity is inconsistent in that it gives rise to gravitational anomalies [93]. In order to cancel such anomalies non-abelian gauge fields have to be present with appropriate couplings to the antisymmetric tensor such that a Green-Schwarz mechanism can be employed [94]. Such additional states can only appear in the twisted sectors of the orbifold theory which are located at the orbifold fixed points $x^{11} = 0$ and $x^{11} = L/2$. However, due to the Z_2 symmetry, these two 10-dimensional hyperplanes have to contribute equally to the anomaly. This can only be achieved for a gauge group which is a product of two factors and thus $E_8 \times E_8$ with one E_8 factor on each hyperplane is the only consistent candidate for such a theory [87]. Just as in the type-IIA case one has $R_{11} = g_{\rm H}^{3/2} L/2\pi$ and thus weak coupling corresponds to small R_{11} and the two 10-dimensional hyperplanes sit close to each other; in the strong-coupling limit the two 10-dimensional hyperplanes move far apart (to the end of the world). Using the previous terminology the heterotic $E_8 \times E_8$ string theory can be viewed as M-theory compactified on T^1/Z_2.
- There is an immediate corollary of the dualities discussed so far. The strong-coupling limit of the type-IIA string compactified on $K3$ is simultaneously

governed by M-theory compactified on $K3 \times T^1$ and the heterotic string compactified on T^4. Since there is a T^1 in both theories one concludes that the strong-coupling limit of the heterotic string compactified on T^3 is governed by M-theory compactified on $K3$ [33]. From (3.17) and (3.19) we learn that indeed the moduli spaces of both theories agree if the heterotic dilaton is related to the overall size of the $K3$. A detailed comparison of the respective effective actions reveals that the strong-coupling limit on the heterotic side corresponds to the large-radius limit of the $K3$ on the M-theory side [33].

- The exact same argument can be repeated in $D = 5$. The strong-coupling limit of the type-IIA string compactified on (a $K3$-fibred) Y_3 is simultaneously governed by M-theory compactified on $Y_3 \times T^1$ and the heterotic string compactified on $K3 \times T^2$. By the same argument used above one concludes that the strong-coupling limit of the heterotic string compactified on $K3 \times T^1$ is governed by M-theory compactified on (a $K3$-fibred) Y_3 [95]. In this case the heterotic dilaton is directly related to the size of the $\mathbf{P}^1$ base of the $K3$-fibration.
- A slightly more involved analysis is necessary to conclude that the strong-coupling limit of the IIB string compactified on $K3$ is governed by M-theory compactified on T^5/Z_2 [96]. Compactifying 11-dimensional supergravity on the orbifold T^5/Z_2 one obtains the chiral $(2,0)$ supergravity with one gravity multiplet and five tensor multiplets from the untwisted sector. The twisted sector is again inferred by anomaly cancellation and provides 16 further tensor multiplets. The weakly-coupled type-IIB theory on $K3$ corresponds to a 'smashed' T^5/Z_2 where the 32 fixed points degenerate into 16 pairs and the 16 tensor multiplets are equally distributed among those pairs. The 81 scalars from the NS-NS sector combine with the 24 scalars from the R-R sector to form the moduli space [52, 97, 98]

$$\mathcal{M} = \frac{\mathrm{SO}(5,21)}{\mathrm{SO}(5)\times\mathrm{SO}(21)}\Big/\,\Gamma_U, \tag{3.39}$$

with a U-duality group $\Gamma_U = \mathrm{SL}(5,21,Z)$. The local structure of this moduli space is already fixed by supergravity (cf. (3.24)) while the global structure follows from M-theory.

Let us summarize the nontrivial compactifications of M-theory:

$$\begin{array}{lcl}
\mathrm{M}/T^1 & \to & \mathrm{IIA} \\
\mathrm{M}/T^1/Z_2 & \to & \mathrm{HE8} \\
\mathrm{M}/K3 & \to & \mathrm{H}/T^3 \\
\mathrm{M}/T^5/Z_2 & \to & \mathrm{IIB}/K3 \\
\mathrm{M}/Y_3 & \to & \mathrm{H}/K3\times T^1
\end{array}$$

Theories whose strong-coupling limit is governed by M-theory are marked with an 'M' in Fig. 5.

3.3.5. *F-theory*

As we discussed previously the type-IIB theory in 10 spacetime dimensions is believed to have an exact $\mathrm{SL}(2,Z)$ quantum symmetry which acts on the complex scalar $\tau = e^{-2\phi} + i\phi'$, where ϕ and ϕ' are the two scalar fields of type-IIB theory (c.f. Table 13). This fact led Vafa to propose that the type-IIB string could be viewed as the toroidal compactification of a twelve-dimensional theory, called F-theory, where τ is the complex structure modulus of a two-torus T^2 and the Kähler-class modulus is frozen [99]. Apart from having a geometrical interpretation of the SL$(2, Z)$ symmetry this proposal led to the construction of new, nonperturbative string vacua in lower space-time dimensions. In order to preserve the SL$(2, Z)$ quantum symmetry the compactification manifold cannot be arbitrary but has to be what is called an elliptic fibration. That is, the manifold is locally a fibre bundle with a two-torus T^2 over some base B but there are a finite number of singular points where the torus degenerates. As a consequence nontrivial closed loops on B can induce a SL$(2, Z)$ transformation of the fibre. This implies that the dilaton is not constant on the compactification manifold, but can have SL$(2, Z)$ monodromy [100]. It is precisely this fact which results in nontrivial (nonperturbative) string vacua inaccessible in string perturbation theory.

F-theory can be compactified on elliptic Calabi–Yau manifolds and each of such compactifications is conjectured to capture the nonperturbative physics of an appropriate string vacua. One finds:

- The IIB string in $D = 10$ can be viewed as F-theory compactified on T^2 with a frozen Kähler modulus.
- F-theory compactified on an elliptic $K3$ yields an 8-dimensional vacuum with 16 supercharges which is quantum equivalent to the heterotic string compactified on T^2 [99, 101].
- F-theory compactified on an elliptic Calabi–Yau threefold has 8 unbroken supercharges and is quantum equivalent to the heterotic string compactified on $K3$ [54]. In fact there is a beautiful correspondence between the heterotic vacua labelled by the instanton numbers (n_1, n_2) and elliptically fibred Calabi-Yau manifolds with the base being the Hirzebruch surfaces $\mathbb{F}_{n_2-12}$ (we have chosen $n_2 \geq n_1$) [54].
- Finally, the heterotic string compactified on a Calabi–Yau threefold Y_3 is quantum equivalent to F-theory compactified on an elliptic Calabi–Yau fourfold [102]. Calabi-Yau fourfolds are Calabi-Yau manifolds of complex dimension four and holonomy group SU(4).

Let us summarize the nontrivial compactifications of F-theory:

$$\begin{array}{lcl} \mathrm{F}/T^2 & \to & \mathrm{IIB} \\ \mathrm{F}/K3 & \to & \mathrm{H}/T^2 \\ \mathrm{F}/Y_3 & \to & \mathrm{H}/K3 \\ \mathrm{F}/Y_4 & \to & \mathrm{H}/Y_3 \end{array}$$

The theories governed by F-theory are marked with an 'F' in Fig. 5.

3.3.6. *Summary of all strong-coupling limits.*

So far we tried to systematically discuss the different possible strong-coupling limits of string theories along with the relevant examples. In this final section let

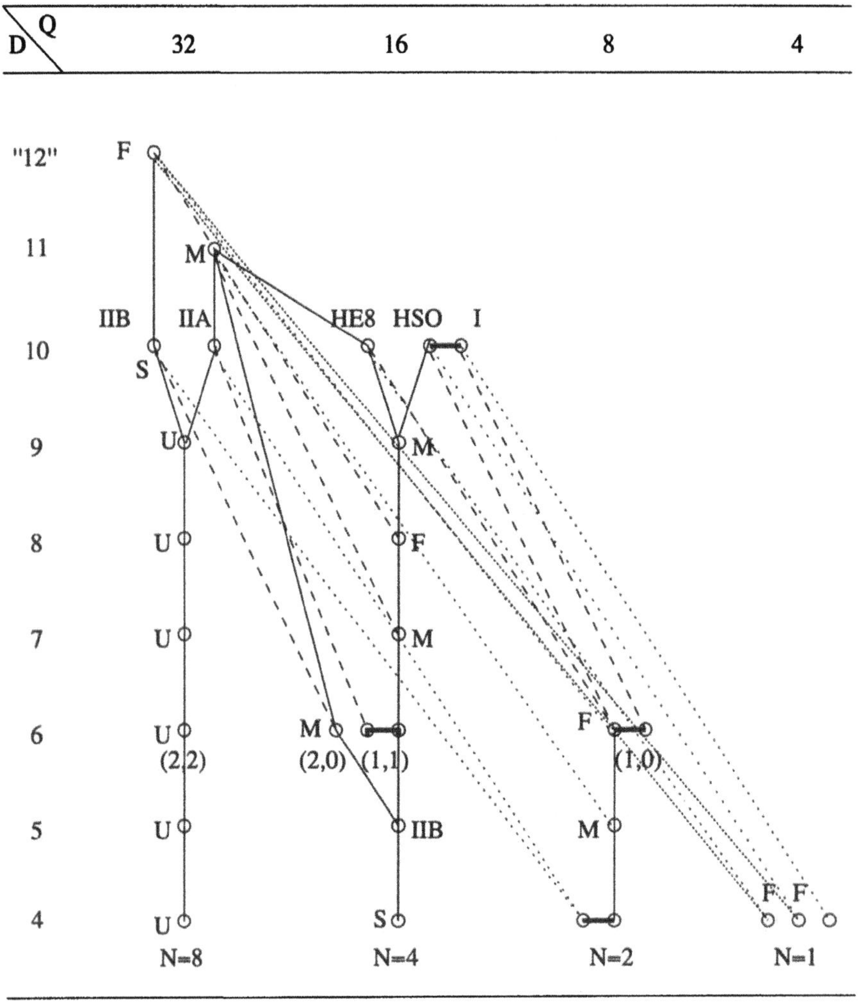

Figure 5. The distinct string theory and their strong-coupling limit. As in Fig. 4 the solid line (–) denotes toroidal compactification, the dashed line (– –) denotes $K3$ compactifications and the dotted line (···) denotes Y_3 compactifications. The fine-dotted line (··) denotes Y_4 compactifications while the horizontal bar (▬) indicates a string-string duality. The theories marked with a 'U' ('S') have a U–duality (S–duality); the strong-coupling limit of the theories marked by 'M' ('F') are controlled by M-theory (F-theory). With a slight abuse of convention, we also denote the two orbifold compactification $\mathrm{M}/T^1/Z_2$ and $\mathrm{M}/T^5/Z_2$ by a solid line.

us one more time summarize all strong-coupling relations but now organized by the number of supercharges. The following discussion is visualized in Fig. 5.

$Q = 32$. Theories with $Q = 32$ have a unique massless multiplet which contains all scalar fields on an equal footing and does not single out a string coupling constant. As a consequence there is a discrete symmetry group Γ_U (listed in Table 15) which leads to global identifications in the moduli space and in any given region a different scalar plays the role of the perturbative expansion parameter. A special situation occurs in $D = 10$ where the type-IIB string has $\Gamma_U = \Gamma_S = \mathrm{SL}(2, Z)$ while the strong-coupling limit of the type-IIA string cannot be a string theory but is something new – M-theory – whose low-energy limit is 11-dimensional supergravity.

$Q = 16$. Theories with $Q = 16$ contain, besides the gravitational supermultiplet, also a set of vector supermultiplets. The gravitational multiplet always contains one scalar (cf. Table 11), which can be uniquely identified to play the role of the coupling constant. In $D = 10$ the heterotic SO(32) string and the type-I string are S-dual while the strong-coupling limit of the $E_8 \times E_8$ string is governed by M-theory compactified on an orbifold T^1/Z_2. In $D = 9$ the two heterotic theories are perturbatively equivalent and their strong coupling limit is governed both by M-theory and type-I. In $D = 8$ the strong coupling limit is governed both by F-theory and type-I; in $D = 7$ the strong-coupling limit is governed by M-theory compactified on $K3$. In $D = 6$ the heterotic string compactified on T^4 and the type-IIA string compactified on $K3$ are S-dual and the strong-coupling limit of type-IIB compactified on $K3$ is captured by M-theory compactified on T^5/Z_2. In $D = 5$ the strong-coupling limit of the heterotic string compactified on T^5 is governed by type-IIB compactified on $K3 \times T^1$ [33]. In $D = 4$ the antisymmetric tensor and the dilaton can be combined into one complex scalar with an appropriate action of $\Gamma_U = \Gamma_S = \mathrm{SL}(2, Z)$.

$Q = 8$. Theories with $Q = 8$ supercharges exist in $D = 6$ and below. In $D = 6$ the heterotic $E_8 \times E_8$ and the heterotic SO(32) string are perturbatively equivalent and their nonperturbative regime is governed by F-theory compactified on elliptic Calabi–Yau threefolds. Furthermore, there also exists an S-duality with type-I compactified on $K3$. In $D = 5$ the strong-coupling limit is controlled by M-theory while in $D = 4$ the heterotic string compactified on $K3 \times T^2$ is S-dual to the type-II string compactified on Y_3.

$Q = 4$. Finally, theories with $Q = 4$ only exist in $D = 4$. Both heterotic string theories are non-perturbatively equivalent to F-theory compactified on an elliptic Calabi–Yau fourfold Y_4, while the strong-coupling limit of the type-I theory is not yet completely understood. Some of the type-I models seem to be S-dual to the heterotic vacua [103]. It might well be that all three theories are part of a larger moduli space.

Acknowledgements

We would like to thank P. Aspinwall, G.L. Cardoso, V. Kaplunovsky, D. Lüst, T. Mohaupt, H. Nicolai, J. Sonnenschein, S. Theisen, A. Van Proeyen, S. Yankielowicz for the most fruitful collaboration on the subjects presented in these lectures. S.

Förste, C.M. Hofman, B. Kleijn, P. Mayr, E. Rabinovici, M.A. Vasiliev, G. Zwart also helped us with useful comments on the manuscript.

We would also like to thank the organizers of this school for providing such a stimulating environment.

Appendix

A. Field representations

In section 2 we have outlined the derivation of various supermultiplets of states. At the noninteracting level, these states can easily be described in terms of local fields. The purpose of this appendix is to present suitable field representations for the relevant states. With the help of these field representations one can then write down free massless supersymmetric field theories. Interactions can be introduced separately, for instance, by iteration or by some more systematic procedure. We should stress that there are sometimes ambiguities, because different field representation can describe the same massless free states. At the interacting level, these ambiguities will usually disappear. So the proper choice of the field representation may be subtle. Our strategy is to discuss a number of standard field representations, in D spacetime dimensions, with their corresponding free wave equations and exhibit the behaviour of the corresponding states under helicity rotations. The supermultiplets discussed previously can then be converted into supersymmetric actions, quadratic in the fields. For selfdual tensor fields, the action must augmented by a duality constraint on the corresponding field strength.

A.1. GRAVITON FIELDS

The linearized Einstein equation for $g_{\mu\nu} = \eta_{\mu\nu} + \kappa h_{\mu\nu}$ implies that (for $D \geq 3$)

$$R_{\mu\nu} \propto \partial^2 h_{\mu\nu} + \partial_\mu \partial_\nu h - \partial_\mu \partial^\rho h_{\rho\nu} - \partial_\nu \partial^\rho h_{\rho\mu} = 0\,, \tag{A.1}$$

where $h \equiv h_{\mu\mu}$ and $R_{\mu\nu}$ is the Ricci tensor. To analyze the number of states implied by this equation, one may count the number of plane-wave solutions with given momentum q^μ. It then turns out that there are D arbitrary solutions, corresponding to the linearized gauge invariance, $h_{\mu\nu} \to h_{\mu\nu} + \partial_\mu \xi_\nu + \partial_\nu \xi_\mu$, which can be discarded. Many other components vanish and the only nonvanishing ones require the momentum to be lightlike. These reside in the fields h_{ij}, where the components i, j are in the transverse $(D-2)$-dimensional subspace. In addition, the trace of h_{ij} must be zero. Hence, the relevant plane-wave solutions are massless and have polarizations (helicities) characterized by a symmetric traceless 2-rank tensor. This tensor comprises $\frac{1}{2}D(D-3)$, which transform irreducibly under the $\mathrm{SO}(D-2)$ helicity group of transverse rotations. For the special case of $D = 6$ spacetime dimensions, the helicity group is SO(4), which factorizes into two SU(2) groups. The symmetric traceless representation then transforms as a doublet under each of the SU(2) factors and it is thus denoted by (2,2).

As is well known, for $D = 3$ there are obviously no dynamic degrees of freedom associated with the gravitational field. When $D = 2$ there are again no dynamic degrees of freedom, but here (A.1) should be replaced by $R_{\mu\nu} = \frac{1}{2} g_{\mu\nu} R$.

A.2. ANTISYMMETRIC TENSOR GAUGE FIELDS

Antisymmetric tensor gauge fields have field-strength tensors which are antisymmetric and of rank $p+2$. They satisfy field equations and Bianchi identities, generalizations of the Maxwell equations, which read

$$\partial_{[\mu_1} F_{\mu_2 \cdots \mu_{p+3}]} = 0 \,, \qquad \partial_\mu F^{\mu\nu_1 \cdots \mu_{p+1}} = 0 \,. \tag{A.2}$$

A trivial example is the case $p = -1$, which describes an ordinary scalar field. For $p = -1$ the solution of the first equation of (A.2) yields $F_\mu = \partial_\mu \phi$, so that the second equation yields the Klein-Gordon equation for ϕ. Another example is the case of a vector gauge field, which corresponds to $p = 0$, where (A.2) are just the Maxwell equations.

There are two ways of dealing with (A.2). One is to solve the first equation in terms of an antisymmetric tensor gauge field $A_{\mu_2 \ldots \mu_{p+1}}$ of rank $p+1$,

$$F_{\mu_1 \cdots \mu_{p+2}} = (p+2)\, \partial_{[\mu_1} A_{\mu_2 \cdots \mu_{p+2}]} \,, \tag{A.3}$$

and then impose the second equation. The alternative is to first solve the second equation in terms of an antisymmetric gauge field $B_{\nu_1 \cdots \nu_{D-p+3}}$ of rank $D-p+3$,

$$F_{\mu_1 \cdots \mu_{p+2}} = \frac{1}{(D-p-1)!}\, \varepsilon_{\mu_1 \cdots \mu_{p+2} \rho \nu_1 \cdots \nu_{D-p-3}}\, \partial^\rho B^{\nu_1 \cdots \nu_{D-p-3}} \,, \tag{A.4}$$

after which one imposes the first equation. The second procedure coincides with the first one, but it is based on the dual field strength defined by

$$F_{\mu_1 \cdots \mu_{p+2}} = \frac{1}{(D-p-2)!}\, \varepsilon_{\mu_1 \cdots \mu_{p+2} \nu_1 \cdots \nu_{D-p-2}}\, \tilde{F}^{\nu_1 \cdots \nu_{D-p-2}} \,, \tag{A.5}$$

which can be written as

$$\tilde{F}_{\nu_1 \cdots \nu_{D-p-2}} = (D-p-2)\, \partial_{[\nu_1} B_{\nu_2 \cdots \nu_{D-p-3}]} \,. \tag{A.6}$$

For $\tilde{F}$ the two equations (A.2) are interchanged and the solution in terms of $B_{\nu_1 \ldots \nu_{p'+1}}$, with $p' + p = D - 4$ is the dual formulation of the one in terms of $A_{\mu_1 \ldots \mu_{p+1}}$. As is well known, in a so-called "first-order" formulation it is possible to have a Lagrangian that encompasses both descriptions.

Let us now examine the plane-wave solutions for the equations (A.2). We will be somewhat more explicit here and start from a decomposition of $F_{\mu_1 \cdots \mu_{p+2}}$ in the momentum representation, with a fixed momentum vector q^μ. Introducing $D-2$ transverse polarization vectors ϵ^i_μ, with $i = 1, 2, \ldots, D-2$, and an additional vector $\bar{q}^\mu = (-q^0, \vec{q})$, we decompose the field strength according to

$$\begin{aligned} F_{\mu_1 \ldots \mu_{p+2}}(q) \;\propto\;& a_{i_1 \cdots i_{p+2}}(q)\, \epsilon^{i_1}_{[\mu_1} \cdots \epsilon^{i_{p+2}}_{\mu_{p+2}]} \\ &+ \left(b_{i_1 \cdots i_{p+1}}(q)\, \bar{q}_{[\mu_1} + c_{i_1 \cdots i_{p+1}}(q)\, q_{[\mu_1} \right) \epsilon^{i_1}_{\mu_2} \cdots \epsilon^{i_{p+1}}_{\mu_{p+2}]} \\ &+ d_{i_1 \ldots i_p}(q)\, \epsilon^{i_1}_{[\mu_1} \cdots \epsilon^{i_p}_{\mu_p}\, \bar{q}_{\mu_{p+1}}\, q_{\mu_{p+2}]} \,. \end{aligned} \tag{A.7}$$

Imposing (A.2) yields

$$a_{i_1\cdots i_{p+2}}(q) = b_{i_1\cdots i_{p+1}}(q) = d_{i_1\cdots i_p}(q) = 0\,, \qquad q^2\, c_{i_1\cdots i_{p+1}}(q) = 0\,, \tag{A.8}$$

so that the dynamic degrees of freedom are massless and reside in the antisymmetric $(p+1)$-th rank tensors $c_{i_1\cdots i_{p+1}}(q)$ living in the transverse $(D-2)$-dimensional space. Hence the number of degrees of freedom is equal to $(D-2)!/[(p+1)!(D-p-3)!]$, which is, as expected, invariant under $p \to p' = D-4-p$.

If $D = 2 \bmod 4$ and $p+1 = \frac{1}{2}(D-2)$, it is possible to restrict the tensor $F_{\mu_1\cdots\mu_{p+2}}$ to be selfdual or antiselfdual, viz.

$$F_{\mu_1\cdots\mu_{p+2}} = \pm\frac{1}{(p+2)!}\,\varepsilon_{\mu_1\cdots\mu_{2p+4}}\; F^{\mu_{p+3}\cdots\mu_{2p+4}}\,. \tag{A.9}$$

For such tensors the two equations (A.2) are no longer independent. The above duality condition on the field strength induces a corresponding $(D-2)$-dimensional duality condition (but now in the space of transverse momenta, which is Euclidean) on the coefficients $c_{i_1\cdots i_{p+1}}(q)$,

$$c_{i_1\cdots i_{p+1}}(q) = \mp\frac{1}{(p+1)!}\,\varepsilon_{i_1\cdots i_{2p+2}}\; c^{i_{p+1}\cdots i_{2p+2}}(q)\,. \tag{A.10}$$

Consequently the number of independent solutions associated with the antisymmetric tensor is reduced by a factor 2. For $D=6$ where the helicity group factorizes, the representation of the (anti-)selfdual tensor gauge fields, correspond to (3,1) and (1,3). In $D = 10$ the (anti-)selfdual tensors correspond to the $\mathbf{35}_s$ and $\mathbf{35}_c$ representations.

A.3. SPINOR FIELDS

Consider a spinor $u(q)$ in D space-time dimensions, satisfying the massless Dirac equation (in momentum space),

$$\not{q}\,u(q) = 0, \tag{A.11}$$

The Dirac equation implies that $q^2 = 0$. Using the same manipulations as those leading to (2.3), we rewrite the Dirac equation as

$$q^0\Big(\mathbf{1} - \tilde\Gamma_D\tilde\Gamma_\perp\Big)u(q) = 0\,, \tag{A.12}$$

where $\tilde\Gamma_D$ and $\tilde\Gamma_\perp$ were defined in section 2.2.

In odd dimensions $\tilde\Gamma_D$ is proportional to the unit matrix, so that the above condition determines that the spinors are reduced to a subspace where $\tilde\Gamma_\perp = \pm\mathbf{1}$. For even dimensions the states constitute a spinor representation of the helicity group whose chirality is related to the D-dimensional chirality of the spinor field. For instance, for $D=6$ dimensions a chiral spinor will transform under only one of the SU(2) groups of the helicity group, so we have either (2,1) of (1,2). For

Majorana-Weyl spinors in $D = 10$, the states transform as $\mathbf{8}_c$ or $\mathbf{8}_s$, depending on the chirality of the spinor field.

A.4. GRAVITINO FIELDS

The gravitino field is a vector-spinor ψ_μ and acts as the gauge field of local supersymmetry transformations. Free gravitini satisfy the Rarita-Schwinger equation

$$\Gamma^\mu(\partial_\mu\psi_\nu - \partial_\nu\psi_\mu) = 0\,. \tag{A.13}$$

To examine the nature of plane-wave solutions, we again consider the momentum representation and decompose $\psi(q)$ as

$$\psi_\mu(q) = u_i(q)\,\epsilon^i_\mu + v(q)\,\bar q_\mu + w(q)\,q_\mu\,, \tag{A.14}$$

where the coefficient functions $u_i(q)$, $v(q)$ and $w(q)$ are spinors. The field equation (A.13) takes the form

$$\not{q}u_i(q)\,\epsilon^i_\nu - [\not{\epsilon}^i u_i(q) - \not{\bar q}v(q)]\,q_\nu - \not{q}v(q)\,\bar q_\nu = 0\,, \tag{A.15}$$

where $\not{\epsilon}^i = \epsilon^i_\mu\Gamma^\mu$. The spinor $w(q)$, which is subject to gauge transformations $\delta\psi_\mu = \partial_\mu\epsilon$, is not determined by the gauge invariant field equation (A.13) and can be discarded. The remaining spinors $u_i(q)$ and $v(q)$ satisfy

$$\not{q}u_i(q) = \not{q}v(q) = 0\,, \qquad \not{\bar q}v(q) = \not{\epsilon}^i u_i(q)\,. \tag{A.16}$$

Multiplying the last equation with $\not{q}$ and using the first two equations and $\epsilon^i\cdot q = 0$, one derives $q\cdot\bar q v(q) = 0$. Hence $v(q) = 0$, so that we are left with two equations for $u_i(q)$,

$$\not{q}u_i(q) = \not{\epsilon}^i u_i(q) = 0\,. \tag{A.17}$$

Hence the gravitino states transform under transverse rotations according to the highest helicity representation contained in the product of a vector and a spinor representation.

For instance, in $D = 6$ dimensions $u_i(q)$ transforms as a chiral vector-spinor, which is a product of (2,2) with (2,1) or (1,2). This product decomposes into $(3,2) + (1,2)$, or $(2,3) + (2,1)$, respectively. The second representation is again suppressed by virtue of the second condition in (A.17), so that we are left with (3,2) or (2,3).

In $D = 10$ spacetime dimensions a chiral gravitino field $u_i(q)$ transforms as a chiral vector-spinor, which constitutes a tensor product $\mathbf{8}_v \times \mathbf{8}_c$ (or $\mathbf{8}_v \times \mathbf{8}_s$, depending on the chirality). According to the multiplication rules (2.6), this product decomposes into $\mathbf{8}_s + \mathbf{56}_s$ (or, $\mathbf{8}_c + \mathbf{56}_c$). However, the second equation in (A.17), which is SO(8) covariant, imposes eight conditions thus suppressing the $\mathbf{8}_s$ or $\mathbf{8}_c$ representation. Consequently chiral gravitini transform according to the $\mathbf{56}_s$ or $\mathbf{56}_c$ representations of SO(8).

B. Coupling constants of low-energy effective field theories

In section 2 we discussed various field theories that play a role as effective low-energy field theories for superstrings. The effective field theories can be rigorously derived from the underlying string theory and in this process the free parameters of the field theories are expressed in terms of the parameters of the string theory itself. The purpose of this appendix is to briefly recall the various possibilities of deriving the low-energy effective action.

One method to obtain the low-energy effective action is known as the 'S-matrix approach', which was pioneered in [39]. Here one computes physical scattering amplitudes in both string theory and the low-energy field theory and demands their equality in the limit $p \ll M_s$ where p is the characteristic momentum of the scattering process. This method is carried out most conveniently in the Einstein frame. Alternatively one can use the 'σ-model approach' which was pioneered in [104]. One imposes conformal invariance on the 2-dimensional σ-model specified by the action (3.8). This requirement leads to field equations in spacetime which coincide with the field equations obtained from an action in the string frame. (The σ-model approach is not applicable to all string compactifications.)

Let us first outline how the relation (3.5) emerges in the S-matrix approach. From (3.2) and (3.8) we know that the dilaton couples to the topology of the world sheet, so that in leading order (genus-0), the N-particle S-matrix elements are proportional to g_s^{N-2} multiplied by an appropriate power of α', in accordance with dimensional counting. On the other hand, the corresponding S-matrix elements when calculated from the effective field theory, are expressed in terms of Newton's constant. Comparing the S-matrix elements one obtains (suppressing numerical factors) the relation (3.5),

$$(\kappa_D^2)^{\text{physical}} = \kappa_D^2\, \lambda^{(2-D)/2} = \alpha'^{(D-2)/2}\, e^{2\langle\phi\rangle}\,, \tag{B.1}$$

with λ given in the Einstein frame.

Note that the parameter κ_D^2 is not determined by (B.1) in agreement with our previous arguments that it is intrinsically undetermined. There are basically two ways to proceed: First one may *choose* the constant κ_D^2 to be Newton's constant. This implies that one has to expand the metric around $g_{\mu\nu} = \eta_{\mu\nu}$ in the Einstein frame, so that $\lambda = 1$. This is a convenient setting, which is most commonly used (see, e.g. [39]) and which leads to $\kappa_D^2 = (\alpha')^{(D-2)/2} g_s^2$. However, this choice implies that a coupling constant (κ_D^2) in the effective Lagrangian depends on a parameter (g_s) that arises as the vacuum-expectation value of the dilaton. Alternatively one could insist that any dependence on g_s only arises as a result of the explicit couplings of the dilaton field in the effective Lagrangian. Or in other words, no *parameters* of the effective action are chosen to explicitly depend on g_s. In the Einstein frame this requires to expand the metric around $g_{\mu\nu} = \lambda\,\eta_{\mu\nu}$ with $\lambda^{(D-2)/4} = g_s$, while in the string frame an expansion around $g_{\mu\nu} = \eta_{\mu\nu}$ is necessary. In both frames one obtains $\kappa_D^2 = \alpha'^{(D-2)/2}$ with no explicit dependence on g_s. In the string frame, this effective action coincide with the one obtained by the σ-model approach.[35] In the Einstein frame this choice is somewhat awkward and rarely used.

[35] Observe that the σ-model approach does not insist on a particular ground-state value for the metric. Since it derives the effective action by integration of the field equations, it determines the Lagrangian only up to an overall constant.

As a further illustration of the two different parameter choices let us consider higher order gravitational interactions which generically arise in string theory. For example, the string calculations (still to leading order in g_s) of the S-matrix of graviton-graviton scattering give rise to contributions that require an effective interaction quartic in the Riemann tensor

$$\mathcal{L}_{\rm eff} = \frac{1}{\kappa_D^2}\sqrt{-g}\,{\rm e}^{-2\phi}\Big[R + A\,\alpha'^3\left(R_{abcd}R^{cdef}R_{efgh}R^{ghab} + \cdots\right)\Big]. \tag{B.2}$$

We have displayed $\mathcal{L}_{\rm eff}$ is the string frame and the higher-order terms depend on a dimensionless constant A which is independent of g_s. In the Einstein frame the dilaton factor in front of the Ricci scalar is removed by a Weyl transformation which also changes the coupling in front of the R^4-terms into a factor $\exp(-12\phi/(D-2))$. Expanding the metric in the Einstein frame around $g_{\mu\nu} = \eta_{\mu\nu}$ and comparing the relevant S-matrix elements to the string calculation, one finds that $\kappa_D^2 = (\alpha')^{(D-2)/2}\, g_s^2$ while $A = g_s^{12/(D-2)}$ [39].[36] Again, the dependence on g_s cannot be tied to the presence of a dilaton interaction in neither one of the two terms. However, if one expands the Einstein metric around $g_{\mu\nu} = g_s^{-4/(D-2)}\,\eta_{\mu\nu}$ (or equivalently, the metric in the string frame around $g_{\mu\nu} = \eta_{\mu\nu}$) one finds $\kappa_D^2 = (\alpha')^{(D-2)/2}$ and A = constant – both couplings independent of g_s. This form of the parameters is also obtained in the σ-model approach where the R^4-term arises as a 4-loop counterterm [105].

The final point of this appendix concerns the dilaton in arbitrary spacetime dimensions. It can be defined as the field in the σ-model action (3.8) taken in D dimensions. The corresponding vertex operator is composed only out of operators of the spacetime sector of the conformal field theory (CFT) and no operators in the 'internal CFT'. Let us denote the dilaton defined in this manner by $\phi^{(D)}$. This definition has the virtue that $\phi^{(D)}$ is invariant under T-duality transformations of the D-dimensional theory which originates from the existence of equivalence classes of the internal CFT. The same is true for the graviton and the antisymmetric tensor, whose vertex operators are similarly composed solely out of the spacetime sector of the CFT.

Compactification of the low-energy effective actions relates the dilatons of different dimensions by a volume-dependent factor of the compactification manifold Y and the metric associated with the compactified dimensions. More precisely, starting in $D = 10$ one finds (in the string frame)

$$\frac{1}{\kappa_{10}^2}\,{\rm e}^{-2\phi^{(10)}}\,V_n\sqrt{\det g_n} = \frac{1}{\kappa_{10-n}^2}\,{\rm e}^{-2\phi^{(10-n)}}\,, \tag{B.3}$$

where n is the dimension of Y, V_n is the volume of the n-dimensional compactified coordinates and g_n is the metric associated with the compactified dimensions.[37]

[36] Note that the relation with the dilaton field used in the second reference of [39] is given by $\phi - \langle\phi\rangle = \sqrt{2}\kappa D$ (in 10 spacetime dimensions).

[37] To be precise, V_n is the (higher-dimensional) analog of the length L introduced in section 2.5. The geodesic volume which generalizes R_{11} in (2.44) is instead proportional to $V_n\sqrt{\det g_n}$.

The latter is directly related to the vacuum-expectation value of certain moduli fields. Furthermore, the space-time part of the metric (in the string frame) is left unchanged in the parametrization used in (B.3) and the D-dimensional quantities are defined by

$$\frac{1}{\kappa_{10-n}^2} = \frac{V_n}{\kappa_{10}^2}, \qquad \phi^{(10-n)} = \phi^{(10)} - \tfrac{1}{4}\log\det g_n\,. \tag{B.4}$$

The D-dimensional string metric, the dilaton and the antisymmetric tensor are invariant under T-duality. For the perturbative dualities one can demonstrate this fact by performing a dimensional reduction on the 10-dimensional supergravity field theory (for instance on the Lagrangian (2.38)) from 10 to $10-n$ dimensions. Using the arguments of section 2.5 one establishes the existence of a rank-n group of invariances that leaves the string metric, the dilaton (defined according to (B.4)) and the antisymmetric tensor field invariant. However, the original 10-dimensional dilaton $\phi^{(10)}$ transforms under these symmetries.

References

1. B. de Wit and J. Louis, *"Duality and Supersymmetry"*, to appear in the Proceedings of the *Trieste Spring School on String Theory, Gauge Theory and Quantum Gravity, 1997.*
2. For an overview of results in supersymmetry and supergravity, see, for example, S. Ferrara (ed.), *"Supersymmetry"*, Vol. 1& 2, North-Holland/World Scientific (1989); A. Salam and E. Sezgin (eds.), *"Supergravities in Diverse Dimensions"*, Vol. 1& 2, North-Holland/World Scientific (1989).
   There are many references on the structure of supermultiplets. A few of them are, W. Nahm, *"Supersymmetries and their representations"*, Nucl. Phys. **B135** (1978) 149;
   S. Ferrara, C.A. Savoy and B. Zumino, *"General massive multiplets in extended supersymmetry"*, Phys. Lett. **100B** (1981) 393;
   S. Ferrara and C.A. Savoy, *"Representations of extended supersymmetry on one- and two-particle states"*, in *Supergravity '81*, eds. S. Ferrara and J.G. Taylor, Cambridge University Press (1982), p 47.
3. B. de Wit, A. Tollstén and H. Nicolai, *"Locally supersymmetric $D = 3$ non-linear sigma models"*, Nucl. Phys. **B392** (1993) 3, hep-th/9208074.
4. E. Cremmer, *"Supergravities in 5 dimensions"*, in *Superspace & Supergravity*, p. 267, eds. S.W. Hawking and M. Roček, Cambridge University Press (1981).
5. E. Cremmer, B. Julia and J. Scherk, *"Supergravity theory in 11 dimensions"*. Phys. Lett. **76B** (1978) 409.
6. R. Slansky, *"Group theory for unified model building"*, Phys. Rep. **79** (1981) 1.
7. L. Brink, J. Scherk and J.H. Schwarz, *"Supersymmetric Yang-Mills Theories"*, Nucl. Phys. **B121** (1977) 77;
   F. Gliozzi, J. Scherk and D. Olive, *"Supersymmetry, Supergravity Theories and the Dual Spinor Model"*, Nucl. Phys. **B122** (1977) 253.
8. M.B. Green and J.H. Schwarz, *"Superstring interactions"*, Phys. Lett. **122B** (1983) 143;
   J.H. Schwarz and P.C. West, *"Symmetries and Transformation of Chiral $N = 2$ $D = 10$ Supergravity"*, Phys. Lett. **126B** (1983) 301;
   J.H. Schwarz, *"Covariant Field Equations of Chiral $N = 2$ $D = 10$ Supergravity"*, Nucl. Phys. **B226** (1983) 269;

P. Howe and P.C. West, *"The Complete $N = 2$ $D = 10$ Supergravity"*, Nucl. Phys. **B238** (1984) 181.

9. M. Huq and M. A. Namazie, *"Kaluza–Klein supergravity in ten dimensions"*, Class. Q. Grav. **2** (1985) 293;
F. Giani and M. Pernici, *"$N = 2$ supergravity in ten dimensions"*, Phys. Rev. **D30** (1984) 325;
I.C.G. Campbell and P.C. West, *"$N = 2$ $D = 10$ nonchiral supergravity and its spontaneous compactification."* Nucl. Phys. **B243** (1984) 112.
10. M. Dine, P. Huet and N. Seiberg, *"Large and small radius in string theory"*, Nucl. Phys. **B322** (1989) 301.
11. J. Dai, R.G. Leigh, J. Polchinski, *"New connections between string theories"*, Mod. Phys. Lett. **A4** (1989) 2073.
12. E. Bergshoeff, C. Hull and T. Ortin, *"Duality in the type-II superstring effective action"*, Nucl. Phys. **B451** (1995) 547, hep-th/9504081.
13. P.K. Townsend, *"A new anomaly free chiral supergravity theory from compactification on $K3$"*, Phys. Lett. **139B** (1984) 283.
14. C. Fronsdal, *"Massless Fields with Integer Spin"*, Phys. Rev. **D18** (1978) 3264;
J. Fang and C. Fronsdal, *"Massless Fields with Half Integral Spin"*, Phys. Rev. **D18** (1978) 3630;
J. Schwinger, *Particles, Sources and Fields*, Addison Wesley (1970):
T. Curtright, *"Massless Field Supermultiplets with Arbitrary Spin"*, Phys. Lett. **85B** (1979) 219.
15. M.A. Vasiliev, *"Higher-Spin Gauge Theories in Four, Three and Two Dimensions"*, Int. J. Mod. Phys.**D5** (1996) 763, hep-th/9611024.
16. W. Nahm, *"Supersymmetries and their representations"*, Nucl. Phys. **B135** (1978) 149.
17. C. Aragone and S. Deser, *"Consistency Problems of Hypergravity"*, Phys. Lett. **B86** (1979) 161;
F.A. Berends, J.W. van Holten, P. van Nieuwenhuizen, and B. de Wit, *"On spin-5/2 gauge fields"*, J. Phys. **A13** (1980) 1643;
B. de Wit and D.Z. Freedman, *"Systematics of higher-spin gauge fields"*, Phys. Rev. **D21** (1980) 358.
18. P.K. Townsend, *"Four Lectures on M-theory"*, lectures given at the 1996 ICTP Summer School in High Energy Physics and Cosmology, Trieste, hep-th/9612121; *"M-theory from its Superalgebra"*, hep-th/9712004.
19. S. Ferrara, *"Supergravities in Diverse Dimensions and their Central Extensions"*, Nucl. Phys. (Proc. Suppl.) **55B** (1997) 145;
L. Andrianopoli, R. D'Auria and S. Ferrara, *"U-Duality and Central Charges in Various Dimensions Revisited"*, hep-th/9612105.
20. P.K. Townsend and P. van Nieuwenhuizen, *"Geometrical Interpretation of Extended Supergravity"*, Phys. Lett. **67B** (1977) 439;
A.H. Chamseddine and P.C. West, *"Supergravity as a Gauge Theory of Supersymmetry"*, Nucl. Phys. **B129** (1977) 39.
21. D.N. Page, *"Classical stability of round and squashed seven-spheres in eleven-dimensional supergravity"*, Phys. Rev. **D28** (1983) 2976.
22. M.J. Duff and K.S. Stelle, *"Multi-membrane solutions of D=11 supergravity"*, Phys. Lett. **B253** (1991) 113
23. R. Güven, *"Black p-brane solutions of D=11 supergravity theory"*, Phys. Lett. **B276** (1992) 49.
24. For a review see, for example, M.J. Duff, R.R. Khuri and J.X. Lü, *"String solitons"*, Phys. Rept. **259** (1995) 213, hep-th/9412184;
M. Duff, *"Supermembranes"*, hep-th/9611203;
K.S. Stelle, *"Lectures on Supergravity p-Branes"*, lectures given at the 1996 ICTP Summer School in High Energy Physics and Cosmology, Trieste, hep-th/9701088.
25. B. Biran, F. Englert, B. de Wit and H. Nicolai, *"Gauged N=8 supergravity and its*

breaking from spontaneous compactification", Phys. Lett. **124B** (1983) 45.
26. For a reprint collection and a review of Kaluza-Klein theory, see, for instance, M.J. Duff, B.E.W. Nilsson and C.N. Pope, *"Kaluza-Klein Supergravity"*, Phys. Rep. **130** (1986) 1;
T Appelquist, A. Chodos and P.G.O. Freund, *"Modern Kaluza-Klein Theories"*, Addison-Wesley (1987).
27. B. de Wit, F. Vanderseypen and A. Van Proeyen, *"Symmetry structures of special geometries"*, Nucl. Phys. **B400** (1993) 463, hep-th/9210068.
28. E. Cremmer, *"N=8 Supergravity"*, in *Unification of Fundamental particle Interactions*, p. 137, eds. S. Ferrara, J. Ellis, P. van Nieuwenhuizen, Plenum (1980);
B. Julia, *"Group Disintegrations"*, in *Superspace & Supergravity*, p. 331, eds. S.W. Hawking and M. Roček, Cambridge University Press (1981).
29. S. Ferrara, J. Scherk and B. Zumino, *"Algebraic Properties of Extended Supersymmetry"*, Nucl. Phys. **B121** (1977) 393;
E. Cremmer, J. Scherk and S. Ferrara, *"SU(4) Invariant Supergravity Theory"*, Phys. Lett. **74B** (1978) 61;
B. de Wit, *"Properties of SO(8)-extended supergravity"*, Nucl. Phys. **B158** (1979) 189;
M.K. Gaillard and B. Zumino, *"Duality rotations for interacting fields"*, Nucl. Phys. **B193** (1981) 221;
B. de Wit and H. Nicolai, *"N=8 Supergravity"*, Nucl. Phys. **B208** (1982) 323.
30. E. Cremmer and B. Julia, *"The $N = 8$ supergravity theory. I. The Lagrangian"*, Phys. Lett. **80B** (1978) 48.
31. B. de Wit and A. Van Proeyen, *"Potentials and symmetries of general gauged $N = 2$ supergravity - Yang-Mills models"*, Nucl. Phys. **B245** (1984) 89.
32. M. de Roo, *"Matter Couplings in $N = 4$ Supergravity"*, Nucl. Phys. **B255** (1985) 515.
33. E. Witten, *"String theory dynamics in various dimensions"*, Nucl. Phys. **B443** (1995) 85, hep-th/9503124.
34. C.M. Hull and P.K. Townsend, *"Unity of superstring dualities"*, Nucl. Phys. **B438** (1995) 109, hep-th/9410167.
35. P.K. Townsend, *"The eleven-th dimensional supermembrane revisited"*, Phys. Lett. **B350** (1995) 184, hep-th/9501068, *"D-branes from M-branes"*, Phys. Lett. **B373** (1996) 68, hep-th/9512062.
36. A.H. Chamseddine, *"Interacting supergravity in ten dimensions: the role of the six-index gauge field"*, Phys. Rev. **D24** (1981) 3065;
E. Bergshoeff, M. de Roo, B. de Wit and P. van Nieuwenhuizen, *"Ten-dimensional Maxwell-Einstein supergravity, its currents, and the issue of its auxiliary fields"*, Nucl. Phys. **B195** (1982) 97;
E. Bergshoeff, M. de Roo and B. de Wit, *"Conformal supergravity in ten dimensions"*, Nucl. Phys. **B217** (1983) 143.
G. Chapline and N.S. Manton, *"Unification of Yang-Mills theory and supergravity in ten dimensions"*, Phys. Lett. **120B** (1983) 105.
37. For a review of perturbative string theory see, for example, M. Green, J. Schwarz and E. Witten, *"Superstring theory"*, Vol. 1&2, Cambridge University Press, (1987);
D. Lüst and S. Theisen, *"Lectures on string theory"*, Springer, (1989);
J. Polchinski, *"What is String Theory"*, Les Houches lectures 1994, hep-th/94111028;
E. Kiritsis, *"Introduction to Superstring Theory"*, hep-th/9709062.
38. For a review see, for example, A.A. Tseytlin, in *Superstrings '89*, eds. M. Green, R. Iengo, S. Randjbar-Daemi, E. Sezgin and A. Strominger, World Scientific (1990).
C. Callan and L. Thorlacius, in *Particles, Strings and Supernovae*, eds. A. Jevicki and C.-I. Tan World Scientific, (1989).
39. D. Gross and E. Witten, *"Superstring Modification of Einstein's Equations"*, Nucl. Phys. **B277**, (1986) 1;

D. Gross and J. Sloan, *"The Quartic Effective Action for the Heterotic String"*, Nucl. Phys. **B291**,(1987) 41.

40. For a review see, for example, J. Polchinski, *"TASI lectures on D-branes"*, hep-th/9611050;
C. Bachas, *"(Half) a Lecture on D-branes"*, in *Gauge Theories, Applied Supersymmetry and Quantum Gravity II*, ed. A. Sevrin, K.S. Stelle K. Thielemans and A. Van Proeyen, Imperial College Press, 1997, hep-th/9701019.
41. L.J. Dixon and J.A. Harvey, *"String theories in ten-dimensions without space-time supersymmetry"*, Nucl. Phys. **B274**, (1986) 93.
42. O. Bergman and M.R. Gaberdiel, *"A Non-Supersymmetric Open String Theory and S-Duality "*, Nucl. Phys. **B499** (1997) 183, hep-th/9701137;
J.D. Blum and K.R. Dienes, *" Strong/Weak Coupling Duality Relations for Non-Supersymmetric String Theories"*, hep-th/9707160.
43. A. Schellekens and N. Warner, *"Anomalies and modular invariance in string theory"*, Phys. Lett. **177B** (1986) 317.
44. For a review, see P. Aspinwall, *"$K3$ surfaces and string duality"*, hep-th/9611137.
45. For a review, see, for example, S. Hosono, A. Klemm and S. Theisen, *"Lectures on mirror symmetry"*, hep-th/9403096;
B. Greene, *"String Theory on Calabi-Yau Manifolds"*, hep-th/9702155.
46. For a review, see, for example, A. Giveon, M. Porrati and E. Rabinovici, *"Target space duality in string theory"*, Phys. Rept. **244** (1994) 77, hep-th/9401139.
47. K.S. Narain, *"New heterotic string theories in uncompactified dimensions < 10"*, Phys. Lett. **169B** (1986) 41;
K.S. Narain, M.H. Samadi and E. Witten, *"A note on toroidal compactification of heterotic string theory"*, Nucl. Phys. **B279** (1987) 369.
48. P. Ginsparg, *"Comment on toroidal compactification of heterotic superstrings"*, Phys. Rev. **D35** (1987) 648.
49. P. Hořava, *"Strings on world sheet orbifolds"*, Nucl. Phys. **B327** (1989) 461; *"Background duality of open string models"*, Phys. Lett. **B231** (1989) 251.
50. N. Seiberg, *"Observations on the moduli space of superconformal field theories"*, Nucl. Phys. **B303** (1988) 286.
51. P.S. Aspinwall and D.R. Morrison, *"String theory on $K3$ surfaces"*, hep-th/9404151.
52. L. Romans *"Self-duality for interacting fields: covariant field equations for six-dimensional chiral supergravities*, Nucl. Phys. **B276** (1986) 71.
53. M.B. Green, J. Schwarz and P. C. West, *"Anomaly-free chiral theories in six dimensions"*, Nucl. Phys. **B254** (1985) 327;
J. Erler, *"Anomaly cancellation in six dimensions"*, J. Math. Phys. **35** (1994) 1819, hep-th/9304104;
J. Schwarz, *"Anomaly-free supersymmetric models in six dimensions"*, Phys. Lett. **B371** (1996) 223, hep-th/9512053.
54. D.R. Morrison and C. Vafa, *"Compactifications of F-theory on Calabi-Yau threefolds - I"*, Nucl. Phys. **B473** (1996) 74, hep-th/9602111; *"Compactifications of F-theory on Calabi-Yau threefolds - II"*, Nucl. Phys. **B476** (1996) 437, hep-th/9603161.
55. M. Berkooz, R. Leigh, J. Polchinski, J. Schwarz, N. Seiberg and E. Witten, *"Anomalies, Dualities, and Topology of $D = 6, N = 1$ superstring vacua"*, Nucl. Phys. **B475** (1996) 115, hep-th/9605184.
56. G. Aldazabal, A. Font, L.E. Ibañez and F. Quevedo, *"Heterotic/heterotic duality in $D = 6, 4$"*, Phys. Lett. **B380** (1996) 33, hep-th/9602097;
for a review see, for example, L.E. Ibañez and A.M. Uranga, *"$D = 6, N = 1$ String Vacua and Duality"*, hep-th/9707075.
57. M. Bianchi and A. Sagnotti, *"On the systematics of open string theories"*, Phys. Lett. **B247** (1990) 517; *"Twist symmetry and open string Wilson lines"*, Nucl. Phys. **B361** (1991) 519;
A. Sagnotti, *"A note on the Green-Schwarz mechanism in open string theories"*,

Phys. Lett. **B294** (1992) 196;
for a review see, for example, A. Sagnotti, *"Surprises in open-string perturbation theory"*, hep-th/9702093.
58. E.G. Gimon and J. Polchinski, *"Consistency conditions for orientifolds and D-manifolds"*, Phys. Rev. **D54** (1996) 1667, hep-th/9601036.
59. L. Dixon, *"Some Worldsheet Properties of Superstring Compactifications, on Orbifolds and Otherwise"*, in Superstrings, Unified Theories, and Cosmology ed. G. Furlan et al., World Scientific (1988);
P. Candelas, M. Lynker and R. Schimmrigk, *"Calabi–Yau manifolds in weighted $P(4)$"*, Nucl. Phys. B341 (1990) 383;
B. Greene and R. Plesser, *"Duality in Calabi–Yau Moduli Space"*, Nucl. Phys. B338 (1990) 15.
60. P. Candelas and X. de la Ossa, *"Moduli space of Calabi-Yau manifolds"*, Nucl. Phys. **B355** (1991), 455.
61. A. Strominger, *"Special Geometry"*, Commun. Math. Phys. **133** (1990) 163.
62. For recent reviews on special geometry, see, for example, B. de Wit and A. van Proeyen, *"Special geometry and symplectic transformations"*, Nucl. Phys. B (Proc. Suppl.) **45B,C** (1996) 196, hep-th/9510186;
A. van Proeyen, *"Vector multiplets in $N = 2$ supersymmetry and its associated moduli spaces"*, hep-th/9512139;
B. Craps, F. Roose, W. Troost and A. Van Proeyen, Nucl. Phys. **B503** (1997) 565, hep-th/9703082.
63. S. Cecotti, S. Ferrara and L. Girardello, *"Geometry of type-II superstrings and the moduli of superconformal field theories"*, Int. J. Mod. Phys. **A4** (1989), 2475.
64. P. Candelas, G. Horowitz, A. Strominger and E. Witten, *"Vacuum Configurations for Superstrings"*, Nucl. Phys. **B258** (1985) 46.
65. For a collection of reprints see, for example, *"String Theory in four Dimensions"*, ed. M. Dine, North-Holland, 1988.
For a review see, for example, F. Quevedo, *"Lectures on Superstring Phenomenology"*, hep-th/9603074.
66. C. Angelantonj, M. Bianchi, G. Pradisi, A. Sagnotti and Y. Stanev, *"Chiral Asymmetry in Four-Dimensional Open-String Vacua"*, Phys. Lett. **B385** (1996) 96, hep-th/9606169.
67. For a review see, for example, J. Schwarz, *"Lectures on superstring and M-theory dualities"*, hep-th/9607021;
A. Sen, *"Unification of string dualities"*, hep-th/9609176;
W. Lerche, *"Introduction to Seiberg-Witten Theory and its Stringy Origin"*, Nucl. Phys. Proc. Suppl. **55B** (1997) 83, hep-th/9611190;
S. Förste and J. Louis, *" Duality in String Theory"*, in *Gauge Theories, Applied Supersymmetry and Quantum Gravity II*, ed. A. Sevrin, K.S. Stelle K. Thielemans and A. Van Proeyen, Imperial College Press, 1997, hep-th/9612192;
C. Vafa, *"Lectures on Strings and Dualities"*, hep-th/9702201;
A. Klemm, *"On the Geometry behind N=2 Supersymmetric Effective Actions in Four Dimensions"*, hep-th/9705131;
E. Kiritsis, *"Introduction to non-perturbative String Theory"*, hep-th/9708130;
W. Lerche, *"Recent Developments in String Theory"*, hep-th/9710246.
68. A. Dabholkar, *"Ten dimensional heterotic string as a soliton"*, Phys. Lett. **B357** (1995) 307, hep-th/9506160.
69. C.M. Hull, *"String-string duality in ten-dimensions"*, Phys. Lett. **B357** (1995) 545, hep-th/9506194.
70. J. Polchinski and E. Witten, *"Evidence for heterotic - type-I duality"*, Nucl. Phys. **B460** (1996) 525, hep-th/9510169.
71. M.J. Duff and J.X. Lü, *"Black and super p-branes in diverse dimensions"*, Nucl. Phys. **B416** (1994) 301;
M.J. Duff and R. Minasian, *"Putting string/string duality to the test"*, Nucl. Phys.

B436 (1995) 261, hep-th/9406198.

72. A. Sen, *"String-string duality conjecture in six dimensions and charged solitonic strings"*, Nucl. Phys. **B450** (1995) 103, hep-th/9504027.
73. J.A. Harvey and A. Strominger, *"The heterotic string is a soliton"*, Nucl. Phys. **B449** (1995) 456, hep-th/9504047.
74. S. Kachru and C. Vafa, *"Exact results for $N = 2$ compactifications of heterotic strings"*, Nucl. Phys. **B450** (1995) 69, hep-th/9505105.
75. S. Ferrara, J.A. Harvey, A. Strominger and C. Vafa, *" Second-quantized mirror symmetry"*, Phys. Lett. **B361** (1995) 59, hep-th/9505162.
76. A. Klemm, W. Lerche and P. Mayr, *"K3 fibrations and heterotic type-II string duality"*, Phys. Lett. **B357** (1995) 313, hep-th/9506112.
77. C. Vafa and E. Witten, *"Dual string pairs with $N = 1$ and $N = 2$ supersymmetry in four-dimensions"*, hep-th/9507050.
78. P.S. Aspinwall and J. Louis, *"On the ubiquity of K3 fibrations in string duality"*, Phys. Lett. **B369** (1996) 233, hep-th/9510234.
79. V. Kaplunovsky, J. Louis and S. Theisen, *"Aspects of duality in $N = 2$ string vacua"*, Phys. Lett. **B357** (1995) 71, hep-th/9506110;
I. Antoniadis, E. Gava, K.S. Narain and T.R. Taylor, *"$N = 2$ type-II - heterotic duality and higher derivative F-terms"*, Nucl. Phys. **B455** (1995) 109, hep-th/9507115;
G. Aldazabal, A. Font, L .E. Ibañez and F. Quevedo, *"Chains of $N = 2, D = 4$ heterotic type-II duals"*, Nucl. Phys. **B461** (1996) 85, hep-th/9510093;
G. Lopes Cardoso, G. Curio, D. Lüst, T. Mohaupt and S.-J. Rey, *"BPS spectra and non-perturbative couplings in $N = 2,4$ supersymmetric string theories"*, Nucl. Phys. **B464** (1996) 18, hep-th/9512129;
P. Candelas and A. Font, *"Duality Between the Webs of Heterotic and Type II Vacua*, hep-th/9603170;
P. Berglund, S. Katz, A. Klemm and P. Mayr, *"New Higgs transitions between dual $N = 2$ string models"*, Nucl. Phys. **B483** (1997) 209, hep-th/9605154;
M. Bershadsky, K. Intriligator, S. Kachru, D.R. Morrison, V. Sadov and C. Vafa, *"Geometric singularities and enhanced gauge symmetries"*, Nucl. Phys. **B481** (1996) 215, hep-th/9605200;
J. Louis, J. Sonnenschein, S. Theisen and S. Yankielowicz, *"Nonperturbative properties of heterotic string vacua"*, Nucl. Phys. **B480** (1996) 185, hep-th/9606049;
B. de Wit, G. Lopes Cardoso, D. Lüst, T. Mohaupt and S.-J. Rey, *"Higher order gravitational couplings and modular forms in $N = 2$, $D = 4$ heterotic string compactifications"*, Nucl. Phys. **B481** (1996) 353hep-th/9607184;
for a review see, J. Louis and K. Förger, *"Holomorphic couplings in string theory"*, Nucl. Phys. (Proc. Suppl.) **55B** (1997) 33, hep-th/9611184.
80. I. Antoniadis, C. Bachas, C. Fabre, H. Partouche and T.R. Taylor, *"Aspects of Type I-Type II-Heterotic Triality in Four Dimensions"*, Nucl. Phys. **B489** (1997) 160, hep-th/9608012;
for a review see, for example, I. Antoniadis, H. Partouche and T.R. Taylor, *"Lectures on heterotic-type-I duality"*, hep-th/9706211.
81. C. Montonen and D. Olive, *"Magnetic monopoles as gauge particles?"*, Phys. Lett. **72B** (1977) 117;
E. Witten and D. Olive, *"Supersymmetry algebras that include topological charges"*, Phys. Lett. **78B** (1978) 97.
82. J. Schwarz, *"An $SL(2, Z)$ multiplet of type-IIB superstrings"*, Phys. Lett. **B360** (1995) 13, hep-th/9508143.
83. E. Witten, *"Bound states of strings and p-branes"*, Nucl. Phys. **B460** (1996) 335, hep-th/9510135.
84. For a review see, J.A. Harvey *"Magnetic monopoles, duality and supersymmetry"*, hep-th/9603086.
85. A. Font, L.E. Ibañez, D. Lüst and F. Quevedo, *"Strong–weak coupling duality and*

nonperturbative effects in string theory", Phys. Lett. **B249** (1990) 35.
86. A. Sen, *"SL(2, Z) duality and magnetically charged strings"*, Int. J. Mod. Phys. A8 (1993) 5079, hep-th/9302038; *"Dyon-monopole bound states, selfdual harmonic forms on the multi-monopole moduli space, and SL(2, Z) invariance in string theory"*, Phys. Lett. **B329** (1994) 217, hep-th/9402032;
J. Schwarz and A. Sen, *"Duality symmetric actions"*, Nucl. Phys. **B411** (1994) 35, hep-th/9304154; *"Duality symmetries of 4-D heterotic strings"*, Phys. Lett. **B312** (1993) 105, hep-th//9305185.
87. P. Hořava and E. Witten, *"Heterotic and type-I string dynamics from eleven-dimensions"*, Nucl. Phys. **B460** (1996), 506, hep-th/9510209; *"Eleven-dimensional supergravity and a manifold with boundary"*, Nucl. Phys. **B475** (1996)94, hep-th/9603142.
88. T. Banks, W. Fischler, S.H. Shenker and L. Susskind, *"M Theory As A Matrix Model: A Conjecture"*, Phys. Rev. **D55** (1997) 5112, hep-th/9610043.
89. M. Claudson and M.B. Halpern, *"Supersymmetric ground state wave functions"*, Nucl. Phys. **B250** (1985) 689;
R. Flume, *"On quantum mechanics with extended supersymmetry and nonabelian gauge constraints"*, Ann. Phys. **164** (1985) 189;
M. Baake, P. Reinicke, and V. Rittenberg, *"Fierz identities for real Clifford algebras and the number of supercharges"*, J. Math. Phys. **26** (1985) 1070.
90. E.A. Bergshoeff, E. Sezgin and P.K. Townsend, *"Supermembranes and eleven-dimensional supergravity"*, Phys. Lett. **B189** (1987) 75; *"Properties of the eleven-dimensional supermembrane theory"*, Ann. Phys. **185** (1988) 330.
91. B. de Wit, J. Hoppe and H. Nicolai, *"On the quantum mechanics of supermembranes"*, Nucl. Phys. **B305** (1988) 545.
92. For a review, see, for example, R. Dijkgraaf, E. Verlinde and H. Verlinde, *"Notes on Matrix and Micro Strings"*, hep-th/9709107;
N. Seiberg, *"Why is the Matrix Model correct?"*, hep-th/9710009;
A. Bilal, *"M(atrix) Theory: a Pedagogical Introduction"*, hep-th/9710136;
T. Banks, *"Matrix Theory"*, hep-th/9710231;
D. Bigatti and L. Susskind, *"Review of Matrix Theory"*, hep-th/9712072.
93. L. Alvarez-Gaumé and E. Witten, *"Gravitational anomalies"*, Nucl. Phys. **B234** (1984) 269.
94. M.B. Green and J. Schwarz, *"Anomaly cancellations in supersymmetric $D = 10$ gauge theory and superstring theory"*, Phys. Lett. **149B** (1984) 117.
95. I. Antoniadis, S. Ferrara and T.R. Taylor, *"$N = 2$ Heterotic Superstring and its Dual Theory in Five Dimensions"*, Nucl. Phys. **B460** (1996) 489, hep-th/9511108.
96. K. Dasgupta and S. Mukhi, *"Orbifolds of M theory"*, Nucl. Phys. **B465** (1996) 399, hep-th/9512196;
E. Witten, *"Five-branes and M-theory on an orbifold"*, Nucl. Phys. **B463** (1996) 383, hep-th/9512219.
97. P.S. Aspinwall and D.R. Morrison, *"U-duality and integral structures"*, Phys. Lett. **B355** (1995) 141, hep-th/9505025.
98. C.M. Hull, *"String dynamics at strong coupling"*, Nucl. Phys. **B468** (1996) 113, hep-th/9512181.
99. C. Vafa, *"Evidence for F theory"*, Nucl. Phys. **B469** (1996) 403, hep-th/9602022.
100. B.R. Greene, A. Shapere, C. Vafa and S.-T. Yau, *"Stringy cosmic strings and noncompact Calabi-Yau Manifolds"*, Nucl. Phys. **B337** (1990) 1.
101. A. Sen, *"F-theory and orientifolds"*, hep-th/9605150.
102. E. Witten, *"Non-Perturbative Superpotentials In String Theory"*, Nucl. Phys. **B474** (1996) 343, hep-th/9604030;
S. Sethi, C. Vafa and E. Witten, *"Constraints on Low-Dimensional String Compactifications"*, Nucl. Phys. **B480** (1996) 213, hep-th/9606122;
I. Brunner and R. Schimmrigk, *"F-Theory on Calabi-Yau Fourfolds"*, Phys. Lett. **B387** (1996) 750, hep-th/9606148;

R. Donagi, A. Grassi, E. Witten, *"A Non-Perturbative Superpotential With E_8 Symmetry"*, Mod. Phys. Lett. **A11** (1996) 2199, hep-th/9607091;
P. Mayr, *"Mirror Symmetry, N=1 Superpotentials and Tensionless Strings on Calabi-Yau Four-Folds"*, Nucl. Phys. **B494** (1997) 489, hep-th/9610162;
I. Brunner, M. Lynker and R. Schimmrigk, *"Unification of M- and F-Theory Calabi-Yau Fourfold Vacua"*, Nucl. Phys. **B498** (1997) 156, hep-th/9610195;
M. Bershadsky, A. Johansen, T. Pantev, V. Sadov and C. Vafa, *"F-theory, Geometric Engineering and N=1 Dualities"*, hep-th/9612052;
A. Klemm, B. Lian, S.-S. Roan and S.-T. Yau, *"Calabi-Yau fourfolds for M- and F-Theory compactifications"*, hep-th/9701023;
M. Bershadsky, A. Johansen, T. Pantev and V. Sadov, *"On Four-Dimensional Compactifications of F-Theory"*, hep-th/9701165;
G. Curio and D. Lüst, *"A Class of N=1 Dual String Pairs and its Modular Superpotential"*, hep-th/9703007;
B. Andreas, G. Curio and D. Lüst, *"N=1 Dual String Pairs and their Massless Spectra"*, Nucl. Phys. **B507** (1997) 175, hep-th/9705174;
W. Lerche, *"Fayet-Iliopoulos Potentials from Four-Folds"*, hep-th/9709146.

103. Z. Kakushadze, *"Aspects of N=1 Type I-Heterotic Duality in Four Dimensions"*, hep-th/9704059;
Z. Kakushadze and G. Shiu, *"A Chiral N=1 Type I Vacuum in Four Dimensions and Its Heterotic Dual"*, Phys. Rev. **D56** (1997) 3686, hep-th/9705163;
Z. Kakushadze and G. Shiu, *"4D Chiral N=1 Type I Vacua with and without D5-branes"*, hep-th/9706051.

104. C.G. Callan, D. Friedan, E. Martinec and M. Perry, *"Strings in Background Fields"*, Nucl. Phys. **B262** (1985) 593.

105. M. Grisaru, A. van de Ven and D. Zanon, *"Four Loop Divergences for the $N = 1$ Supersymmetric Nonlinear Sigma Model in Two-Dimensions"*, Nucl. Phys. **B277** (1986) 409.

SEIBERG-WITTEN THEORY AND S-DUALITY

TOHRU EGUCHI
Department of Physics, Faculty of Science,
University of Tokyo, Tokyo, Japan 113

1. Introduction

In these lectures I am going to give an elementary introduction to the Seiberg-Witten theory, i.e. the exact solutions of $N = 2$ supersymmetric Yang-Mills theories. Seiberg-Witten solution [1, 2] is the first exact solution in strongly-coupled quantum gauge theories in 4-dimensions. The solution was first constructed for $SU(2)$ Yang-Mills theory with $0 \leq N_f \leq 4$ hypermultiplets in vector representations. It was subsequently generalized to the case of classical gauge groups $SU(N), SO(N), Sp(N)$ [3, 4, 5, 7, 8, 9, 10] and provides a wealth of information on the strong coupling dynamics of gauge theories. There are excellent pedagogical reviews on Seiberg-Witten theory [11, 12, 13].

Let us first recall that there exist two types of supermultiplets in $N = 2$ supersymmetric filed theories: vector and hypermultiplets. An $N = 2$ vector-multiplet consists of a gauge field A^a_μ, two Weyl spinors λ^a, ψ^a and a complex scalar field ϕ^a. Here the suffix 'a' runs over the generators of the gauge group G, $a = 1, \cdots, dimG$. An $N = 2$ vector-multiplet decomposes into a sum of an $N = 1$ vector-multiplet made of (A^a_μ, λ^a) and a chiral-multiplet made of (ϕ^a, ψ^a).

On the other hand an $N = 2$ hypermultiplet consists of a pair of $N = 1$ chiral fields (q^i, ψ^i_q) and $(\tilde{q}^j, \tilde{\psi}^j_q)$ which belong to a representation R of G ($i = 1, 2 \cdots, dim(R)$) and its conjugate R^* ($j = 1, 2, \cdots, dim(R^*)$), respectively. In the following we restrict ourselves to the case of $G = SU(2)$ with R in the vector representation.

$N = 1$ vector and chiral superfields in an $N = 2$ vector-multiplet are denoted as W^a_α and A^a, respectively. $N = 1$ chiral fields of an $N = 2$ hypermultiplet are denoted as Q^i and $\tilde{Q}^j$. Under the $U(1)_R$ rotation they transform as

$$U(1)_R : \quad \begin{cases} \Phi(\theta) \to e^{2i\alpha}\Phi(e^{-i\alpha}\theta), & W_\alpha(\theta) \to e^{i\alpha}W_\alpha(e^{-i\alpha}\theta) \\ Q \to Q(e^{-i\alpha}\theta), & \tilde{Q} \to \tilde{Q}(e^{-i\alpha}\theta). \end{cases} \tag{1.1}$$

On the other hand under the 3rd component of the $SU(2)_R$ fields transform as

$$U(1)_J : \quad \begin{cases} \Phi(\theta) \to \Phi(e^{-i\alpha}\theta), & W_\alpha(\theta) \to e^{i\alpha}W_\alpha(e^{-i\alpha}\theta) \\ Q \to e^{i\alpha}Q(e^{-i\alpha}\theta), & \tilde{Q} \to e^{i\alpha}\tilde{Q}(e^{-i\alpha}\theta). \end{cases} \tag{1.2}$$

L. Baulieu et al. (eds.), Strings, Branes and Dualities, 103–120.

In the following we first consider the case of $SU(2)$ gauge theory without matter hypermultiplets $N_f = 0$. If we denote the scalar field in the vector-multiplet by a 2×2 matrix $\phi = \sum_a \phi^a\sigma_a/2$, the potential energy of the scalar fields is expressed as

$$V(\phi) = \frac{1}{g^2} Tr\left[\phi, \phi^\dagger\right]^2. \tag{1.3}$$

Thus the potential has a flat direction

$$\langle\phi^3\rangle = a \neq 0, \quad \langle\phi^1\rangle = \langle\phi^2\rangle = 0. \tag{1.4}$$

When the scalar field develops a vacuum value $a \neq 0$, the gauge symmetry breaks down to $U(1)$ and the theory is in the Coulomb phase. The vacuum value a can take an arbitrary value and the theory possesses an infinitely degenerate vacua. The vacuum value becomes the moduli of the $N = 2$ gauge theory. It turns out that the moduli space of $N = 2$ theory has the structure of a special Kähler geometry with its Kähler potential written in terms of some holomorphic function.

On the other hand if there exists matter in the theory, scalar fields in the matter hypermultiplets could develop vacuum values $\langle q^i\rangle \neq 0$ which in general completely break gauge invariance. A theory with vacua $\langle q^i\rangle \neq 0$ is called in the Higgs phase. Due to Yukawa-like couplings between scalars of vector and hypermultiplets, Coulomb and Higgs phase mutually exclude each other. Moduli space of the Higgs phase has a structure of a hyper-Kähler manifold. Hyper-Kähler manifolds are not deformed by quantum corrections and do not possess interesting dynamics in the present model. We will study the structure of the Coulomb phase in the following.

Let us introduce a gauge invariant "order parameter" $u = Tr\phi^2$. Then the complex u-plane becomes the moduli space. Classically $u = a^2/2$, however, this relation is modified at a quantum level. As is well-known, the nature of the quantum theory strongly depends on the sign of the beta function. In $N = 2$ supersymmetric theories a perturbative beta function is given only by the one-loop contribution. A well-known formula for the one-loop beta function is given by

$$\beta = \frac{-bg^3}{16\pi^2}, \quad b = \frac{11}{3}c_2(G) - \sum_{R_S}\frac{1}{6}T(R_S) - \sum_{R_F}\frac{2}{3}T(R_F) \tag{1.5}$$

$$T(R)\delta^{ab} = TrR^aR^b, \quad c_2(G)\delta^{ab} = f_{acd}f_{bcd}, \tag{1.6}$$

where $c_2(G)$ is the value of the 2nd-order Casimir invariant for the group G. R^a are representation matrices and $T(R)$ is the so called Dynkin index for the representation R. In eq.(1.5) $S(F)$ stands for the contributions of the scalar (spinor) fields (coefficient is 1/6 (2/3) for a real scalar (spinor) field). In the case of $SU(N)$, $c_2(SU(N)) = T(\text{adjoint rep}) = N$ and $T(\text{vector rep}) = 1/2$.

An $N = 2$ vector-multiplet consists of a vector, 2 real spinors and 1 complex scalar fields all in the adjoint representation. Thus in the case of $SU(N_c)$ gauge group

$$b = \left(\frac{11}{3} - \frac{1}{6}\times 2 - \frac{2}{3}\times 2\right)N_c = 2N_c. \tag{1.7}$$

Similarly an $N = 2$ hypermultiplet with 2 complex scalars and 2 real spinors in the vector or adjoint representation of $SU(N_c)$ yield

$$\text{vector representation}: \quad b = -\left(\frac{1}{6}\times 4 + \frac{2}{3}\times 2\right)\times\frac{1}{2} = -1, \tag{1.8}$$

$$\text{adjoint representation}: \quad b = -\left(\frac{1}{6}\times 4 + \frac{2}{3}\times 2\right)\times N_c = -2N_c. \tag{1.9}$$

Therefore the beta function for an $SU(N_c)$ gauge theory coupled to N_f vector matter or one adjoint matter is given by

$$\begin{cases} \beta = \frac{(N_f - 2N_c)g^3}{16\pi^2}: & N_f \text{ vector matter,} \\ \beta = 0: & \text{one adjoint matter.} \end{cases} \tag{1.10}$$

Thus the theory is asymptotically free in the range

$$0 \le N_f < 2N_c. \tag{1.11}$$

The theory with $N_f = 2N_c$ vector matter or one adjoint matter has a vanishing beta function. Thus these are ultra-violet finite theories. It is known that the theory with an adjoint matter has in fact an $N = 4$ supersymmetry.

2. $SU(2)$ gauge theory without matter

Let us first look at the dynamics in the Coulomb branch of $N = 2$ $SU(2)$ pure gauge theory. When the scalar field develops a vacuum value $\phi^3 \neq 0$, the gauge fields A^1_μ, A^2_μ (together with its fermionic partners) become massive while the $U(1)$ sector of the theory $(A^3_\mu, \phi^3, \lambda^3, \psi^3)$ remains massless. Due to asymptotic freedom regions of large vacuum value $u = Tr\phi^2$ correspond to the weak coupling regime. In the limit of $u \to \infty$ A^1_μ, A^2_μ become very heavy and almost decouple from the theory. Thus it seems reasonable to rewrite the theory in this region in terms of the remaining massless degrees of freedom. Let us drop the suffix 3 of the $U(1)$ fields $(A^3_\mu, \phi^3, \lambda^3, \psi^3)$ and write their $N = 1$ gauge and chiral superfield as W_α, A, respectively. Then the most general effective action for the $N = 2, U(1)$ gauge theory is written as

$$S = \frac{1}{4\pi} Im\Big(\int d\theta^4 \bar{A}\frac{\partial F}{\partial A} + \frac{1}{2}\int d\theta^2 \frac{\partial^2 F}{\partial A^2} W_\alpha W^\alpha\Big) \tag{2.1}$$

Here $F(A)$ is the prepotential of the theory and is an holomorphic function of the chiral field A. In (2.1) the kinetic term of the chiral field A and the coefficient of the gauge kinetic term are both described in terms of a single function F. This constraint is due to $N = 2$ supersymmetry. The Kähler potential of the chiral field is given by

$$K = \frac{1}{4\pi} Im\left(\bar{A}\frac{\partial F}{\partial A}\right) \tag{2.2}$$

(2.2) is the characteristic equation of the special Kähler geometry.

In the case of the standard action for the $N = 2\, U(1)$ gauge field the prepotential is simply quadratic in A

$$F = \frac{1}{2}\tau_{cl}A^2, \qquad \tau_{cl} = \frac{\theta}{2\pi} + \frac{4\pi i}{g^2}, \qquad K = \frac{1}{4\pi} Im\tau \cdot A\bar{A} \tag{2.3}$$

Here θ denotes the vacuum angle. Then the above action (2.1) reduces to the sum of free kinetic terms of $A_\mu, \lambda, \psi, \phi$ and the θ term. In the present situation we originally had an interacting $SU(2)$ gauge theory and have eliminated the charged fields to recast the theory in terms of massless neutral fields. Then all the renormalization effects due to massive fields are encoded in the prepotential $F(A)$. Let us introduce a field-dependent coupling constant by

$$\tau(A) \equiv \frac{\partial^2 F(A)}{\partial A^2} = \frac{\theta(A)}{2\pi} + \frac{4\pi i}{g^2(A)}. \tag{2.4}$$

Practically superfield A is identified as a (vacuum value of the scalar component) and thus $F(A) = F(a)$.

Due to the asymptotic freedom one expects that F approaches its weak coupling behavior (2.3) at large $|u|$. We can then use the value of the one-loop beta function and obtain

$$F(a) \approx \frac{ia^2}{2\pi}\log\frac{a^2}{\Lambda^2} + \sum_{k=1} F_k \left(\frac{\Lambda}{a}\right)^{4k} a^2. \tag{2.5}$$

Here Λ denotes the mass scale of the theory generated by the dimensional transmutation and the sum over k is the non-perturbative instanton contributions. In fact the perturbative part gives

$$\tau(a) = \frac{\partial^2 F}{\partial a^2} \approx \frac{2i}{\pi}\log\frac{a}{\Lambda}, \qquad \frac{1}{g(a)^2} \approx \frac{1}{2\pi^2}\log\frac{a}{\Lambda}. \tag{2.6}$$

This reproduces the beta function (1.10).

In (2.5) the k-instanton contribution is proportional to a^{-4k} and violates the $U(1)_R$ charge conservation by $-8k$. This is due to the $U(1)_R$ axial anomaly

$$\partial_\mu J^\mu_R = -\frac{2\times 2}{16\pi^2} F^a_{\mu\nu}\, {}^*F^{a\mu\nu} \tag{2.7}$$

which yields

$$\Delta R = -8 \times \text{instanton number}. \tag{2.8}$$

Instanton expansion coefficients F_k may be evaluated directly by perturbation theory around instanton backgrounds [14, 15, 16]. However, the calculation will soon become prohibitively difficult as one goes to higher orders in the expansion. The analysis of the strong coupling region $|u| \leq \Lambda^2$ needs a powerful new technique which goes beyond perturbation theory.

In order to determine the behavior of the prepotential F in the strong coupling region one can use the following information:

(1) F has the asymptotic behavior (2.5) at $|u| \to \infty$.
(2) F is an holomorphic function of a.
(3) Imaginary part of the second derivative of F is proportional to the coupling constant and hence must be positive definite.

The second derivative of a holomorphic function is holomorphic. If the function does not have singularities other than the one at ∞, then the positivity (3) becomes violated because of the property of a holomorphic function (mean-value theorem). Thus F must have at least one additional singularity in the strong coupling region. It turns out, however, if there exists only one extra singularity in the strong coupling region, the behavior of the theory around the singularity is in conflict with physical expectations. Thus there exist at least two singularities in the strong coupling regime.

What then happens at these singularities in the complex u-plane where the description by the effective Lagrangian (2.1) breaks down? Effective action has been written down based on the assumption that the $N = 2$ $U(1)$ vector-multiplet is the only massless degrees of freedom in the Coulomb phase. Thus if a new massless degrees of freedom appears at some point in the u-plane, description of the theory by means of the $U(1)$ multiplet becomes inadequate and the effective Lagrangian necessarily develops a singularity. Seiberg and Witten made a brilliant assumption that in the strong coupling region of the $N = 2$ gauge theory solitons (rather than the elementary particles) become massless at singular points.

We now recall that there exists a special class of solitons, i.e. BPS states in supersymmetric field theories with extended supersymmetries. BPS states saturate the Bogomolnyi bound on the soliton mass and form "small" representations of extended supersymmetry algebra. Since the BPS and non-BPS states belong to different representations and have different number of degrees of freedom, BPS states can not be converted into non-BPS states by quantum corrections and vice versa. In the case of $N = 2$ supersymmetric gauge theory 't Hooft-Polyakov monopoles appear as the BPS states. At the classical level masses of the BPS states with n_e electric and n_m magnetic charges are given by

$$M_{BPS} = \sqrt{2}\left|n_e a + n_m \frac{4\pi i}{g^2} a\right| \qquad n_e, n_m \in \mathbf{Z}^2 \tag{2.9}$$

(Note that the vacuum value a of the Higgs field is related to its conventional value v by $a = gv$). We recall that in classical theory $F = \tau a^2/2$ and hence $4\pi i a/g^2 = \partial F/\partial a$ (when $\theta = 0$). At the quantum level the RHS of (2.9) is modified due to quantum effect. Let us define the dual scalar field as

$$a_D = \frac{\partial F}{\partial a}. \tag{2.10}$$

Then the quantum version of (2.9) is given by

$$M_{BPS}(u) = \sqrt{2}\,|n_e a(u) + n_m a_D(u)| \tag{2.11}$$

In (2.11) a, a_D are regarded as functions of u.

Let us suppose that the function $a_D(u)$ vanishes at $u = u_0$. Then at u_0 monopoles become massless and the effective action develops a singularity. In order to restore a regular description we have to incorporate the monopole degrees of freedom explicitly in the effective Lagrangian. It is known that monopoles occur in hypermultiplets in $N = 2$ theory. Let us denote monopole fields by the chiral superfields $M, \tilde{M}$. We also perform a duality transformation of the $U(1)$ vector-multiplet so that it has a local coupling to the monopole fields. Then around the singularity $u = u_0$ we have an effective action

$$S = \frac{1}{4\pi}\left(\int d^4\theta Im\Big(h_D(A_D)\bar{A}_D\Big) + \frac{1}{2}\int d^2\theta Im\Big(\tau_D W_{\alpha D}W^{\alpha D}\Big)\right) + \int d^4\theta M^\dagger e^{V_D} M + \int d^4\theta \tilde{M} e^{-V_D}\tilde{M}^\dagger \tag{2.12}$$

Here h_D and τ_D are defined by

$$h(A) \equiv A_D, \qquad h_D(A_D) \equiv -A, \tag{2.13}$$

$$\frac{-1}{\tau(A)} = \frac{-1}{\frac{\partial h(A)}{\partial A}} = \frac{\partial h_D(A_D)}{\partial A_D} = \tau_D(A_D) \equiv \frac{\theta_D}{2\pi} + \frac{4\pi i}{g_D^2}. \tag{2.14}$$

The kinetic term of the chiral field of (2.1) is rewritten in (2.12) as

$$Im\int d^4\theta h(A)\bar{A} = Im\int d^4\theta A_D\bar{A} = Im\int d^4\theta h_D(A_D)\bar{A}_D \tag{2.15}$$

V_D ($W_{\alpha D}$) denotes the dual gauge superfield containing the dual photon which couples to the monopoles in a standard manner. Thus (2.12) describes a dual $N = 2$ QED where a magnetic photon interacts with the magnetically charged matter field. τ_D encodes the renormalization of the magnetic charge due to the vacuum polarization by monopoles.

It turns out that the positions of two strong coupling singularities in the u-plane are given by Λ^2 and $-\Lambda^2$ (after breakdown by anomaly the Z_2 subgroup $u \to -u$ of $U(1)_R$ symmetry survives which maps two singularities into each other). Thus at $u = \Lambda^2$ a_D vanishes and monopole becomes massless. On the other hand at $u = -\Lambda^2$, $a_D - a$ vanishes and the dyons become massless. Near the dyon point the theory is described by a Lagrangian similar to (2.12) with dyon hypermultiplet replacing the monopole and a mixture of a photon with its dual replacing the magnetic photon. Around these singularities we can compute the one-loop beta function and determine the behavior of the prepotential. At the monopole point, for instance, renormalization of the magnetic charge g_D is given by $a_D\partial\tau_D/\partial a_D = -8\pi i/g_D^3 \times a_D\partial g_D/\partial a_D = -2\times 8\pi i/16\pi^2$. This leads to $\tau_D \approx \frac{-i}{\pi}\log a_D$. We find

At monopole point :

$$\tau_D \approx \frac{-i}{\pi}\log a_D, \quad a_D \approx const(u-\Lambda^2),$$

$$a(u) \approx a_0 + \frac{i}{\pi}a_D\log a_D \tag{2.16}$$

At dyon point :

$$\tau_D \approx \frac{-i}{\pi}\log(a_D - a), \quad a_D - a \approx const(u + \Lambda^2),$$

$$a \approx a_0 + \frac{i}{\pi}(a_D - a)\log(a_D - a) \tag{2.17}$$

From these formula we can determine the monodromy of a, a_D as one goes around the singularities

At monopole point :

$$a_D \to a_D, \qquad a \to a - 2a_D$$

$$\begin{pmatrix} a_D \\ a \end{pmatrix} \longrightarrow \begin{pmatrix} 1 & 0 \\ -2 & 1 \end{pmatrix}\begin{pmatrix} a_D \\ a \end{pmatrix}, \quad M_{\Lambda^2} \equiv \begin{pmatrix} 1 & 0 \\ -2 & 1 \end{pmatrix}$$

At dyon point :

$$a_D \to -a_D + 2a, \qquad a \to -2a_D + 3a$$

$$\begin{pmatrix} a_D & a \end{pmatrix} \longrightarrow \begin{pmatrix} a_D & a \end{pmatrix}\begin{pmatrix} -1 & 2 \\ -2 & 3 \end{pmatrix}, \quad M_{-\Lambda^2} \equiv \begin{pmatrix} -1 & 2 \\ -2 & 3 \end{pmatrix}$$

If we introduce the S, T matrices which generate $SL(2, \mathbf{Z})$, $M_{\pm\Lambda^2}$ are represented as

$$S = \begin{pmatrix} 0 & 1 \\ -1 & 0 \end{pmatrix}, \quad T = \begin{pmatrix} 1 & 1 \\ 0 & 1 \end{pmatrix}, \tag{2.18}$$

$$M_{\Lambda^2} = ST^2S^{-1}, \quad M_{-\Lambda^2} = (TS)T^2(TS)^{-1} \tag{2.19}$$

Monodromy at ∞ is given by

$$a \approx \sqrt{2u} \longrightarrow -a, \quad a_D = \frac{\partial F}{\partial a} \approx \frac{2ia}{\pi}\log\frac{a}{\Lambda} + i\frac{a}{\pi} \longrightarrow -a_D + 2a$$

$$M_\infty = \begin{pmatrix} -1 & 2 \\ 0 & -1 \end{pmatrix} = PT^{-2}, \quad P \equiv \begin{pmatrix} -1 & 0 \\ 0 & -1 \end{pmatrix} \tag{2.20}$$

Complex u-plane has the topology of a sphere. One may verify

$$M_{\Lambda^2}M_{-\Lambda^2} = M_\infty \tag{2.21}$$

and see that the monodromies consistently fit into each other.

Now it is possible to present an exact solution to $N = 2$ pure $SU(2)$ gauge theory. One introduces the so-called Seiberg-Witten curve C and differential

$$\text{curve } C: \qquad y^2 = (x^2 - \Lambda^4)(x - u), \tag{2.22}$$

$$\text{differential}: \qquad \lambda = \frac{\sqrt{2}}{2\pi}\sqrt{\frac{x-u}{x^2-\Lambda^4}}\,dx. \tag{2.23}$$

These curves and differentials are designed so that their period integrals produce the functions $a(u)$ and $a_D(u)$

$$a(u) = \oint_\alpha \lambda = \frac{\sqrt{2}}{\pi}\int_{-\Lambda^2}^{\Lambda^2}\sqrt{\frac{x-u}{x^2-\Lambda^4}}\,dx,$$

$$a_D(u) = \oint_\beta \lambda = \frac{\sqrt{2}}{\pi} \int_{\Lambda^2}^{u} \sqrt{\frac{x-u}{x^2-\Lambda^4}} dx \tag{2.24}$$

where α, β are the standard homology cycles of a torus. At generic values of u the cubic equation $(x^2 - \Lambda^4)(x-u) = 0$ has three distinct roots $x_i = u, \Lambda^2, -\Lambda^2$ $(i = 1, 2, 3)$. When some of these roots coincide, the curve becomes degenerate and we have a singular situation. Thus we have the monopole $(u = \Lambda^2)$ and dyon singularity $(u = -\Lambda^2)$ in the present case. In general the discriminant Δ of a cubic equation $x^3 + fx + g = 0$ is given by $\Delta = 4f^3 + 27g^2$ and the positions of the singularities in the u-plane are located by $\Delta(u) = 0$.

It is easy to recover the behavior at the three singularities (2.16),(2.17), (2.20) from (2.24). In fact the above periods obey a differential equation with regular singular points (Picard-Fuchs equation)

$$\frac{d^2\Omega(u)}{du^2} + \frac{\Omega(u)}{4(u^2 - \Lambda^4)} = 0. \tag{2.25}$$

a and a_D are two independent solutions of this equation. (2.25) is nothing but the Gauss hypergeometric equation with singular pints at $u = \pm\Lambda^2$ and ∞. Expansions (2.5), (2.16),(2.17) are just the standard Taylor expansions around singular points. It is easy to see that the indices of the equation are degenerate at each singularity, which corresponds to the non-vanishing beta function and the appearance of the logarithmic solution.

A special feature of (2.25) is the absence of the first-derivative term. Then the Wronskian $a(u)\partial a_D(u)/\partial u - a_D(u)\partial a(u)/\partial u$ becomes a constant. Using $a_D = \partial F/\partial a$ we can integrate this equation and obtain [17]

$$a\frac{\partial F}{\partial a} - 2F = const \times u. \tag{2.26}$$

The value of *const* can be fixed by comparing the behavior of LHS and RHS at $u = \infty$. It turns out that it is proportional to the coefficient of one-loop beta function, $const = ib/2\pi$ (b=4 in the present case). The above equation shows the breakdown of scale invariance of the prepotential due to non-vanishing beta function. Generalization of (2.26) is known to hold in all $N = 2$ supersymmetric gauge theories including massless matter fields [18, 19].

3. $SU(2)$ gauge theory with matter

Let us next turn to the case of $SU(2)$ gauge theory coupled to matter in the vector representations. As it turns out that the ultraviolet-finite theory $N_f = 4$ is somewhat special, we restrict ourselves to the case of $1 \le N_f \le 3$ in the following. A special feature of the $N = 2$ gauge theory coupled to matter is the existence of Yukawa-like interaction between scalars of the vector and hypermultiplets. The superpotential consists of Yukawa-type term and bare mass term of matter fields

$$W = \sum_i \sqrt{2}\tilde{M}_i A M_i + \sum_i m_i \tilde{M}_i M_i. \tag{3.1}$$

Thus in the Coulomb phase with the vacuum value $\phi^3 = a$ the matter field acquires an effective mass $m_i \pm \frac{1}{\sqrt{2}}a$.

Let us first look at the case of large masses $m_i \gg \Lambda$ and study regions in the moduli space around $u \approx m_i^2$ $(i = 1, \cdots, N_f)$. These are in the weak coupling regime where the analysis based on the bare Lagrangian is reliable. Then there exists a massless particle at each $u = m_i^2$ $(i = 1, \cdots, N_f)$ with the bare mass m_i being canceled by the vacuum value a. Therefore, when masses are large, there are N_f additional singular points in the complex u-plane at large $|u|$. We call these singularities as the squark singularities. If the mass m_i goes to ∞, the corresponding singularity moves away to ∞ in the u-plane and we eventually loose one flavor, $N_f \to N_f - 1$. Thus the squark singularity describes an elementary particle which becomes massless at $u = m_i$. On the other hand if the masses m_i decrease, squark singularities move into the region of smaller values of $|u|$. In the limit of zero mass squark singularities will be located in the strong coupling region $|u| \approx \Lambda^2$ and appear as massless solitons. We will later discuss how an elementary particle could covert itself into a soliton as the mass is varied.

Let us first look at the massless case. In the massless limit there exists a global discrete symmetry (remnant of the $U(1)_R$ symmetry broken by anomaly) which acts on the u-plane and determines the the location of the singularities. In the following we list the global symmetry, electric and magnetic charges (n_m, n_e) carried by the strong- coupling singularities and their monodromies:

	global symmetry	monodromies	
			(3.2)
$N_f = 0$	Z_2		
	$(1,0),(1,2);$	$STS^{-1},(T^2S)T(T^2S)^{-1}$	(3.3)
$N_f = 1$	Z_3		
	$(1,0),(1,1),(1,2);$	$STS^{-1},(TS)T(TS)^{-1},(T^2S)T(T^2S)^{-1}$	(3.4)
$N_f = 2$	Z_2		
	$(1,0),(1,1);$	$ST^2S^{-1},(TS)T^2(TS)^{-1}$	(3.5)
$N_f = 3$	none		
	$(1,0),(2,1);$	$ST^4(S)^{-1},(ST^2S)T(ST^2S)^{-1}$	(3.6)

(In the case $N_f = 0$ the normalization of electric charge is changed by a factor 2 from the previous section). Note that in the massless limit the theory with N_f flavors has a global $SO(2N_f)$ symmetry (**2** and **2*** reps. of $SU(2)$ are equivalent and thus M_i and $\tilde{M}_i$ form a $2N_f$ dimensional rep. of $SO(2N_f)$) and the singularities fall into representations of $SO(2N_f)$. Both of the singularities in $N_f = 2$ theory have multiplicity 2 and form reps. $(\mathbf{2},\mathbf{1}),(\mathbf{1},\mathbf{2})$ of $SO(4)$ while those of $N_f = 3$

have multiplicity 4 and 1 and form **4** and **1** reps. of $SO(6)$. Note also that the monodromy around a singularity of multiplicity k with charge $(n_m, n_e) = (1, n)$ is given by $(T^n S)T^k(T^n S)^{-1}$ in the above table. We can also check that the strong-coupling monodromies add up to the weak-coupling one at ∞, $\prod_i M_i = M_\infty$ where $M_\infty = PT^{-(4-N_f)}$.

One of the curious features of the massless solution is the appearance of so-called jumping lines in the u-plane. A jumping line is defined by

$$\{a_D(u)/a(u) = \text{real} \mid u \in \mathbf{C}\}. \tag{3.7}$$

It passes through strong-coupling singularities and divides the u-plane into two, inner and outer, parts with the outer part being accessible to the observation at ∞ while the inner part is not. In the case of $N_f = 1$, for instance, jumping line looks like a circle passing through singularities at $u = u_1, u_2, u_3$ ($u_j = \exp((2j - 1)\pi i/3)(27/4)^{1/3}\Lambda^2/4)$. See Fig.3 [20, 21].

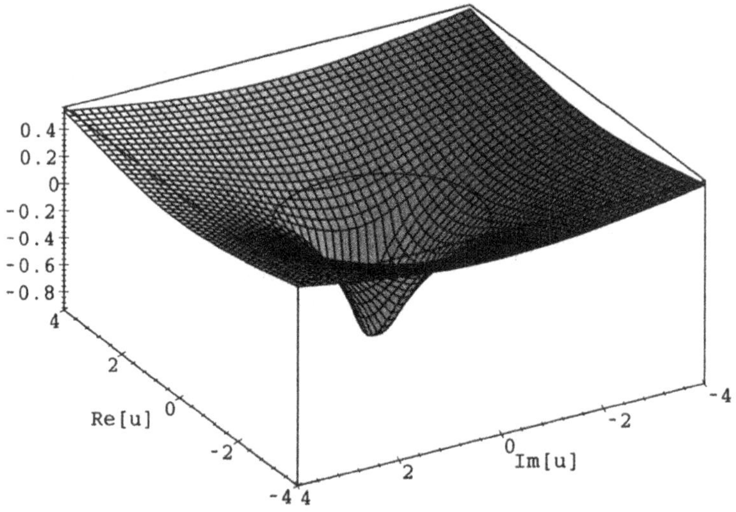

Figure 1. Im $a_D(u)/a(u)$ ($N_f = 1$, $m = 0$)

On the jumping line the distinction between the electric and magnetic charge becomes ambiguous. In the BPS mass formula (2.11) a and a_D span a 2-dimensional lattice whose coordinates define the electric and magnetic charges. When a_D/a is real, the lattice collapses and one can not distinguish n_e from n_m. Thus a soliton may change its identity when it crosses a jumping line. We note that since all the strong-coupling singularities are located on the jumping line, they can be reached from $u = \infty$ without crossing the line. Thus these singularities have well-defined quantum numbers.

Let us, for instance, examine the quantum numbers of three singularities in the $N_f = 1$ case. Seiberg-Witten curve for the massless $N_f = 1$ theory is given by

$$y^2 = x^2(x - u) - \frac{\Lambda^6}{64}. \tag{3.8}$$

Its discriminant reads as $\Delta = 256u^3 + 27\Lambda^6$ and hence the singularities are located at $u_2 = -(27/4)^{1/3}\Lambda^2/4, u_1 = \exp(-2i\pi/3)u_2, u_3 = \exp(2i\pi/3)u_2$ in the u-plane. We denote the zeros of the cubic polynomial $x^2(x-u) - \Lambda^6/64$ in the x-plane as x_1, x_2, x_3. See Fig.3.

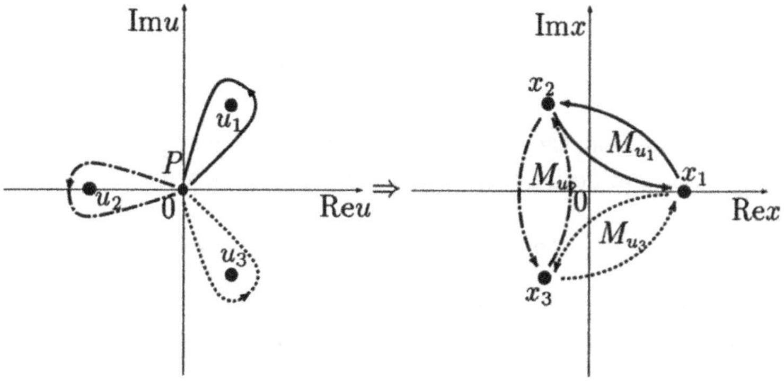

Figure 2. Singularities in the u and x plane

We also define three cycles in the x-plane, α, β, γ which go around a pair of points among x_1, x_2, x_3. See Fig.3.

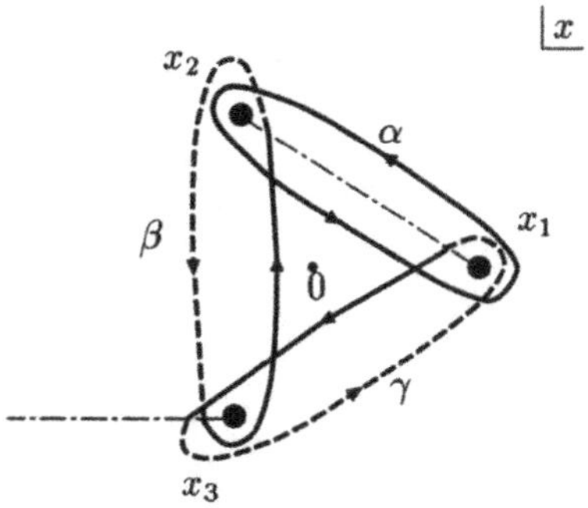

Figure 3. α, β, γ cycles

Note that the sum of these cycles vanish $\alpha + \beta + \gamma = 0$. We then consider contours C_i which starts from $u = 0$, goes around u_i and returns to $u = 0$. When one travels along C_1, for instance, x_1 and x_2 are interchanged while x_3 remains invariant as in Fig.3. Then it is easy to see that by going along C_i three cycles transform as

$$C_1 : \begin{cases} \alpha & \to & \alpha \\ \beta & \to & \beta + \alpha, \\ \gamma & \to & \gamma - \alpha \end{cases} \quad C_2 : \begin{cases} \alpha & \to & \alpha - \beta \\ \beta & \to & \beta, \\ \gamma & \to & \gamma + \beta \end{cases} \quad C_3 : \begin{cases} \alpha & \to & \alpha + \gamma \\ \beta & \to & \beta - \gamma. \\ \gamma & \to & \gamma \end{cases} \tag{3.9}$$

If we take α, β as the basis of cycles, we obtain the monodromy matrices

$$M_{u_1} = \begin{pmatrix} 1 & 0 \\ 1 & 1 \end{pmatrix}, M_{u_2} = \begin{pmatrix} 1 & -1 \\ 0 & 1 \end{pmatrix}, M_{u_3} = \begin{pmatrix} 0 & -1 \\ 1 & 2 \end{pmatrix}. \tag{3.10}$$

Monodromy at ∞ is then given by

$$M_\infty = M_{u_1} M_{u_2} M_{u_3} = \begin{pmatrix} -1 & -3 \\ 0 & -1 \end{pmatrix}. \tag{3.11}$$

On the other hand, from the known weak coupling behavior of a, a_D, $a \sim \sqrt{u}$, $a_D \sim 3ia/2\pi \log u$, one can derive their monodromy property at ∞

$$a_D \to -a_D + 3a, \qquad a \to -a. \tag{3.12}$$

Hence a is identified as the period around the cycle β and a_D as the period around the cycle $-\alpha$. If we rewrite the monodromy matrices in the basis of a_D and a, they read as

$$M_{u_1} = \begin{pmatrix} 1 & 0 \\ -1 & 1 \end{pmatrix}, M_{u_2} = \begin{pmatrix} 1 & 1 \\ 0 & 1 \end{pmatrix}, M_{u_3} = \begin{pmatrix} 0 & 1 \\ -1 & 2 \end{pmatrix}. \tag{3.13}$$

In order to define the charges of the singularities u_i properly we next consider paths extending from infinity to u_i as in Fig.3.

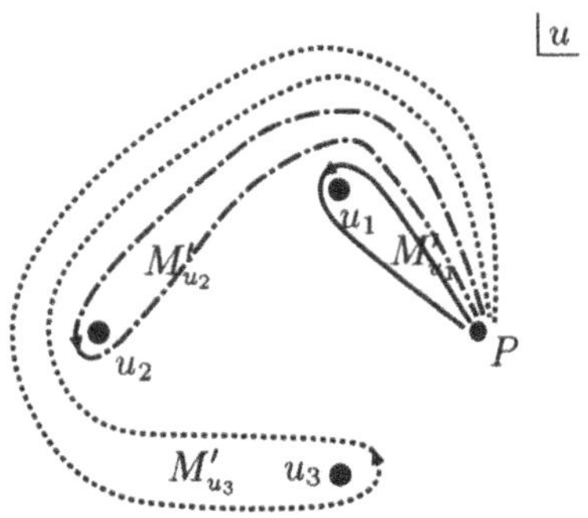

Figure 4. Charge of solitons

These paths avoid entering into the area inside the jumpling line. They generate monodromies as

$$M'_{u_1} = M_{u_1} = \begin{pmatrix} 1 & 0 \\ -1 & 1 \end{pmatrix}, \tag{3.14}$$

$$M'_{u_2} = M_{u_1} M_{u_2} M_{u_1}^{-1} = \begin{pmatrix} 2 & 1 \\ -1 & 0 \end{pmatrix}, \tag{3.15}$$

$$M'_{u_3} = (M_{u_1} M_{u_2}) M_{u_3} (M_{u_1} M_{u_2})^{-1} = \begin{pmatrix} 3 & 4 \\ -1 & -1 \end{pmatrix} \tag{3.16}$$

By considering eigenvectors with eigenvalue 1 we deduce charges of massless solitons: (1,0) at $u = u_1$, (1,1) at $u = u_2$, (1,2) at $u = u_3$, respectively. This reproduces the result of (3.4).

Let us now generalize our discussions and consider the case of massive matter fields. Seiberg-Witten curves for massive theories are constructed by requiring the correct singularity structure and decoupling limit $m_i \to \infty$. They are given by [2]

$$N_f = 1: \quad y^2 = x^2(x-u) + \frac{1}{4}m\Lambda_1^3 x - \frac{1}{64}\Lambda_1^6, \tag{3.17}$$

$$N_f = 2: \quad y^2 = (x^2 - \frac{1}{64}\Lambda_2^4)(x-u) + \frac{1}{4}m_1 m_2 \Lambda_2^2 x - \frac{1}{64}(m_1^2 + m_2^2)\Lambda_2^4, \tag{3.18}$$

$$\begin{aligned} N_f = 3: \quad y^2 &= x^2(x-u) - \frac{1}{64}\Lambda_3^2(x-u)^2 - \frac{1}{64}(m_1^2+m_2^2+m_3^2)\Lambda_3^2(x-u) \\ &+\frac{1}{4}m_1m_2m_3\Lambda_3 x - \frac{1}{64}(m_1^2m_2^2 + m_2^2m_3^2 + m_3^2m_1^2)\Lambda_3^2. \end{aligned} \tag{3.19}$$

Here the Λ_{N_f} stands for the Λ parameter of the N_f-flavor theory and obeys the decoupling condition

$$\lim_{m_i \to \infty} m_i \Lambda_{N_f}^{4-N_f} = \Lambda_{N_f-1}^{4-(N_f-1)}. \tag{3.20}$$

In fact in the above (3.17)-(3.19) a curve of N_f-theory reduces to that of (N_f-1)-theory in the decoupling limit $m_i \to \infty$.

Discriminants of the curves read as

$$N_f = 1: \quad \Delta_1 = 27\Lambda_1^6 + 256\Lambda_1 m^3 - 288\Lambda_1^3 mu - 256m^2u^2 + 256u^3, \tag{3.21}$$

$$N_f = 2: \quad \Delta_2 = (\Lambda_2^2 + 8m^2 - 8u)^2(\Lambda_2^4 - 64\Lambda_2^2m^2 + 16\Lambda_2^2 u + 64u^2), \tag{3.22}$$

$$\begin{aligned} N_f = 3: \quad \Delta_3 &= (\Lambda_3 m + 8m^2 - 8u)^3(3\Lambda_3^3 m + 24\Lambda_3^2 m^2 + 2048\Lambda_3 m^3 \\ &+84\Lambda_3^2 u - 768\Lambda_3 mu - 2048u^2). \end{aligned} \tag{3.23}$$

Here we have put masses to be equal $m_i = m$ for simplicity. When matter fields have the same mass, the flavor symmetry of the theory becomes $SU(N_f)$ and the singularities fall into their representations. In fact in the $N_f = 2$ case the singularity at $u = m^2 - \Lambda_2^2/8$ has the multiplicity 2 and hence forms the **2** rep. of $SU(2)$. Similarly the singularity at $u = m^2 - \Lambda_3 m/8$ in $N_f = 3$ theory is a **3** rep. of $SU(3)$.

Seiberg-Witten differentials are now given by

$$N_f = 1: \quad \lambda = \frac{-\sqrt{2}}{4\pi}\frac{y dx}{x^2}, \tag{3.24}$$

$$N_f = 2: \quad \lambda = \frac{-\sqrt{2}}{4\pi}\frac{y dx}{(x^2 - \frac{1}{64}\Lambda_2^4)}, \tag{3.25}$$

$$N_f = 3: \quad \lambda = \frac{\sqrt{2}}{\pi\Lambda_3} dx \log\left(y + i\frac{\Lambda_3}{8}\left(x - u + \frac{1}{2}\sum_i m_i^2 - \frac{32}{\Lambda_3^2}x^2\right)\right). \tag{3.26}$$

A new aspect of massive theory is that the differential has poles with non-zero residues proportional to the mass of matter fields. In the cases of $N_f = 1, 2$ residues are given by

$$2\pi i Res(\lambda, x = 0) \qquad = \pm\tfrac{m}{\sqrt{2}}, \tag{3.27}$$

$$2\pi i Res(\lambda, x \quad = \pm\Lambda^2/8) = \pm\tfrac{1}{2}\tfrac{m_1 \mp m_2}{\sqrt{2}}. \tag{3.28}$$

Then the values of the periods of λ depends explicitly on the choice of contours (rather than their homology classes) and the integral picks up an additive factor when the contour is deformed across a pole. This is in accord with the BPS mass formula in the massive theory which has an additional term proportional to the mass of matter fields

$$M_{BPS} = \sqrt{2}\left| n_e a(u) + n_m a_D(u) + \sum_i S_i \frac{m_i}{\sqrt{2}} \right| \tag{3.29}$$

Quantum numbers S_i appear in the central charge of $N = 2$ algebra when there exist Abelian conserved quantities. $S_i = 1$ $(1/2)$ for a soliton in the singlet (spinor) rep. of flavor symmetry. Monodromy transformation in massive theory then takes a form

$$\begin{pmatrix} \frac{m}{\sqrt{2}} \\ n_m \\ n_e \end{pmatrix} \longrightarrow M \begin{pmatrix} \frac{m}{\sqrt{2}} \\ n_m \\ n_e \end{pmatrix} \tag{3.30}$$

where M is a 3×3 matrix.

In the case of massive $N_f = 1$ theory the sum of periods around the cycles α, β, γ is now given by $\alpha + \beta + \gamma = 2m/\sqrt{2}$ (for an integral of λ around a cycle we use the same notation as the cycle itself). Cycles transform when one goes around the trajectories C_i

$$C_1 : \begin{cases} \alpha & \to \quad \alpha \\ \beta & \to \quad \beta + \alpha, \end{cases} \quad C_2 : \begin{cases} \alpha & \to \quad \alpha - \beta \\ \beta & \to \quad \beta, \end{cases} \quad C_3 : \begin{cases} \alpha & \to \quad 2\frac{m}{\sqrt{2}} - \beta \\ \beta & \to \quad \alpha + 2\beta - 2\frac{m}{\sqrt{2}}. \end{cases} \tag{3.31}$$

It is easy to see that monodromy matrices are now given by

$$M_{u_1} = \begin{pmatrix} 1 & 0 & 0 \\ 0 & 1 & 0 \\ 0 & 1 & 1 \end{pmatrix}, M_{u_2} = \begin{pmatrix} 1 & 0 & 0 \\ 0 & 1 & -1 \\ 0 & 0 & 1 \end{pmatrix}, M_{u_3} = \begin{pmatrix} 1 & 0 & 0 \\ 2 & 0 & -1 \\ -2 & 1 & 2 \end{pmatrix}. \tag{3.32}$$

Let us next go back to the discriminant (3.21) and follow the motion of singularities as the mass m is varied. When the mass is increased (in the real positive direction), the singularity u_2 moves to the right and eventually goes away to ∞ as $u_2 \approx m^2$ while the other two singularities u_1, u_3 remain in the strong coupling region. They turn into $N_f = 0$ singularities in the decoupling limit

$$\Delta_1 \approx (u - m^2)(u^2 - m\Lambda_1^3) = (u - m^2)(u^2 - \Lambda_0^4). \tag{3.33}$$

Thus u_2 is the squark singularity in this case (which of the three singularities becomes the squark depends on the direction along which m moves to ∞). Then we have an apparent paradox: the singularity at u_2 was originally a dyon with charges (1,1) but now in the weak coupling region it behaves like an elementary particle. How a soliton throws away its magnetic charge when it moves out to infinity? The precise mechanism of this metamorphosis seems not well understood, however, it is obviously related to the existence of the jumping line. In Fig.3 we have defined the charge of the singularity u_2 making use of a contour which avoids entering into the region inside the jumping line. On the other hand when the singularity is in the weak coupling region close to ∞, we may define the charge of the singularity simply by a contour as in Fig.3.

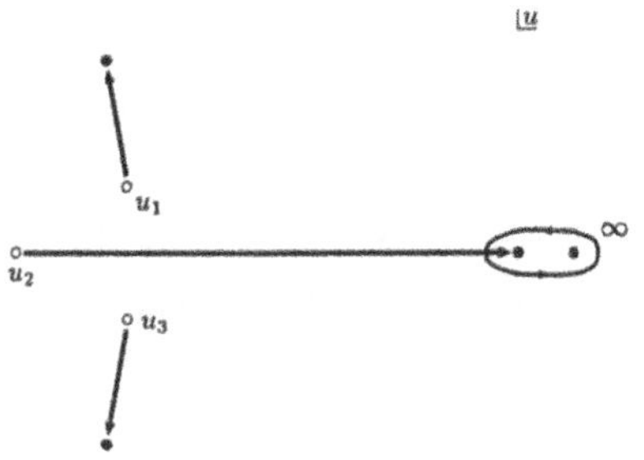

Figure 5. Charge of squark

The monodromy matrix at ∞ is given by

$$M_\infty = M_{u_1} M_{u_2} M_{u_3} = \begin{pmatrix} 1 & 0 & 0 \\ 4 & -1 & -3 \\ 2 & 0 & -1 \end{pmatrix}. \tag{3.34}$$

The period a transforms as $a \to -a$ under $u \to \exp(2\pi i)u$ and thus corresponds to an eigenvector of (3.34) with an eigenvalue -1. It is easy to see that the eigenvector is given by $(-1, 0, 1)$ and hence $a = \beta - m/\sqrt{2}$ or $\beta = a + m/\sqrt{2}$. Thus the cycle β which gave rise to a massless dyon in the strong-coupling region now behaves like an elementary particle at infinity. Flip in the quantum number has taken place due to the existence of jumping line.

4. 4-dimensional conformal field theory

Another interesting feature of the $SU(2)$ theory with matter is the existence of superconformal points. Four-dimensional conformal field theory with $N = 2$ supersymmetry was first discovered by Argyres and Douglas in the $SU(3)$ theory [22]. In this section, however, I will follow the approach of [23] and locate the superconformal points in $SU(2)$ theories coupled to matter.

We have seen in the previous section that when we increased the value of the mass, the would-be squark singularity passed in between the other singularities and moved into the weak coupling region. Hence if we fine-tune the value of the mass the squark singularity could collide with the other singularities and generates a new phase of $N = 2$ field theory.

We know that the locations of singularities are given by the zeros of the discriminant $\Delta(u, m)$. We may then fine-tune m to a critical value m^* so that $\Delta(u, m^*)$ has a multiple root at $u = u^*$. Values of u^*, m^* are easily found by calculating the discriminant of $\Delta(u, m)$ as a function of u. At these critical points Seiberg-Witten curves become completely degenerate

$$N_f = 1 : y^2 \approx (x - x^*)^3, \quad m^* = \frac{3\Lambda}{4}, \; u^* = \frac{3\Lambda^2}{4}, \; x^* = \frac{\Lambda^2}{4}, \tag{4.1}$$

$$N_f = 2 : y^2 \approx (x - x^*)^3, \quad m^* = \frac{\Lambda}{2}, \; u^* = \frac{3\Lambda^2}{8}, \; x^* = \frac{\Lambda^2}{8}, \tag{4.2}$$

$$N_f = 3 : y^2 \approx (x - x^*)^3, \quad m^* = \frac{\Lambda}{8}, \; u^* = \frac{\Lambda^2}{32}, \; x^* = \frac{\Lambda^2}{64}. \tag{4.3}$$

Thus both α and β cycles vanish at these points. In order to see in more detail what is happening here let us look at the differential equations satisfied by the periods a, a_D. Since in the case of massive theories monodromy is described by 3×3 matrices, Picard-Fuchs equation becomes a 3rd order differential equations

$$D(u)\Omega'''(u) + P(u)\Omega''(u) + Q(u)\Omega'(u) = 0. \tag{4.4}$$

We always have a trivial solution $\Omega = const$ which corresponds to a cycle surrounding the simple pole. The coefficient functions are given by

$$N_f = 1 :$$

$$\begin{cases} P(u) = (81\Lambda^6 - 384\Lambda^3 m^3 - 2048 m^4 u + 3840 m^2 u^2 - 1536 u^3), \\ Q(u) = 8(-9\Lambda^3 m - 32 m^4 + 72 m^2 u - 24 u^2), \\ D(u) = (4m^2 - 3u)(27\Lambda_1^6 + 256\Lambda_1^3 m^3 - 288\Lambda_1^3 m u - 256 m^2 u^2 + 256 u^3). \end{cases} \tag{4.5}$$

$$N_f = 2 :$$

$$\begin{cases} P(u) = 64(15\Lambda^4 m^2 - 96\Lambda^2 m^4 - 2\Lambda^4 u + 48\Lambda^2 m^2 u \\ \qquad\qquad -256 m^4 u - 32\Lambda^2 u^2 + 448 m^2 u^2 - 128 u^3), \\ Q(u) = 16(-\Lambda^4 - 24\Lambda^2 m^2 - 128 m^4 - 16\Lambda^2 u + 320 m^2 u - 64 u^2), \\ D(u) = (\Lambda^2 + 8m^2 - 8u)(\Lambda^2 - 16 m^2 + 8u)(\Lambda^4 - 64\Lambda^2 m^2 + 16\Lambda^2 u + 64 u^2). \end{cases} \tag{4.6}$$

$$N_f = 3 :$$

$$\begin{cases} P(u) = 4(-7\Lambda^4 m^2 - 208\Lambda^3 m^3 - 4288\Lambda^2 m^4 + 73728\Lambda m^5 - 32\Lambda^3 m u \\ \quad +2560\Lambda^2 m^2 u - 40960\Lambda m^3 u + 131072 m^4 u - 64\Lambda^2 u^2 + 16384\Lambda m u^2 \\ \quad -212992 m^2 u^2 + 32768 u^3), \\ Q(u) = 256(7\Lambda^2 m^2 - 8\Lambda m^3 + 256 m^4 + 32\Lambda m u - 704 m^2 u + 64 u^2), \\ D(u) = (\Lambda m + 8m^2 - 8u)(\Lambda m - 16 m^2 + 4u)(3\Lambda^3 m + 24\Lambda^2 m^2 \\ \qquad\qquad +2048\Lambda m^3 + 8\Lambda^2 u - 768\Lambda m u - 2048 u^2). \end{cases} \tag{4.7}$$

In the $N_f = 1$ case, for instance, at the critical value $m = m^*$ we have

$$\Omega'''(u) + \frac{1}{(u-u^*)}\Omega''(u) - \frac{1}{36(u-u^*)^2}\Omega'(u) = 0. \tag{4.8}$$

Now the exponents of Ω around $u = u^*$, $\Omega(u) \approx (u-u^*)^\delta$ are non-degenerate and given by $\delta = \frac{7}{6}, \frac{5}{6}, 0$. Thus the degeneracy among the exponents are lifted and both of the solutions are power-behaved: $a(u)$, $a_D(u) \approx (u-u^*)^{5/6}$ ($(u-u^*)^{7/6}$ component vanishes faster at $u \approx u^*$). The logarithmic solution has now disappeared. (Explicit evaluation of periods at critical points is discussed in [24]).

We may interpret this as the vanishing of the beta function and the restoration of the scale invariance at the critical point. As we noted before, both α and β cycles vanish at the critical point. In general a cycle corresponds to a massless soliton with some charge (n_m, n_e) and the intersection number of cycles on the torus is proportional to the "cross-product" of the charges $n_m n'_e - n_e n'_m$. When the cross-product does not vanish, two set of charges represents solitons which are mutually non-local to each other (there are no Lagrangians where both of the fields have a local description). Thus the general criterion for the appearance of conformal invariant theories is the coexistence of massless solitons with mutually non-local charges [22]. If we normalize the scaling dimensions of periods a, a_D as 1, the scaling dimension of the order parameter u becomes 6/5 in $N_f = 1$ theory. Similarly we find that the exponents of the order parameter are given by 4/3 (3/2) in $N_f = 2, 3$ theories.

Detailed physical nature of these superconformal theories is yet unknown, however, one may attempt to classify them by means of their global symmetries and critical exponents. In [25] a detailed study has been made by making use of all solutions of $N = 2$ Yang-Mills theories with classical gauge groups and fine-tuning parameters to critical points. It turns out that the critical points obtained from $SU(n+1), SO(2n+1), Sp(2n)$ gauge theories have the same exponents and global symmetries and thus seem to define identical conformal field theories. On the other hand $SO(2n)$ and E_n gauge theories lead to different exponents. Thus we have some evidence for an A-D-E type classification of $N = 2$ conformal field theories. The critical exponents are given by

$$A_n: \quad 2\frac{e+1}{h+2}, \quad e = 1, 2, \cdots, n, \qquad h = n+1 \tag{4.9}$$

$$D_n: \quad 2\frac{e+1}{h+2}, \quad e = 1, 3, \cdots, 2n-3, n-1 \qquad h = 2n-2 \tag{4.10}$$

$$E_6: \quad 2\frac{e+1}{h+2}, \quad e = 1, 4, 5, 7, 8, 11, \qquad h = 12 \tag{4.11}$$

$$E_7: \quad 2\frac{e+1}{h+2}, \quad e = 1, 5, 7, 9, 11, 13, 17, \qquad h = 18 \tag{4.12}$$

$$E_8: \quad 2\frac{e+1}{h+2}, \quad e = 1, 7, 11, 13, 17, 19, 23, 29, \qquad h = 30 \tag{4.13}$$

In all these cases e runs over the Dynkin exponents of the group and h is the dual Coxeter number. $SU(2)$ theory with $N_f = 1, 2$ belongs to the class A_2, A_3 above and $SU(2)$ theory with $N_f = 3$ belongs to D_4.

Acknowledgements

I would like to thank organizers of the Cargèse summer school for their kind hospitality. I also would like to thank D. Hanawa for helpful discussions.

References

1. N. Seiberg and E. Witten, Nucl. Phys. **B426** (1994) 19.
2. N. Seiberg and E. Witten, Nucl. Phys. **B431** (1994) 484.
3. P. Argyres and A. Faraggi, Phys. Rev. Lett. **74** (1995) 3931.
4. A. Klemm, W. Lerche, S. Theisen and S. Yankielowicz, Phys. Lett. **B344** (1995) 169.
5. A. Hanany and Y. Oz, Nucl. Phys. **B452** (1995) 283: A. Hanany, Nucl. Phys. **B466** (1996) 85.
6. A. Brandhuber and K. Landsteiner, Phys. Lett. **B358** (1995) 73.
7. P.C. Argyres, R.N. Plesser and A.D. Shapere, Phys. Rev. Lett. **75** (1995) 1699.
8. J.A. Minahan and D. Nemeschansky, Nucl. Phys. **B464** (1996) 3.
9. P.C. Argyres and A.D. Shapere, Nucl. Phys. **B461** (1996) 437.
10. U.H. Danielsson and B. Sundborg, Phys. Lett. **370B** (1996) 83.
11. A. Bilal, *Duality in N=2 SUSY SU(2) Yang-Mills Theory: A pedagogical introduction to the work of Seiberg and Witten*, hep-th/9601007.
12. S.V. Ketov, *Solitons, monopoles and duality: from sine-Gordon to Seiberg-Witten*, hep-th/9611209.
13. L. Alvarez-Gaumé and S. F. Hassan *An introduction to S-Duality in N=2 Supersymmetric Gauge Theory*, (A pedagogical review of the work of Seiberg and Witten), hep-th/9701069.
14. H. Aoyama, T. Harano, M. Sato and S. Wada, Phys.Lett. B388 (1996) 331: T. Harano, M. Sato, Nucl.Phys. B484 (1997) 167.
15. K. Ito and N. Sasakura, Phys. Lett. **B382** (1996) 95, Nucl.Phys. B484 (1997) 141.
16. N.Dorey, V.V.Khoze and M.P.Mattis, Phys.Rev. D54 (1996) 2921, Phys.Lett. B388 (1996) 324, Phys.Rev. D54 (1996) 7832.
17. M. Matone, Phys. Lett. **B357** (1995) 342.
18. J. Sonnenschein, S. Theisen and S. Yankielowicz, Phys.Lett. B367 (1996) 145.
19. T. Eguchi and S.-K. Yang, Mod. Phys. Lett. A11 (1996) 131.
20. D. Hanawa, Master's thesis, Univ. of Tokyo, 1996.
21. A. Bilal and F. Ferrari, Nucl. Phys. B480 (1996) 589.
22. P.C. Argyres and M. Douglas, Nucl. Phys. **B448** (1995) 93.
23. P.C. Argyres, R.N. Plesser, N. Seiberg and E. Witten, Nucl.Phys. B461 (1996) 71.
24. T. Masuda and H. Suzuki, Nucl.Phys. B495 (1997) 149.
25. T. Eguchi, K. Hori, K. Ito and S.K. Yang, Nucl. Phys. **B471** (1996) 430: T. Eguchi and K. Hori, in *The Mathematical Beauty of Physics*, eds. J.M. Drouffe and J.B. Zuber, World Scientific 1997.

DUALITIES IN THE CLASSICAL SUPERGRAVITY LIMITS†

Dualisations, dualities and a détour via 4k+2 dimensions

B. L. JULIA
Laboratoire de Physique théorique de l'Ecole Normale Supérieure, CNRS
24 rue Lhomond 75005 Paris FRANCE.

1. Introduction

The duality symmetries are invariances of equations of motion and even sometimes, for instance in odd dimensions, symmetries of a suitable action. The case of $4k+2$ dimensions seems to be related to classical Lie-Poisson actions leading to Quantum groups, indeed the latter were discovered in 2 dimensional theories. The duality group gets its name from some of its elements that are actually Hodge dualities like in Dirac's famous analysis of the exchange of electricity and magnetism in four dimensions which permutes strong and weak coupling expansions. This is also the nature of so-called discrete S-dualities in quantum four dimensional heterotic string theories as discussed by A. Sen. The $SL(2,\mathbb{R})$ symmetry in type IIB theory in 10 dimensions is also called S duality as it exchanges weak and strong string couplings. The string coupling is affected via the dilaton field. In four dimensions these symmetries involve Hodge duality in the target spacetime.

In string theory the so-called T-dualities do exchange strong and weak couplings of the sigma model on the worldsheet. They can be seen as generalisations of the Kramers-Wannier symmetry of the Ising model that permutes inverse temperature (or equivalently the euclidean period of time) and its inverse. T-dualities may also be realised as Hodge dualities but on the 2d worldsheet, as was shown by Buscher when the target space admits isometries. A generalisation called Lie-Poisson T-duality relaxes slightly the latter condition.

Three years ago Hull and Townsend unified the two main kinds of dualities (S and T) inside a much larger if conjectural discrete U-duality group of the quantum supergravity and string theories. These classical theories are now believed to be inequivalent limits of a quantum model called M-theory after Witten's discovery

†After the original lectures were delivered at this Cargèse school in May 1997, various developments have been presented at the Neuchâtel Workshop "Quantum aspects of Gauge theories, Supersymmetry and Unification" in September 1997 and at the Trieste Conference on "Superfivebranes and Physics in 5+1 dimensions" in April 1998.

L. Baulieu et al. (eds.), Strings, Branes and Dualities, 121–139.

of the correspondence between both formulations. The string coupling constant becomes geometrical, essentially equal to the length of the compactification circle along the eleventh dimension. An important point is that all solitonic excitations should be included in the quantum theory, as well as their duals which include fundamental strings in ten dimensions or fundamental membranes in 11.

Now the proposed U-duality groups are discrete subgroups of the duality symmetries of the equations of motion of maximally dualised supergravities which have been known since about 1980. We shall use the same letter for the discrete subgroup and the Lie group when it is clear from the context which one it is. The first classical supergravities under consideration were the fully dualised toroidal dimensional reductions of 11 dimensional (ie type II or maximal or $N_4 = 8$) supergravity. We shall only briefly mention their truncations to pure ($N_4 \leq 6$) supergravities. Type I supergravities possess also interesting duality symmetries of the classical equations of motion first studied by Chamseddine [1]; considered together with those of type II [2] they strongly suggest that the two simply laced hyperbolic Kac-Moody algebras of maximal rank (equal to 10) should appear as symmetries of some huge space covering the set of unidimensional classical solutions of supergravity theories. Although there are only indications for that yet, let us give a name to these hyperbolic Kac-Moody groups: E_{10} and HD_{10} (called overextended D_8 in [2]), they correspond to type II resp. type I. It would be interesting to accommodate the heterotic theories in the hyperbolic game [3]. The situation there is still moving with hyperbolic Kac-Moody algebras and generalised Kac-Moody algebras (including Borcherds algebras) appearing in toroidal compactifications.

What has been established and extensively studied is the occurrence of infinite dimensional symmetry groups in the reduction to two dimensions, there the affine Kac-Moody extension $G^{(1)}$ enlarges G, the corresponding maximal U-duality group in three dimensions. In two dimensions scalar potentials (fields) are dual to scalar fields, as a result the exchange between gauge and internal (rigid) symmetries does not take place; the duality group acting on the appropriate set of fields covering the set of solutions becomes infinite dimensional. It is also an important problem to describe precisely and uniformly, namely for all dimensions, what the discrete (infinite) groups of U duality are. They are most probably groups over $\mathbb{Z}$, the rational integers, as defined by Chevalley. This conjecture is nicely compatible with the observation that any $\mathbb{R}$ factor group of the classical duality group has no infinite discrete duality analogue and disappears from the U-duality group at the quantum level.

In section 2 we shall recall the "silver rules" of supergravities and the building blocks of the U-groups. The latter use three ingredients: the dilatonic rescaling symmetry one obtains in 10 dimensions, the "Ehlers phenomenon" where a scaling symmetry becomes a whole $SL(2)$ and finally the fusing together of such $SL(2)$'s with the expected symmetry $GL(D-d)$ that comes from the dimensional reduction from D to d dimensions into a simple rigid internal U-duality group. Given G the U-duality group, the three silver rules are that the scalar fields parameterise a symmetric space of the noncompact type $K\backslash G$ where K is the maximal compact subgroup of G, that equations simplify dramatically upon restoring a local gauge invariance under K and finally that in even dimensions $d = 2f$ the equations of motion of the field strengths of order f and their Bianchi identities are unified in a twisted self-duality equation.

$$*S.\mathcal{V}.\mathcal{F} = \mathcal{V}.\mathcal{F}$$

where S is an invariant operator acting on the appropriate representation of K and where $\mathcal{V}$ is the coset representative of the scalar field transforming under G on the right contragrediently to $\mathcal{F}$ on the left, and transforming also under the local gauge group K on the left. The case of timelike compactification is somewhat different [4] but it is important for Euclidean signature, see [5]. The maximal compact subgroup is replaced in general by a noncompact subgroup and the quotient has a rather problematic topology.

In section 3 we proceed to study the effects of dualisations beginning with a comparison between pure gravity reduced from 4 to 3 dimensions and its dualised theory with Ehlers symmetry acting in a local way as a rigid $SL(2,\mathbb{R})$ invariance of the action. In the theory of integrable Hamiltonian systems, Baecklund transformations (for instance Miura transformations) may exchange solutions of one system with those of another. Here dualisations are discrete duality transformations, essentially Legendre transformations, that modify the (perturbative) field contents of the action and the analogy is useful. Then we analyse in detail the case of maximal supergravity, also in three dimensions. It may seem at first that as soon as one leaves the simple case of maximally dualised theories one is in danger of loosing oneself among all the possibilities. It is not so, the key choice is to pick a grading of the root space, for instance along a particular root, one can then show how undualisation of some scalar fields belonging to the corresponding highest level (or levels) do actually reduce the U-symmetry by changing the dimension of (and simultaneously contracting ie partially abelianising) the remaining subalgebra. In passing we note that large abelian subalgebras of dualities do occur and in fact those of maximal dimension as classified by Malcev [6] for the case of complex Lie algebras (or their normal real forms) do appear. In the case at hand there can be 36 commuting generators in the maximally noncompact (so-called split or normal) real form of E_8: $E_8(+8)$. This is to be contrasted with the compact form situation where the maximal abelian dimension is the rank.

In the fourth section we shall try to learn about the higher form fields from the Dynkin diagram of the U-group. It turns out that one can read off from the diagram the number of forms of various degrees, because they belong to very specific fundamental representations of G. In the case of maximal supergravities the Dynkin diagram reflects the possibility of dualisation between higher form fields by the existence of some outer automorphism. In general and most importantly it suggests that one should unify internal rigid symmetries with diffeomorphisms (or at least the $GL(D-d)$ subgroup) in a larger group of rank at least eleven (or twelve?). This can be applied to type IIA, IIB or I. In fact we find a purely group theoretical version of the Horava Witten orbifolding relating type II and type I string theories, it corresponds to using a Cartan involution of the U-duality in any dimension. The heterotic duality groups are interesting too, they are non-split real forms in which one still recognizes the expected linear diffeomorphism symmetries, but they will be treated separately. Note that the non-maximal supergravities also lead to non-split real forms of the duality groups [7].

Then we shall present the general analysis of duality symmetries in curved space with either Lorentzian or Euclidean signature. The matter will be taken in an N-plet of middle rank ($f = d/2$)-forms plus sufficiently many scalar fields. Since our paper [8] appeared on the hep-th archive we realised that the general Lorentzian case seems to have been investigated by Tanii [9], some partial results on the 2 dimensional case were also obtained in [11]. The 4 dimensional case was systematically analysed in [10] but as one might expect the $4k + 2$ dimensional case is quite different. The fact that self-duality becomes possible is not directly relevant here. We have also constructed constrained actions that preserve the U-symmetry and allow to simplify the computations by doubling the set of fields (third silver rule).

In the last section I shall summarize two research projects I have been working on for the past few years. On the one hand I want to stress the importance of the existence of "complementary " classical limits appropriate to different experimental situations. This may lead to some clarification of the formulation of Quantum Mechanics. There is no classical world only classical approximations: the limit of the Planck constant tending to zero is to be defined by a dimensionless criterion but more importantly by specifying what is being kept fixed. Particle limit and wave or classical field limit are in *duality*. On the other hand I would like to point out more publicly than before the intimate relation between intersection theory, fermionisation and charge quantization of dyons. The fundamental differences among even dimensions between $d = 4k$ and $d = 4k + 2$ will be discussed. The idea is that well defined statistics is what saves locality in theories of extended objects, it is associated to a charge quantization condition in two dimensions. But the latter has to be symmetric to allow for chiral (self-dual) particles [12] [13]. We show that higher dimensional fermionisation requires a détour through 6 (or 10) dimensions or twisting in 4d by some internal symmetry in the sense of section 2. In guise of conclusion I give a preview of a forthcoming paper where it is shown that indeed the twisted self-duality equation is universal excluding for the time being the graviton. There is a candidate for F-duality that brings us into the realm of graded superalgebras.

2. Silver rules

In the quest for supergravity actions the use of extra-dimensions of space made the deformation or so-called Noether method significantly simpler by reducing the number of scalar fields to zero in the best cases and thus avoiding nonpolynomial expressions. The idea of higher dimensions is in fact quite natural in the context of extended Poincaré superalgebras, in particular the doubling of the number of fermionic charges can result from the addition of 2 dimensions. Interest in four-dimensional supergravity with the maximal possible number of supercharges (32) led us to consider ten dimensions. In fact we went directly to eleven dimensions with Majorana spinors as suggested by the spectrum analysis by Nahm at the linear level but also by practical considerations. No scalar fields are left in eleven dimensions and the spectrum is amazingly simple: a graviton, a gauge three-form and the Rarita-Schwinger field. Furthermore dimensional reduction is easier than its converse group disintegration [7]. For instance the theory with 24 supercharges in 6 dimensions predicted in [7] was only constructed 16 years later [14]. Returning to eleven dimensions, the requirement of local supersymmetry leads to a unique action for this set of classical fields. After toroidal compactification one obtains the expected $N = 8$ supergravity in 4 dimensions but also peculiar duality symmetries. Indeed in dimension d one discovered (maximally) noncompact symmetry groups $E_{11-d}(11-d)$ of the toroidally reduced theories; the number between parentheses is the real rank to be defined in the next paragraph.

We found ourselves in a situation similar to that of general relativists in front of the Ehlers $SL(2,\mathbb{R})$ symmetry or its extension for stationary electrovacs $SU(2,1)$. In the absence of conceptual understanding we began a systematic analysis of the symmetry Lie algebras for all dimensions and all number of supersymmetries. This led to a list of noncompact symmetric spaces, ie cosets of the form $K\backslash G$ of the U-symmetry groups G by their maximal compact subgroups K. The real rank of the coset is the maximal dimension of the subspace of a Cartan subalgebra orthogonal to the compact directions (for short noncompact Cartan generators). The real rank r is equal to the rank l for a normal (also called split) real form, this is the case for the descendants of type I or II supergravities in 10 dimensions. It turns out that in all cases the real rank increases by one upon each step of dimensional reduction. One can also check that $d + l - N_4' = 4$ is constant in the disintegration triangle [7] for type II(A) and other pure (in four dimensions) supergravities in various dimensions; N_4' is the number of supersymmetries in four dimensions except for the maximal case where it is equal to seven. For Chamseddine's type I disintegration column, the constant is equal to 7 instead of 4. Now the scalar fields parameterize this coset and other fields transform as representations of K or if one prefers of G through the scalar group element which intertwines between K and G (we do not consider spinor fields here). It is quite suggestive to think of (the scalars as) a moving frame exchanging the (K) tangent space-Lorentz indices with (G) world indices.

The split or maximally noncompact real form of a complex simple Lie group is the generalisation to a general Lie group of $SL(2,\mathbb{R})$ inside $SL(2,\mathbb{C})$. One simply defines the split form as generated by a number of copies of $SL(2,\mathbb{R})$ equal to the rank l and a finite set of relations, the Serre relations, which restrict their simple and multiple commutators. The same relations hold over the real numbers for the split form as over the complex for the complex form. Another characteristic of the split form is that it admits a Cartan subalgebra of $l = r$ commuting, noncompact but ad-diagonalisable generators. The rest of the generators come in real pairs associated to opposite roots and closing on Cartan generators to form again copies of $SL(2,\mathbb{R})$. In the non-split case, the rank l is strictly larger than the number r of linearly independent noncompact Cartan generators. However if we relax the diagonalisability assumption the number a of linearly independent commuting generators might be strictly larger than r and even than l. We shall see examples of this in the next section for real split forms. The split real form is also characterized by having the smallest compact subalgebra and consequently the largest dimension for the coset space $K\backslash G$.

This symmetric space structure $K\backslash G$ for the moduli space of scalar fields in the maximally dualised form of pure supergravity theories constitutes the first silver rule. Some of the gauge forms descending either from the gauge three form or from the metric, namely the Kaluza-Klein one forms, can be simultaneously dualised when their potentials can be covered with derivatives despite the presence of the Chern-Simons-like term in 11 dimensions. In fact all those potentials that are of degree $d-2$ can be dualised into scalars in the case of pure supergravities, this we shall call maximal dualisation in the scalar sector. That particular form of the action has the maximal rigid internal symmetry.

The second silver rule is the generalisation of the orthonormal moving frame technique to internal symmetries. In General Relativity the metric is given on a manifold, it does not depend on any frame choice but its components depend on coordinate choices in a tensorial way. Alternatively the same local information can be obtained from a frame of orthonormal 1-forms, but they are only defined up to a gauge Lorentz transformation. In our scalar manifold case, $K\backslash G$ can be parameterised either as a manifold by ignoring the coset structure or alternatively as the base of a principal bundle the group G itself. Then the subgroup action (on the left here) is a gauge symmetry that compensates for the extra freedom that was added to make the G symmetry manifest. Typically, our algebraic power being limited, the coset representatives we encounter first are very often in the triangular or Borel and more generally in the solvable K-gauge [15]. One reason is that in that gauge the exponential parameterisation is bijective but more important probably is the fact that unipotent elements and their inverse are parameterised polynomially. The search for hidden symmetries is consequently the restoration of the gauge freedom or at least, and this is easier, the enlargment of the Borel symmetry to the full G invariance. For instance in two dimensions only the second process has been realised yet.

The third silver rule of supergravity is that the middle degree field strengths are self-dual; more precisely their equations of motion and their Bianchi identities can be unified in a covariant set of Bianchi identities for a doubled set of fields restricted by a twisted self-duality condition so that the original n second order equations have been encoded into a set of $2n$ first order ones. This was first discovered in 4 dimensions with Minkowskian signature such that the square of the Hodge dual is equal to minus one: clearly the self-duality needs a twist in the form of a G invariant operator S acting on the representation $2n$ and of square equal to minus one in four dimensions in order to compensate for the previous minus sign. As an illustration let us recall that maximal supergravity in 4 dimensions has 28 vectors, with their dual potentials this makes 56. But this is precisely the dimension of the fundamental representation of E_7, it is symplectic and can be extended to a representation of $SP(56, \mathbb{R})$.

It is an easy exercise to check that the square of the Hodge operator is equal to $(-1)^{(s-t)/2}$. Note that the difference $s-t$ also appears in the classification of Majorana spinors. In fact instantons require Euclidean (or (2,2)) signature in four dimensions. But supersymmetry can provide some internal degrees of freedom to compensate for the otherwise devastating minus sign, in other words it recalls its higher (10 or 6) dimensional origin to allow for generalised or we shall say twisted self-duality (TS). Twist can also be invoked to allow for generalised spin structure or to define a generalised Majorana condition. This self-duality is a feature of toroidal compactifications as the appearance of bare potentials rather than field strengths associated to nonabelian gauge theories or simply (charged) matter couplings seems to ruin the possibility of dualisation.

We should be slightly more careful though as we know examples of dual pairs of theories one of which has bare potentials. For instance the Freedman-Townsend-Thierry-Mieg dual sigma models are theories of two forms in four dimensions with bare potentials. This is compatible with dualisation and even twisted self-duality [8], [16]. Indeed, in the most favorable case, dualisation can use either one of two routes: one either introduces a Lagrange multiplier for the field strength Bianchi identity which becomes the dual potential, this is possible provided no potential appears in the Lagrangian or in the Bianchi identity, or one may use another first order formalism where now both the original potentials and field strengths are considered as independent variables. In the sigma model case the first route is taken when going from the scalar field description towards the two-form version but the reverse route is of the second type. This reciprocity of the two routes occurs in general. In the dual sigma model case the alternative paths are not allowed. Furthermore doubling does not mean one can actually find a dual Lagrangian as the example of 11 dimensional supergravity shows [16] and [17]. There exists a doubled TS (twisted self-dual) formalism but the 3-form cannot be integrated out to give the dual 6-form theory.

Let us now return to the general discussion of the third silver rule of supergravity. The structure is a mixture of K and G group theory. The potentials form a G-multiplet but the field strength combinations that are self-dual are the K-multiplet. The key equation was already mentioned, it is $*S.\mathcal{V}.\mathcal{F} = \mathcal{V}.\mathcal{F}$. This is to be supplemented by the G tensor equation $\mathcal{F} = d\mathcal{A}$. This coexistence of differentiation in curved space (G representation) and self-duality in flat or tangent space (K representation) seems to be quite general, it holds for instance also for various (super)brane actions and for Born-Infeld theories.

There are important remarks to make at this point. Firstly S is an operator from the K representation to itself. Secondly one should point out that there have been two ways to write this self duality equation, we just gave the second one, the first one was given in [18] and references therein: one constructs the analog of the metric (the monodromy in integrable systems). In the simple case of $SO\backslash SL$ we may define using a SO-invariant positive definite metric

$$\mathcal{M}_{MN} := \mathcal{V}^{t\,A}_{M}\eta_{AB}\mathcal{V}^{B}_{N}.$$

Then the TS equation reads

$$*\Omega^{PM}\mathcal{M}_{MN}\mathcal{F}^{N} = \mathcal{F}^{P}.$$

The invariant tensor Ω is an invariant of the noncompact group G, for instance the symplectic form of E_7 in its 56 dimensional representation in the case of four spacetime dimensions.

How can we relate this formula to the previous one using K-tensors? It does not seem to have been written up in general in the litterature but we need to establish

$$\mathcal{V}^A_M \Omega^{MN} \mathcal{V}^{tB}_N = S^A_C \eta^{-1CB} := \omega^{AB}$$

or alternatively

$$\mathcal{V}^{t\,-1} = \omega^{-1}.\mathcal{V}.\Omega$$

choosing the identity matrix for $\mathcal{V}$ we must identify ω and Ω. Things becomes even simpler if we use the operator S and rewrite the above formula as

$$\sigma(\mathcal{V}) = S^{-1}\mathcal{V}S$$

where σ is the Cartan involution[1] The Cartan involution is "inner" in this representation, the condition for involutivity (that S^2 be central) is automatic here. This property of σ has interesting consequences to which we hope to return in the future. For a general group, following [19], we must introduce $\sigma(\mathcal{V}^{-1})\mathcal{V}$ to replace $\eta^{-1}\mathcal{M}$.

There is no one-to-one relation between the symmetry property of the twist-form ω and the parity of $(s-t)/2$. The latter is related to reality properties whereas the former is a linear question. The symmetry of the twist can be found by inspection, surprisingly it does not depend on the signature of spacetime. One instance of this "signature blindness", namely the fact that duality transformations are canonical in $4k$ dimensions and not in $4k+2$ dimensions in Euclidean signature as well as in the Minkowskian one, was actually discovered some time ago[2]. Let us recall also that the fact that dualities are not canonical in 2 dimensions was realised before Quantum groups were invented see for instance [20]. In $4k$ Lorentzian dimensions one has no real "self-dual" tensor but there is a Euclidean one and twist is antisymmetric, in $4k+2$ dimensions on the other hand twist is symmetric. The existence of a symmetric resp. antisymmetric bilinear, invariant form for an irreducible complex representation of a group that is equivalent to its contragredient (here by conjugation by ω) makes it a "real" resp. "quaternionic" representation [21].

The analysis is complicated here by the fact that as we shall see the subgroup H that replaces K in the Euclidean case is non-compact. The properties of the twist ω, the invariant metric η and the operator S are intertwined in the relation $S := \omega.\eta$. The origin of the two different symmetries of ω has not been completely clarified yet, but the detailed study of supergravities is quite suggestive. We shall review the systematic analysis of [8] in section 4. For comparison, let us recall that half-spinor representations in even Lorentzian dimensions are complex if $d = 4k$, real

[1] Let us recall that the maximal compact subgroup of a real simple Lie group can be defined as the fixed point set of an involution σ, the Cartan involution of G. In the simple case under consideration $\sigma(\mathcal{V}) := \eta^{-1}.\mathcal{V}^{t\,-1}.\eta$.

[2] in a discussion with S. Deser in Sept. 1996.

if $d = 8k+2$ and quaternionic if $d = 8k+6$. The distinction between quaternionic and real spinors is irrelevant for quadratic expressions like the Ramond-Ramond bosons under consideration here. More relevant is the fact that the only symplectic fundamental representation of $SL(D, \mathbb{C})$ is the self-dual tensor for $D = 4k+2$ [21]. Let us also note that the fundamental representations of $SP(2N, \mathbb{C})$ and of $SO(2N, \mathbb{C})$ are obviously respectively symplectic and real.

Let us now review how U-dualities grow upon toroidal dimensional reduction. In eleven dimensions the classical equations of motion admit an engineering scaling symmetry because the dimensionful coupling constant appears as an overall factor in the action. In 10 dimensions this leads to an $\mathbb{R}$ internal and rigid symmetry. Technically one must specify an (Einstein) frame. This noncompact symmetry rescales the dilaton or the length of the compactification circle. Beyond this one generator there is an automatic $SL(11-d)$ internal symmetry directly originating from the diffeomorphism symmetries of our starting point. The second fact to notice is that in 8 dimensions the scaling group becomes an $SL(2, \mathbb{R})$ factor. This is also the dimension in which the 4-form field strength can be self-dual. More surprising still is the fusion into a simple Lie group of both ingredients, $SL(2, \mathbb{R})$ and $SL(11-d)$ below 8 dimensions.

This discussion can be adapted to the type I family. The starting point is in 10 dimensions [1] with one dilatonic symmetry. Now in 8 dimensions a second $SL(2, \mathbb{R})$ factor appears beyond the expected $SL(10-8, \mathbb{R})$, it incorporates one of the dilatonic symmetries of the 9 dimensional model. In lower dimensions the duality symmetries are the now familiar $SO(10-d, 10-d) \times \mathbb{R}$ enlarging $SL(10-d, \mathbb{R})$. In 4 dimensions the last dilatonic subgroup becomes a full $SL(2, \mathbb{R})$ factor by dualisation of the two-form potential. Finally in three dimensions the group becomes simple namely $SO(8, 8)$.

Until now we focussed our attention on the maximally dualised theories. But if instead of deciding to lower as much as possible the degrees of the forms in the action by dualising, when possible, any k-form field strength to the lower degree ($k' = d-k$)-form we decide not to do it or to undualise some of the latter, the internal symmetry group action on the scalar fields that are being undualised is transmuted into a gauge group for the dual forms, at least above 2 dimensions. For the two-dimensional case see for instance [22] and references therein. The purpose of the next section is to discuss more systematically these choices (to dualise or not to dualise).

3. Dualisations

It may be advisable to begin with the simplest typical example rather than a list of definitions and formal properties. So let us follow J. Ehlers who recognized a $U(1)$ duality symmetry of Einstein's vacuum solutions in 4 dimensions with one non-null Killing vector. One can view the situation as a fibration of spacetime over a three dimensional set of orbits. One question is whether the base inherits a geometrical structure namely a metric from the original Minkowskian manifold. Locally the answer is yes, it is the beautiful work of Kaluza (who actually developed the idea in 5 dimensions), it was completed by O. Klein (after Quantum Mechanics), Jordan, Thiry ... In fact one may see that the Killing orbits being non-null the distribution of normal hyperplanes is transverse to the fibration, equivariant under the isometries by construction and thus defines a principal abelian connection (locally in all directions). Furthermore the length of the Killing vector defines a scalar field C on the base. The construction of the geometry on the base is given in [23]. The case of null Killing vectors is much more subtle see [24] and references therein.

Now as we started from 4 dimensional spacetime the two degrees of freedom of the graviton have become a scalar field plus one polarisation of a vector gauge field (the connection we just presented). What are the symmetries? From 4d diffeomorphism invariance we expect a remaining scaling invariance of the cyclic (ie now internal) direction and of course the Maxwell-Weyl gauge invariance of the connection. However it turns out that the connection 1-form can be dualised, in other words it can be exchanged for a dual scalar field B by a Legendre transform. Then the miracle in today's state of affairs is that the two scalar fields do form a couple of coordinates for the Poincaré upper-half-plane. In other words they parameterise the symmetric Riemannian non-compact space $SO(2)\backslash SL(2,\mathbb{R})$. Ehlers considered the $SO(2)$ subgroup of $SL(2,\mathbb{R})$ but the full group is now called Ehlers' group. The action is pointwise as in a sigma model and rigid ie. independent on the position. But that means that it is actually non-local in terms of the original scalar C and the original connection. Furthermore the other two generators are the expected scaling and a new shift of the second scalar B which is defined up to an arbitrary additive constant by the Baecklund type formula:

$$dA \propto *dB\ C^4.$$

See for instance [7] for complete formulas. It is also established there that the Ehlers rotations act at the linearised level as helicity rotations.

To summarise, the maximal dualisation leads to an unexpected symmetry that is nonlocal in terms of the original (non-dualised) fields. The undualisation of the new scalar field to the connection one-form hides the large symmetry but restores the gauge invariance of the connection field.

Our next example will be three dimensional maximal supergravity (type II but for our purpose the fermion fields may be set to zero). The maximally dualised theory is in a topological background gravity and its dynamics is that of a Riemannian symmetric space sigma model of the non-compact type. The 128 bosonic degrees of freedom span $SO(16)\backslash E_8$ or equivalently, but after gauge fixing the $SO(16)$ gauge invariance, they span a Borel (or upper triangular) subgroup. In a maximally noncompact real form the Cartan generators may be chosen noncompact and together with the positive roots they generate a solvable group that is univocally parameterised by the exponential map. The compact generators are the differences of positive root generators and their opposite and disappear in the gauge fixed description. Let us briefly recall the origin of the various scalar fields. The three form in eleven dimension reduces to 8x7x6/3!= 56 scalars plus 28 vectors (to be dualised to scalars) and some non propagating components. The metric leads to 11-3=8 vectors (also dualisable) as well as 36 scalars. It turns out that these fields fit into the graded decomposition of the Borel subalgebra along the simple positive root labeled 10 of the Dynkin diagram of E_8 in Figure 1.

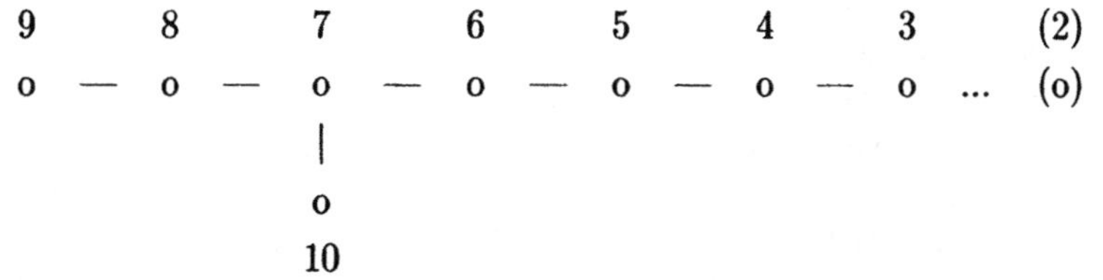

Figure 1: type IIA symmetry, Dynkin diagram of E_8 with affine extension. The label indicates the dimension at which each vertex appears.

Let us recall that in the decomposition of a positive root into a linear combination of simple positive roots with non-negative coefficients there is a $\mathbb{Z}^8$ grading at our disposal. We claim that the simple root corresponding to the vertex 10 must be selected to find the symmetry of the bare reductions from 11 dimensional supergravity doing no dualisation at all. Firstly one must select the grading along that root, and then erase its vertex. Namely there are 28 positive roots of level 0 along that simple root (plus 8 Cartan generators) all inside the Borel subalgebra of $gl(7,\mathbb{R})$, as well as 56 level 1 roots, 28 at level 2 and 8 at level 3. Clearly the level 2 and 3 root generators form an abelian ideal because the level runs only up to 3. This means in practical terms that the corresponding 36 scalars can be undualised to vectors. They actually originated as such from dimensional reduction, and $GL(7,\mathbb{R})$ is the almost obvious symmetry after strict toroidal dimensional reduction.

We just gave the simplest example of the phenomenon: the big Borel duality group of the maximally dualised theory is quotiented by the abelian ideal of the undualised scalar generators to become typically a semidirect product, here (after restoring the full linear group) $GL(7,\mathbb{R})\ltimes\mathbb{R}^{56}$. What has happened is that the 36 arbitrary constants in the definition of the new scalar fields obtained by dualisations are not available before dualisations and their shifts by corresponding group elements drop out because the realisation of the Borel group becomes non faithful; furthermore non commuting generators of E_8, like those at level 1, become commuting after undualisation because their commutators vanish by the previous mechanism. This phenomenon has been uncovered in [8] but abelian ideals have many other uses see [15]. If one wants to restore gauge invariance under the compact subgroup (second silver rule) one may alternatively describe the division by the ideal of the Borel group as a double coset construction for the full U-duality group [8].

Such undualisation along the root 10 of all the E_{11-d} Borel subgroups leads to symmetries of a similar type: semidirect products of the corresponding linear subgroup by some abelian group which are the symmetries obtained from bare dimensional reduction from 11 dimensions. We did not give the details here but the $SL(11-d)$ is visible as a subdiagram of the E_8 Dynkin diagram that grows together with E_{11-d} from the vertex labeled 9 towards the vertex labeled 3 ... For all dimensions one uses abelian ideals, those of the maximal possible dimension occur in various places as one can check following [6] who found them by inspection. A more conceptual discussion will be attempted in a further paper. But let us illustrate their usefulness on more examples.

Let us now turn briefly to three other natural choices of undualisations. We shall consider undualisation of all NS-NS fields (the even forms), it corresponds to exchanging the role of the simple roots 10 and 9 in Figure 1, namely to undualising generators whose root has the highest coefficient along the simple root 9. We could also consider type IIB supergravity in 10 dimensions, reduce to 9 dimensions and below without dualising the 5 form field-strength. This series is obtained by using root 8 (and simultaneously root 10) in a similar way as root 10 or 9 in the previous examples. The manifest and expected symmetry is now only $GL(9-d, \mathbb{R})$. Finally the truncation of E_{11-d} to E_{10-d} can also be obtained by deleting the vertex labelled d at the end of the $SL(11-d)$ line inside (E_{11-d}) that shortens when d increases. The origin of this property is that any vector in dimension $d+1$ gives rise to a new scalar in dimension d. Malcev has shown that the maximal dimension of a maximal abelian subalgebra of $E_7(7)$ is equal to 27, but 27=63-36 which corresponds precisely to the contraction of $E_7(7)$ towards E_6. The rest of the E family gives similar results.

The last series we would like to discuss is the type I family between 10 and 3 dimensions. The diagrams are subdiagrams of Figure 2:

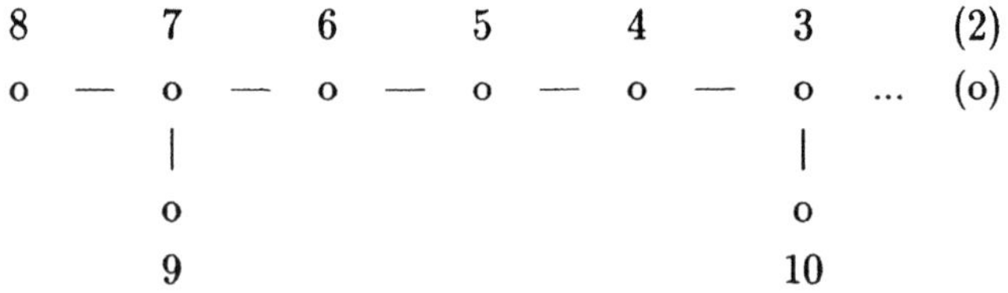

Figure 2: type I symmetry, Dynkin diagram of D_8 with affine extension.

Again the labels denote the dimension at which the Cartan generators appear, sometimes with a full $SL(2, \mathbb{R})$ but not immediately. We may conclude this section with the remark that there is also a couple of undualisations that are easy to describe in the type I case, namely those of the 2-form. They are done by the same type of contraction as for type II by using the simple root labelled 10 in Figure 2.

4. Higher order potentials

In the previous section we have seen how specific locations on the Dynkin diagram of internal dualities correspond to specific sets of dualisations and in the process of studying these we have seen that root labelled d is associated to vectors. We shall now discover, in type II supergravities, that root 10 is associated to the 3-form, root 8 to the 4-form and root 9 to the 2-form potentials. Let us consider for instance dimension 7, there are 4 vectors from the metric plus 6 from the 3-form, together they build the representation 10 of $SL(5,\mathbb{R})$ which corresponds to the fundamental dominant weight on vertex $d = 7$. Similarly in dimension 6 the 3-form can (and should) be dualised to a 16th vector to implement the symmetry $SO(5,5)$ and again this is the fundamental half spinorial representation associated to vertex $d = 6$. If the 3-form is associated to vertex 10 we might guess that the symmetry of the Dynkin diagram in 6 dimensions is responsible for the possibility of dualising it away.

In dimension 2 one knows that the U-symmetry of type I supergravity is the affine $D_8^{(1)}$. Note that there is a Weyl reflection that exchanges the highest root (it is the opposite of the root labelled (2) corresponding to the affine extension) with the simple positive root labelled 3. Roots 3 and (2) are connected by a simple line and so belong to a $SL(3,\mathbb{R})$ subgroup. This is related to the fact that 1-forms can be dualised to scalars in 3 dimensions. In fact for any dimension, and also in type I theories, 1-form potentials in the maximally dualised form belong to the representation of highest weight equal to the fundamental dominant weight along the last root of the $SL(10-d)$ line (root labelled $d, d \leq 9$ for type I) at least *before they can be dualised away.*

Scalars correspond to the adjoint representation of highest weight equal to the highest root. Their dualisability, which is a spacetime property, somehow can be seen in the internal symmetry Dynkin diagram as the contiguity of the highest weights. Any weight is Weyl conjugate to a dominant weight but not necessarily to a fundamental one. Conjugacy seems to be a necessary condition for the dualisability. For higher forms, the critical dimensions for which the degree of forms can be lowered (3 for vectors to scalars, 5 for 2-forms to 1-forms etc...) are most visible on the Dynkin diagram when dualisability corresponds to the existence of an outer automorphism.

Type II theories provide more examples of this. I have checked them one by one! The last vertex of the Dynkin diagram to appear (label d) is always associated to the highest weight of the vector representation (it is the fundamental weight for that vertex), the vertex labeled 9 in Figure 1 gives the highest weight of the 2-forms and similarly the vertex labeled 10 there corresponds to the 3-form. Note that E_{11-d} has outer automorphisms visible as symmetries of its Dynkin diagram whenever k-forms and $(k' = d-2-k)$-forms are in duality, this symmetry exchanges the locations of the corresponding (fundamental) weights in all 9 cases of duality for degrees of potential-forms between 0 and 4!

Let us choose dimension 6 for instance, the group G is $SO(5,5)$ and the 2-forms can be selfdual under the symmetry of the diagram that exchanges the two half-spinor ends, ie the vertices 10 and 6 and leaves invariant the vertex 9. These vertices correspond to the fundamental weights of the 3-form, dual 1-form and self-dual 2-forms respectively. The scalars in the adjoint correspond to the lowest root whose vertex is adjacent to the vertex 8 related to the potentially dual 4-form as we saw at the end of the previous section by a Weyl reflection. Let us note in passing the strange rule that when the affine diagram is cyclic ($SL(p,\mathbb{R})^{(1)}\ p \geq 3$) the sum of the degrees of the potential forms whose vertex is attached to the (affine) extension vertex is its dual degree; for example in dimension 8 scalars are dual to 6 forms and 6=4+2, and similarly in dimension 7.

Let us now remark that the outer automorphisms of the internal Lie algebra of dualities g should be defined as simultaneously dualising the world indices. This is quite easy to implement by similar symmetries of the Dynkin diagram of $SL(d)$. It is well known that the fundamental representations of $SL(d)$ are from one end to the other the antisymmetric powers of the vector representation in increasing order: the vector, then the antisymmetric twice contravariant tensor etc... The symmetry under the exchange of the two ends of $SL(d)$ is nothing but duality provided we consider field strengths. By this we mean that the tensor character of the field strength should define the fundamental weight and consequently the vertex to be dualised. Under dimensional reduction and its inverse group disintegration [7] the linear group is shared in a d-dependent way between internal and external spaces. Let us denote by a star the scaling factor $\mathbb{R}$ of symmetry, the general picture for type IIA is given in Figure 3.

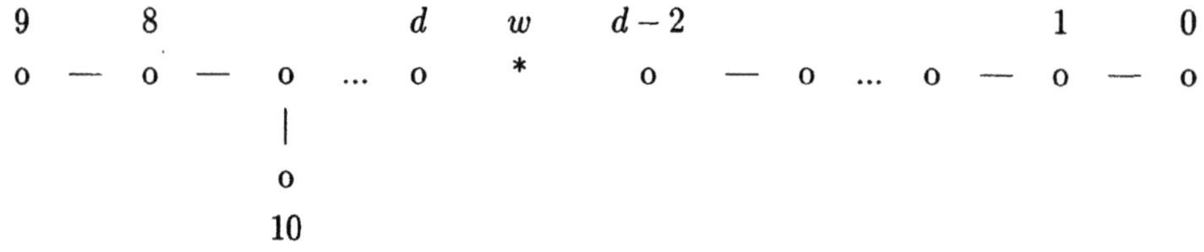

Figure 3: type II F''_{11} subgroups of F-symmetry
The vertex labeled w represents scaling symmetry

The diagram obtained by adding two bonds to Figure 3 to connect a regular $SL(2,\mathbb{R})$ replacing the $\mathbb{R}$ factor to the rest of the diagram contains all the above diagrams in any dimension. Its rank is eleven let us call it F''_{11}, it has an analogue F'_{11} for type I. Note that $SL(12,\mathbb{R})$ cannot be found in them by naive inspection. However it could still be included in there, like $SL(9,\mathbb{R})$ inside E_8. But $GL(12,\mathbb{R})$ is unlikely to be included and this could be related to the same property of F-theory. Note that supersymmetry implies that the twelfth dimension would be timelike and hence the subgroup of $SL(12,\mathbb{R})$ would be $SO(10,2)$ [25] so it may be not so surprising that the $SL(12)$ is hidden.

F''_{11} contains of course E_{10}. HD_{10} lies analogously in F'_{11} and M-ology suggests that types I and II should be unified to make the full F-group. In fact it is encouraging to discover that orbifolding by an involution (a Cartan involution) leads in all dimensions from the type II to the type I U-duality. The invariant set of the involution is given by erasing the vertex 9 of Figure 1 and adding vertex 10 of Figure 2. This group theoretical construction should be compared to the geometrical orbifolding of [26]. It is tantalising now to propose that the one-loop diagram obtained either by attaching through a new single vertex (as the simplest choice) vertices 8 and 10 of Figure 2 or vertices 9 and 3 of Figure 1 or 3 should be studied more carefully. This diagram has rank 12 and a $\mathbb{Z}_2$ symmetry.

In summary dualisability seems to require the existence of (outer) involutions of the F-duality group of a very special type. The F-group should be universal but then split into a product of spacetime symmetry by an internal symmetry factor in a type- and d-dependent way.

Let us now review the possible duality symmetries of a Lagrangian field theory in curved space with scalar fields and a N-plet of middle rank ($f = d/2$) forms. In [10] it was established that in Minkowskian signature the set of equations of motion for the N vector fields and sufficiently many scalar fields are invariant at most under the symplectic group $SP(2N, \mathbb{R})$. Then the energy-momentum tensor is automatically invariant. In fact if there is no scalar compensator like in the free Maxwell theory the available linear invariance is under $GL(N, \mathbb{C})$. So there remains some work to be done to clarify the situation with a few scalars. For the free Maxwell case we should use the invariance of the energy-momentum tensor as an input instead of deriving it as in [10] but even then the discussion may be more involved. In [8] the analysis was extended to all dimensions higher than two and also to Euclidean signature. It is assumed that the half dimension forms occur at most quadratically in the action and again that there are enough scalar fields to be able to reduce the symmetry as in [10].

The Minkowskian analysis seems to have been done a long time ago as we mentioned but the Euclidean case is interesting because it illustrates the fact that a change of (non-degenerate) signature does not change the duality group; it changes, in general, the coset however. The non-compactness of the subgroup H that replaces K implies some chaos. What we have shown is that the case of dimension $d = 4k$ resembles that of 4 dimensions but that of $4k + 2$ dimensions however admits maximal duality group of the type $O(N, N)$ with compact subgroup (Lorentzian case) $O(N)^2$ resp. subgroup $H = O(N, \mathbb{C})$. For completeness let us mention that in $4k$ dimensions $K = U(N)$ is similarly replaced in the Euclidean signature by $H = GL(N, \mathbb{R})$.

This analysis was motivated in part by my desire to clarify the difference between two-dimensional quantisation of single charges by requiring locality (symmetric quadratic form) and the four-dimensional Dirac-Schwinger-Zwanziger antisymmetric quantisation condition. As we have seen this alternation between symmetry and antisymmetry of the quadratic invariant ω for every other dimension can be checked by inspection and is rather robust as it does not depend on the signature of spacetime. The supersymmetric origin of the supergravity bosonic actions is not relevant here but let us recall that dualities are often coupled to chiral rotations in such models. Now it has been noticed repeatedly that the first nontrivial example of what we would like to call a quantisation dimension namely 6 is closely related to quaternions and hyperKaehler structures [27]. But k was arbitrary in the previous discussion so we should not restrict ourselves to the supersymmetry domain.

One conclusion at this stage is that, as one had maybe anticipated, the symmetry must be defined by a very basic counting, namely parity of the half-dimension $f := d/2$; this number precisely determines the symmetry or antisymmetry of the intersection form of the corresponding cohomology. An important shift of one occurs though as $f = 1$ in the symmetric case here, we shall elaborate on it later, it is related to the time extension.

Let us finish this section by giving some technical advice for the handling of non-manifestly duality invariant actions. It was inspired by [28] and consists in doubling as usual the set of forms of degree f as we explained above and in keeping along the TS constraint which is also invariant. This leads to nicer couplings to scalar fields [8] for instance and to a derivation of the Noether current of duality which had to be guessed in [10].

5. Complementarity, dyons, TS and F-duality and conclusion

Let us now elaborate on the relationship between charge quantisation conditions for dyons in 4 dimensions and for chiral fermions in 2d. I noticed in March 1996 that the famous Skyrme minus sign, namely the fermionic character of his vertex operator had to be understood in units where $h = 2\pi$ as

$$e^{2i\pi(eg'+e'g)} = -1$$

This should have been (and maybe has been) analysed long ago. The analogy between chirality of spinors and helicity of vectors is a Lorentz fact. As we mentioned above supersymmetric dualities act simultaneously on the fermionic fields by chiral transformations. As it appears from [13] chiral fermions are obtained by a vertex construction if their charge obeys the quantisation condition: $e = g = \sqrt{s}/2$, where s is an integer (a squarefree integer for the algebra of observables to be maximal). The model is the chiral $U(1)$ current algebra. The idea is that a nonlocal expression can only satisfy causality, or fermionic causality, if it obeys a quantisation condition à la Dirac. In the monopole case one wants the Dirac string to be invisible, in the fermionisation problem one wants the Mandelstam-Skyrme string to be almost invisible, giving only a minus sign but actually an unavoidable one. In the Sine-Gordon case the quantisation is the celebrated $\beta^2 = 4\pi$ relation. It could be rewritten, compare [12], as the relation $eg' = \frac{1}{4}$ but the symmetry of the quadratic form was overlooked there and generalisations attempted in 4 dimensions. We now know much more about affine Kac-Moody algebras (then 2 years old in Mathematics and to be born in Physics). The finiteness of the Schwinger term in 1 (null) dimension or in 1+1 dimensions is the possibility of central extension of Loop groups. A lot of effort has been devoted to find higher dimensional generalisations but divergences have spoiled the game. Yet non-relativistic versions of 2+1 or 3+1 dimensional fermionisation have been proposed since 1975 in the presence of gauge fields. The role of the Dirac quantisation condition appeared immediately through the formula for angular momentum, which gives straightforwardly the minus sign of the Schwinger-Zwanziger condition. It is very natural to compare this to the Bohr-Rosenfeld uncertainty relation for electromagnetic fields. This is in 4 dimensions but free field theory and hence (bad)-divergence free.

Interestingly enough the quantisation condition for extended objects (p-branes) with $p \geq 2$ were given in 1985-86 without adressing the question of the sign. It was thus very exciting to check that in 6d supergravities ($N_4 = 8$ and $N_4 = 6$ [7]) the duality symmetry was not symplectic but pseudo-Riemannian. Since then a few papers have appeared to fill the gap mentioned above by direct arguments and to verify the sign difference between $4k$ and $4k+2$ dimensions for the generalised Dirac-Schwinger-Zwanziger condition [29]. One may remark that these papers exclude the fermionic case. Now dimensional reduction techniques are available to exploit a desired fermionisation that could be done in six dimensions without any internal symmetry, so as to do it in lower dimensions with an automatic twist. The simplest candidate is a two-form (self-dual if one prefers) in 6d and in fact such a theory has become quite fashionable in the meantime starting with [30]. We can check that the fibration of the 6d Minkowskian space on a torus over either Euclidean or Minkowskian base gives in both cases the $SL(2,\mathbb{Z})$ modular group as duality group on the base for Maxwell theory. It was clear already from the disintegration magic of [7] that the rather fundamental nature of the so-called ADE structure seemed to play a role in supergravity theory. Since then it appeared in CFT, in the applications of singularity theory (where it was already a dominant character) to string theory, etc... Its home, intersection theory seems to be the most important tool for fermionisation and charge quantisation problems [31].

The second project I have been interested in that relates to dualities is the full analysis of the 1975 observation that classical field theory is obtained by letting Planck's constant tend to zero but hiding some of them in rebaptised parameters like E=e/h and M=m/h. The upper case parameters are relevant to the wave world in which one does not count quanta. Note that e/m=E/M, so the anniversary of the electron as a particle should have been next year instead of last year. Now this relates to dualities because in this (electric) classical field theory limit one can study magnetic particles and their solitonic "quantised" charge. Clearly the dual theory, ignoring its strong coupling for the moment, has magnetic waves and electric particles at the new classical level. One message is that the words "classical limit" are so ambiguous that they confuse everybody. Let us mention in passing that non-relativistic limits are equally ambiguous as anybody at ease with Unit Systems can testify. How do classical people measure Planck's constant? They in fact combine two complementary limits to measure for instance two vertices of the [e, E or g, α] triangle. The validity of a particular classical limit is to be decided case by case and one may be better than another. Is it true that the collection of all classical limits contains all the information about the quantum theory? This is postulated in the usual interpretation of Quantum Mechanics, but can we make this precise and restrict the set of limits that is necessary? For instance non-relativistically invariant classical limits are sometimes necessary for instance to obtain particles rather than quanta in order to avoid the Klein paradox. Similarly non-relativistically invariant limits of supergravity theories might be useful for some perturbative computations.

To conclude with less ambitious goals let me sketch some results of [16] in order to motivate the reader to dig deeper into the present lines. It turns out that indeed the TS idea can be implemented for all differential forms of the type II supergravity theories in all dimensions between 3 and 11. The structure is particularly simple of course in 11 dimensions but the equations can be rewritten in a dimension independent way. It applies to sigma models both gauge fixed and not gauge fixed. The restriction to toroidal compactifications remains and should be relaxed in the future. Our present reduction of type II to type I suggests that this might be doable. Bosonic strings and heterotic strings should be reconciled with the present group theoretical approach that smells of integrable systems, in a loose sense as chaos generally follows non-compact groups.

Acknowledgements I would like to thank P. Windey for inviting me to present some of these results at the NATO Advanced Study Institute on "Strings, Branes, and Dualities". Parts 3 and 4 benefited from many discussions with E. Cremmer, part 5 from encouragements from and discussions with D. Buchholz, K. Fredenhagen and J. Roberts in 1995 and 1996 resp. C. Callan in 1975 and H. B. Nielsen in 1993 and 1994. I am also grateful for useful comments to C. Bachas, Y. Benoist, D. Bernard, M. Duflo, M. Henneaux, C. Hull, H. Nicolai, D. Olive, M. Trigiante. This work was supported in part by EEC under TMR contracts ERBFMRX-CT96-0045 and -0012 and by the A. von Humboldt foundation.

References

1. A. H. Chamseddine, Nucl. Phys. **B185** (1981) 403.
2. B. Julia (1982) in *Lectures in Applied Mathematics*, AMS-SIAM, **vol. 21** (1985) 355.
3. J.A. Harvey and G. Moore, Nucl. Phys. **B463** (1996) 315.
4. C. Hull and B. Julia, hep-th/9803239, *Duality and Moduli Spaces for Time-Like Reductions.*
5. G. Gibbons and S. Hawking, Com. Math. Phys. **66** (1979) 291. E. Cremmer et al. hep-th/9803259.
6. A.I. Malcev, *Commutative subalgebras of semi-simple Lie algebras*, Izv. Akad. Nauk SSR, Ser. Mat. **9** (1945) 291 (Russian); translation No. 40, series 1, American Mathematical Society (English).
7. B. Julia in *Supergravity and Superspace*, S.W. Hawking and M. Rocek, C.U.P. Cambridge, 1981.
8. E. Cremmer, B. Julia, H. Lu and C. Pope, *Dualisation of dualities I*, Nucl. Phys. **B522** (1998) to appear, hep-th/9710119.
9. Y. Tanii, Phys. Lett. **B145** (1984) 197 and unpublished.
10. M.K. Gaillard and B. Zumino, Nucl. Phys. **B193** (1981) 221.
11. S. Cecotti, S. Ferrara and L. Girardello, Nucl. Phys. **B308** (1988) 436 (quoted in M.K. Gaillard and B. Zumino, hep-th/9712103).
12. R.F. Streater and I.F. Wilde, Nucl. Phys. **B24** (1970) 561.
13. D. Buchholz, G. Mack and A. Todorov, Nucl. Phys. B (Proc. Suppl.) **5B** (1988) 20.
14. R. D'Auria, C. Kounnas and S. Ferrara, hep-th/9711048.
15. M. Trigiante, hep-th/9801144.
16. E. Cremmer, B. Julia, H. Lu and C. Pope, *Dualisation of dualities II*, in preparation.
17. I. Bandos, N. Berkovits and D. Sorokin, hep-th/9711055.
18. E. Cremmer and B. Julia, Nucl. Phys **B156** (1979) 141.
19. P. Breitenlohner and D. Maison in *"Solutions of Einstein's equations: Techniques and Results"*, Lect. Notes in Physics **205**, ed. C. Hoenselaers and W. Dietz, Springer 1984.
20. M.C. Davies et al., Phys. Lett. **119B** (1982) 187.

21. J. Tits, *Tabellen zu den einfachen Lie Gruppen*, Lect. Notes in Math. **40** (1967) Springer.
22. D. Bernard and B. Julia, hep-th/9712254.
23. R. Geroch, J. Math. Phys. **12** (1971) 918.
24. B. Julia and H. Nicolai, **B439** (1995) 291.
25. B. Julia, in *Proc. M. Grossmann Second Meeting on General Relativity* (1979),p. 79, R. Ruffini ed. North Holland (1982) and in *Proc. Conf. on Math. Problems in Th. Physics*, Lausanne 1979, ed. K. Osterwalder, Lecture Notes in Physics **116** (1980) 368.
26. P. Horava and E. Witten, Nucl. Phys. **B460** (1996) 506.
27. T. Kugo and P. Townsend, Nucl. Phys. **B221** (1983) 357.
28. E. Bergshoeff, H.J. Boonstra and T. Ortin, *S duality and dyonic p-brane solutions in type II string theory*, Phys. Rev. **D53** (1996) 7206: hep-th/9508091.
29. S. Deser et al., hep-th/9702184.
M.S. Bremer et al., hep-th/9710244.
S. Deser et al.,hep-th/9712189.
S. Deser, M. Henneaux and A. Schwimmer, hep-th/9803106.
see also M. Perry and J. Schwarz, hep-th/9612191.
30. E. Verlinde, hep-th/9506011, see also E. Witten, hep-th/9505186.
31. B. Julia, in preparation.

M-THEORY FROM ITS SUPERALGEBRA

P.K. TOWNSEND
DAMTP, UNIVERSITY OF CAMBRIDGE
SILVER ST., CAMBRIDGE CB3 9EW, U.K.

Lecture 1: M-branes and supersymmetry

About twenty years ago, it was pointed out that D=11 is the maximum dimension in which one can expect to find an interacting supersymmetric field theory [1]. Its existence, and uniqueness subject to standard assumptions, was established with the construction of D=11 supergravity [2]. The linearized theory can be deduced from an analysis of the standard D=11 super-Poincaré algebra and it will prove worthwhile to first present a variant of this analysis. Since the commutators involving Lorentz generators are determined by the Lorentz indices it is convenient to concentrate on the subalgebra of supertranslations spanned by the 11-momentum P_M and the 32-component Majorana spinor charge Q_α. Translation invariance implies that P and Q commute, leaving the anticommutator

$$\{Q_\alpha, Q_\beta\} = (C\Gamma^M)_{\alpha\beta} P_M \tag{1.1}$$

as the only non-trivial relation, where Γ^M are the Dirac matrices and C is the (real, antisymmetric) charge conjugation matrix. There are actually two inequivalent representations of the Dirac matrices. They differ according to whether the product of all 11 of them is 1 or -1. The choice is arbitrary and we shall suppose that

$$\Gamma^0\Gamma^1\cdots\Gamma^9\Gamma^\natural = 1\,. \tag{1.2}$$

The symbol $\natural$ is used here to denote the number 10. It is convenient to have a single symbol for this number because we shall be using a convention in which Γ^{10} will mean the product $\Gamma^1\Gamma^0$. More generally, the antisymmetrized product of n Dirac matrices will be written as $\Gamma^{M_1\cdots M_n}$. Thus (1.2) is equivalent to $\Gamma^\natural = \Gamma^{0\cdots 9}$ (also known as Γ_{11} in the context of D=10 IIA supergravity).

Let us suppose that there exists some quantum theory that realizes the algebra (1.1) as an asymptotic symmetry, e.g. of the S-matrix. The asymptotic metric is assumed to be the D=11 Minkowski metric and the vacuum is assumed to be annihilated by all supersymmetry charges. Consider a state preserving some non-zero fraction ν of the supersymmetry of the vacuum. This state will be annihilated by some combination of supersymmetry charges and the expectation value of $\{Q,Q\}$

L. Baulieu et al. (eds.), Strings, Branes and Dualities, 141–177.

will be a real symmetric positive semi-definite matrix with $32 \times \nu$ zero eigenvalues. In particular, its determinant will vanish. From (1.1) we see that this will happen when

$$0 = \det \Gamma \cdot P = (P^2)^{16} \,. \tag{1.3}$$

Thus, given the algebra (1.1), we deduce that the only states preserving a non-zero fraction of the supersymmetry, apart from the vacuum, are those for which the 11-momentum is null. To determine the fraction ν we note that if P is null we can choose a frame in which (the metric convention is 'mostly plus' and P^0 is the energy)

$$P_M = \frac{1}{2}(\pm 1; 1, 0, \ldots, 0) \,. \tag{1.4}$$

We may also choose the Majorana (real) representation of the Dirac matrices for which $C = \Gamma^0$. The algebra (1.1) is then

$$\{Q_\alpha, Q_\beta\} = \frac{1}{2}(1 \mp \Gamma_{01})_{\alpha\beta} \tag{1.5}$$

where $\Gamma_{01} = \Gamma_0\Gamma_1$, and similarly for other products of Dirac matrices. Clearly, eigenspinors of $\{Q, Q\}$ with zero eigenvalue satisfy

$$\Gamma_{01}\epsilon = \pm\epsilon \,. \tag{1.6}$$

Since Γ_{01} squares to the identity its eigenvalues are ± 1, and since it is also traceless precisely half are $+1$ and half -1. Thus, the space of solutions to (1.6) is 16-dimensional and we therefore deduce that $\nu = 1/2$. *There are no other possibilities allowed by the supersymmetry algebra (1.1).*

It is natural to associate the $\nu = 1/2$ states for which $P^2 = 0$ with massless particles. Classically, and in the absence of interactions, a massless particle in a D=11 Minkowski vacuum is described by the action

$$S[X; e] = \int d\tau \, e^{-1} g \tag{1.7}$$

where the 'einbein' $e(\tau)$ is an independent worldline scalar density and g is the metric on the worldline induced from spacetime, i.e. $g = \dot{X}^2$. In the D=11 Minkowski vacuum, this action is invariant under translations of X. The corresponding Noether charge is the null 11-momentum P. To incorporate supersymmetry we must replace the translation-invariant differential dX by the supertranslation-invariant differential $\omega = dX - i\bar{\theta}\Gamma d\theta$, so that $g = \omega^2$. The worldline 'fields' of the resulting 'superparticle' action are therefore the bosons $X^M(\tau)$ and the fermions $\theta_\alpha(\tau)$. The fermion 'fields' θ_α transform inhomogeneously under supersymmetry ($\theta \to \theta + \epsilon$) and can be interpreted as Nambu-Goldstone (NG) variables for broken supersymmetry. This would appear to imply that all supersymmetry is broken but the massless superparticle action has a fermionic gauge invariance, known as 'κ-symmetry', that allows half the fermions to be 'gauged away'. There are therefore only 16 NG variables, corresponding to $\nu = 1/2$. On the one hand this symmetry appears miraculous since we made no attempt to build it into the construction. On the other hand there would have been a contradiction with our previous analysis had it not been present. In any case, only half of the 32 components of $\theta(\tau)$ are physical. These 16 physical fermion fields may be split into 8 creation operators and 8 annihilation operators. Quantization of the massless superparticle then leads to linearized field equations for a supermultiplet of D=11 supersymmetry with 2^8 components, of which 128 are bosons and 128 fermions. Detailed investigation reveals that the Lorentz representations are those of the graviton supermultiplet of D=11 supergravity, i.e. the graviton, gravitino, and 3-form gauge potential A. In other words, quantization of the massless D=11 superparticle yields the linearized field equations of D=11 supergravity.

By including interactions we arrive at the full non-linear field equations of D=11 supergravity. We may now ask whether these equations admit solutions preserving some fraction of the supersymmetry. The procedure for doing this is to seek D=11 spacetimes admitting Killing spinor fields, i.e. spinors statisfying the first order differential equation

$$\delta_\epsilon \psi \equiv \mathcal{D}\epsilon = 0\,, \tag{1.8}$$

where ψ is the one-form gravitino field and $\delta_\epsilon \psi$ its supersymmetry variation with spinor parameter $\epsilon(x)$. This variation defines a covariant exterior derivative $\mathcal{D}$. In principle, this operator includes terms involving contractions of Dirac matrices with the 4-form field strength $F = dA$. These terms will eventually prove to be of crucial importance but they play no role in the present discussion. Setting $F = 0$, the general 11-metric admitting Killing spinors may be shown to take the 'M-wave' form [3]

$$ds^2 = dudv + K(\mathbf{x}, u)du^2 + d\mathbf{x}\cdot d\mathbf{x}\,, \tag{1.9}$$

where $\mathbf{x}$ are Cartesian coordinates for $\mathbb{E}^9$ and K is an arbitrary function of u but is harmonic on $\mathbb{E}^9$. This metric is asymptotically flat provided that $K \to 0$ as $|\mathbf{x}| \to \infty$. In this case there is an asymptotically flat region in which P is defined by an ADM-type formula, and one can verify that $P^2 = 0$. If the above metric is used to compute $\mathcal{D}$ one finds that the condition (1.12) reduces to $\Gamma_v \epsilon = 0$, which (setting $v = t + x^1$) is equivalent to (1.6). Actually, the spinor ϵ here is a field but it is assumed to be constant at infinity and this constant spinor may be identified with a constant zero-eigenvalue eigenspinor of $\{Q, Q\}$.

Given a solution of D=11 supergravity one may ask for the effective action determining the dynamics of small fluctuations about it. This dynamics is governed by the Nambu-Goldstone variables associated with the symmetries broken by the solution and these variables are fields on the orbits of Killing vector fields of the background. The M-wave breaks translational invariance in the $(u, \mathbf{x})$ directions (generically) and $1/2$ of the supersymmetry. It has $\partial/\partial v$ as its only Killing vector field. We conclude that the effective action for small fluctuations is a functional of the 'fields' $\big(u(v), \mathbf{x}(v); \lambda(v)\big)$, where λ represents the 16 NG fermions of $1/2$-broken supersymmetry. Symmetry considerations now imply that the action must be the light-cone gauge-fixed version of the massless superparticle. We have now arrived back at our starting point in the analysis of $\nu = 1/2$ realizations of the standard D=11 superymmetry algebra.

About ten years ago it was realized that it is possible to couple D=11 supergravity to a closed supermembrane [4]. The membrane action has the form

$$S = -T_2 \int vol(g) + Q_2 \int \mathcal{A} \tag{1.10}$$

where $vol(g)$ is the worldvolume density in the induced metric g and $\mathcal{A}$ is the pull-back to the worldvolume of the 3-form potential A. The constant T_2 is the surface tension, while Q_2 is a membrane charge. The supermembrane action is formally the same, but $vol(g)$ and $\mathcal{A}$ are induced from superspace. Actually, the supermembrane action of [4] is the special case of (1.10) in which $Q_2 = T_2$. It turns out that precisely in this case the action is κ-symmetric, i.e. it has a fermionic gauge invariance that allows half of the worldvolume fermion fields to be 'gauged away'. More precisely, the supermembrane action is κ-symmetric if the background satisfies the field equations of D=11 supergravity. This allows, in particular, a D=11 vacuum background, in which the worldvolume fermions can be interpreted as the NG variables associated with the breaking by the membrane of 1/2 the supersymmetry of the vacuum. This presents us with a paradox because we have just shown that the D=11 supersymmetry algebra (1.1) allows only one object to have this property, a massless particle. The resolution of this paradox is that the supersymmetry algebra is modified in the presence of a membrane. In the D=11 Minkowski vacuum the supermembrane action is supertranslation invariant and there therefore exist conserved Noether charges P_M and Q_α. However, these charges do not obey the algebra (1.1). The reason for this is that, unlike the massless superparticle action, the supermembrane action is not *manifestly* supersymmetric: the coupling to the 3-form potential A does not vanish in the D=11 Minkowski vacuum but instead reduces to a Wess-Zumino term for the super-Poincaré group. This term (which is defined up to the addition of an exact 3-form) can be chosen to be translation invariant but it then changes under a supersymmetry transformation by the addition of an exact 3-form and this leads to a modification of the algebra of the supertranslation Noether charges [5]. The modified algebra is

$$\{Q_\alpha, Q_\beta\} = (C\Gamma^M)_{\alpha\beta} P_M + \frac{1}{2}(C\Gamma_{MN})_{\alpha\beta} Z^{MN} , \tag{1.11}$$

where

$$Z^{MN} = Q_2 \int dX^M \wedge dX^N , \tag{1.12}$$

and the integral is taken over the 2-cycle occupied by the membrane in spacetime. The 2-form charge Z is not central in the super-Poincaré algebra because it does not commute with Lorentz transformations, but this is to be expected of a charge carried by an extended object[1]. The charge Q_2 is thus analogous to the string winding number in string theory; it vanishes unless the two-cycle is non-contractible.

[1]The fact that p-form charges are carried by p-branes was the main result of [5] but the charges themselves were considered previously on mathematical grounds [6], in the context of super Yang-Mills theory [7, 8], and in some other 'p-brane-inspired' generalizations of the supersymmetry algebra [9, 10]. A related modification of the worldvolume supersymmetry algebra of gauge-fixed super p-branes was found in [11].

The case of an infinite planar membrane can be dealt with by considering it as a limit of one wrapped on a large torus. In this limit P^0 and some components of Z are infinite but the tension T_2 and the charge Q_2 remain finite.

Essentially, the supermembrane is an exotic form of matter. As for any other form of matter, a membrane will produce long-range gravitational and other fields, allowing the tension T_2 and charge Q_2 to be detected as surface integrals at infinity. In the case of an infinite planar membrane, 'infinity' should be interpreted to mean 'transverse spatial infinity', i.e. a large distance limit in non-parallel directions. Transverse spatial infinity is topologically $\mathbb{R}^2 \times S^7$, but the asymptotic translational invariance in directions parallel to the membrane allows us to reduce integrals for total charges to integrals over S^7 for charge densities. The energy density is the surface tension T_2, which is given by a modification of the usual ADM formula [12]. The membrane charge density is given by

$$Q_2 = \Omega_7^{-1} \int_{S^7} \star F \tag{1.13}$$

where $F = dA$ is the 4-form field strength and Ω_7 the volume of the unit 7-sphere. The integral should be evaluated at infinity because, in contrast to electrodynamics, the field equation for A is not $d \star F = 0$, but rather

$$d[\star F + F \wedge F] = 0\,, \tag{1.14}$$

where the second term is due to the Chern-Simons (CS) term in the action. Thus, $\star F$ is not necessarily a closed 7-form. We shall consider the implications of this below, but $d \star F$ will vanish asymptotically in the circumstances described above.

From (1.12) we see that a membrane in the 12 plane is associated with non-zero Z_{12}. Let us again choose the Majorana representation of the Dirac matrices in which $C = \Gamma^0$. Then, for a static membrane, the algebra (1.1) becomes

$$\{Q, Q\} = P^0 + \Gamma^{012} Z_{12}\,. \tag{1.15}$$

Now, Q is real in the Majorana representation, so the left hand side is manifestly positive. Since the sign of Z_{12} can be flipped by replacing the membrane by an anti-membrane we must have $P^0 \geq 0$. If $P^0 = 0$ we have the vacuum. Otherwise $P^0 > 0$ and we derive the Witten-Olive-type bound $P^0 \geq |Z_{12}|$, which is equivalent to

$$T_2 \geq |Q_2|\,. \tag{1.16}$$

A stable membrane is expected to saturate this bound, so the case in which the bound is saturated, i.e. $T_2 = |Q_2|$, is of particular importance. In this case the anticommutator (1.15) becomes

$$\{Q, Q\} = P^0[1 \mp \Gamma^{012}]\,. \tag{1.17}$$

Spinors ϵ satisfying

$$\Gamma^{012}\epsilon = \pm\epsilon \tag{1.18}$$

are eigenspinors of $\{Q, Q\}$ with zero eigenvalue. Since Γ^{012} squares to the identity, and is traceless, the dimension of the zero-eigenvalue eigenspace of $\{Q, Q\}$ is 16.

In other words, a membrane saturating the bound (1.9) preserves 1/2 the supersymmetry of the vacuum [5]. This fact is directly related to the κ-symmetry of the supermembrane action.

The bound (1.16) can also be derived [13], subject to standard assumptions, via a modification of the Gibbons-Hull bound on the mass of charged black holes in General Relativity[2]. In this approach, the solutions saturating the bound are, as in the M-wave case, those admitting Killing spinor fields satisfying (1.8), but now one must allow for a 4-form field strength F consistent with the asymptotic form required by a non-zero membrane charge Q_2. In this way one obtains the following 1/2 supersymmetric membrane solution of D=11 supergravity [14]:

$$\begin{aligned} ds^2 &= H^{-2/3} ds^2(\mathbb{E}^{2,1}) + H^{1/3} ds^2(\mathbb{E}^8) \\ F &= vol(\mathbb{E}^{2,1}) \wedge dH^{-1} \end{aligned} \tag{1.19}$$

where $H(\mathbf{x})$ is a harmonic function on $\mathbb{E}^8$. The choice

$$H = 1 + \sum_{s=1}^{n} \frac{|q_s|}{|\mathbf{x} - \mathbf{x}_s|^6} \tag{1.20}$$

leads to a metric that is 'transverse-asymptotically' flat and which can be interpreted as n parallel membranes with charges q_s at 'centres' $\mathbf{x} = \mathbf{x}_s$, $(s = 1, \ldots, n)$. The singularities at the centres of the metric are actually degenerate Killing horizons. The maximal analytic extension is analogous to the extreme Reissner-Nordstom (RN) multi black hole solution of Maxwell-Einstein theory in D=4; there is a singular timelike membrane source behind each horizon [15]. The existence of multi-membrane solutions indicates that the static force between parallel membranes cancels because of a balance between the attraction due to gravity and the electrostatic-type repulsion due to the 3-form potential. This balance is possible only when $T_2 = |Q_2|$. Again, there is a close analogy here to extreme RN black holes.

It took almost a further ten years before it was appreciated that there is yet another term in the $\{Q, Q\}$ anticommutator. Consider the surface integral

$$Q_5 = \Omega_4^{-1} \int_{S^4} F\,, \tag{1.21}$$

where Ω_4 is the volume of the unit 4-sphere. Because of the Bianchi identity $dF = 0$, this integral is homotopy invariant: if it is non-zero the 4-sphere must surround a magnetically-charged topological defect of the 3-form potential A. This defect must be 5-dimensional, and the 4-sphere can be taken to be a sphere of fixed radius in a 5-dimensional space transverse to the 6-dimensional worldvolume

[2]Actually, in the conventions used in [13], the bound is saturated when $T_2 = (1/2)|Q_2|$, and there is a similar factor of 1/2 for the fivebrane to be discussed below. To retain these factors would suggest a level of attention to the consistency of conventions that has not been attempted here. Various factors in formulas appearing in these lectures have therefore been set to one on the grounds that this is the correct factor for some convention.

of a fivebrane defect. This is just an application to D=11 and $p = 2$ of the well-known fact that the magnetic dual of a p-brane in a D-dimensional spacetime is a $(D-p-4)$-brane. Assuming boundary conditions appropriate to an infinite planar fivebrane one can derive the bound $T_5 \geq |Q_5|$ on the fivebrane tension. The configuration saturating the bound is [16]

$$\begin{aligned} ds^2 &= H^{-1/3} ds^2(\mathbb{E}^{(5,1)}) + H^{2/3} ds^2(\mathbb{E}^5) \\ F &= *dH \end{aligned} \tag{1.22}$$

where $*$ is the Hodge dual on $\mathbb{E}^5$ and $H(\mathbf{x})$ is a harmonic function on $\mathbb{E}^5$. The choice

$$H = 1 + \sum_{s=1}^{n} \frac{|p_s|}{|\mathbf{x} - \mathbf{x}_s|} \tag{1.23}$$

can be interpreted as n parallel fivebranes with charges p_s at positions $\mathbf{x} = \mathbf{x}_s$, $(s = 1, \ldots, n)$ in $\mathbb{E}^5$. Like the membrane solution of D=11 supergravity, this fivebrane solution admits 16 Killing spinor fields and so preserves half the supersymmetry. The singularities of H are again just Killing horizons of the metric, but in this case there is no singularity behind the horizon; the maximal analytic extension is geodesically complete [13].

The fact that $dF = 0$ means that the 4-sphere surrounding the fivebrane can be deformed at will without changing the value of the charge Q_5. We do not even have to worry about passing through singularities of F because, as just mentioned, there are none. One consequence of this is that a fivebrane carrying non-zero Q_5 charge must be closed, i.e. it cannot have a boundary. If there were a boundary then the 4-sphere of the Q_5 integral could be pushed past it and shrunk to a point, so Q_5 would vanish, contrary to hypothesis. A similar conclusion would hold for the membrane if the equation of motion for the 3-form potential were $d{\star}F = 0$, but as this is not the equation of motion the conclusion is modified. A closer analysis [17] shows that the membrane may have a boundary on a fivebrane[3]. We pass over the details here as we shall eventually recover the result from a different approach.

By analogy with the membrane, one would expect the fivebrane charge Q_5 to be the magnitude of a 5-form charge in the D=11 supertranslation algebra. In other words, taking both the membrane and the fivebrane into account we expect that (1.11) should be replaced by [21]

$$\{Q_\alpha, Q_\beta\} = (\Gamma^M C)_{\alpha\beta} P_M + \frac{1}{2}(\Gamma_{MN} C)_{\alpha\beta} Z^{MN} + \frac{1}{5!}(\Gamma_{MNPQR} C)_{\alpha\beta} Y^{MNPQR} \tag{1.24}$$

where Y is the 5-form charge. One indirect argument for its presence is that the fivebrane solution could not otherwise be half-supersymmetric. To see that this *is* possible in the presence of a 5-form charge, let us associate a static fivebrane in the 12345 5-plane with non-zero Y_{12345}, and define $q_5 = Y_{12345}/P^0$. The $\{Q, Q\}$ anticommutator is then

$$\{Q, Q\} = P^0 \left[1 + \Gamma^{012345} q_5\right]. \tag{1.25}$$

[3]A conclusion originally arrived at in [18, 19], following the discovery that Type II superstrings can end on D-branes [20].

As before, we deduce that $|q_5| \leq 1$. We may identify $|q_5|$ with the ratio of the fivebrane's charge Q_5 to its tension T_5, so the bound is equivalent to $T_5 \geq |Q_5|$. When this bound is saturated, half of the eigenvalues of $\{Q,Q\}$ vanish so that half the supersymmetry is preserved.

If the fivebrane is to be associated with a 5-form charge in the supersymmetry algebra then its worldvolume action must contain a coupling to a 6-form potential $\mathcal{B}$ induced from a 6-form B on superspace. It is natural to identify B as the magnetic potential dual to A, but the presence of the CS term in the D=11 supergravity action seems to prevent the dualization of A. However, the A field equation allows us to write

$$[\star F + F \wedge F] = dB\,, \tag{1.26}$$

and there is a superfield version of this [22]. Thus, the required superspace 6-form potential exists for on-shell supergravity backgrounds[4]. The on-shell restriction is no disadvantage since this is in any case an expected consequence of κ-symmetry.

At this point one might suppose that the superfivebrane action is just a higher-dimensional generalization of the supermembrane action, but this is not the case. In both cases the worldvolume field theory must, after (partial) gauge fixing, have 16 (linearly realized) supersymmetries (counting each component of the supercharge separately). Massless supermultiplets of such supersymmetry algebras have 8 boson degrees of freedom, per worldvolume point (16 in phase space). In the membrane case the worldvolume fields are the maps from the worldvolume to superspace so that on fixing the 3 worldvolume diffeomorphisms we are left with 8 physical scalar fields describing transverse fluctations of the membrane. For the fivebrane a similar count yields 5 scalars describing fluctations transverse to the fivebrane. We therefore need 3 additional boson degrees of freedom, which must be provided by other boson field(s) needed to complete a supermultiplet of N=2 D=6 supersymmetry. The only possibility is the antisymmetric tensor multiplet of chiral (2,0) D=6 supersymmetry [24, 25] (originally identified as the one containing the worldvolume fields of the IIA fivebrane [26][5]). As the terminology suggests, the (2,0) antisymmetric tensor multiplet has a 2-form gauge potential U. Its field strength $H = dU$ is self-dual. This makes the construction of an action difficult, even at the linearized level. Some of these difficulties are intrinsic and lead to unavoidable complications in the quantum theory [27]. One could ask only for field equations, and these were presented in [28] as superfield constraints on the extrinsic supergeometry. One could also abandon manifest six-dimensional Lorentz covariance, and an action of this type was given in [29]. However, it is possible to find a manifestly Lorentz invariant superfivebrane action [30], at the cost of having to include inverses of worldvolume scalars. These three formulations have been shown to be classically equivalent [31, 32], and the Noether charges of the

[4]Both A and B appear off-shell in a new formulation of D=11 supergravity [23].

[5]In that case only four of the five scalars were interpreted as NG variables representing transverse fluctuations, but the presence of the fifth scalar suggests an 11-dimensional interpretation. In fact, the fivebrane solution of D=10 IIA supergravity *is* the fivebrane solution of D=11 supergravity; one has only to reinterpret the fields.

Lorentz covariant action have been shown to obey the algebra (1.24), with both 5-form and 2-form charges [33].

The action of [30] provides a simple route to the hamiltonian formulation, some aspects of which will be explained here for later use. This formulation is especially simple in a vacuum background. Omitting fermions, the phase-space Lagrangian density is [34]

$$\begin{aligned}\mathcal{L} &= P\cdot\dot{X} + \Pi^{ab}\dot{U}_{ab} + \lambda_a\partial_b\Pi^{ab} - s^a\big(P\cdot\partial_a X - V_a\big)\\ &\quad + \sigma_{ab}(\Pi^{ab} + \frac{1}{4}\tilde{\mathcal{H}}^{ab}) - \frac{1}{2}v\big[(P - g^{ab}V_a\partial_b X)^2 + \det(g+\tilde{H})\big] \qquad (1.27)\end{aligned}$$

where g is here the 'worldspace' 5-metric of the fivebrane and

$$\begin{aligned}\tilde{\mathcal{H}}^{ab} &= \frac{1}{6}\varepsilon^{abcde}H_{cde}\\ \tilde{H}_{ab} &= \frac{1}{\sqrt{\det g}}g_{ac}g_{ad}\tilde{\mathcal{H}}^{cd}\\ V_f &= \frac{1}{24}\varepsilon^{abcde}H_{abc}H_{def} \qquad (1.28)\end{aligned}$$

Note that the Gauss law constraint on the electric 2-form Π (imposed by the Lagrange multiplier λ_a) becomes equivalent to the Bianchi identity $dH = 0$ on using the constraint imposed by the Lagrange multiplier σ_{ab}. This is how the self-duality of the worldvolume 3-form H is incorporated in the phase-space action, which depends only the worldspace components of H.

At this point it may seem that a new term in the D=11 supersymmetry algebra is found every ten years and that it might therefore be prudent to wait another ten years to see what happens. While there will undoubtedly be many advances in our understanding of M-theory over the next ten years, the addition of another term to the D=11 supersymmetry algebra is not likely to be one of them. The total number of entries of the real symmetric 32×32 matrix $\{Q,Q\}$ is 528. This is the same as the total number of components of P, Z and Y. To see that this is no accident we first note that the matrices $C\Gamma^{M_1\cdots M_n}$ are either symmetric or antisymmetric, depending on the value of n. Because of (1.2), we need consider only $n\le 5$ and of these only the $n = 1,2,5$ matrices are symmetric. Thus, (1.24) is the most general D=11 super-translation algebra with one D=11 spinor charge. The matrix $\{Q,Q\}$ may, initially, be considered as an element in the Lie algebra of the group $Sp(32;\mathbb{R})$. This has an $SO(1,10)$ subgroup with respect to which the 528-dimensional adjoint representation of $Sp(32;\mathbb{R})$ has the following decomposition

$$\mathbf{528}\to\mathbf{11}\oplus\mathbf{55}\oplus\mathbf{462}\,. \qquad (1.29)$$

This is just the decomposition provided by the charges P, Z and Y. In fact, because the bosonic charges are all assumed to be abelian, the algebra spanned by $(Q;P,Y,Z)$, is a contraction of $OSp(1|32;\mathbb{R})$.

The algebra (1.29) was called the 'M-theory algebra' in [35] because it encodes many of the important features of M-theory. The most significant is that it shows that, in addition to M-waves, both membranes and fivebranes, henceforth to be called M-2-branes and M-5-branes, can also preserve 1/2 supersymmetry. In this sense, and also because they are related by dualities (in a way to be explained below) the M-Wave, M-2-brane and M-5-brane should be considered on a similar footing. However, these are not the only ingredients. In compactified spacetimes there are additional possibilities, as we now discuss.

Lecture 2: More branes from M-theory

Membranes and fivebranes are associated with the *spatial* components of the charges Z and Y in the M-theory superalgebra. Having introduced these charges we should now consider what might be the significance of their time components. This is most straightforward for the 5-form charge Y, so we consider it first. Let us suppose that all bosonic charges other than P^0 and $Y_{0789\natural}$ vanish, and set $\tilde{q}_5 = Y_{0789\natural}/P^0$. Then

$$\{Q_\alpha, Q_\beta\} = P^0\left[1 - \Gamma_{789\natural}\,\tilde{q}_5\right]. \tag{2.1}$$

Again, $\Gamma_{789\natural}$ squares to the identity and has zero trace, and therefore has eigenvalues ± 1. Positivity implies the bound $|\tilde{q}_5| \leq 1$ and configurations which saturate the bound preserve 1/2 supersymmetry and are associated with the constraint

$$\Gamma^{789\natural}\epsilon = \pm\epsilon\,. \tag{2.2}$$

This constraint is not obviously associated with a brane, but it is equivalent to

$$\Gamma^{0123456}\epsilon = \pm\epsilon \tag{2.3}$$

which suggests a 6-brane. There is no 7-form gauge potential of D=11 supergravity to couple to a 6-brane but if we compactify in the $\natural$ th direction on S^1 we obtain a Kaluza-Klein (KK) vector potential C in D=10, with 6-form dual. Thus, an object that is magnetically charged with respect to C is a 6-brane. This 6-brane has an M-theory interpretation [36] as a KK monopole [37, 38].

To see this in more detail we need the KK ansatz relating the bosonic fields of D=11 supergravity with those of D=10 IIA supergravity. Setting $y = X^\natural$, the KK ansatz for the bosonic IIA string-frame fields is

$$\begin{aligned} ds^2_{11} &= e^{-\frac{2}{3}\phi(x)}ds^2_{10} + e^{\frac{4}{3}\phi(x)}\left(dy + dx^\mu C_\mu(x)\right)^2 \\ A_{11} &= A(x) + dy \wedge B(x)\,. \end{aligned} \tag{2.4}$$

where we now suppose that y is periodically identified with period 2π. The 10-metric, 'dilaton' field ϕ and two-form potential B are the massless NS-NS fields of IIA superstring theory, while the D=10 gauge potentials A and C are the massless R-R fields. Defining $G = dC$ one finds the following 6-brane solution of the IIA supergravity field equations [39]

$$\begin{aligned} ds^2_{10} &= H^{-\frac{1}{2}}ds^2(\mathbb{E}^{(6,1)}) + H^{\frac{1}{2}}d\mathbf{x}\cdot d\mathbf{x} \\ e^{-4\phi} &= H^3 \\ G &= *dH\,, \end{aligned} \tag{2.5}$$

where H is a harmonic function on $\mathbb{E}^3$ (with coordinates $\mathbf{x}$) and $*$ indicates the Hodge dual on $\mathbb{E}^3$. The singularities of H are the positions of parallel 6-branes; far away from these singularies the metric will be asymptotic to the D=10 Minkowski metric if we choose H such that $H \to 1$ as $|\mathbf{x}| \to \infty$. From the ansatz (2.4) we can read off the corresponding D=11 supergravity solution. It has $F = 0$ and 11-metric

$$ds^2 = ds^2(\mathbb{E}^{(6,1)}) + H\, d\mathbf{x}\cdot d\mathbf{x} + H^{-1}\left(dy + \mathbf{C}\cdot d\mathbf{x}\right)^2 . \tag{2.6}$$

Given the assumed asymptotic property of H, the 11-metric is asymptotic to the KK vacuum $\mathbb{E}^{(9,1)} \times S^1$ as $|\mathbf{x}| \to \infty$, confirming the KK interpretation.

If we set $|\mathbf{x}| = r$ and choose

$$H = 1 + \frac{1}{2r} \tag{2.7}$$

then the 11-metric (2.6) is just the product of $\mathbb{E}^{(6,1)}$ with the Euclidean-Taub-NUT (ETN) 4-metric. The dimensional reduction to D=10 yields the metric of a single 6-brane located at the origin of $\mathbb{E}^3$. However, whereas the 6-brane solution is genuinely singular at $r = 0$, the corresponding singularity of the 11-metric is just a coordinate singularity, provided that y is identified with period 2π. To see this we note that $H \sim 1/2r$ near the singularity, so the metric near the singularity is

$$\begin{aligned} ds_4^2 &\sim \frac{1}{2r}(dr^2 + r^2 d\Omega_2^2) + 2r(dy + C)^2 \\ &\sim \frac{dr^2}{2r} + \frac{r}{2}\left[d\Omega_2^2 + 4(dy + C)^2\right] \end{aligned} \tag{2.8}$$

where $d\Omega_2^2$ is the $SO(3)$-invariant metric on the 2-sphere. Setting $\rho = \sqrt{2r}$ we have

$$ds_4^2 \sim d\rho^2 + \rho^2\left[d\Omega_2^2 + 4(dy + C)^2\right] . \tag{2.9}$$

Now, provided that $y \sim y + 2\pi$, the hypersurfaces of constant ρ are 3-spheres, each being a $U(1)$ bundle over S^2 (the Hopf fibration), so the asymptotic metric near the singularity of H is just the metric on $\mathbb{E}^4$ in polar coordinates. The 'M-monopole' metric (2.6) therefore interpolates between the 'M-theory vacuum' $\mathbb{E}^{(10,1)}$ near $r = 0$ and the KK vacuum $\mathbb{E}^{(9,1)} \times S^1$ near $r = \infty$. The M-2-brane and M-5-brane solutions of D=11 supergravity similarly interpolate between maximally-supersymmetric vacuum solutions [24], with the difference that the 'M-theory vacuum' is at now at infinity and one finds either the $adS_4 \times S^7$ or the $adS_7 \times S^4$ KK vacuum near the singularities of H (i.e. at the horizons).

Having now dealt with the significance of M-theory configurations carrying the time-component of the 5-form charge Y it remains for us to consider the significance of the time component of the two-form charge Z. If we suppose that $Z^{0\natural} = \mp P^0$ with all other charges vanishing then we find that $\{Q, Q\} = P^0(1 \mp \Gamma_\natural)$. Configurations preserving 1/2 supersymmetry are now associated with spinors ϵ satisfying the constraint $\Gamma_\natural \epsilon = \pm\epsilon$. Again, this is not immediately interpretable as the condition imposed by the presence of a brane but it is equivalent to

$$\Gamma^{0123456789}\epsilon = \pm\epsilon \tag{2.10}$$

which suggests a 9-brane. It seems that this must be interpreted as a constraint associated with a (9+1)-dimensional boundary of the 11-dimensional spacetime, or 'M-boundary', as in the Hořava-Witten (HW) description of the heterotic string [40]. Note that the D=11 constraint (2.10) is equivalent to the D=10 chirality constraint $\Gamma_{11}\epsilon = \pm\epsilon$, so the field theory on the M-boundary boundary is indeed chiral. The D=10 interpretation of the M-boundary is as the Minkowski vacuum of the heterotic string theory (which might equally well be called the IIA 9-brane [41]). The connection to *bone fide* branes becomes clearer if we instead consider $Z^{09} = P^0$ but still compactify to the IIA theory in the $\natural$th direction. The spinor constraint associated with 1/2 supersymmetric configurations is then equivalent to

$$\Gamma_{012345678}\Gamma_{11}\epsilon = \pm\epsilon\,, \tag{2.11}$$

which is naturally associated with an 8-brane; in fact, it is the IIA D-8-brane constraint [42].

We have now discussed the significance of all charges in the M-theory algebra. We have seen that the time components of the 2-form and 5-form charges can be interpreted as those carried by branes of the S^1-compactified M-theory, alias IIA superstring theory. All other IIA branes can be obtained directly from the M-wave, M-2-brane and M-5-brane. Consider, for example, the M-wave. Compactifying M-theory in one of the nine transverse ($\mathbf{x}$) directions we obviously get a similar wave solution of IIA supergravity, but if we choose K to be u-independent then we can also consider compactification in the direction parametrized by $y = v - u$. Defining $t = v + u$, the M-wave 11-metric can then be written as

$$\begin{aligned} ds^2_{11} &= -dt^2 + dy^2 + K(\mathbf{x})(dt - dy)^2 + d\mathbf{x}\cdot d\mathbf{x} \\ &= -H^{-\frac{1}{2}}\big[H^{-\frac{1}{2}}dt^2 - H^{\frac{1}{2}}d\mathbf{x}\cdot d\mathbf{x}\big] + H\big[dy + (H^{-1}-1)dt\big]^2 \end{aligned} \tag{2.12}$$

where

$$H = 1 + K\,. \tag{2.13}$$

We may now use the ansatz (2.4) to deduce that the following IIA configuration is a solution of IIA supergravity

$$\begin{aligned} ds^2_{10} &= -H^{-\frac{1}{2}}dt^2 + H^{\frac{1}{2}}d\mathbf{x}\cdot d\mathbf{x} \\ e^{\frac{4}{3}\phi} &= H \\ G &= dt\wedge dH^{-1}\,. \end{aligned} \tag{2.14}$$

Note that $H \to 1$ at infinity because $K \to 0$, so this solution is asymptotically flat. In fact, it is a kind of extreme black hole although the isolated singularities of H are not horizons but rather genuine singularities of the metric, which must be resolved by going beyond IIA supergravity. In the context of IIA superstring theory the solution (2.14) gives the long-range fields of the D-0-brane. We note here that the condition (1.6) imposed on Killing spinors by the presence of an M-wave in D=11 becomes, in D=10, the condition

$$\Gamma_0\Gamma_{11}\epsilon = \pm\epsilon\,, \tag{2.15}$$

which is characteristic of a D-0-brane. Since a D-0-brane is 1/2 supersymmetric its mass is simply related to its charge, which is quantized as a consequence of the Dirac-Nepomechie-Teitelboim (DNT) quantization condition between 0-branes and their 6-brane magnetic duals. In the context of M-theory this quantization condition is due to the compactness of the 11th dimension. The quantization condition on the 6-brane charge is then just what is needed to ensure the non-singularity of the M-monopole metric in D=11, as discussed above.

Let us now turn to the membrane solution (1.19). We can write it as

$$\begin{aligned} ds^2_{11} &= H^{\frac{1}{3}}\left[H^{-1}ds^2(\mathbb{E}^{(1,1)}) + ds^2(\mathbb{E}^8)\right] + H^{-\frac{2}{3}}dy^2 \\ F &= \left[vol(\mathbb{E}^{(1,1)}) \wedge dH^{-1}\right] \wedge dy \end{aligned} \tag{2.16}$$

The ansatz (2.4) then leads to the following string solution of IIA supergravity

$$\begin{aligned} ds^2_{10} &= H^{-1}ds^2(\mathbb{E}^{(1,1)}) + ds^2(\mathbb{E}^8) \\ e^{-2\phi} &= H \\ F_{(3)} &= vol(\mathbb{E}^{(1,1)}) \wedge dH^{-1} \end{aligned} \tag{2.17}$$

where $F_{(3)} = dB$ is a 3-form field strength. This gives the long range fields of an infinite straight IIA superstring, which may therefore be interpreted as the D=11 membrane wrapped on the KK circle [43].

If we choose the harmonic function in (1.19) to be independent of one of the $\mathbb{E}^8$ coordinates, which we now call y, then we can instead rewrite the membrane 11-metric as

$$ds^2_{11} = H^{-\frac{1}{6}}\left[H^{-\frac{1}{2}}ds^2(\mathbb{E}^{(2,1)}) + H^{\frac{1}{2}}ds^2(\mathbb{E}^7)\right] + H^{\frac{1}{3}}dy^2 \tag{2.18}$$

where H is now harmonic on $\mathbb{E}^7$. From the ansatz (2.4) we can then read off a membrane solution of IIA supergravity. Note that it has a similar form to the 6-brane and 0-brane solutions already given. In fact, all are special cases of a general type II supergravity D-p-brane solution (even p for IIA and odd p for IIB) for which the 10-metric and dilaton take the form

$$\begin{aligned} ds^2_{10} &= H^{-\frac{1}{2}}ds^2(\mathbb{E}^{(p,1)}) + H^{\frac{1}{2}}ds^2(\mathbb{E}^{9-p}) \\ e^{4\phi} &= H^{3-p} \end{aligned} \tag{2.19}$$

where H is a harmonic function on $\mathbb{E}^{9-p}$. The metric is asymptotically flat (for $p \leq 6$) if $H \to 1$ at infinity, and if we assume isolated singularities then these can be interpreted as the positions of parallel p-branes in static equilibrium. Clearly, the $p \geq 7$ cases are special because H is logarithmic for $p = 7$ and linear for $p = 8$. The 8-brane case of (2.19) is a solution of the 'massive' IIA supergravity theory which is not directly obtainable by reduction of D=11 supergravity[6].

[6]This is perhaps not unexpected in view of the association made above of the 8-brane charge with a D=11 spacetime boundary because a boundary is determined by imposing boundary conditions rather than by solving local field equations. Perhaps the distinction between these two aspects of traditional physics is blurred in M-theory.

Turning finally to the 5-brane solution (1.22) we may either wrap it on S^1, which yields the D-4-brane solution, i.e. the $p=4$ case of (2.19), or we may simply reduce it to D=10, as done above for the membrane, to get the following IIA 5-brane solution

$$\begin{aligned} ds_{10}^2 &= ds^2(\mathbb{E}^{(5,1)}) + H ds^2(\mathbb{E}^4) \\ e^{2\phi} &= H \\ F_{(3)} &= *dH \end{aligned} \tag{2.20}$$

where H is now harmonic on $\mathbb{E}^4$ and $*$ indicates the Hodge dual on $\mathbb{E}^4$. This is the magnetic dual of the IIA string solution. Unlike the string solution, however, the dual 5-brane solution is non-singular in the sense that the metric is geodesically complete, assuming of course that H has only point singularities and is such that $H \to 1$ at infinity. Near a singularity at $r=0$ in $\mathbb{E}^5$, where r is a radial coordinate, we have $H \sim 1/r^3$ so that the distance to the singularity from $r=R$ is

$$ds \sim \int_0^R \frac{dr}{r^{3/2}} = \infty \, . \tag{2.21}$$

This is the distance on a spacelike geodesic. Because of the direct product structure of the metric it suffices to consider spacelike geodesics. Thus, the singularities of H are at infinite affine parameter on all geodesics and the IIA 5-brane *string-frame* metric is therefore geodesically complete [39]. As the 5-brane is the magnetic dual of the 'fundamental' string it is perhaps not surprising that it should be non-singular. Note, however, that the non-singularity is achieved by a very different mechanism to that of the M-5-brane, for which the singularities of H are horizons at finite affine parameter.

We have now seen all the p-brane solutions of D=11 supergravity and IIA D=10 supergravity. The relation between them is nicely summarized by the reduction to D=10 of the M-theory superalgebra. Simply decomposing the $SO(10,1)$ representations into $SO(9,1)$ representations we obtain the algebra

$$\begin{aligned} \{Q_\alpha, Q_\beta\} &= (C\Gamma^\mu)_{\alpha\beta} P_\mu + (C\Gamma_{11})_{\alpha\beta} P_{11} + (C\Gamma^\mu \Gamma_{11})_{\alpha\beta} Z_\mu \\ &+ \frac{1}{2}(C\Gamma^{\mu\nu})_{\alpha\beta} Z_{\mu\nu} + \frac{1}{4!}(C\Gamma^{\mu\nu\rho\sigma}\Gamma_{11})_{\alpha\beta} Y_{\mu\nu\rho\sigma} \\ &+ \frac{1}{5!}(C\Gamma^{\mu\nu\rho\sigma\lambda})_{\alpha\beta} Y_{\mu\nu\rho\sigma\lambda} \, . \end{aligned} \tag{2.22}$$

We can associate the space components of P_μ with IIA-waves, or supergravitons of IIA supergravity. The central charge P_{11} is carried by D-0-branes which (together with their threshold bound states) are IIA-superstring manifestation of the KK-modes arising from the compactification from D=11 [36, 44]. Each space component of Z_μ is a charge carried by a 'fundamental' IIA superstring, the component depending upon the orientation of the string. All other charges are similarly carried by other IIA p-branes.

We conclude this lecture with a brief digression from the main line of our argument. Recall that *isolated* singularities of the harmonic function H in the M-KK-monopole 11-metric (2.6) are just coordinate singularities, although they are genuine singularities of the IIA 6-brane 10-metric. The singularity in D=10 can be viewed as arising from an illegitimate neglect of the 11th dimension. But what about *non-isolated* singularities? Consider the harmonic function

$$H = 1 + \frac{1}{2|\mathbf{x}-\mathbf{a}|} + \frac{1}{2|\mathbf{x}+\mathbf{a}|}, \tag{2.23}$$

which represents two 'parallel' M-KK-monopoles, i.e. that D=11 supergravity configuration reducing in D=10 to two parallel IIA 6-branes. At the two singularities of H the KK circle contracts to a point. The two M-KK-monopoles are therefore connected by a homology 2-sphere with azimuthal angle y and poles at $\mathbf{x} = \pm\mathbf{a}$. If we now take $|\mathbf{a}| \to 0$ then this homology 2-sphere shrinks to a point and the metric acquires a genuine curvature singularity.

To resolve this singularity one needs to take into account the fact that membranes may wrap around the 2-sphere. The total energy of such a membrane is proportional to the area A of the 2-sphere, at least for the large area limit in which the semiclassical description of the M-2-brane is valid. One would normally expect this semiclassical result to be modified by corrections that are small for large A but become dominant as $A \to 0$. However, supersymmetry ensures that this does not happen, and wrapped M-2-branes become massless as $A \to 0$ [45]. This was originally shown for K_3 compactifications of M-theory, following the suggestion in [44] that the symmetry enhancement expected on the basis of duality with the heterotic string should be associated with collapsing 2-cycles of K_3. But the geometry near a collapsing two-cycle of K_3 is the same as the geometry near a pair of nearly-coincident parallel M-theory monopoles[7]. The geometry in the K_3 case is that of the Eguchi-Hanson instanton which differs from two-centre ETN metric only in the absence of the '1' term in the harmonic function H. This difference is insignificant near the centres of the metric when $|\mathbf{a}| \to 0$. From the D=10 perspective, the wrapped membranes are strings stretched between two D-6-branes and the massless states that appear in the coincidence limit are the string states that lead to an enhancement to $U(2)$ of the $U(1) \times U(1)$ gauge symmetry on the 6-branes's common worldvolume [47]. Thus, the short distance, or UV, singularity of the classical supergravity solution that occurs when $|\mathbf{a}| = 0$ is due to an illegitimate neglect of massless supermembrane states. The inclusion of these additional massless states resolves the singularity. We see that at least some of the UV singularities of supergravity are resolved in M-theory by relating them to the IR physics of massless particles coming from membranes. The resolution of UV singularities of quantum gravity by new IR physics on branes is embodied in the M(atrix) model approach to M-theory, which will not be discussed here (and for which we refer to [48] for a recent review). Instead we shall continue to concentrate on what can be learnt from classical solutions of D=11 supergravity and, more abstractly, the M-theory superalgebra.

[7]This point has been made independently by Sen [46].

Lecture 3: Dualities

There is a web of dualities connecting M-theory with both the Type IIA and Type IIB superstring theories, a summary of which may be found in my previous lectures on M-theory [35]. The relation between M-theory and the Type IIA superstring theory will be called 'M-duality'; we have just seen how some aspects of this relation are encoded in the respective supersymmetry algebras. We are now going to see how the 'T-duality' connecting the IIA and IIB superstring theories is similarly encoded.

We begin by reconsidering the IIA algebra of (2.22) in a form in which the D=10 Majorana supercharge Q is decomposed into the sum of two Majorana-Weyl supercharges $Q^{\pm}$ of opposite chirality, i.e.

$$Q^{\pm} = \mathcal{P}^{\pm} Q \qquad \mathcal{P}^{\pm} \equiv \frac{1}{2}(1 \pm \Gamma_{11}). \tag{3.1}$$

The IIA supersymmetry algebra (2.22) becomes

$$\begin{aligned} \{Q^+_\alpha, Q^+_\beta\} &= (C\mathcal{P}^+\Gamma^\mu)_{\alpha\beta}(P+Z)_\mu + \frac{1}{5!}(C\Gamma^{\mu\nu\rho\sigma\lambda})_{\alpha\beta} Y^+_{\mu\nu\rho\sigma\lambda} \\ \{Q^-_\alpha, Q^-_\beta\} &= (C\mathcal{P}^-\Gamma^\mu)_{\alpha\beta}(P-Z)_\mu + \frac{1}{5!}(C\Gamma^{\mu\nu\rho\sigma\lambda})_{\alpha\beta} Y^-_{\mu\nu\rho\sigma\lambda} \\ \{Q^+_\alpha, Q^-_\beta\} &= (C\mathcal{P}^+)P_{11} + \frac{1}{2}(C\mathcal{P}^+\Gamma^{\mu\nu})_{\alpha\beta} Z_{\mu\nu} \\ &\quad + \frac{1}{4!}(C\mathcal{P}^+\Gamma^{\mu\nu\rho\sigma})_{\alpha\beta} Y_{\mu\nu\rho\sigma}\,, \end{aligned} \tag{3.2}$$

where $Y^{\pm}$ are (anti)self-dual 5-form charges. The p-forms occuring in the last anticommutator are the charges carried by the D-p-branes of type IIA superstring theory. There are manifestly charges for $p = 0, 2, 4$, but there are also charges for $p = 6, 8$ coming from the time components of $Y_{\mu\nu\rho\sigma}$ and $Z_{\mu\nu}$, respectively. These D-brane charges couple to the massless R-R fields of IIA superstring theory and so are also called the (IIA) 'R-R charges'. All other charges are 'NS-NS charges' because they couple to massless fields in the NS-NS sector of the superstring theory. The latter include the 1-form charge carried by the IIA superstring itself and *two* 5-form charges, counting the algebraically irreducible self-dual and anti-self-dual 5-forms separately. The combination $Y = Y^+ + Y^-$ is the 5-form charge descending from the 5-form in D=11 and is therefore the charge carried by the magnetic 5-brane dual to the IIA string. This is called the NS-5-brane or the 'solitonic' 5-brane (S-5-brane). The other combination $\tilde{Y} = Y^+ - Y^-$ is the 5-form charge associated to IIA-KK-monopoles.

The IIA algebra is invariant under the transformation for which Q^- and all RR charges change sign. If we 'mod out' by this symmetry we arrive at the N=1 D=10 supersymmetry algebra of the heterotic string (since this is the N=1 superstring without a RR sector). This heterotic string supersymmetry algebra, is equivalent to

$$\{Q^+_\alpha, Q^+_\beta\} = (C\mathcal{P}^+\Gamma^\mu)_{\alpha\beta}(P+Z)_\mu + \frac{1}{2.5!}(C\Gamma^{\mu\nu\rho\sigma\lambda})_{\alpha\beta}(Y+\tilde{Y})_{\mu\nu\rho\sigma\lambda}\,. \tag{3.3}$$

Note that this is invariant under the interchange of, say, P_9 with Z_9. For this to represent a symmetry of the heterotic string theory the spectra of these two operators would have to coincide. This is not normally the case but if the 9-direction is a circle, of radius R, then the spectra of P_9 and Z_9 are isomorphic, the isomorphism involving the transformation $R \to 1/R$. In other words, the spectrum of the heterotic string theory compactified on a circle of radius R is identical to that of the same string theory compactified on a circle of radius $1/R$ because the transformation $R \to 1/R$ exchanges the KK modes, i.e. the spectrum of P_9, with the string winding modes, i.e. the spectrum of Z_9. In fact, the two heterotic string theory compactifications are equivalent to all orders in string perturbation theory; they are said to be 'T-dual'.

To see whether this remains true of the fully non-perturbative theory one must consider the effects of the T-duality transformation on the non-perturbative spectrum. For example, having chosen X^9 as the parameter of a KK circle of radius R, we have KK-monoples with p-form charges $\tilde{Y}$ proportional to R. We also have a KK tower of 5-branes with charges Y proportional to $1/R$. Now, just as compactification from D=11 to D=10 allows the M-KK-monopole to be interpreted as a 6-brane, so compactification from D=10 to D=9 allows a D=10 KK-monopole to be interpreted as a 5-brane. This is nominally a 5-brane in D=9 but it can be re-interpreted in D=10 as one of the KK tower of 5-branes. Given that the KK-monopole charge was, say, $\tilde{Y}_{06789}$, the charge on the T-dual 5-brane will be Y_{12345}. In fact, for each KK-monopole with charge $\tilde{Y} = \frac{1}{2}(Y^+ - Y^-)$ we have a T-dual 5-brane with charge $Y = \frac{1}{2}(Y^+ + Y^-)$, so at least this sector of the non-perturbative spectrum is invariant under $R \to 1/R$ provided that we also take $Y^- \to -Y^-$ (for all 09 and 8-space components). Since only Y^+ appears in the heterotic algebra (3.3), this algebra is invariant, consistent with the non-perturbative validity of heterotic T-duality.

So far, T-duality can be summarized by the statement that T-duality in the 9-direction effects the transformation

$$(P - Z)_9 \to -(P - Z)_9, \qquad Y^-_{0abcd9} \to -Y^-_{0abc9}, \qquad Y^-_{abcde} \to -Y^-_{abcde} \tag{3.4}$$

where $a, b, c, d, e = 1, 2, \ldots, 8$. The IIA superstring theory cannot be T-dual to itself because the full IIA supersymmetry algebra is definitely not invariant under this transformation. Of course, in performing this exchange we are just relabelling the generators that span the algebra, but it is a relabelling that destroys the manifest D=10 Lorentz covariance. One might suppose that manifest Lorentz covariance could be restored only by reversing the interchange but, remarkably, there exists another way. We first note that (3.4) changes some signs, in a non-Lorentz covariant fashion, in the $\{Q^-, Q^-\}$ anticommutator. We can reverse these sign changes by defining a new charge $\tilde{Q}^+$ by

$$Q^- = \tilde{Q}^+ \Gamma_9 \,. \tag{3.5}$$

The new charge is chiral rather than antichiral because multiplication by Γ^9 changes the chirality. We now find that

$$\{\tilde{Q}^+_\alpha, \tilde{Q}^+_\beta\} = (C\mathcal{P}^+\Gamma^\mu)_{\alpha\beta}(P - Z)_\mu + \frac{1}{5!}(C\Gamma^{\mu\nu\rho\sigma\lambda})_{\alpha\beta} V^+_{\mu\nu\rho\sigma\lambda} \tag{3.6}$$

where V^+ is a new self-dual 5-form such that

$$V^+_{mnpqr} = Y^-_{mnpqr} \qquad (m,n,p,q,r = 0,1,\ldots,8), \tag{3.7}$$

the other components being determined by (anti)self-duality. The remaining anti-commutator with the RR charges becomes

$$\begin{aligned}\{\tilde{Q}^+_\alpha, \tilde{Q}^+_\beta\} &= (C\mathcal{P}^+\Gamma^\mu)_{\alpha\beta}\tilde{Z}_\mu + \frac{1}{3!}(C\mathcal{P}^+\Gamma^{m u\nu\rho})_{\alpha\beta}W_{\mu\nu\rho} \\ &\quad + \frac{1}{5!}(C\Gamma^{\mu\nu\rho\sigma})_{\alpha\beta}\tilde{V}^+_{\mu\nu\rho\sigma\lambda}\end{aligned} \tag{3.8}$$

where

$$\begin{aligned}\tilde{Z}_\mu &= (Z_{m9}, P_{11}) \\ W_{\mu\nu\rho} &= (Y_{mnp9}, Z_{mn}) \\ \tilde{V}^+_{mnpq9} &= Y_{mnpq}\,.\end{aligned} \tag{3.9}$$

We have now arrived at the other D=10 N=2 supersymmetry algebra, the IIB algebra, which has two chiral supercharges $Q^I = (Q^+, \tilde{Q}^+)$. The $\{Q,Q\}$ anticommutator can be written in the form

$$\begin{aligned}\{Q^I_\alpha, Q^J_\beta\} &= (C\mathcal{P}^+\Gamma^\mu)_{\alpha\beta}(\delta^{IJ}P_\mu + \sigma_3^{IJ}Z_\mu + \sigma_1^{IJ}\tilde{Z}_\mu) \\ &\quad + \frac{1}{3!}\sigma_2^{IJ}(C\mathcal{P}^+\Gamma^{\mu\nu\rho})_{\alpha\beta}W_{\mu\nu\rho} + \frac{1}{5!}\delta^{IJ}(C\Gamma^{\mu\nu\rho\sigma\lambda})_{\alpha\beta}K^+_{\mu\nu\rho\sigma\lambda} \\ &\quad + \frac{1}{5!}(C\Gamma^{\mu\nu\rho\sigma\lambda})_{\alpha\beta}(\sigma_3^{IJ}V^+_{\mu\nu\rho\sigma\lambda} + \sigma_1^{IJ}\tilde{V}^+_{\mu\nu\rho\sigma\lambda})\end{aligned} \tag{3.10}$$

where

$$K^+ = Y^+ - V^+ \tag{3.11}$$

The process by which we arrived at this algebra suggests, correctly [49, 50], that the T-dual of the IIA superstring theory is the IIB superstring theory, and vice-versa. T-duality relates the IIA theory compactified on a circle of radius R to the IIB-theory compactified on a circle of radius $1/R$. Of course, neither theory is D=10 Lorentz covariant for finite R or $1/R$, but D=10 Lorentz covariance is recovered in either of the two limits $R \to 0$ or $R \to \infty$. *We also learn from the T-duality map between the two Type II algebras how the branes of one theory are to be interpreted in the other one.*

For example, from the fact that the transverse 1-form charge Z_m in the IIA algebra is mapped to the same charge in the IIB algebra we learn that T-duality in a direction perpendicular to the IIA string transforms it into the IIB string, and vice-versa. This can be verified by inspection of the solutions representing the long-range fields of the IIA and IIB strings. In fact, the solution is the same for both IIA and IIB, and it is also a solution of the effective supergravity field equations of the heterotic string. The (string-frame) 10-metric of this common string solution is given in (2.17). T-duality in a direction perpendicular to the string requires that we compactify one of the directions of the transverse 8-space, which becomes $\mathbb{E}^7 \times S^1$.

The absence of any power of the harmonic function multiplying the metric on this transverse space shows that the 10-metric is invariant under the inversion of the radius of the S^1 factor, apart from a possible constant rescaling of the coordinates to achieve a standard identification of the S^1 coordinate. In the heterotic case we interpret this to mean that the heterotic string is mapped into itself. In the Type II cases we must interpret it to mean that the IIA string is mapped to the IIB string, and vice-versa. Similarly, the fact that Z_9 in the IIA(B) algebra is mapped to P_9 in the IIB(A) algebra shows that T-duality in a direction parallel to a IIA(B) string results in a IIB(A)-wave. This is slightly less straightforward to verify at the level of supergravity solutions, essentially because the wave has an off-diagonal metric. Since all fields involved are common to the Type II and heterotic strings the details can be found in most other accounts of T-duality and will therefore be omitted here.

Turning to the 5-form charges in the Type II algebras we see that T-duality of a IIA-KK-monopole in a 'parallel' direction yields a IIB-KK-monopole, where 'parallel' means in a direction other than that 'occupied' by the ETN 4-metric of the KK-monopole. This can also be seen from inspection of the IIA-KK-monopole metric. The D=11 KK-monopole metric is given in (2.6). Using the dimensional reduction ansatz (2.4) we see that the IIA dilaton is constant and the IIA 10-metric is

$$ds_{10}^2 = ds^2(\mathbb{E}^{5,1}) + H d\mathbf{x}\cdot d\mathbf{x} + H^{-1}(dy + C)^2 \,. \tag{3.12}$$

The only non-zero fields in the complete solution are those in common with IIB supergravity, so the same solution also serves as the IIB-KK-monopole. The absence of any power of the harmonic function multiplying the 6-dimensional 'worldvolume' factor confirms that T-duality in these directions just takes the IIA-KK-monopole into the IIB-KK-monopole. If, on the other hand, we T-dualize in the y-direction then the H^{-1} factor is inverted and we begin to see the emergence of a 5-brane metric. This is confirmed by inspection of the T-duality map between the 5-form charges in the Type II algebras.

We now have a more or less complete set of rules for T-duality of objects carrying NS-NS charges. The results are essentially the same as those for the heterotic string with the difference that each time we T-dualize we move from the IIA(B) to the IIB(A) theory. The principal novelty of T-duality in the Type II context is its effect on the D-branes carrying the RR charges [42]. When the T-duality map (3.9) between RR charges is interpreted in terms of branes we see, for example, that T-duality in a direction parallel to a D-p-brane results in a $D-(p-1)$-brane in the dual theory. This can be understood as follows. In performing the T-duality we first compactify on a circle of large radius R. Saying that this direction is parallel to the p-brane amounts to saying that the p-brane is wrapped around this direction. As we take $R \to 0$ the p-brane becomes, effectively, a $(p-1)$-brane. *A priori*, we might expect to have to interpret this as a $(p-1)$-brane in a D=9 theory but T-duality allows us to re-interpret it as a $(p-1)$-brane in the T-dual D=10 theory. Conversely, T-duality in a direction orthogonal to a RR p-brane results in a RR $(p+1)$-brane in the T-dual theory. We could also have considered T-duality in a direction that is neither parallel nor perpendicular to a given brane, but we shall ignore this (inessential) complication here.

The T-duality map between RR p-branes in the IIA and IIB theories is nicely reflected in the form of the 10-metric (2.19) of the corresponding supergravity solutions. A T-duality transformation takes the radius of a circle to the inverse radius. If one supposes that the circle in question is parameterized by one of the cartesian space coordinates of either $\mathbb{E}^{(p,1)}$ (i.e. a direction parallel to the p-brane) or $\mathbb{E}^{9-p}$ (i.e. a direction orthogonal to the p-brane) then the inversion of the radius either takes one factor of $H^{-\frac{1}{2}}$ to $H^{\frac{1}{2}}$ or vice-versa, respectively. Thus the p-brane metric is taken either to the $(p-1)$-brane metric or to the $(p+1)$-brane metric. We have still to consider the other fields of the D-brane supergravity solutions but these merely confirm the result suggested by the T-duality transformation of the metric, which is the same result as we deduced above from the T-duality map between the Type II supersymmetry algebras.

We are now in a position to see how some of the various branes of M-theory and Type II string theories are related to each other by dualities. Let us start from an M-wave in the $\natural$ direction. It is convenient to represent this by the array

$$MW: \quad - \quad - \quad - \quad - \quad - \quad - \quad - \quad - \quad - \quad \natural$$

Reduction to D=10 in the $\natural$ direction yields the D-0-brane, which we represent by the array

$$D0: \quad - \quad - \quad - \quad - \quad - \quad - \quad - \quad - \quad -$$

T-duality in the 1-direction now yields the D-string oriented in the 1-direction

$$D1: \quad 1 \quad - \quad - \quad - \quad - \quad - \quad - \quad - \quad -$$

Further T-duality in the 2-direction yields the D-2-brane

$$D2: \quad 1 \quad 2 \quad - \quad - \quad - \quad - \quad - \quad - \quad -$$

This can be lifted to D=11 to the M-2-brane

$$M2: \quad 1 \quad 2 \quad - \quad - \quad - \quad - \quad - \quad - \quad - \quad -$$

Instead, we may continue to T-dualize the D-2-brane until arriving at the D-4-brane

$$D4: \quad 1 \quad 2 \quad 3 \quad 4 \quad - \quad - \quad - \quad - \quad -$$

which we may then lift in the $\natural$ th direction to arrive at an M-5-brane in the 1234$\natural$ 5-plane

$$M5: \quad 1 \quad 2 \quad 3 \quad 4 \quad - \quad - \quad - \quad - \quad - \quad \natural$$

If we instead continue to T-dualize to the D-6-brane

$$D6: \quad 1 \quad 2 \quad 3 \quad 4 \quad 5 \quad 6 \quad - \quad - \quad -$$

then we may lift in the $\natural$ th direction to an M-KK-monopole. We represent this by the array

$$MKK: \quad 1 \quad 2 \quad 3 \quad 4 \quad 5 \quad 6 \quad - \quad - \quad - \quad \times$$

where the cross indicates the compact KK-circle of the KK-monopole. In carrying out these steps at the level of supergravity solutions we start from the harmonic function K of the M-wave. At the first step this is converted into the harmonic function H of the D-0-brane in the way indicated earlier, and this harmonic function then appears in all the subsequent dualized solutions. However, certain steps require H to be independent of the coordinate in the T-duality direction. For example, in passing from the D-0-brane to the D-2-brane H goes from a harmonic function on $\mathbb{E}^9$ to one on $\mathbb{E}^7$. Re-interpreting this as a solution of D=11 supergravity then yields the M-2-brane solution but with H harmonic on $\mathbb{E}^7$ instead of $\mathbb{E}^8$. On the other hand, if we instead continue to T-dualize to the D-4-brane then H is reduced to being harmonic on $\mathbb{E}^5$ but when the D-4-brane solution is re-interpreted in D=11 we recover the general M-5-brane solution.

The D-6-brane is the magnetic dual of the D-0-brane. If we continue to T-dualize we arrive at the IIB 7-brane and IIA 8-brane (the 'high-branes') for which there are no obvious electric duals. The asymptotic behaviour of the 7-brane and 8-brane supergravity solutions are also special. In the 7-brane case the function H is harmonic on $\mathbb{E}^2$, so point singularities are conical singularities and the energy density per unit 7-volume is logarithmically divergent. This case will not be discussed in these lectures. In the 8-brane case H is harmonic on $\mathbb{E}^1$; allowing for point singularities means that it is piecewise linear. The 8-brane configuration of IIA supergravity is actually not a solution of the standard IIA field equations but rather of a 'massive' variant with a cosmological constant. The 8-brane is effectively a IIA domain wall separating regions of different cosmological constant [51, 52]. Formally, we may continue to T-dualize to arrive at the IIB D-9-brane. The function H is now constant, so the D-9-brane solution is just flat D=10 Minkowski space. It might appear from this that, unlike all other D-branes, the D-9-brane does not break 1/2 supersymmetry. But we still have the constraint

$$\Gamma^{0123456789}\epsilon = \pm\epsilon \tag{3.13}$$

which is equivalent to the chirality constraint $\Gamma_{11}\epsilon = \pm\epsilon$. It follows that the D-9-brane must be interpreted as the Minkowski vacuum spacetime of a D=10 string theory with N=1 supersymmetry. Since the D-9-brane is a D-brane, strings may end on it. These strings can only be those of the Type I string theory, because this is the only D=10 superstring theory with both closed and open strings.

One potential problem with the attempt to interpret the Type I string theory in this way is that the 8-brane carries a RR charge, for which the lines of force must either terminate on an anti 8-brane or go off to infinity. As long as the one transverse direction is non-compact the force lines may go off to infinity but to T-dualize in this last direction we must first compactify it. Since the lines of force cannot now wander off to infinity the total 8-brane charge must vanish. If D-branes were the only objects to carry RR charge then the net number of 8-branes minus anti 8-branes would have to vanish. Such a configuration would relax to one with neither 8-branes nor anti 8-branes because the anti 8-branes would attract the 8-branes and annihilate them. In this case the initial IIA configuration preserving supersymmetry would be a IIA KK-vacuum, for which the T-dual is a IIB KK vacuum, rather than a D-9-brane. Thus, to arrive at the Type I string theory by T-duality of the IIA theory 8-brane requires an additional ingredient. This ingredient, the orientifold 8-plane, is best understood from the other direction, i.e. by T-duality of the Type I string [50].

The existence of the Type 1 string is suggested by the IIB supersymmetry algebra since there are *two* ways to truncate the latter to an N=1 supersymmetry algebra, corresponding to the following two involutions of the IIB algebra:

$$\begin{aligned} Q^+ &\to \sigma_3 Q^+ \\ \tilde{Z} &\to -\tilde{Z} \\ W &\to -W \\ \tilde{V}^+ &\to -\tilde{V}^+ \end{aligned} \tag{3.14}$$

and

$$\begin{aligned} Q^+ &\to \sigma_1 Q^+ \\ \tilde{Z} &\to -Z \\ W &\to -W \\ \tilde{V}^+ &\to -V^+ . \end{aligned} \tag{3.15}$$

We can obtain an N=1 supersymmetry algebra by 'modding out' by either of these involutions. In the first case we set $\tilde{Q}^+ = 0$ to recover the heterotic string algebra of (3.3). In the second case, we must set $Q^+ = \tilde{Q}^+$. We then find an algebra isomorphic to (3.3) but with the NS 1-form Z replaced by the RR 1-form $\tilde{Z}$ and the NS self-dual 5-form Y^+ replaced by the RR self-dual 5-form $\tilde{V}^+$. The fact that there is now no NS string charge means that what was originally the fundamental type IIB string has become a string without the 2-form charge needed to prevent it from breaking. This fits the description of the Type 1 string. To confirm it we must examine the consequences of the Z_2 identification on the IIB string worldsheet fields.

If the IIB supercharges are expressed as Noether charges in terms of worldsheet fields of the IIB superstring, then the Z_2 transformation (3.15) is effected by a worldsheet parity transformation $\sigma \to -\sigma$. This is a symmetry of the IIB superstring action provided all worldsheet scalars X^μ are indeed scalars, rather than pseudo-scalars. Thus, the identification under worldsheet parity leads to the Type 1 supersymmetry algebra and the worldsheet scalar fields are now subject to $X^\mu(\sigma) = X^\mu(-\sigma)$. If we now T-dualize in, say, the 9-direction then X^9 becomes a pseudoscalar and the identification under worldsheet parity leads to $X^9(-\sigma) = -X^9(\sigma)$. In particular $X^9(0) = -X^9(0)$ so the KK circle in the 9-direction is actually the interval $S^1/\mathbb{Z}_2$. The 8-planes $X^9(0) = 0$ are effectively boundaries of the 9-dimensional space of the D=10 string theory T-dual to the Type 1 string theory[8]. These are the orientifold 8-planes. Open strings, which previously had endpoints in spacetime with $SO(32)$ Chan-Paton factors, now end on any of 16 8-branes. The RR charge of these 8-branes is cancelled by the RR charge

[8]This has been called the Type I' theory but a possibly better terminology is Type IA, in which case the Type I string could be renamed Type 1B.

of the orientifold planes. Much more could be said about the Type 1 string theory but we have now learnt what we can from the supersymmetry algebra.

We have one more duality transformation to consider before we can relate all M-theory or Type II branes to each other. This is IIB S-duality [53, 44]. We first note that the IIB algebra is invariant under the cyclic group generated by the transformation

$$Q \to e^{\frac{i\pi}{4}\sigma_2} Q \tag{3.16}$$

on the two supercharges, combined with the transformation

$$(Z, X) \to (X, -Z) \qquad (V, U) \to (U, -V) \tag{3.17}$$

on the bosonic charges. This group is a subgroup of an $Sl(2, R)$ symmetry group of IIB supergravity. We see from its action on the IIB charges that it interchanges the IIB string with the D-string and the IIB solitonic 5-brane with the D-5-brane. The IIB 3-brane is invariant. The group is actually of order 4 because it squares to $-1 \in Sl(2, R)$ but this is equivalent to the identity element of $PSl(2, R)$, which is the isometry group of the target space of the scalar fields of IIB supergravity. Thus, there is a Z_2 action on the moduli space of the IIB theory exchanging NS-branes with RR-branes. This is clearly non-perturbative within the context of Type IIB superstring theory, but is believed to be a Z_2 subgroup of a $PSl(2, Z)$ S-duality group of the fully non-perturbative IIB theory[9]. Here we shall need only the Z_2 subgroup which we shall also call S-duality.

The effects of S-duality on IIB supergravity solutions are most easily seen in the Einstein-frame because the Einstein-frame metric (defined by the absence of a power of the dilaton multiplying the $\sqrt{-g}R$ term in the action) is $Sl(2; R)$ invariant. The relation between the Einstein-frame metric ds_E^2 and the string frame metric is

$$ds^2 = e^{\frac{1}{2}\phi} ds_E^2 \, . \tag{3.18}$$

An S-duality transformation is then achieved by changing the sign of the dilaton while simultaneously exchanging the two 3-form field strengths in the way indicated by the exchange (3.17) of the 1-form charges. Thus, the effect of S-duality on the string-frame metric and dilaton is

$$ds^2 \to e^{-\phi} ds^2 \, , \qquad \phi \to -\phi \, . \tag{3.19}$$

Let us consider the effect of an S-duality transformation on the IIB string solution, which is the same as the IIA string solution of (2.17). In particular, the metric and dilaton are given by

$$\begin{aligned} ds_{10}^2 &= H^{-1} ds^2(\mathbb{E}^{(1,1)}) + ds^2(\mathbb{E}^8) \\ e^{-\phi} &= H^{\frac{1}{2}} \end{aligned} \tag{3.20}$$

[9]It is possible that there exists a new superstring theory containing both the IIB string and the D-string in which the $PSl(2, Z)$ symmetry is visible in perturbation theory [54].

The transformation (3.19) then yields

$$\begin{aligned} ds^2_{10} &= H^{-\frac{1}{2}} ds^2(\mathbb{E}^{(1,1)}) + H^{\frac{1}{2}} ds^2(\mathbb{E}^8) \\ e^{-\phi} &= H^{-\frac{1}{2}} \end{aligned} \tag{3.21}$$

which is precisely the metric and dilaton of the D-string. We thus confirm the previous conclusion that IIB S-duality takes the IIB string to the D-string. One can similarly confirm that IIB S-duality takes the IIB NS-5-brane to the D-5-brane.

By means of the three dualities, 'M', 'T' and 'S', we can go, in the way described above, from any of the 1/2-supersymmetric 'objects' of M-theory or Type II superstring theories to any other one. M-theory is therefore a theory of a single object with various dual manifestations, no one of which is sufficient by itself. The Type 1 and heterotic string theories might at first appear to stand apart, but the objects within these theories preserving 1/2 of the supersymmetry of the D=10 N=1 Minkowski vacuum can be viewed as special cases of 1/4 supersymmetric 'intersecting brane' configurations of M-theory or Type II supersring theory. For example, the D-5-brane of Type I superstring theory can be viewed as a IIB D-5-brane inside a IIB 9-brane. As another example, the heterotic string can be viewed as the boundary of an M-2-brane on an M-boundary. Neither of these cases is, strictly speaking, an example of 'intersecting' branes but it is convenient to consider them under this rubric because they are dual (at least formally) to cases to which the term has its obvious meaning. We shall now consider some of these cases in more detail.

Lecture 4: Intersecting M-branes

Each of the 'basic', 1/2 supersymmetric, objects of M-theory or Type II superstring theory, with a given orientation, is associated with a constraint of the form $\Gamma\epsilon = \epsilon$ for some traceless product Γ of Dirac matrices with the property that $\Gamma^2 = 1$. Given two such objects we have two matrices with these properties. Let us call them Γ and Γ'. Let ζ and ζ' be the charge/tension ratios of the objects associated, respectively, with Γ and Γ'. Then the $\{Q,Q\}$ anticommutator takes the form

$$\{Q,Q\} = P^0\left[1 + \zeta\Gamma + \zeta'\Gamma'\right]. \tag{4.1}$$

Positivity imposes a bound on ζ and ζ', but the form of this bound depends on whether the two matrices Γ and Γ' commute or anticommute.

If $\{\Gamma,\Gamma'\} = 0$ then

$$(\zeta\Gamma + \zeta'\Gamma')^2 = \zeta^2 + \zeta'^2 , \tag{4.2}$$

so the bound is $\zeta^2 + \zeta'^2 \leq 1$, which is equivalent to a bound of the type

$$T \geq \sqrt{Z^2 + Z'^2} \tag{4.3}$$

where Z and Z' are the charges of the two branes. Since the right hand side is strictly greater than $|Z| + |Z'|$, unless either Z or Z' vanishes, a configuration saturating this bound must be a 'bound state' with strictly positive binding energy. It is associated with a constraint of the form $\Gamma''\epsilon = \epsilon$ where

$$\Gamma'' = \cos\vartheta\,\Gamma + \sin\vartheta\,\Gamma' \tag{4.4}$$

for some angle ϑ. Since Γ'' is traceless and squares to the identity, the 'bound state' is another configuration preserving 1/2 supersymmetry. A trivial example is provided by the matrices $\Gamma = \Gamma_{01}$ and $\Gamma' = \Gamma_{02}$ associated with M-waves in the 1-direction and 2-direction, respectively. In this case Γ'' is clearly associated with a wave in some intermediate direction. To call this a 'bound state' is clearly an abuse of terminology (hence the quotes) but the term can be understood in its usual sense in less trivial cases. An example is provided by

$$\Gamma = \Gamma_{012} \qquad \Gamma' = \Gamma_{012345} \tag{4.5}$$

In this case, the matrix Γ'' can be associated with a bound state of an M-5-brane with an M-2-brane. From the perspective of the M-5-brane's effective worldvolume field theory, the M-2-brane charge is the magnitude of the topological 2-form charge

$$Z^{MN} = \int dX^M \wedge dX^N \wedge H \tag{4.6}$$

where $H = dU$ is the 3-form field strength and the integral is over the M-5-brane's 'worldspace'.

Given that there is an M-5-brane/M-2-brane bound state, the fact that its binding energy is strictly positive suggests that there is an attractive force between a M-2-brane and an M-5-brane when one is parallel to the other and separated by some distance. The M-2-brane would then be attracted to the core of the M-5-brane, thereby lowering the energy until the energy bound is saturated. In this case, one would not expect to find the relative separation of the branes entering as a free parameter in the supergravity solution representing the long-range fields of the M-5-brane/M-2-brane bound state. On the other hand, the 1/2 supersymmetry leads one to expect that it should be possible to superpose parallel M-5-branes carrying the same M-2-brane charge. These considerations, together with the fact that the solution must reduce to that of the M-5-brane for zero M-2-brane charge, suggest that the supergravity solution will again depend on a single harmonic function. There is indeed a D=11 supergravity solution with these properties. The 11-metric of this solution is [55]

$$\begin{aligned} ds^2_{11} &= H^{1/3}\left(\sin^2\vartheta + H\cos^2\vartheta\right)^{1/3}\Big[H^{-1}ds^2(\mathbb{E}^{(1,2)}) + \\ &\quad +\left(\sin^2\vartheta + H\cos^2\vartheta\right)^{-1}ds^2(\mathbb{E}^3) + ds^2(\mathbb{E}^5)\Big] \end{aligned} \tag{4.7}$$

where H is harmonic on $\mathbb{E}^5$. As expected, it depends on the angle ϑ which is arbitrary, classically, but is restricted in the quantum theory because of the DNT quantization condition satisfied by the M-2-brane and M-5-brane charges. This metric interpolates between that of the M-2-brane and that of the M-5-brane; the same is true of the complete solution.

Given a 1/2-supersymmetric M-2-brane/M-5-brane bound state various others follow by duality. Compactifying to D=10 on one of the M-2-brane directions we arrive at the IIA configuration represented by the array

$$\begin{array}{lcccccccccc} D4: & 1 & 2 & 3 & 4 & - & - & - & - & - \\ F1: & 1 & - & - & - & - & - & - & - & - \end{array}$$

where 'F1' stands for 'Fundamental string'. T-dualizing in the 4-direction we obtain a similar 1/2 supersymmetric bound state of a 'Fundamental' IIB string with a D-3-brane. This can be converted by S-duality into a bound state of a D-string with a D-3-brane because the latter is S-self-dual. We now have the IIB array

$$\begin{array}{lcccccccccc} D3: & 1 & 2 & 3 & - & - & - & - & - & - \\ D1: & 1 & - & - & - & - & - & - & - & - \end{array}$$

T-duality in the 2 and 3 directions converts this into a bound state of a D-string with a fundamental string, as required by the $Sl(2;Z)$ duality of IIB superstring theory [56, 47].

Let us now turn to the case in which Γ and Γ' commute. In this case they may be simultaneously diagonalized with eigenvalues ± 1. It immediately follows that positivity implies the bound $|\zeta| + |\zeta'| \leq 1$, which is equivalent to a bound of the form

$$T \geq |Z| + |Z'| . \tag{4.8}$$

When this bound is saturated we can rewrite (4.1) as

$$\{Q, Q\} = 2P^0 [\zeta\mathcal{P} + \zeta'\mathcal{P}'] . \tag{4.9}$$

where $\mathcal{P} = (1/2)(1 - \Gamma)$ and $\mathcal{P}' = (1/2)(1 - \Gamma')$. Since the projectors $\mathcal{P}$ and $\mathcal{P}'$ commute, a zero eigenvalue eigenspinor of $\{Q, Q\}$ must be annihilated by both of them, i.e. it must satisfy the joint conditions

$$\Gamma\epsilon = \epsilon \qquad \Gamma'\epsilon = \epsilon . \tag{4.10}$$

Provided that the product $\Gamma\Gamma'$ is traceless (a condition that is always met), the commuting matrices Γ and Γ' can be brought to the diagonal form

$$\begin{aligned} \Gamma &= \text{diag.}(\overbrace{1, \cdots, 1}^{16}, \overbrace{-1, \cdots, -1}^{16}), \\ \Gamma' &= \text{diag.}(\overbrace{1, \cdots, 1}^{8}, \overbrace{-1, \cdots, -1}^{8}, \overbrace{1, \cdots, 1}^{8}, \overbrace{-1, \cdots, -1}^{8}), \end{aligned} \tag{4.11}$$

from which it is evident that the constraints (4.10) preserve 1/4 supersymmetry.

An example of a 1/4 supersymmetric configuration is provided by the orthogonal intersection, on a point, of two M-2-branes, such that $\Gamma = \Gamma_{012}$ and $\Gamma' = \Gamma_{034}$ [57]. This can be represented by the array

$$\begin{array}{lcccccccccc} M2: & 1 & 2 & - & - & - & - & - & - & - & - \\ M2: & - & - & 3 & 4 & - & - & - & - & - & - \end{array}$$

Most other 1/4 supersymmetric configurations of orthogonally intersecting branes may be obtained from this one by various 'duality chains'. For example, compactifying on the $\natural$ th direction we obtain a similar configuration of two intersecting D-2-branes. Then, T-dualizing in the 5 and 6 directions we arrive a configuration of intersecting D-4-branes

$$\begin{array}{lccccccccc} D4: & 1 & 2 & - & - & 5 & 6 & - & - & - \\ D4: & - & - & 3 & 4 & 5 & 6 & - & - & - \end{array}$$

which may lifted to D=11 to yield a configuration of two M-5-branes intersecting on a 3-brane [57]

$$\begin{array}{lcccccccccc} M5: & 1 & 2 & - & - & 5 & 6 & - & - & - & \natural \\ M5: & - & - & 3 & 4 & 5 & 6 & - & - & - & \natural \end{array}$$

In an alternative notation, the initial configuration of intersecting M-2-branes is denoted by $(0|M2, M2)$ and the final configuration of intersecting M-5-branes by $(3|M5, M5)$. The duality chain from one to the other is then indicated as follows

$$(0|M2, M2) \stackrel{M}{\to} (0|D2, D2) \stackrel{T^2}{\to} (2|D4, D4) \stackrel{M}{\to} (3|M5, M5)\,. \qquad (4.12)$$

From the same starting point we can compactify instead on the 2-direction to obtain a IIA string intersecting a D-2-brane. Relabelling the directions, we have

$$\begin{array}{lccccccccc} F1: & 1 & - & - & - & - & - & - & - & - \\ D2: & - & 2 & 3 & - & - & - & - & - & - \end{array}$$

T-dualizing in the 4 and 5 directions we obtain an intersection of a IIA string with a D-4-brane, which is a reduction to D=10 of

$$\begin{array}{lcccccccccc} M2: & 1 & - & - & - & - & - & - & - & - & 10 \\ D2: & - & 2 & 3 & 4 & 5 & - & - & - & - & 10 \end{array}$$

i.e. the intersection on a string of an M-2-brane with an M-5-brane. This sequence of steps can be denoted by the following duality chain

$$(0|M2, M2) \stackrel{M}{\to} (1|F1, D2) \stackrel{T^2}{\to} (1|F1, D4) \stackrel{M}{\to} (1|M2, M5)\,. \qquad (4.13)$$

The intersection of the M-2-brane with the M-5-brane can be interpreted as a coincidence of two membrane boundaries, one of an 'incoming' M-2-brane and the other of an 'outgoing' M-2-brane. This interpretation is possible because, as mentioned previously, the M-2-brane can end on an M-5-brane. Given this, it then follows that the array representing the intersection of the IIA string with the D-2-brane, for example, can be reinterpreted as the coincidence of two endpoints of IIA strings. More generally, all 'brane-boundary' possibilities, including Type II strings ending on D-branes, follow by duality from the possibility of an M-2-brane ending on an M-5-brane, so we may consider this to be the key case to understand. It does not follow from the D=11 superalgebra alone, but it will be deduced below from considerations related to the worldvolume supersymmetry algebra of the M-5-brane.

For terminological convenience we shall regard brane boundaries as special cases of intersections. It should also be stated that 'intersections' include cases in which there is merely an 'overlap'. This point may be illustrated by the case of two 'intersecting' M-2-branes. There is a genuine intersection only if the positions of the two M-2-branes in the 'overall' transverse 6-space coincide; otherwise they might be said to be 'overlapping'. However, the fact that the configuration saturates an energy bound of the form (4.8) implies the absence of a force between the two M-2-branes and hence that the distance L of separation between them in the overall transverse 6-space is a free parameter. In particular, we may choose $L = 0$, so a genuine intersection is included as a special case. This is equally true for any of the 1/4 supersymmetric configurations represented by the above arrays.

A further implication of the 'no force' condition is that one can obtain a supergravity solution representing a 1/4 supersymmetric intersecting brane configuration by a type of superposition, which is summarized by the 'harmonic function rule' [58, 59]. The above M-brane intersections will serve to illustrate the rule. We shall only consider its application to the 11-metric. In the $(0|M2, M2)$ case this is

$$ds^2_{11} = (H_1H_2)^{1/3}\left[(H_1H_2)^{-1}dt^2 + H_1^{-1}|dz_1|^2 + H_2^{-1}|dz_2|^2 + ds^2(\mathbb{E}^6)\right] \tag{4.14}$$

where z_1 and z_2 are complex coordinates parametrizing the two orthogonal 2-planes occupied by the two membranes. This 11-metric is essentially determined by the requirement that if either $H_1 - 1$ or $H_2 - 1$ vanishes we recover the 11-metric of a single M-2-brane. The other fields are determined by the same requirement. The fact that the solution depends on *two* harmonic functions is directly related to the no-force condition implied by 1/4 supersymmetry. The only subtlety lies in the fact that to solve the field equations of D=11 supergravity we must restrict the variables on which the harmonic functions H_1 and H_2 depend; for the moment we postpone discussion of this point.

A feature of the 11-metric (4.14) (shared by the 11-metric (4.7) representing a 1/2 supersymmetric non-marginal bound state) is that, apart from an overall conformal factor, there is a factor of H_i^{-1} for each term in the metric corresponding to a direction 'occupied' by the i'th brane. Applying these principles to the $(3|M5, M5)$ case we find that

$$\begin{aligned} ds^2_{11} &= (H_1H_2)^{2/3}\Big[(H_1H_2)^{-1}ds^2(\mathbb{E}^{(1,3)}) + H_1^{-1}|dz_1|^2 \\ &\qquad + H_2^{-1}|dz_2|^2 + ds^2(\mathbb{E}^3)\Big] \end{aligned} \tag{4.15}$$

where the complex variables z_1 and z_2 now parametrize the two 'relative transverse' spaces, i.e. the 2-space in each M-5-brane transverse to the 3-brane intersection. A similar application to the $(1|M2, M5)$ case yields

$$ds^2_{11} = H_1^{1/3}H_2^{2/3}\Big[(H_1H_2)^{-1}ds^2(\mathbb{E}^{(1,1)}) + H_1^{-1}dx^2$$

$$+H_2^{-1}|dq|^2 + ds^2(\mathbb{E}^4)\Big] \tag{4.16}$$

where x is a (real) coordinate parametrizing the relative transverse space of the membrane and q is a quaternonic coordinate parametrizing the relative transverse space of the M-5-brane. In all of these cases, the harmonic functions must be taken to be independent of all but the overall transverse coordinates. In other words, the D=11 supergravity solutions of this form are necessarily translational invariant in all directions tangent to a participating brane. This is an appropriate restriction if we wish to consider toroidal compactification in which each p-brane is wrapped on a p-cycle, since we may then essentially read off the solution for an extreme black hole of the compactified supergravity theory. Otherwise the restriction is not appropriate but it seems to be the best that can be achieved by superposition; it is not excluded that there exists a more general class of solution but if so it will not be found by the harmonic function rule[10].

Given two commuting matrices Γ and Γ' associated with a $1/4$ supersymmetric configuration of two intersecting branes the product $\Gamma\Gamma'$ is also traceless and squares to the identity. It also commutes with both Γ and Γ'. Thus it is always possible to add an additional brane 'for free' in the sense that the configuration still preserves $1/4$ supersymmetry. The $1/4$ supersymmetric configuration of two orthogonally intersecting M2-branes will illustrate the point. Since both $\Gamma_{012}\epsilon = \epsilon$ and $\Gamma_{034}\epsilon = \epsilon$ it is automatic that $\Gamma_{1234}\epsilon = -\epsilon$. In this case the 'third brane' is actually an M-KK-monopole, or rather an anti M-KK-monopole; note that if we wish to maintain $1/4$ supersymmetry then we are not free to choose the orientation of the 'third brane'. The $1/4$ supersymmetric configuration of two M-2-branes and an M-KK-monopole may be represented by the array

$M2:$	1	2	–	–	–	–	–	–	–	–
$M2:$	–	–	3	4	–	–	–	–	–	–
$MKK:$	$\times$	–	–	–	5	6	7	8	9	$\natural$

where $\times$ indicates the KK circle. Reducing to D=10 in this direction and relabelling the result yields

$F1:$	1	–	–	–	–	–	–	–	–
$D2:$	–	2	3	–	–	–	–	–	–
$D6:$	–	–	–	4	5	6	7	8	9

T-dualizing in the 4 and 5 directions now yields

$F1:$	1	–	–	–	–	–	–	–	–
$D4:$	–	2	3	4	5	–	–	–	–
$D4:$	–	–	–	–	–	6	7	8	9

which may be lifted to D=11 as

$M2:$	1	–	–	–	–	–	–	–	–	$\natural$
$M5:$	–	2	3	4	5	–	–	–	–	$\natural$
$M5:$	–	–	–	–	–	6	7	8	9	$\natural$

[10]We refer to [60] for a review of recent progress in this direction.

The M-2-brane now intersects each of two M-5-branes on a string. Alternatively, we may view the configuration as one in which a membrane is 'stretched' between two M-5-branes, in such a way that it has a string boundary on each. If we remove the membrane we obtain a new 1/4 supersymmetric configuration, $(1|M5, M5)$, in which two M-5-branes 'intersect' on a string although, as before, there is a genuine intersection only when the separation in the 1 direction vanishes. The harmonic function rule applied to the $(1|M5, M5)$ configuration yields

$$ds_{11}^2 = (H_1H_2)^{2/3}\left[(H_1H_2)^{-1}ds^2(\mathbb{E}^{(1,1)}) + H_1^{-1}|dq_1|^2 + H_2^{-1}|dq_2|^2 + dx^2\right] \quad (4.17)$$

where q_1 and q_2 are quaternionic coordinates parametrizing the two relative transverse spaces to the string intersection in the M-5-branes, and x is the coordinate in which the two M-5-branes are, in principle, separated. In fact, for this to be a solution of D=11 supergravity both harmonic functions must be taken to be *independent of the overall transverse dimension* x; instead, H_1 is harmonic in q_2 and H_2 is harmonic in q_1 [59]. Again, there may be a more general solution in which the M-5-branes are localized in the x direction but, if so, it is not given by the harmonic function rule.

When the branes of a 1/4 supersymmetric 'intersecting brane' configuration actually intersect, the intersection should appear in the worldvolume field theory of each as a 1/2 supersymmetric worldvolume 'soliton' of some kind. This is one way of understanding how 1/4 supersymmetric intersecting brane configurations arise. A simple example of this point of view is provided by consideration of the 'intersecting brane' configuration $(0|D0, D4)$ obtained from $(0|M2, M2)$ by the duality chain

$$(0|M2, M2) \overset{M}{\to} (0|D2, D2) \overset{T^2}{\to} (0|D0, D4)\,. \quad (4.18)$$

The worldvolume field theory for a single D-4-brane is a (4+1)-dimensional maximally supersymmetric Dirac-Born-Infeld theory. In the IIA Minkowski vacuum, and assuming a Minkowski worldvolume (i.e. choosing the 'static' gauge and setting to zero all worldvolume scalars describing transverse fluctuations), the bosonic Lagrangian reduces to the pure Born-Infeld form

$$\mathcal{L} = -\sqrt{-\det\left(\eta + F\right)} \quad (4.19)$$

where η is the D=5 Minkowski metric and F is the BI 2-form. We now set the electric components of F to zero, in which case we can interpret $\mathcal{L}$ as minus the energy density $\mathcal{E}$. The 5×5 determinant also reduces to minus the 4×4 determinant of the matrix $(1+F)$, where F is now the magnetic 2-form component of the BI field strength. Thus $(a, b = 1,2,3,4)$

$$\begin{aligned}\mathcal{E}^2 &= \det(\delta_{ab} + F_{ab}) \\ &= (1 \pm \frac{1}{4}\mathrm{tr}\ \mathrm{F}\tilde{\mathrm{F}})^2 - \frac{1}{4}\mathrm{tr}(\mathrm{F} \pm \tilde{\mathrm{F}})^2 \end{aligned} \quad (4.20)$$

where $\tilde{F}$ is the worldspace Hodge dual of F. The trace is over the worldspace indices, i.e. $\mathrm{trF}^2 = \mathrm{F}_{ab}\mathrm{F}^{ba} \leq 0$, but the notation can be extended to include a trace over $su(n)$ indices, as appropriate to the $U(n)$ gauge theory on n coincident D-4-branes. In either case, th we deduce that [61]

$$\mathcal{E} \geq 1 \pm \frac{1}{4}\mathrm{trF\tilde{F}} \tag{4.21}$$

with equality when $F = \tilde{F}$. The total energy E, relative to the worldvolume vacuum, is just the integral of $\mathcal{E} - 1$ over the worldspace. Thus we derive the bound

$$E \geq |Z| \tag{4.22}$$

where Z is the topological charge

$$Z = \frac{1}{4}\mathrm{tr}\int \mathrm{F\tilde{F}}\,. \tag{4.23}$$

The bound is saturated by solutions of $F = \tilde{F}$. These are just multi-instantons in the non-abelian case, which is appropriate to coincident D-4-branes. In the abelian case any solution must have a singular $U(1)$ gauge potential, although the energy will remain finite for finite charge. Since not much is known about abelian BI instantons, let us concentrate on the non-abelian case. The soliton energy is independent of the instanton size, which is therefore a modulus of the solution. There is nothing to prevent us from shrinking the instanton to zero size. In fact, if we were to allow a slow time variation of the instanton parameters then the instanton could shrink to zero size in *finite* time. This is a reflection of the fact that the natural metric on instanton moduli space is geodesically incomplete, indicating that new physics is needed to determine what happens when an instanton shrinks to zero size. The new physics is provided by the interpretation of the (4+1)-dimensional Minkowski vacuum as the worldvolume of a D-4-brane: when the instanton shrinks to zero size it simply leaves the worldvolume as a D-0-brane [62].

The existence of a 1/2 supersymmetric soliton in the worldvolume field theory of the D-4-brane, or multi D-4-brane, is suggested by the presence of a central charge in its five-dimensional worldvolume supersymmetry algebra. This charge has a natural interpretation as the momentum in an additional space direction. In fact, the algebra is the reduction of the D=6 (2,0) supersymmetry algebra, and the (scalar) central charge arising in this reduction is precisely the 6-component of the 6-momentum. This is as expected because the D-4-brane has an M-theory interpretation as an S^1-wrapped M-5-brane, for which the (gauge-fixed) worldvolume field theory has D=6 (2,0) supersymmetry. Thus, M-theory predicts the existence of a tower of KK quantum states in the D-4-brane worldvolume field theory [63, 64] with one such state for each integer value of the soliton charge, alias instanton number. This amounts to a prediction of a marginal bound state in the system of n D-0-branes and a D-4-brane for each n. Any such bound state must indeed be marginal, and there is good evidence that the prediction is correct [63].

We have just invoked the presence of a central charge in the worldvolume supersymmetry algebra of the D-4-brane to explain the presence of a marginal 1/2 supersymmetric bound state with a D-0-brane, for which the long range supergravity fields are those of a 1/4 supersymmetric $(0|D0, D4)$ 'intersecting brane' configuration. However, as by now should be clear, there is no reason to concentrate exclusively on *scalar* charges; we should consider all charges. Since the D-4-brane is a wrapped M-5-brane it is more economical to consider the p-form charges in the M-5-brane's worldvolume supersymmetry algebra. Allowing for all possible p-form charges, we have [65]

$$\{Q^I_\alpha, Q^J_b\} = \Omega^{IJ} P_{[\alpha\beta]} + Y^{[IJ]}_{[\alpha\beta]} + Z^{(IJ)}_{(\alpha\beta)} \tag{4.24}$$

where $\alpha, \beta = 1, \ldots, 4$ is a spinor index of the Lorentz group $SU^*(4) \cong Spin(5,1)$ and $I = 1, \ldots, 4$ is an index of the internal 'R-symmetry' group $Sp(2)$, with Ω^{IJ} being its invariant antisymmetric tensor. The spinor supercharge therefore has 16 complex components but is subject to a 'symplectic Majorana' condition that reduces the number of independent components by a factor of 2. The Y-charge is a worldvolume 1-form and the Z-charge a worldvolume self-dual 3-form. For simplicity, we shall consider here only the space components of these charges, which would be carried by worldvolume p-branes for $p = 1$ and $p = 3$, respectively. The existence of these worldvolume branes could be anticipated from the fact that there exist 1/4 supersymmetric intersecting brane configurations of M-theory in which an M-5-brane intersects another M-brane on a string (the $(1|M2, M5)$ configuration) or a 3-brane (the $(3|M5, M5)$ configuration).

It might appear that there are more p-brane charges than are required because the $Sp(2)$ representations of these charges are

$$Y(p = 1) : \mathbf{5} \qquad Z(p = 3) : \mathbf{10}\,. \tag{4.25}$$

However, $Sp(2)$ is isomorphic to $Spin(5)$, which we may interpret as the rotation group in the 5-space transverse to the M–brane worldvolume in the D=11 spacetime. In this case, the $Sp(2)$ representations simply provide the information needed for the *spacetime* interpretation of the worldvolume branes as intersections with other branes. Specifically, the **5** and **10** representations can be interpreted as, respectively, a 1-form and a 2-form in the transverse 5-space. A p-form charge that is a transverse q-form is naturally interpreted as the charge of a (p+q)-brane in spacetime with a p-brane intersection with (in this case) the M-5-brane. We thereby recover the from the worldvolume supersymmetry algebra the spacetime interpretation of the 1-brane and 3-brane on the M-5-brane that we earlier deduced by consideration of the spacetime supersymmetry algebra [66].

Just as the scalar charge Z in the D-4-brane's worldvolume superalgebra was expressible as the topological charge integral (4.23), so the worldvolume charges Y and Z in the M-5-brane's worldvolume superalgebra must be expressible as integrals of worldvolume fields over the transverse subspace of worldspace. Let us concentrate on the string charge. Its magnitude must be expressible as an integral over the 4-dimensional subspace of worldspace transverse to the string; let us call this w_4. Since Y is a 5-vector in the transverse subspace of spacetime, it must depend on one of the five worldvolume scalar fields $\mathbf{X}$ describing fluctuations of the M-5-brane in these directions; the choice of scalar corresponds to a choice of direction of the 5-vector. These considerations imply that Y must take the form [61]

$$Y = \int_{w_4} H \wedge dX \tag{4.26}$$

where H is the closed 3-form field strength of the 2-form worldvolume gauge potential and X is the scalar.

If we set to zero all other physical scalars then, in the static gauge, a static configuration is one for which

$$X^M = (t, \sigma^a, X(\sigma), 0, \ldots, 0) \tag{4.27}$$

where (t, σ^a) are the worldvolume coordinates. The conjugate momentum is

$$P_M = (-\mathcal{E}, V_a, 0, 0, \ldots, 0) \tag{4.28}$$

where $\mathcal{E}$ is the energy density, V_a is the worldspace vector density introduced in (1.28), and we have used the fact that the diffeomorphism constraint reduces in static gauge, and for static configurations, to $P_a = V_a$. Under these conditions the hamiltonian constraint of (1.28) implies that

$$\mathcal{E}^2 = 1 + (\partial X)^2 + \frac{1}{2}|\tilde{\mathcal{H}}|^2 + |\tilde{\mathcal{H}} \cdot \partial X|^2 + |V|^2 \tag{4.29}$$

where the vertical bars indicate contraction with the kronecker delta. Here we are following [61], in which it is pointed out that this expression for the energy density of static worldvolume field configurations can be rewritten as

$$\mathcal{E}^2 = \left|\zeta^a \pm \tilde{\mathcal{H}}^{ab}\partial_b X\right|^2 + 2\left|\partial_{[a}X\zeta_{b]} \mp \frac{1}{2}\delta_{ac}\delta_{bd}\tilde{\mathcal{H}}^{cd}\right|^2 + \frac{1}{2}(\zeta^a\partial_a X)^2 \tag{4.30}$$

where ζ is a constant unit 5-vector. One can deduce from this (e.g. by making a particular choice for ζ) that

$$\mathcal{E} - 1 \geq \pm i_\zeta(\star H dX) \tag{4.31}$$

where $\star$ is the Hodge dual on worldspace and i_ζ indicates contraction with the (constant) vector field ζ. The inequality is saturated when

$$H = \pm i_\zeta(\star dX) \qquad \mathcal{L}_\zeta X = 0 \tag{4.32}$$

where $\mathcal{L}_\zeta$ is the Lie derivative with respect to ζ. Note that these conditions imply that $\mathcal{L}_\zeta H = 0$ (and hence that $V_a = 0$). Since the field configuration saturating the bound is translational invariant in the ζ direction we should integrate the energy density over the subspace w_4 orthogonal to this direction and interpret the result as a string tension T. The result is that

$$T \geq |Y| \tag{4.33}$$

where Y is the topological charge of (4.26), and the bound is saturated when

$$H = \pm * dX \tag{4.34}$$

where it is now understood that all fields are defined on w_4 and $*$ is the Hodge dual in this space.

Since $dH = 0$, we learn from (4.34) that X is harmonic on w_4. Isolated singularities of this harmonic function represent infinite straight string defects in the M-5-brane [67]. We may assume that X vanishes at spatial infinity, so that if the only singularity is at the origin then $X = X(r)$ where

$$X(r) = \frac{1}{r^2}\,. \tag{4.35}$$

As we approach the singularity at $r = 0$, the position of the M-5-brane in the transverse 5-space extends further and further along the X-axis, thereby by creating an infinite 'spike' in this direction[11]. The 'spike' is actually a 'ridge' due to the translational invariance in the ζ-direction. To determine the nature of this ridge we note that the topological charge Y may be rewritten as the small radius limit of a surface integral over a 3-sphere centred on the origin. Since X is constant over this integration surface we then have

$$Y = \big[\lim_{\delta\to 0} X(\delta)\big]\int_{S^3} H\,. \tag{4.36}$$

The integral of H is what one might naively have taken to be the string charge. We see that the charge is instead the product of the naive charge with an infinite factor, since $X(\delta)$ diverges as $\delta \to 0$. However, this divergence is exactly what we would expect if the string is the boundary of a membrane of constant surface tension $T_2 = \int_{S^3} H$. Thus, the 'ridge' of the solution is actually a membrane with its boundary on the M-5-brane. Remarkably, the worldvolume string soliton of the M-5-brane provides its own *spacetime* interpretation as the boundary of an M-2-brane.

Epilogue

We have seen that an enormous amount of information about M-theory and its duality relations to superstring theories is encoded in the M-theory and Type II supersymmetry algebras. We have also begun to see how 1/4-supersymmetric intersections of the 'basic', 1/2-supersymmetric, branes are encoded in their worldvolume supersymmetry algebras. The next step would be to consider supersymmetry fractions less than 1/4; for example 1/8, which corresponds to N=1 in D=4. This and other fractions are realized by certain non-orthogonal intersections and some triple intersections. Having classified possible fractions of supersymmetry associated with intersecting brane configurations our next step would be to determine the supersymmetric field theories on the intersections. It seems possible that all supersymmetric field theories may be obtainable this way. However, there is much

[11]This feature of 1/2 supersymmetric worldvolume solitons was first noted in the context of D-branes [68, 69] and extended to the string soliton in the M-5-brane in [67]. The presentation here follows [61].

less that can be said about these further developments from the algebraic viewpoint advocated here, at least for the present, so this is a good place to end these lectures.

Acknowledgements

The written version of these lectures was completed while on leave at the University of Barcelona. I thank the faculty of Physics for their hospitality and am grateful for the support of the Iberdrola *Profesor Visitante* program. In addition, I thank my recent collaborators Eric Bergshoeff, Martin Cederwall, Phillip Cowdall, Jerome Gauntlett, Gary Gibbons, Joachim Gomis, Chris Hull, George Papadopoulos and Dmitri Sorokin for conversations related to work reviewed here.

References

1. W. Nahm, *Supersymmetries and their representations*, *Nucl. Phys.* **B135** (1978) 149.
2. E. Cremmer, B. Julia and J. Scherk, *Supergravity theory in 11 dimensions*, *Phys. Lett.* **76B** (1978) 409.
3. C.M. Hull, *Exact p-p wave solutions of eleven-dimensional supergravity*, *Phys. Lett.* **139B** (1984) 39.
4. E. Bergshoeff, E. Sezgin and P.K. Townsend, *Supermembranes and 11-dimensional supergravity*, *Phys. Lett.* **189B** (1987) 75.
5. J. A. de Azcárraga, J.P. Gauntlett, J.M. Izquierdo and P.K. Townsend, *Topological extensions of the supersymmetry algebra for extended objects*, *Phys. Rev. Lett.* **63** (1989) 2443.
6. J.W. van Holten and A. Van Proeyen, *N=1 supersymmetry algebras in D=2, D=3, D=4 mod 8*, *J. Phys.* **A 15** (1982) 3763.
7. H. Nicolai and A. van Proeyen, *Off-shell representations with central charges for ten-dimensional super Yang-Mills Theory*, *Nucl. Phys.* **B203** (1982) 510.
8. P. Zizzi, *Antisymmetric tensors in supersymmetry algebras and spontaneous compactification*, *Phys. Lett.* **149B** (1984) 333.
9. E. Bergshoeff and E. Sezgin, *Super p-brane theories and new spacetime superalgebras*, *Phys. Lett.* **354B** (1995) 256.
10. E. Sezgin, *The M-algebra*, *Phys. Lett.* **392B** (1997) 323.
11. J. Hughes and J. Polchinski, *Partially broken global supersymmetry and the superstring*, *Nucl. Phys.* **B278** (1986) 147.
12. J.X. Lu, *ADM masses for black strings and p-branes*, *Phys. Lett.* **313B** (1993) 29.
13. G.W. Gibbons, G.T. Horowitz and P.K. Townsend, *Higher dimensional resolution of dilaton black hole singularities*, *Class. Quantum Grav.* **12** (1995) 297.
14. M.J. Duff and K.S. Stelle, *Multi-membrane solutions of D=11 supergravity*, *Phys. Lett.* **253B** (1981) 113.
15. M.J. Duff, G.W. Gibbons and P.K. Townsend, *Macroscopic superstrings as interpolating solitons*, *Phys. Lett.* **332 B** (1994) 321.
16. R. Güven, *Black p-brane solutions of D=11 supergravity*, *Phys. Lett.* **276B** (1992) 49.
17. P.K. Townsend, *Brane Surgery*, *Nucl. Phys. (Proc. Suppl.)* **B58** (1997) 163-175; hep-th/9609217.
18. A. Strominger, *Open p-branes*, *Phys. Lett.* **383B** (1996) 44.
19. P.K. Townsend, *D-branes from M-branes*, *Phys. Lett.* **373B**, (1996) 68.
20. J. Polchinski, *Dirichlet branes and Ramond-Ramond charges*, *Phys. Rev. Lett.* **75** (1995) 184.

21. P.K. Townsend, *p-brane democracy*, in *Particles, Strings and Cosmology*, eds. J. Bagger, G. Domokos, A Falk and S. Kovesi-Domokos (World Scientific 1996), hep-th/9507048.
22. A. Candiello and K. Lechner, *Duality in supergravity theories*, *Nucl. Phys.* **B412** (1994) 479.
23. I. Bandos, N. Berkovits and D. Sorokin, *Duality-symmetric eleven-dimensional supergravity and its coupling to M-branes*, hep-th/9711055.
24. G.W. Gibbons and P.K. Townsend, *Vacuum interpolation in supergravity via super p-branes*, *Phys. Rev. Lett.* **71** (1993) 3754.
25. D. Kaplan and J. Michelson, *Zero modes for the D=11 membrane and fivebrane*, *Phys. Rev.* **D 53** (1996) 3474.
26. C. Callan, J. Harvey and A. Strominger, *Worldbrane actions for string solitons*, *Nucl. Phys.* **B367** (1991) 60.
27. E. Witten, *Fivebrane effective action in M-theory*, hep-th/9610234.
28. P.S. Howe and E. Sezgin, *D=11, p=5*, *Phys. Lett.* **394B** (1997) 62.
29. M. Aganagic, J. Park, C. Popescu and J.H. Schwarz, *Worldvolume action of the M-theory five-brane*, *Nucl. Phys.* **B496** (1997) 191.
30. I. Bandos, K. Lechner, A. Nurmagambetov, P. Pasti, D. Sorokin and M. Tonin, *Covariant action for the super-five-brane of M-theory*, *Phys. Rev. Lett.* **78** (1997) 4332.
31. P.S. Howe, E. Sezgin and P.C. West, *Covariant field equations of the M-theory five-brane*, *Phys. Lett.* **399B**, (1997) 49.
32. I. Bandos, K. Lechner, A. Nurmagambetov, P. Pasti, D. Sorokin and M. Tonin, *On the equivalence of different formulations of the M-theory fivebrane*, *Phys. Lett.* **408B** (1997) 135.
33. D. Sorokin and P.K. Townsend, *M-theory superalgebra from the M-5-brane*, hep-th/9708003.
34. E. Bergshoeff and P.K. Townsend, *Phys. Lett.* **412 B** (1997) 265.
35. P.K. Townsend, *Four Lectures on M-theory*, hep-th/9612121.
36. P.K. Townsend, *The 11-dimensional supermembrane revisited*, *Phys. Lett.* **350B** (1995) 184.
37. R. Sorkin, *Kaluza-Klein monopole*, *Phys. Rev. Lett.* **51** (1983) 87.
38. D.J. Gross and M. Perry, *Magnetic monopoles in Kaluza-Klein theory*, *Nucl. Phys.* **B226** (1983) 29.
39. G.T. Horowitz and A. Strominger, *Black strings and p-branes*, *Nucl. Phys.* **B360** (1991) 197.
40. P. Hořava and E. Witten, *Heterotic and Type I string dynamics from eleven dimensions*, *Nucl. Phys.* **B460** (1996) 506.
41. C.M. Hull, *Gravitational duality, branes and charges*, hep-th/9705162.
42. J. Polchinski, *Tasi lectures on D-branes*, hep-th/9611050.
43. M.J. Duff, P.S. Howe, T. Inami and K.S. Stelle, *Superstrings in D=10 from supermembranes in D=11*, *Phys. Lett.* **191B** (1987) 70.
44. E. Witten, *String theory dynamics in various dimensions*, *Nucl. Phys.* **B443** (1995) 85.
45. C.M. Hull and P.K. Townsend, *Enhanced gauge symmetries in superstring theories*, *Nucl. Phys.* **B451**, (1995) 525.
46. A. Sen, *Dynamics of multiple Kaluza-Klein monopoles in M- and String theory*, hep-th/9707042.
47. E. Witten, *Bound states of strings and p-branes*, *Nucl. Phys.* **B460** (1996) 335.
48. T. Banks, *Matrix Theory*, hep-th/9710231.
49. M. Dine, P. Huet and N. Seiberg, *Large and small radius in string theory*, *Nucl. Phys.* **B322** (1989) 301.
50. J. Dai, R. Leigh and J. Polchinski, *New connections between string theories*, *Mod. Phys. Lett.* **A4** (1989) 251.
51. J. Polchinski and E. Witten, *Evidence for Heterotic-Type I string duality*, *Nucl.*

Phys. **B460** (1996) 525.
52. E. Bergshoeff, M. de Roo, M.B. Green, G. Papadopoulos and P.K. Townsend, *Duality of Type II 7-branes and 8-branes*, *Nucl. Phys.* **B470** (1996) 113.
53. C.M. Hull and P.K. Townsend, *Unity of superstring dualities*, *Nucl. Phys.* **B438** (1995) 109.
54. M. Cederwall and P.K. Townsend, *The manifestly $Sl(2;\mathbb{Z})$-covariant superstring*, *JHEP* 09, (1997) 003, hep-th/9709002.
55. J.M. Izquierdo, N.D. Lambert, G. Papadopoulos and P.K. Townsend, *Dyonic Membranes*, *Nucl. Phys.* **B460**, (1996) 560.
56. J.H. Schwarz, *An $Sl(2;\mathbb{Z})$ Multiplet of Type II B Superstrings*, *Phys. Lett.* **B360** (1995) 13; Erratum: ibid. **B364** (1995) 252.
57. G. Papadopoulos and P.K. Townsend, *Intersecting M-branes*, *Phys. Lett.* **380B** (1996) 273.
58. A.A. Tseytlin, *Harmonic superpositions of M-branes*, *Nucl. Phys.* **B475** (1996) 149.
59. J.P. Gauntlett, D.A. Kastor and J. Traschen, *Overlapping Branes in M-Theory*, *Nucl. Phys.* **B478** (1996) 544.
60. J.P. Gauntlett, *Intersecting Branes*, hep-th/9705011.
61. J.P. Gauntlett, J. Gomis and P.K. Townsend, *BPS bounds for worldvolume branes*, hep-th/9711205.
62. M. Douglas, *Branes within branes*, hep-th/9512077.
63. M. Douglas, D. Kabat, P. Pouliot and S.H. Shenker, *D-branes and Short Distances in String Theory*, *Nucl. Phys.* **B485** (1997) 85.
64. G. Papadopoulos and P.K. Townsend, *Kaluza-Klein on the brane*, *Phys. Lett.* **393B** (1997) 59.
65. P.S. Howe, N.D. Lambert and P.C. West, *The threebrane soliton of the M-5-brane*, hep-th/9710033.
66. E. Bergshoeff, J. Gomis and P.K. Townsend, *M-brane intersections from worldvolume superalgebras*, hep-th/9711043.
67. P.S. Howe, N.D. Lambert and P.C. West, *The self-dual string soliton*, hep-th/9709014.
68. C.G. Callan and J.M. Maldacena, *Brane dynamics from the Born-Infeld action*, hep-th/9708147.
69. G.W. Gibbons, *Born-Infeld particles and Dirichlet p-branes*, hep-th/9709027.

LECTURES ON HETEROTIC–TYPE I DUALITY

I. ANTONIADIS AND H. PARTOUCHE
TH-Division, CERN
CH-1211 Geneva 23, Switzerland

Centre de Physique Théorique, Ecole Polytechnique
F-91128 Palaiseau cedex, France

AND

T.R. TAYLOR
Department of Physics, Northeastern University
Boston, MA 02115, U.S.A.

1. Introduction

In the last couple of years, there has been remarkable progress in understanding non-perturbative aspects of string theories. The new ingredient is the discovery of various duality symmetries; these, in particular, invert the coupling constant and exchange perturbative states with non-perturbative solitonic excitations. Since inversion of the coupling involves the Planck constant $\hbar$, duality symmetries can be realized only at the quantum level. Once the dual models are identified, non-perturbative dynamics of one model can be studied by means of the standard perturbative techniques in the other model.

Dualities are known under the names of S, T and U, depending on whether they invert the string coupling in the target space-time, or the coupling of the underlying two-dimensional field theory of the string world-sheet, or exchange these two, respectively [1]. One of the main consequences of string dualities is that all known, apparently distinct, superstring theories correspond to different perturbative expansions of the same underlying theory, the so-called M-theory [2], whose low-energy limit is eleven-dimensional supergravity. In fact, there are five consistent superstring theories in ten dimensions: two type II theories of closed strings, IIA and IIB, which have two supersymmetries in ten dimensions, of opposite and same chirality, respectively; two heterotic closed string theories with $N = 1$ supersymmetry and $SO(32)$ or $E_8 \times E_8$ gauge groups; and the type I theory of open and closed strings with $N = 1$ supersymmetry and $SO(32)$ gauge group [3].

It was already known for quite some time that the two type II theories, as well as the two heterotic ones, are related by T-duality, which is an exact perturbative

L. Baulieu et al. (eds.), Strings, Branes and Dualities, 179–199.

symmetry after compactifying to lower dimensions. For instance, upon compactification to nine dimensions on a circle of radius R, type IIA and type IIB theories become equivalent by inverting R:

$$R \overset{T}{\longleftrightarrow} \frac{\alpha'}{R}, \tag{1.1}$$

where α' is the Regge slope. In four dimensions, the two theories are also identified by mirror symmetry with respect to the internal six-dimensional Calabi–Yau manifold. T-duality can then be used to compute world-sheet instanton corrections by using algebraic geometry in the mirror (dual) theory [4]. Similarly, the two heterotic theories are equivalent but T-duality is more complicated due to the presence of Wilson lines.

It has recently been realized that, in ten dimensions, there are two fundamental non-perturbative S-dualities. On the one hand, the heterotic and type I $SO(32)$ theories are equivalent by inverting the string coupling [2, 5]:

$$\lambda \overset{S}{\longleftrightarrow} \frac{1}{\lambda}. \tag{1.2}$$

We recall that the string coupling is given by the vacuum expectation value (vev) of the dilaton field: $\lambda = \langle e^{\phi} \rangle$. On the other hand, type IIB is self-dual under an $SL(2;\mathbf{Z})$ S-duality [6] acting on the complex field $\chi + i\lambda$, where χ is the Ramond–Ramond (RR) scalar. In this way, the two $N = 2$ and three $N = 1$ theories are separately connected. Their mutual connection is believed to arise from U-duality via M-theory [2, 7]. In fact, the coupling constants of the type IIA and heterotic $E_8 \times E_8$ strings can be identified with the radius of the eleventh dimension of the M-theory compactified on S^1 and $S^1/\mathbf{Z}_2$, respectively:

$$\lambda = (R_{11}/l_P)^{3/2}, \tag{1.3}$$

where l_P is the Planck length in eleven dimensions.

The purpose of these notes is to describe the main features, consequences and tests of heterotic–type I string duality. In Section 2, we give a brief introduction to type I strings, D-branes and orientifolds. In Section 3, we derive the main duality conjectures in dimensions 10, 6 and 4 by arguments based on the effective field theory. In Section 4, we discuss duality in six dimensions with $N = 1$ supersymmetry, while in Section 5, we study four-dimensional compactifications with $N = 2$ supersymmetry. In Section 6, we present general results for the one-loop corrections to the $N = 2$ prepotential and vector moduli metrics. We then perform various duality tests in the Higgs (Section 7.1) and Coulomb (Section 7.2) phase of $(T^4/\mathbf{Z}_2) \times T^2$ compactifications. Finally, in Section 8, we compare the infinite series of couplings corresponding to higher-derivative F-terms.

2. Type I Strings and D-Branes

Type I strings arise as world-sheet orbifolds (orientifolds) of the world-sheet left–right-symmetric IIB theory by modding it out with respect to the involution Ω that exchanges left- and right-movers [8, 9]. The projection in the "untwisted" closed string sector introduces unoriented closed strings and has the effect of symmetrizing the Neveu–Schwarz–Neveu–Schwarz (NSNS) sector and antisymmetrizing the RR sector. As a result, the number of supersymmetries is reduced by half. In ten dimensions, the invariant (bosonic) massless states are the dilaton and graviton of the NSNS sector and the 2-index antisymmetric tensor of the RR sector.

Open strings appear as type IIB strings that close up to an Ω reflection, in a way similar to "twisted" states in orbifold constructions:

$$\Omega X(\sigma,\tau) \equiv X(-\sigma,\tau) = X(\sigma,\tau)\,. \tag{2.1}$$

Indeed, the mode expansion of an Ω-invariant free string satisfying Eq. (2.1) is

$$\begin{aligned} X(\sigma,\tau) &= x + 2\alpha' p\tau \\ &+ i\sqrt{2\alpha'}\sum_{m\neq 0}\frac{\alpha_m}{m}\cos(m\sigma)e^{-im\tau}\,, \end{aligned} \tag{2.2}$$

which is the usual open string with Neumann boundary conditions $\partial_\sigma X|_{\sigma=0,\pi} = 0$. Open strings carry at their ends Chan–Paton charges that give rise to a gauge group. From the orientifold point of view, they play the role of fixed points. Their multiplicity is constrained by the tadpole cancellation which replaces (oriented) closed string modular invariance. In ten dimensions, one finds 32 real charges leading to an $SO(32)$ gauge group.

The modern approach to type I theory emphasizes the role of p-brane solitons of the corresponding type II string theory. In particular, the multiplicity of (complex) charges corresponds to the number of the so-called Dp-branes where open strings can end. Dp-branes provide sources for the RR type II (p+1)-potentials [10]. An open string with one end (say at $\sigma = 0$) on a Dp-brane has Neumann boundary conditions for X^μ with $\mu = 0, 1, \ldots$,p and Dirichlet boundary conditions along the perpendicular directions, $\partial_\tau X^\mu|_{\sigma=0} = 0$ for $\mu = \mathrm{p}+1, \ldots, 9$. These conditions fix the position of the string end in the space orthogonal to the p-brane, allowing it to move freely only in the directions parallel to the p-brane. According to D-brane terminology, strings with one end on a Dp-brane and the other one on a Dq-brane are called pq strings.

Type IIB has even-form RR potentials that can couple to Dp-branes with p odd. The Ω projection leaves invariant 2 mod 4 forms coupled to D1-, D5- and D9-branes only. In ten dimensions, Lorentz invariance allows only D9-branes with multiplicity 16 fixed by the tadpole cancellation. However after compactification to six or four dimensions, D5-branes can also appear.

Following the discussion above, the 1-loop partition function receives contributions from four genus-one surfaces: the torus ($\mathcal{T}$), the Klein bottle ($\mathcal{K}$), the annulus ($\mathcal{A}$) and the Möbius strip ($\mathcal{M}$). It takes the generic form:

$$Z_{1\text{-loop}} = \frac{1}{2}(\mathcal{T}+\mathcal{K}) + \frac{1}{2}(\mathcal{A}+\mathcal{M})\,, \tag{2.3}$$

where the two terms in brackets correspond to closed (untwisted) and open (twisted) states. In particular, $\mathcal{K}$ and $\mathcal{M}$ describe the propagation of closed and open unoriented strings, imposing the Ω projection in the two sectors of the theory.

T-duality can be incorporated into the open string framework with the main effect of interchanging Neumann (N) with Dirichlet (D) boundary conditions. In fact, imposing NN conditions in some compact direction $X \equiv X + 2\pi R$ yields

$$X(\sigma,\tau) = X_L(\tau+\sigma) + X_R(\tau-\sigma)\,, \tag{2.4}$$

and gives the expression (2.2) with the quantized momentum $p = n/R$. In analogy with closed strings, T-duality transforms $X_L \to X_L$, $X_R \to -X_R$, and inverts the radius R to $\tilde{R} = \alpha'/R$ (1.1). It follows that $X \to \tilde{X}$, with the dual coordinate

$$\begin{aligned} \tilde{X}(\sigma,\tau) &= X_L(\sigma+\tau) - X_R(\sigma-\tau)\,, \\ &= \tilde{x} + 2nR\sigma + i\sqrt{2\alpha'}\sum_{l\neq 0}\frac{\alpha_l}{l}\sin l\sigma\,. \end{aligned} \tag{2.5}$$

We see that $\tilde{X}(0,\tau) = \tilde{x}$ and $\tilde{X}(\pi,\tau) = \tilde{x} + 2\pi nR$, which means that the string is wrapped n times around the circle and has Dirichlet boundary conditions on a D-brane at $\tilde{X} = \tilde{x}$. Thus a Dp-brane compactified on a circle is exchanged under T-duality with a D(p-1)-brane obtained by wrapping around the compact dimension.

Furthermore, under T-duality, Ω is mapped to $\tilde{\Omega} = \Omega\mathcal{R}$ where $\mathcal{R}$ is the $\mathbf{Z}_2$ orbifold transformation $X \to -X$. Indeed, starting from Neumann boundary conditions for the compact coordinate X (2.4) and performing a T-duality transformation, one finds that the dual coordinate $\tilde{X}$ (2.5) is antiperiodic under the exchange of left- and right-movers.

Finally, T-duality exchanges Wilson lines of the gauge fields associated with Neumann boundary conditions with the D-brane positions in the compactified space. In fact, in the presence of Wilson lines $a_{i,j}$ corresponding to an NN (traceless) gauge group generator $T_{(i,j)} = \mathrm{diag}(0\ldots 1_i \ldots -1_j \ldots 0)$, on a circle, the compact momentum is shifted as:

$$\frac{n}{R} \to \frac{n + a_i - a_j}{R}\,. \tag{2.6}$$

Using the same line of reasoning as before, it is easy to see that the dual coordinate satisfies $\tilde{X}(\pi,\tau) - \tilde{X}(0,\tau) = 2\pi(n + a_i - a_j)R$. This shows that up to a common shift, a_i's represent positions of the open string ends, or equivalently the location of D-branes.

3. Effective Field Theories

Effective Lagrangians, and most of all their dependence on the dilaton fields, provide a powerful guide into various duality symmetries. Let ϕ_n denote the dilaton field in n non-compact dimensions. Focusing on the Einstein and Yang–Mills kinetic terms of the heterotic $SO(32)$ and type I string effective Lagrangians we have

$$\mathcal{L} = e^{-2\phi_{10}}\left\{\frac{1}{2}R + \frac{1}{4}F^2 \,\bigg|\, \frac{1}{e^{\phi_{10}}} + \ldots\right\}, \tag{3.1}$$

where the heterotic and type I gauge kinetic factors are assembled together as upper and lower entries, respectively. These Lagrangian terms contain also implicit dimensionful factors, which are all equal to 1 in the units of α'.

After compactification from 10 to $D \leq 10$ dimensions on the same manifold of volume V_{10-D},

$$\mathcal{L} = e^{-2\phi_{10}} V_{10-D} \left\{ \frac{1}{2} R + \frac{1}{4} F^2 \left| \begin{matrix} 1 \\ e^{\phi_{10}} \end{matrix} \right. + \dots \right\} . \tag{3.2}$$

The above expressions allow the following identification of dilatons in dimension D:

$$e^{-2\phi_D} \equiv e^{-2\phi_{10}} V_{10-D} , \tag{3.3}$$

in terms of which we have

$$\mathcal{L} = e^{-2\phi_D} \left\{ \frac{1}{2} R + \frac{1}{4} F^2 \left| \begin{matrix} 1 \\ e^{\phi_D} V_{10-D}^{1/2} \end{matrix} \right. + \dots \right\} . \tag{3.4}$$

Finally, we go to the Einstein frame by rescaling the metric $g_{\mu\nu} \to g_{\mu\nu}\, e^{4\phi_D/(D-2)}$ to obtain

$$\mathcal{L} = \frac{1}{2} R + \frac{1}{4} F^2 \left| \begin{matrix} e^{\frac{-4}{D-2}\phi_D} \\ e^{\frac{D-6}{D-2}\phi_D} V_{10-D}^{1/2} \end{matrix} \right. + \dots \tag{3.5}$$

For $D = 10$, the identification of the two Lagrangians implies $\phi_{10}^H = -\phi_{10}^I$; the two string theories are equivalent under S-duality. In each case, since the gravitational coupling constant is $\kappa_{10}^2 = e^{2\phi_{10}} \alpha'^4$ [2, 5],

$$\lambda_{10} \leftrightarrow 1/\lambda_{10} \quad ; \quad \alpha' \leftrightarrow \lambda_{10} \alpha' . \tag{3.6}$$

A derivation of type I and heterotic theories from eleven dimensions provides more insight into their duality. One can see that M-theory compactified on $S^1/\mathbf{Z}_2 \times S^1$ and $S^1 \times S^1/\mathbf{Z}_2$ gives rise to the heterotic and type I strings compactified on S^1 [7]. Here, $\mathbf{Z}_2$ acts as the reflection on the eleventh and tenth coordinate, respectively, changing also the sign of the 3-form. Moreover in the second case, one can show that the $\mathbf{Z}_2$ symmetry is identical to $\Omega\mathcal{R}$ which, as shown in Section 2, is T-dual to the world-sheet involution of type IIB, allowing construction of type I strings from type IIA. The corresponding identifications are [7]:

$$\lambda^I = \frac{R_{11}^M}{R_{10}^M} = 1/\lambda^H , \tag{3.7}$$

$$R_{10}^I = \frac{1}{R_{10}^M \sqrt{R_{11}^M}} \quad , \quad R_{10}^H = \frac{1}{R_{11}^M \sqrt{R_{10}^M}} ,$$

where $R_{10,11}^M$, R_{10}^I and R_{10}^H are the compact radii in the respective theories. Thus, S-duality is now recovered from U-dualities of M-theory compactified on a torus whose complex structure (3.7) is identified with the coupling constants.

For $D = 6$, one derives from Eq. (3.5) a U-duality relation $e^{-2\phi_6^H} = V_4^I$. Moreover, using Eqs. (3.3) and (3.6), one obtains the same identification with the indices H and I exchanged, hence

$$V_4 \leftrightarrow 1/\lambda_6^2 . \tag{3.8}$$

For $D = 4$, Eq. (3.5) implies a weak–weak coupling duality with $e^{-2\phi_4^H} = e^{-\phi_4^I}(V_6^I)^{1/2}$. Using Eqs. (3.3) and (3.6), one also obtains $V_6^H = e^{-3\phi_4^I}(V_6^I)^{-1/2}$ and similar relations with the indices H and I exchanged. Finally, we get [11]

$$\lambda_4^2 \leftrightarrow \lambda_4 V_6^{-1/2} \quad ; \quad V_6 \leftrightarrow \lambda_4^{-3} V_6^{-1/2} . \tag{3.9}$$

It is interesting that the ten-dimensional strong–weak coupling duality between heterotic and type I strings leads in lower dimensions to relations that hold when both theories are weakly coupled and thus, can be tested at the perturbative level. These relations can also be used to study non-perturbative effects in one theory by using the perturbative description of the dual theory. For instance, world-sheet instanton contributions of the characteristic magnitude $e^{-V_4^{1/2}/\alpha'}$ or $e^{-V_6^{1/3}/\alpha'}$ are mapped through the relations (3.8) and (3.9) to non-perturbative effects of order e^{-1/λ_6} and e^{-1/λ_4}, respectively. Since λ_4^H is also the gauge coupling, such world-sheet instantons on the type I side translate on the heterotic theory to stringy non-perturbative effects of order $e^{-1/g}$, which are much stronger than ordinary field-theoretical instantons.

4. Type I–Heterotic Duality in Six Dimensions

The first non-trivial test of heterotic–type I duality is for six-dimensional vacua with $N = 1$ supersymmetry, obtained upon compactification of the ten-dimensional theories on $K3$ manifolds [12].

We start by recalling some basic facts about $D = 6$, $N = 1$ supersymmetry. Besides the gravitational multiplet, which contains 12 bosonic and 12 fermionic degrees of freedom, namely the metric, an anti-self-dual 2-index antisymmetric tensor and the gravitino (with 9, 3 and 12 physical components respectively), there are three different matter multiplets: the tensor (T), the vector (V) and the hyper (H), containing $4+4$ physical states. These are a self-dual 2-form and a real scalar (for T), a vector (for V) and two complex scalars (for H), together with a Weyl spinor. Cancellation of the gravitational anomaly in six dimensions requires the number of massless matter multiplets n_T, n_V and n_H to satisfy the following constraint:

$$29 n_T + n_H - n_V = 273 . \tag{4.1}$$

Perturbative heterotic spectra always contain one tensor multiplet whose scalar component is identified with the six-dimensional dilaton ϕ_6^H. All other moduli belong to hypermultiplets, since vector multiplets contain no scalars. Moreover, Eq. (4.1) fixes the difference between hypers and vectors, $n_H - n_V = 244$. On the other hand, perturbative type I strings may contain several tensor multiplets. Therefore in order to make perturbative duality tests possible, we will restrict ourselves to type I vacua with $n_T = 1$. In contrast to the heterotic case, however, the type I dilaton ϕ_6^I belongs to a hypermultiplet, in agreement with the duality relation (3.8).

At a generic point of the moduli space, all massless hypermultiplets are neutral under the gauge group. Supersymmetry requires that hypermultiplets decouple from the interactions of vectors and tensors, so that the effective action splits into two terms: $S_{eff} = S_{T,V} + S_H$, where $S_{T,V}$ depends on tensors and vectors only, while S_H describes the interactions of hypers. The important point is that the type I string coupling is given by a vev of a hypermultiplet, while the heterotic coupling is determined by a vev of a tensor multiplet. It follows that $S_{T,V}$ receives no quantum corrections on the type I side, while S_H is determined classically on the heterotic side.

Here, we focus only on $S_{T,V}$, which is easier to analyze. In fact, its form is almost completely fixed by supersymmetry and anomaly cancellation. In the Einstein frame, the bosonic part of the Lagrangian becomes [13]:

$$\begin{aligned}\mathcal{L}^{(6)}_{T,V} &= \frac{1}{2}R + \frac{1}{2}\left(\frac{\partial\omega}{\omega}\right)^2 + \frac{1}{4}(v\omega + v'\omega^{-1})F^2 \\ &+ \frac{1}{16}\omega^2(dB - v\Omega)^2 + \frac{v'}{16}B\wedge F\wedge F, \end{aligned} \tag{4.2}$$

where ω is the scalar of the tensor multiplet, B is the 2-form potential, F denotes the field-strengths of gauge fields with Ω the corresponding Chern–Simons terms, v and v' are gauge group-dependent constants, and an implicit summation over gauge fields is understood.

On the heterotic side, $\omega \equiv e^{-\phi_6^H}$, therefore v is given classically by the Kač–Moody level k of the underlying affine Lie algebra, while v' represents a one-loop string correction. The difference $\Delta v'$ between two gauge groups coincides with the corresponding difference of $N=2$ β-functions that one obtains upon further compactification to $D=4$ on a T^2 [14]. This can be compared with the result of the dual type I theory, where $\omega \equiv V_{K3}^{1/2}$, and both v and v' are determined at the tree level.

Gauge fields with $v=0$ are non-perturbative from the heterotic point of view and correspond to parallel 5-branes on the type I side. On the other hand, when $vv' < 0$ for some gauge field, there is a point in the tensor moduli space where the gauge coupling blows up. It has been argued that such a singularity is resolved by the appearance of tensionless strings [15], generated on the type I side by type IIB 3-branes wrapping around a collapsing 2-cycle of $K3$. Note finally that if there is a $D=6$, $N=1$ type II vacuum with only one antisymmetric tensor, v' will be given at the tree level and v at one loop, as required by heterotic–type II S-duality in six dimensions.

We conclude this Section by describing one of the simplest $N=1$ heterotic–type I dual pairs in six dimensions. They are obtained by compactifying the ten-dimensional theories on the $K3$ orbifold $T^4/\mathcal{R}$, where $\mathcal{R}$ is the $\mathbf{Z}_2$ transformation $X^m \to -X^m$ for $m=6,7,8,9$.

On the heterotic side, the $\mathbf{Z}_2$ projection reduces the supersymmetry to $N=1$ and breaks the gauge group $SO(32)$ down to $U(16)$. From the original $SO(32)$ gauge multiplet there remain also two massless hypermultiplets transforming in the $\mathbf{120}_{1/2}$ representation. In addition, there are 4 untwisted neutral hypermultiplets parametrizing the moduli of T^4, as well as 16 twisted ones transforming in the

$\mathbf{16}_{1/4}$ representation. Of course, there is also the standard $N = 1$ supergravity multiplet together with the tensor dilaton multiplet.

This model is dual to the type I theory compactified on the same $T^4/\mathcal{R}$ which can be constructed from type IIB by modding it out by the orientifold group $\{1, \Omega, \mathcal{R}, \Omega\mathcal{R}\}$ [8, 16]. Then, besides the 16 D9–branes, tadpole cancellation requires the presence of 16 D5–branes. The massless spectrum of this $K3$ orientifold model is as follows. In the Ω-twisted sector, open strings satisfy 99 boundary conditions and the $SO(32)$ gauge group is broken by the $\mathcal{R}$ projection to $U(16)_9$. In the $\Omega\mathcal{R}$-twisted sector, one finds 55 strings with both endpoints fixed on $T^4/\mathcal{R}$, and an additional 5-brane gauge group of rank 16. The maximal gauge group is $U(16)_5$ which appears when all 5-branes are located at the same fixed point of the orbifold. There are also two massless hypermultiplets transforming in the $\mathbf{120}_{1/2}$ representation of $U(16)_5$, which parametrize the relative distances between the D5-branes, as mentioned in Section 2. When the 5-branes are pulled apart the gauge group is broken, and the maximum breaking occurs when each 5-brane is located at a separate fixed point; the gauge group is then $U(1)_5^{16}$ and there is no massless matter. There are also 95 + 59 strings which give 16 massless hypermultiplets in the $\mathbf{16}_{1/4}$ representation of $U(16)_9$ [or 1 in the $(\mathbf{16}_{1/4}, \mathbf{16}_{1/4})$ of $U(16)_9 \times U(16)_5$]. Finally, the closed string sector contains the $N = 1$ six-dimensional supergravity multiplet, one tensor multiplet and 20 hypermultiplets, 4 of them from the $\mathcal{R}$-untwisted sector and 16 from the $\mathcal{R}$-twisted one. Out of the four scalar components of each twisted hypermultiplet, three are coming from the symmetrized NSNS sector while the fourth one originates by Poincaré duality from a RR 4-form potential that is equivalent to a scalar in six dimensions.

The above type I massless spectrum is the same as in the heterotic model with the exception of the $U(1)_5^{16}$ gauge multiplets which are non-perturbative on the heterotic side, and 16 hypermultiplets from the $\mathcal{R}$-twisted closed string sector. However, all $U(1)_5$'s are broken due to anomalous couplings of their gauge field strengths with the RR 4-forms, $C_4^i \wedge F_2^i$ [17]. As a result these gauge multiplets become massive, absorbing all 16 twisted hypermultiplets, and the massless spectra of the two theories become identical. The effective Lagrangian that describes the interactions of the 9-brane gauge multiplets with the tensor multiplet is of the form (4.2) with $v = 1$ and $v' = 0$. In Section 7, we will compactify the above dual pair on T^2 and discuss some non-trivial duality tests in four dimensions.

5. Type I–Heterotic Duality in Four Dimensions

When heterotic and type I theories are compactified on $K3 \times T^2$, one obtains $N = 2$ supersymmetric models in four dimensions with $n_V + 3$ vector multiplets and n_H hypermultiplets. The three additional vector multiplets describe the universal sector which arises from the single tensor multiplet and the Kaluza-Klein reduction of the graviton multiplet on T^2. More precisely, the vector field components correspond to three linear combinations of $G_{\mu 4}$, $G_{\mu 5}$, $B_{\mu 4}$ and $B_{\mu 5}$, while the fourth combination can be identified with the graviphoton of the $N = 2$ supergravity multiplet. Similarly, their scalar components are combinations of the three T^2 metric components G_{44}, G_{55} and G_{45}, the internal antisymmetric tensor B_{45}, the four-dimensional dilaton ϕ_4, the $K3$ volume V, and the universal axion α dual to the antisymmetric tensor $B_{\mu\nu}$. In addition, there are $2n_V$ scalar components of the six-dimensional gauge fields $a_{4,5}$ that correspond to the Wilson lines on T^2. The way these fields are arranged into $N = 2$ multiplets is different in the two theories.

We first discuss the case of vanishing Wilson lines. On the heterotic side, the $K3$ volume belongs to a hypermultiplet while the remaining six scalars form the well-known complex fields:

$$S_H = \alpha + ie^{-2\phi_4} \quad , \quad T = B_{45} + i\sqrt{G}$$
$$U = (G_{45} + iG^{1/2})/G_{44}\,, \tag{5.1}$$

where $G = G_{44}G_{55} - G_{45}^2$. On the type I side, the six-dimensional string dilaton $\phi_6 = \phi_4 + (1/2)\ln\sqrt{G}$ remains in a hypermultiplet. A straightforward dimensional reduction of the six-dimensional Lagrangian (4.2) gives [11]

$$\mathcal{L}_V^{(4)} = \frac{1}{2}R + \frac{1}{4}(v\mathrm{Im}S_I + v'\mathrm{Im}S')F^2 + \ldots \tag{5.2}$$

where

$$S_I = \alpha + ie^{-\phi_4}G^{1/4}V^{1/2}\,,$$
$$S' = B_{45} + ie^{-\phi_4}G^{1/4}V^{-1/2}\,. \tag{5.3}$$

It is easy to see that with the above definitions and with U the same as in Eq. (5.1), S_I, S' and U have diagonal kinetic terms and form the scalar components of the three vector multiplets.

As seen in Eq. (5.1) the heterotic dilaton belongs to a vector multiplet, which implies that the hypermultiplet moduli space remains the same as in six dimensions and does not receive any quantum corrections. On the other hand, on the type I side, the four-dimensional dilaton ϕ_4 is a "mixture" of vector and hypermultiplet components, so that the string coupling $e^{\phi_4} = e^{\phi_6}(\mathrm{Im}S_I\mathrm{Im}S')^{-1/4}$. Hence, both hyper and vector sectors can receive quantum corrections once the string coupling combines with other fields to form the full scalar supermultiplet components.

Wilson lines can be turned on along the flat directions of the potential corresponding to the Cartan directions of the gauge group. They can be diagonalized to $a_{4,5}^i$, $i = 1\ldots r$, where r is the rank, and generically break the gauge group to its abelian Cartan subgroup $U(1)^r$. In the presence of Wilson lines, the $N = 2$ special coordinates for the type I theory become:

$$A_i = a_4^i - a_5^i U\,, \quad S_I = S_I|_{A=0} + \sum_i \frac{v_i'}{2} a_5^i A_i\,,$$
$$S' = S'|_{A=0} + \sum_i \frac{v_i}{2} a_5^i A_i\,. \tag{5.4}$$

Similarly, in the heterotic theory, A_i is defined as above, the T modulus is redefined as S', while S_H remains the same as in Eq. (5.1).

In terms of these fields, the tree-level prepotentials that determine the interactions of $N = 2$ vector multiplets in the two theories can be read off from the six-dimensional action (4.2) and take the form [11]:

$$\begin{aligned} \mathcal{F}^{I}_{\text{tree}} &= S_I S' U - \frac{1}{2}\sum_i (v_i S_I + v'_i S') A_i^2 \\ \mathcal{F}^{H}_{\text{tree}} &= S_H T U - \frac{1}{2}\sum_i v_i S_H A_i^2 \,. \end{aligned} \tag{5.5}$$

It follows that the heterotic–type I duality mapping inherited from Eq. (3.8) in $D = 6$, is [11]

$$S_H \leftrightarrow S_I, \quad T \leftrightarrow S', \quad U \leftrightarrow U, \quad A_i \leftrightarrow A_i \,. \tag{5.6}$$

Let us now discuss quantum corrections. As already mentioned above, both prepotentials can be modified quantum mechanically. Perturbative corrections are restricted by continuous Peccei–Quinn symmetries. On the heterotic side, there is a Peccei–Quinn symmetry associated with the universal axion $\alpha = \text{Re}S_H$, which is dual to the antisymmetric tensor. On the type I side, besides a similar symmetry associated with $\text{Re}S_I$, there is a second symmetry due to the fact that $\text{Re}S' = B_{45}$ originates from the RR sector. These symmetries, together with analyticity, imply that the perturbative corrections must be independent of $S_{I,H}$ and S', hence they are entirely due to one-loop effects.

The general forms of the respective prepotentials are:

$$\begin{aligned} \mathcal{F}^{I}(S,S',U,A) &= \mathcal{F}^{I}_{\text{tree}} + f^I(U,A) + \ldots \\ \mathcal{F}^{H}(S,T,U,A) &= \mathcal{F}^{H}_{\text{tree}} + f^H(T,U,A) + \ldots \end{aligned} \tag{5.7}$$

where $f^{I,H}$ are the one-loop corrections and dots represent non-perturbative contributions. In Eq. (5.7) and everywhere below we drop the subscripts I and H referring to S, in view of the identification (5.6). Since the Peccei–Quinn symmetries are broken by instanton effects to discrete shifts of axions, the non-perturbative corrections can be expanded in integer powers of $e^{2\pi i S}$ whose magnitude is suppressed by the instanton action.

As already mentioned before, heterotic–type I duality includes regimes that are weakly coupled on both sides. It is possible to compare prepotentials (as well as other quantities that will be discussed later) already at the perturbative level, in the appropriate limits of moduli parameters. First of all, in order to reach the perturbative limit on type I side, one has to take the limit of large $\text{Im}S$ and $\text{Im}S'$. This is mapped via Eq. (5.6) to the region of large $\text{Im}S$ and $\text{Im}T$ which corresponds to a weakly coupled heterotic theory in the limit of "large" T^2 torus compactification, provided that the limit is taken with $\text{Im}S > \text{Im}S'$. Thus the perturbative heterotic prepotential reproduces its type I counterpart in the region of $K3$ volume $V > 1$, cf. Eq. (5.3). More precisely, from the expansions (5.7), one obtains

$$f^H(T,U,A) \stackrel{T\to i\infty}{\longrightarrow} -\frac{1}{2}\sum_i v'_i T A_i^2 + f^I(U,A), \tag{5.8}$$

with the fields mapped as in Eq. (5.6). Note that a part of the one-loop heterotic prepotential is mapped into the tree-level type I term proportional to v_i', while the remaining part reproduces the type I one-loop correction.

It is also interesting to study the other perturbative branch of the type I theory with $V < 1$, which is mapped in the heterotic theory to the non-perturbative region $\mathrm{Im}T > \mathrm{Im}S \to \infty$. These two regions are related by type I T-duality $V \to 1/V$ which corresponds on the heterotic side to the non-perturbative $S \leftrightarrow T$ exchange. If in addition there exists a type IIA description of the same model, the exact prepotential can be determined classically on the type II side, and this region can be probed directly, providing a perturbative test of type I–type II duality that is non-perturbative from the heterotic point of view.

6. One-loop Corrections

The standard way of deriving the one-loop prepotential in heterotic theory [18, 19] is to extract it from the one-loop corrections to gauge couplings:

$$\frac{4\pi^2}{g^2} = \frac{\pi}{2}\mathrm{Im}S + \Delta^H , \tag{6.1}$$

where Δ^H is the threshold function [20]. Its moduli dependence is governed by [18]

$$\partial_U \partial_{\bar{U}} \Delta^H = b K^{H(0)}_{U\bar{U}} + 4\pi^2 K^{H(1)}_{U\bar{U}} , \tag{6.2}$$

where b is the beta-function coefficient, $K^{H(0)}_{U\bar{U}} = -1/(U-\bar{U})^2$ is the tree-level moduli metric and $K^{H(1)}_{U\bar{U}}$ is the one-loop correction. The first term in the r.h.s. of Eq. (6.2) depends on the gauge group while the second one is a universal, gauge-group–independent contribution. The latter can be used to extract the one-loop prepotential f^H.

A direct string computation gives Δ^H as an integral over the complex Teichmüller parameter $\tau = \tau_1 + \tau_2$ of the world-sheet torus in its fundamental domain [18]:

$$\Delta^H = -\frac{1}{2}\int \frac{d^2\tau}{\tau_2} \bar{\eta}^{-2} \mathrm{Tr}_R F(-)^F \left(Q^2 - \frac{1}{2\pi\tau_2}\right), \tag{6.3}$$

where η is the Dedekind eta function, Q is the gauge group generator, and the trace is over the Ramond sector of the internal $(2,0)$ superconformal theory with $U(1)$-charge operator F. After taking $\partial_U \partial_{\bar{U}}$ derivatives, one finds that the two terms in the trace give rise to the two terms in the r.h.s. of Eq. (6.2), respectively. Using the properties of the underlying superconformal field theory that describes $K3 \times T^2$ compactifications, one can rewrite the integrand as a sum over $N = 2$ BPS states [21]. In particular, the one-loop Kähler metric reads:

$$\begin{aligned} K^{H(1)}_{U\bar{U}} &= -\frac{1}{8\pi^2}\int \frac{d^2\tau}{\tau_2} \partial_U \partial_{\bar{U}} \\ &\left(\sum_{\substack{\text{BPS} \\ \text{hypermultiplets}}} - \sum_{\substack{\text{BPS} \\ \text{vectormultiplets}}} \right) e^{i\pi\tau M_L^2} e^{-i\pi\bar{\tau} M_R^2} , \end{aligned} \tag{6.4}$$

where M_L and M_R denote the contribution to the masses from the left- and right-movers, respectively. In this formula, the supergravity multiplet is counted as a vector multiplet.

One-loop threshold corrections have also been studied in type I theory [22], however their structure is different from the heterotic case. The function Δ^I contains the group-dependent contribution only, proportional to the beta function. The universal term is absent, which means that it is automatically absorbed into the definition of ImS. Hence another procedure is needed to compute the one-loop Kähler metrics. It turns out that the quantity to be examined is the Planck mass [11]. Unlike the heterotic case, the Einstein term receives a one-loop correction:

$$e^{-2\phi}R \overset{\text{1-loop}}{\longrightarrow} (e^{-2\phi}+\frac{\delta}{\sqrt{G}})R. \tag{6.5}$$

The relation between the function δ and the Kähler metric is[1]:

$$K^{I(1)}_{U\bar{U}} = \frac{1}{16\pi \mathrm{Im}S'}\partial_U\partial_{\bar{U}}\delta\,. \tag{6.6}$$

The computation of the one-loop correction to the Planck mass in type I theory has been presented in [11]. The function δ can be determined from a physical amplitude involving one modulus and two gravitons, and receives contributions from the annulus and Möbius strip diagrams for open strings, and from the Klein bottle for closed strings. The result can be expressed as an integral over the real modular parameter of these one-loop surfaces:

$$\frac{\delta}{\sqrt{G}} = -\frac{1}{\pi}\int_0^{\prime\infty}\frac{dt}{t^2}(\mathcal{I}_{\mathcal{A}}Z_{\mathcal{A}}+\mathcal{I}_{\mathcal{M}}Z_{\mathcal{M}}+4\mathcal{I}_{\mathcal{K}}Z_{\mathcal{K}}), \tag{6.7}$$

where $\mathcal{I}_\sigma$ are indices associated to $K3$ which count the open string spinors (propagating on $\sigma=\mathcal{A},\mathcal{M}$) and closed string RR bosons (propagating on $\sigma=\mathcal{K}$) weighted with the fermion-parity operator $(-)^{F_{\rm int}}$, while Z_σ denote the corresponding T^2 partition functions. The prime in the integral indicates that the quadratic divergence in the ultraviolet limit $t\to 0$ has been subtracted, as dictated by the tadpole cancellation [22]. There is no need however for such a regularization at the level of the Kähler metric, where the divergence disappears after taking $\partial_U\partial_{\bar{U}}$ derivatives, c.f. Eq. (6.6).

The above result for the Kähler metric generalizes the heterotic expression obtained from Eq. (6.3) to the type I case. It can be reexpressed in a form similar to (6.4) as a sum over $N=2$ BPS states that originate only from the massless modes in six dimensions [11]:

$$\begin{aligned}\frac{1}{\sqrt{G}}\partial_U\partial_{\bar{U}}\delta &= -\frac{2}{\pi}\int_0^\infty\frac{dt}{t^2}\partial_U\partial_{\bar{U}} \\ &\Bigg(\sum_{\substack{\text{BPS}\\ \text{hypermultiplets}}} - \sum_{\substack{\text{BPS}\\ \text{vectormultiplets}}}\Bigg)e^{-\pi t M^2/2}\,,\end{aligned} \tag{6.8}$$

[1]The present definition of δ differs from its original definition in Ref. [11] by a factor of $\sqrt{G}$, so that δ becomes independent of $\sqrt{G}$.

where the masses M come from the momentum in the internal T^2. The same formula gives type I threshold corrections after inserting the operator tQ^2 inside the sum.

7. Example of Type I–Heterotic Duality

In this Section, we discuss an example of a dual pair with $N = 2$ supersymmetry in four dimensions, obtained by compactifying on T^2 the six-dimensional $K3$ orbifold models described in Section 4. More precisely, we will analyze two cases. The first one corresponds to the Higgs phase, in which the six-dimensional gauge group $U(16)_9$ is completely broken by giving appropriate vev's to the charged hypermultiplets. One thus obtains $n_T = 1$, $n_V = 0$ and $n_H = 244$ consistently with the anomaly cancellation condition (4.1) in $D = 6$. Upon compactification to $D = 4$ one finds the so-called STU model which contains 3 massless vector multiplets (S, T and U in the heterotic notation) and 244 neutral hypermultiplets. The second case corresponds to the Coulomb phase obtained by turning on Wilson lines on T^2. $U(16)_9$ is then broken to $U(1)^{16}$ and all charged hypermultiplets become massive. The resulting four-dimensional massless spectrum contains 19 vector multiplets and 4 neutral hypermultiplets.

7.1. HIGGS PHASE

On the heterotic side, the STU model has been studied extensively in the past because it admits also a type II description [23]. In fact, it can be obtained by compactifying the ten-dimensional type IIA theory on the Calabi–Yau manifold described by the weighted hypersurface of degree 24, $WP_{1,1,2,8,12}(24)$, with Hodge numbers $h_{(1,1)} = 3$ and $h_{(1,2)} = 243$. Since the type II dilaton belongs to a hypermultiplet, on the type II side the exact prepotential can be determined at the classical level. Therefore, this model provides an example of type II–heterotic–type I triality which can be tested in appropriate limits at both perturbative and non-perturbative levels.

Starting from the type II prepotential $\mathcal{F}^{II}$ with the field identification of S, T, U guided by the $K3$ fibration, and taking the limit Im$S \to \infty$ one can perform a perturbative test of heterotic–type II duality [23, 24]:

$$\mathcal{F}^{II}(S,T,U) \overset{S\to i\infty}{\longrightarrow} STU + f^H(T,U)\,, \tag{7.1}$$

where the two terms in the r.h.s. coincide with the tree-level and the one-loop contributions to the heterotic prepotential. A non-perturbative test [23] can also be done by taking the zero-slope limit along the conifold singularity of the Calabi–Yau manifold, reproducing the Seiberg–Witten prepotential of the rigid $SU(2)$ $N = 2$ super-Yang–Mills theory whose perturbative limit is described by the heterotic model on the $T = U$ line of enhanced symmetry.

In order to test heterotic–type I duality, we first recall that the heterotic S, T and U fields are mapped to S, S' and U defined in Eq. (5.3). As explained in Section 5, a perturbative test can be performed by taking the limit of Im$T \to \infty$ of the one-loop heterotic prepotential and comparing it with the type I counterpart.

The one-loop type I prepotential can be reconstructed from the Kähler metric given by Eqs. (6.6,6.8). In the model under consideration, Eq. (6.8) yields

$$\begin{aligned}\partial_U\partial_{\bar U}\delta &= -\frac{2\sqrt{G}}{\pi}\int_0^\infty \frac{dt}{t^2}(244-4)\times\\ &\qquad \partial_U\partial_{\bar U}\sum_{p\in\Gamma_2} e^{-\pi t|p|^2/2}\,, \end{aligned}\tag{7.2}$$

where Γ_2 is the T^2 momentum lattice:

$$p = \frac{m_4 - m_5 U}{\sqrt{2G^{1/2}\mathrm{Im}U}}\,, \tag{7.3}$$

with integer $m_{4,5}$. After performing the t-integration, Eq. (6.6) gives [11]:

$$K^{I(1)}_{U\bar U} = -\frac{15}{\pi^4}\frac{1}{\mathrm{Im}S'}\sum_{m_4,m_5}{}' \frac{1}{|m_4 - m_5 U|^4}\,, \tag{7.4}$$

where the prime means $(m_4, m_5) \neq (0,0)$.

The one-loop prepotential can now be determined by the standard $N = 2$ formula [19]:

$$\begin{aligned}\partial_U^3 f^I(U) &= 2\,\mathrm{Im}S'\,\partial_U^2\partial_{\bar U}(U-\bar U)^3 K^{I(1)}_{U\bar U}\\ &= 4\,E_4(U)\,,\end{aligned}\tag{7.5}$$

with the function

$$E_4(U) = \frac{45}{\pi^4}\sum_{m_4,m_5}{}' \frac{1}{(m_4 - m_5 U)^4}\,. \tag{7.6}$$

This result agrees with the heterotic prepotential [19] in the limit $\mathrm{Im}T \to \infty$:

$$\partial_U^3 f^H(U, \mathrm{Im}T \to \infty) = 4\,E_4(U) = \partial_U^3 f^I(U)\,. \tag{7.7}$$

Note that the perturbative type I computation is valid not only in the region $\mathrm{Im}S > \mathrm{Im}S' \to \infty$ ($K3$ volume $V > 1$), but also in the region $\mathrm{Im}S' > \mathrm{Im}S \to \infty$ ($V < 1$). This is due to the symmetry of the present type I model under T-duality, $V \leftrightarrow 1/V$, which exchanges 9-branes and 5-branes. Although the region of $V < 1$ cannot be reached by means of perturbative expansion from the heterotic side, it can be reached from the type II side by taking the limit $\mathrm{Im}T \to \infty$ before the limit $\mathrm{Im}S \to \infty$. Since the exact prepotential of the STU model, as evaluated on the type IIA side, is known to be symmetric under the exchange of S and T [23], such an order of limits gives $\mathcal{F}^{II} \to STU + f^I(U)$, the same as in the other order.

7.2. COULOMB PHASE

In the Coulomb phase, we find it more convenient to test duality by a direct examination of threshold corrections to the gauge couplings of $U(1)^{15}$ associated with the unbroken Cartan generators of $SU(16)_9$ [25]. After combining Eqs. (6.2) and (6.6) with Eq. (5.8), we see that duality predicts the following large ImT expansion of the heterotic threshold corrections:

$$\begin{aligned}\Delta^H(T,U,A) &= \frac{\pi}{2}v'\mathrm{Im}T + \Delta^I(U,A)\\ &\quad + \frac{\pi}{4\mathrm{Im}T}\delta(U,A) + \ldots\,,\end{aligned} \tag{7.8}$$

up to exponentially suppressed corrections. Recall from the discussion of the six-dimensional model in Section 5 that $v' = 0$.

The type I quantities $\Delta^I(U,A)$ and $\delta(U,A)$ have been discussed for generic models in Section 6. In order to apply Eqs. (6.7) and (6.8) to the model under consideration, we first determine the quantities $\mathcal{I}_\sigma Z_\sigma$ for various surfaces $\sigma = \mathcal{A}, \mathcal{M}, \mathcal{K}$. In the 99 open string sector,

$$\begin{aligned}\mathcal{I}_\mathcal{A} Z_\mathcal{A} &= -2\sum_{a^I+a^J+\Gamma_2} s_{IJ}e^{-\pi t|p|^2/2}\\ \mathcal{I}_\mathcal{M} Z_\mathcal{M} &= -2\sum_{2a^I+\Gamma_2} e^{-\pi t|p|^2/2}\,,\end{aligned} \tag{7.9}$$

where for convenience, we introduced the index $I \equiv i$ or $\bar{\imath}$, with i and $\bar{\imath}$ running over the **16** and $\overline{\mathbf{16}}$ of $SU(16)$, respectively, and $a^{\bar{\imath}} \equiv -a^i$. The matrix s_{IJ} represents the action of the orbifold group $\mathcal{R}$ on the Chan–Paton charges: $s_{IJ} = -1$ or 1, depending on whether I and J belong to the same or conjugate representations of $SU(16)$, respectively. For open string surfaces with boundaries, the momenta of the T^2 lattice (7.3) are shifted in a self-explanatory way by the Wilson lines of the 9-brane group, according to Eq.(2.6); for instance:

$$a^I + \Gamma_2 : m_{4,5} \to m_{4,5} + a^I_{4,5}\,. \tag{7.10}$$

Similarly, in the 95+59 sectors:

$$\mathcal{I}_\mathcal{A} Z_\mathcal{A} = 2\times 16 \sum_{a^I+\Gamma_2} e^{-\pi t|p|^2/2}\,,\quad \mathcal{I}_\mathcal{M} = 0\,. \tag{7.11}$$

There should be no contribution from the 55 and closed string sectors since the 5-brane $U(1)^{16}$ gauge supermultiplets become massive by absorbing 16 twisted hypermultiplets from the closed string sector, as explained in Section 4. However, for a generic position of 5-branes, the 55 sector contributes

$$\mathcal{I}_\mathcal{A} Z_\mathcal{A} + \mathcal{I}_\mathcal{M} Z_\mathcal{M} = -4\times 16\sum_{\Gamma_2} e^{-\pi t|p|^2/2}\,, \tag{7.12}$$

while the closed string sector gives

$$\mathcal{I}_\mathcal{K} Z_\mathcal{K} = 16\sum_{\Gamma_2} e^{-\pi t|p|^2/2}\,, \tag{7.13}$$

due to the RR components of the twisted hypermultiplets. Note that these two contributions cancel each other.

Inserting the above results (7.9-7.13) into Eq. (6.7), one obtains [25]:

$$\delta = \frac{2\sqrt{G}}{\pi}\int_0^{'\infty}\frac{dt}{t^2} \tag{7.14}$$

$$\left\{\sum_{a^I+a^J+\Gamma_2} s_{IJ} + \sum_{2a^I+\Gamma_2} - 16\sum_{a^I+\Gamma_2}\right\} e^{-\pi t|p|^2/2}.$$

A similar formula, obtained by inserting the charge operator tQ^2 in the sum, gives the type I threshold corrections [22]:

$$\Delta^I(U,A_i) = -\frac{1}{8}\int_0^{'\infty}\frac{dt}{t}\left\{\sum_{a^I+a^J+\Gamma_2} s_{IJ}(q_I+q_J)^2 + \right.$$

$$\left.\sum_{2a^I+\Gamma_2}(2q_I)^2 - 16\sum_{a^I+\Gamma_2} q_I^2\right\} e^{-\pi t|p|^2/2}, \tag{7.15}$$

where q_i are the $U(1)$ charges associated to SU(16) Cartan generators in the **16** representation and $q_{\bar{\imath}} = -q_i$.

We now turn to the heterotic side and determine the large ImT behavior of Δ^H given by the general formula of Eq. (6.3). Specifying to the model under consideration:

$$\Delta^H(T,U,A) = -\frac{1}{8}\int\frac{d^2\tau}{\tau_2}\sum_s Z_s(\bar{\tau}) \times$$

$$\sum_{p_L,p_R\in\Gamma_s^{(2,18)}}\left(Q^2 - \frac{1}{2\pi\tau_2}\right) e^{i\pi\tau|p_L|^2} e^{-i\pi\bar{\tau}|p_R|^2}, \tag{7.16}$$

where the first sum is over the $N=2$ sectors of the $T^4/\mathbf{Z}_2$ orbifold, $s = (P,A)$, (A,P), (A,A), where P and A denote periodic and antiperiodic boundary conditions, respectively. Z_s are the corresponding partition functions for all right-movers except for the 18 right-moving momenta that are included in the $\Gamma_s^{(2,18)}$ lattices. In fact, the moduli dependence is due entirely to these lattices. In particular, the left-moving (complex) momenta are:

$$p_L = \frac{1}{\sqrt{2\text{Im}T\text{Im}U}}(m_4' - m_5'U - n_5T - n_4TU), \tag{7.17}$$

with integer $n_{4,5}$ and the momentum numbers shifted appropriately to $m_{4,5}'$ by the 16 Wilson lines.

The limit Im$T \to \infty$ can be taken in the following steps [25]. First, one observes that the contributions of all winding modes are exponentially suppressed, hence one can restrict the lattice summation to $n_4 = n_5 = 0$. Next, by making the change of variable

$$\tau_2 = \frac{t}{4}\mathrm{Im}T\ , \tag{7.18}$$

one can easily show that the integration domain becomes the strip $t \geq 4/\mathrm{Im}T$, $-1/2 \leq \tau_1 \leq 1/2$, up to exponentially small corrections in $\mathrm{Im}T$. Finally, the τ_1 integration selects the states that originate from the massless modes only in $D = 6$.

The leading contribution diverges linearly in $\mathrm{Im}T$, reflecting the quadratic ultraviolet divergence of the integral in the region $t \to 0$, where $\mathrm{Im}T$ acts as a regulator. The coefficient of the divergence determines v' of Eq. (7.8). It is given by the leading term of the expression (7.16) Poisson resummed in the T^2 momentum numbers, with the $\Gamma_s^{(2,18)}$ lattices degenerating to the moduli-independent Γ_s^{16}:

$$\begin{aligned} v' &= -\frac{1}{4\pi}\int\frac{d^2\tau}{\tau_2^2}\sum_s Z_s(\bar\tau) \\ &\quad \sum_{p_R\in\Gamma_s^{16}}\left(Q^2-\frac{1}{2\pi\tau_2}\right)e^{-i\pi\bar\tau|p_R|^2} \end{aligned} \tag{7.19}$$

The above result coincides with the one-loop threshold correction of the six-dimensional heterotic theory. The integral (7.19) can be shown to vanish in the model under consideration, in agreement with the fact that $v' = 0$ on the type I side, as explained in Section 4.

The subleading contribution is T-independent and comes entirely from the Q^2 part of Eq. (7.16). It coincides with the type I expression for Δ^I, Eq. (7.15), after using the condition $\sum_i a^i_{4,5} = 0$ for the Wilson lines corresponding to $SU(16)$ Cartan generators. Similarly, the term proportional to $1/\tau_2$ inside the bracket in Eq. (7.16) reproduces the third term in Eq. (7.8), of order $1/\mathrm{Im}T$, with δ given in Eq. (7.14).

It is interesting to trace the origin of individual type I one-loop contributions to the heterotic side. The contributions of the 99 sector (on the annulus and Möbius strip), corresponding to the first two terms in Eqs. (7.14,7.15), originate on the heterotic side from the untwisted (P,A) orbifold sector, while the contribution of the 95+59 sectors (on the annulus), corresponding to the third term, originate from the twisted (A,P)+(A,A) orbifold sectors.

8. Higher-derivative F-terms

We now consider a class of higher-derivative F-terms in the effective actions which, in $N = 2$ superfield formalism, take the form

$$I_g = \mathcal{F}_g W^{2g}\ , \tag{8.1}$$

with integer $g \geq 0$. Here, W is the Weyl superfield

$$W_{\mu\nu} = F^-_{\mu\nu} - R^-_{\mu\nu\lambda\rho}\theta^1\sigma_{\lambda\rho}\theta^2 + \dots\ , \tag{8.2}$$

which is anti-self-dual in the Lorentz indices. $F^-_{\mu\nu}$ and $R^-_{\mu\nu\lambda\rho}$ are the (anti-self-dual) graviphoton field strength and Riemann tensor, respectively. The couplings $\mathcal{F}_g$ are holomorphic sections of degree $2-2g$ of the vector moduli space, up to a holomorphic anomaly for $g \geq 1$ [26]. They generalize the well known prepotential $\mathcal{F} \equiv \mathcal{F}_0$: $\mathcal{F}_g$ ($g \geq 1$) determines ($2g+2$)-derivative couplings of two gravitons and ($2g-2$) graviphotons together with all interactions related by supersymmetry [27].

On the type II side, $\mathcal{F}_g$ is determined entirely at genus g, while on the heterotic and type I sides these couplings arise at one loop (with additional tree-level contributions to $\mathcal{F}_0$ and $\mathcal{F}_1$). The corresponding perturbative expansions in the two theories are [25]:

$$\begin{aligned} \mathcal{F}_1^H &= 4\pi \mathrm{Im} S + f_1^H(T,U,A) \,, \\ \mathcal{F}_g^H &= \mathcal{F}_g^H(T,U,A) \qquad (g \geq 2) \,, \end{aligned} \tag{8.3}$$

$$\begin{aligned} \mathcal{F}_1^I &= 4\pi(\mathrm{Im} S + v_1' \mathrm{Im} S') + f_1^I(U,A) \,, \\ \mathcal{F}_g^I &= \mathcal{F}_g^H(U,A) \qquad (g \geq 2) \,. \end{aligned} \tag{8.4}$$

In the example of dual pair discussed in Section 7, $v_1' = 1$, as required by the symmetry under type I T-duality.

The standard way of encoding $\mathcal{F}_g$'s is to combine them in the generating function [26]:

$$\mathcal{F}(\lambda) = \sum_{g=1}^{\infty} g^2 \lambda^{2g} \mathcal{F}_g \,. \tag{8.5}$$

Heterotic–type I duality predicts the following asymptotic expansion in the large $\mathrm{Im}T$ limit:

$$\begin{aligned} \mathcal{F}^H(\lambda; S,T,U,A) &= 4\pi\lambda^2(\mathrm{Im} S + \mathrm{Im} T) \\ &+ \mathcal{F}^I(\lambda; U,A) + \frac{2\pi\lambda^2}{\mathrm{Im}T}\delta(U,A) + \ldots \end{aligned} \tag{8.6}$$

The heterotic function $\mathcal{F}^H(\lambda)$ has been computed in Refs. [28, 29, 25] and is given by an integral similar to (7.16) with the operator inside brackets replaced by $-(\lambda^2/4\pi^2)\frac{d^2}{d\lambda^2} G^H(\tilde{\lambda},\tau)$, where G^H is the partition function of space-time coordinates in the presence of a (anti-self-dual) graviphoton field strength background λ, and $\tilde{\lambda} \equiv \bar{p}_L \tau_2 \lambda / \sqrt{2\, \mathrm{Im}T \mathrm{Im}U}$.

The type I generating function $\mathcal{F}^I(\lambda)$ receives two types of contributions. The first one can be obtained from the expression for the threshold corrections by a procedure similar to the heterotic case, and arises from the world-sheet surfaces $\mathcal{A}$, $\mathcal{M}$ and $\mathcal{K}$. The second one arises from the $N=4$ supersymmetric type IIB sector propagating on the world-sheet torus; it contributes to $\mathcal{F}_1^I$ only. The sum of the two is [30, 25]:

$$\mathcal{F}^I(\lambda) = \frac{\lambda^2}{64\pi^2} \int_0^{\prime\infty} \frac{dt}{t} (\mathcal{I}_\mathcal{A} Z_\mathcal{A} + \mathcal{I}_\mathcal{M} Z_\mathcal{M} + 4\mathcal{I}_\mathcal{K} Z_\mathcal{K})$$

$$\times \frac{d^2}{d\lambda^2}\left(\frac{\lambda\pi}{\sin\pi\tilde{\lambda}}\right)^2 \tag{8.7}$$
$$+\frac{\lambda^2}{8}\int_0^{'\infty}\frac{dt}{t}(\mathcal{I}_T Z_T - \mathcal{I}_K Z_K)\,,$$

where the indices $\mathcal{I}_\sigma$ and partition functions Z_σ have been defined in Section 6, with the exception of $\mathcal{I}_T$ and Z_T which are the Witten index (equal to 24 for $K3$) and the torus partition function, respectively. In contrast to the first integrand which is reduced to a sum over $N=2$ BPS states as in Section 6, the second term $(\mathcal{I}_T Z_T - \mathcal{I}_K Z_K)$ receives contributions only from the $\mathcal{R}$-untwisted $N=4$ sector of the theory.

Note that the integration in Eq. (8.7) is infrared singular at $t \to \infty$. The divergence is proportional to λ^2 and affects $\mathcal{F}_1$ only, reproducing the trace anomaly of the effective field theory. Unlike the case of gauge couplings, it is not regulated by non-vanishing Wilson lines.

Specializing to the $K3$ orbifold model discussed in Section 7, in the Higgs phase we have:

$$\begin{aligned}\mathcal{F}^I(\lambda) &= \frac{\lambda^2}{32\pi^2}\int_0^{'\infty}\frac{dt}{t}\left\{240\frac{d^2}{d\lambda^2}\left(\frac{\lambda\pi}{\sin\pi\tilde{\lambda}}\right)^2\right. \\ &\quad \left. +4\pi^2(24-16)\right\}\sum_{p\in\Gamma_2} e^{-\pi t|p|^2/2}\,.\end{aligned} \tag{8.8}$$

Similarly, in the Coulomb phase, using Eqs. (7.9-7.13) and

$$\mathcal{I}_T Z_T = 24\sum_{\Gamma_2} e^{-\pi t|p|^2/2}\,, \tag{8.9}$$

we find:

$$\begin{aligned}\mathcal{F}^I(\lambda) &= \frac{\lambda^2}{32\pi^2}\int_0^{'\infty}\frac{dt}{t}\left\{32\pi^2\sum_{\Gamma_2}\right. \\ &\quad -\left(\sum_{a^I+a^J+\Gamma_2} s_{IJ} + \sum_{2a^I+\Gamma_2} -16\sum_{a^I+\Gamma_2}\right) \\ &\quad \left.\times\frac{d^2}{d\lambda^2}\left(\frac{\pi\lambda}{\sin\pi\tilde{\lambda}}\right)^2\right\} e^{-\pi t|p|^2/2}\,,\end{aligned} \tag{8.10}$$

where we used the same notation as in Eq. (7.14), and $\tilde{\lambda} \equiv \bar{p}t\lambda/\sqrt{32G^{1/2}\mathrm{Im}U}$.

It is easy to see [25, 30], by following the same steps as those described in Section 7 for the threshold corrections, that the above type I expressions (8.8) and (8.10) do appear in the $\mathrm{Im}T \to \infty$ limit of the heterotic (one-loop) generating function $\mathcal{F}^H(\lambda, T, U, A)$, in agreement with Eq. (8.6). The correspondence of individual type I and heterotic terms is as before, with the additional torus contribution to $\mathcal{F}_1^I$ [the first term on the r.h.s. of Eq. (8.10)] originating from the (P,A) orbifold sector on the heterotic side. The leading term $4\pi\lambda^2 \mathrm{Im}T$ in Eq. (8.6) reflects the quadratic ultraviolet divergence of $\mathcal{F}_1^H$. The subleading term of order $1/\mathrm{Im}T$ arises also from $\mathcal{F}_1^H$, and corresponds to the "universal" part of gravitational threshold corrections, reproducing δ of Eqs. (7.2) and (7.14).

Acknowledgements

Research supported in part by the National Science Foundation under grant PHY–96–02074, and by the EEC under the TMR contract ERBFMRX-CT96-0090.

References

1. For recent reviews, see: J.H. Schwarz, hep-th/9607067 and 9607201; J. Polchinski, *Rev. Mod. Phys.* **68** (1996) 1245, hep-th/9607050; R. Dijkgraaf, hep-th/9703136.
2. E. Witten, *Nucl. Phys.* **B443** (1995) 85, hep-th/9503124.
3. M.B. Green, J.H. Schwarz and E. Witten, "Superstring Theory" (Cambridge U. Press, 1987).
4. P. Candelas, X. de la Ossa, A. Font, S. Katz and D.R. Morrison, *Nucl. Phys.* **B416** (1994) 481, hep-th/9308083.
5. J. Polchinski and E. Witten, *Nucl. Phys.* **B460** (1996) 525, hep-th/9510169.
6. C.M. Hull and P.K. Townsend, *Nucl. Phys.* **B438** (1995) 109, hep-th/9410167.
7. P. Horava and E. Witten, *Nucl. Phys.* **B460** (1996) 506, hep-th/9510209.
8. A. Sagnotti, *in Non-Perturbative Quantum Field Theory*, G. Mack et al., eds. (Pergamon Press, Oxford, 1988) p. 521; M. Bianchi and A. Sagnotti, *Nucl. Phys.* **B361** (1991) 519.
9. P. Horava, *Nucl. Phys.* **B327** (1989) 461.
10. J. Polchinski, *Phys. Rev. Lett.* **75** (1995) 4724, hep-th/9510017; for a review, see: J. Polchinski, hep-th/9611050.
11. I. Antoniadis, C. Bachas, C. Fabre, H. Partouche and T.R. Taylor, *Nucl. Phys.* **B489** (1997) 160, hep-th/9608012.
12. M. Berkooz, R.G. Leigh, J. Polchinski, J.H. Schwarz, N. Seiberg and E. Witten, *Nucl. Phys.* **B475** (1996) 15, hep-th/9605184.
13. A. Sagnotti, *Phys. Lett.* **B294** (1992) 196.
14. G. Aldazabal, A. Font, L.E. Ibáñez and F. Quevedo, *Phys. Lett.* **B380** (1996) 33.
15. N. Seiberg and E. Witten, *Nucl. Phys.* **B471** (1996) 121, hep-th/9603003.
16. E.G. Gimon and J. Polchinski, *Phys. Rev.* **D54** (1996) 1667, hep-th/9601038.
17. M. Douglas, hep-th/9512077; M. Douglas and G. Moore, hep-th/9603167.
18. I. Antoniadis, E. Gava and K.S. Narain, *Nucl. Phys.* **B383** (1992) 109; I. Antoniadis, E. Gava, K.S. Narain and T.R. Taylor, *Nucl. Phys.* **B407** (1993) 706; I. Antoniadis and T.R. Taylor, *in String Theory, Quantum Gravity and Unification of Fundamental Interactions*, M. Bianchi et al., eds. (World Scientific, 1993) p.6, hep-th/9301033.
19. I. Antoniadis, S. Ferrara, E. Gava, K.S. Narain and T.R. Taylor, *Nucl. Phys.* **B447** (1995) 35, hep-th/9504034; B. de Wit, V. Kaplunovsky, J. Louis and D. Lüst, *Nucl. Phys.* **B451** (1995) 53.
20. L.J. Dixon, V.S. Kaplunovsky and J. Louis, *Nucl. Phys.* **B355** (1991) 649.
21. J.A. Harvey and G. Moore, *Nucl. Phys.* **B463** (1996) 315, hep-th/9510182.
22. C. Bachas and C. Fabre, *Nucl. Phys.* **B476** (1996) 418, hep-th/9605028.
23. S. Kachru and C. Vafa, *Nucl. Phys.* **B450** (1995) 69, hep-th/9508155; A. Klemm, W. Lerche and P. Mayr, *Phys. Lett.* **B357** (1995) 313, hep-th/9506112.
24. I. Antoniadis, S. Ferrara, E. Gava, K.S. Narain and T.R. Taylor, *Nucl. Phys.(*Proc. Suppl.) **46** (1996) 162; V. Kaplunovsky, J. Louis and S. Theisen, *Phys. Lett.* **B357** (1995) 71.

25. I. Antoniadis, H. Partouche and T.R. Taylor, hep-th/9703076.
26. M. Bershadsky, S. Cecotti, H. Ooguri and C. Vafa, *Nucl. Phys.*B405 (1993) 279, hep-th/9302103; *Comm. Math. Phys.*165 (1994) 311, hep-th/9309140.
27. I. Antoniadis, E. Gava, K.S. Narain and T.R. Taylor, *Nucl. Phys.*B413 (1994) 162, hep-th/9307158.
28. I. Antoniadis, E. Gava, K.S. Narain and T.R. Taylor, *Nucl. Phys.*B455 (1995) 109, hep-th/9507115.
29. J.F. Morales and M. Serone, *Nucl. Phys.*B481 (1996) 389, hep-th/9607193.
30. M. Serone, *Phys. Lett.*B395 (1997) 42, hep-th/9611017; J.F. Morales and M. Serone, hep-th/9703049.

CONNECTIONS BETWEEN M-THEORY AND SUPERSTRINGS†

MICHAEL B. GREEN
Department of Applied Mathematics and Theoretical Physics
Silver Street
Cambridge CB3 9EW, United Kingdom

1. Introduction

Many features of M-theory follow directly from the strong algebraic constraints associated with the symmetries of the theory rather than specific details of any microscopic model, such as the matrix model. Certain of these properties must be inevitable consequences of eleven-dimensional supersymmetry. Others may depend on assuming the validity of the intricate web of duality inter-relationships between the perturbative string theories.

This raises the question of the extent to which the higher order terms in a momentum expansion of the M-theory effective action are determined by eleven-dimensional supergravity compactified in various ways. It has been generally assumed that since perturbative supergravity is terribly divergent in eleven dimensions no interesting consequences follow by considering loop diagrams and that a microscopic theory is needed in order to regularize its divergences. However, this ignores the powerful supersymmetry constraints which are expected to reduce the number of arbitrary constants – but by how much? Furthermore, the eleven-dimensional quantum theory compactified to ten or fewer dimensions is supposed to be precisely equivalent to string theory and its compactifications. What mileage can be gained by exploiting this equivalence as an ansatz? The lowest order terms in the expansion of the M-theory action in powers of the momentum are the terms in the original classical supergravity action of [1]. Certain higher order terms have also been unambigously identified. One example is the $C^{(3)} \wedge X_8$ term [2] that arises as a one-loop effect in type IIA string theory [3] and can also be motivated by requiring the cancellation of chiral anomalies in p-branes [4]. Here $C^{(3)}$ is the

†An expanded version of talks given at the ICTP Conference on Duality Symmetries in String Theory, 7-12 April 1997; STRINGS '97, Amsterdam, June 16-21 1997; CERN Workshop on Non-perturbative Aspects of Strings, Branes and Fields, 8-12 December 1997.

L. Baulieu et al. (eds.), Strings, Branes and Dualities, 201–218.

Ramond–Ramond three-form and X_8 is an eight-form made out of the Riemann curvature.

In the next section I will review the evidence that certain 'protected' terms in the low energy effective type II string action compactified on S^1 are precisely determined by one-loop effects in eleven-dimensional supergravity compactified on T^2. For example the one-loop eleven-dimensional four-graviton amplitude leads to a R^4 term in M-theory [1] which is related by supersymmetry to the $C^{(3)} \wedge X_8$ term. This term can be expressed in the type IIB coordinates by making use of the equivalence of M-theory compactified on T^2 with type IIB string theory compactified on S^1 [5, 6]. This leads to a term of the form $f(\rho, \bar{\rho})R^4$ term in the type IIB action [7] where f is a $SL(2, Z)$-invariant function of the complex IIB scalar field, ρ. The modular properties of the IIB theory are inhereted from the geometric $SL(2, Z)$ of T^2 in M-theory with ρ identified with the complex structure, Ω, of the torus. The interplay between the duality symmetries of nine-dimensional type II string theories and the one-loop eleven-dimensional supergravity amplitude are encoded in f which contains both perturbative terms and non-perturbative D-instanton contributions. Furthermore, even though the eleven-dimensional loop amplitude is cubically divergent the consistent identification of the compactified theory with the IIB theory determines a specific finite renormalized value for the coefficient of the R^4 term in eleven dimensions — this finite value would necessarily arise as the regularized value in any microscopic theory. In fact, since the ultraviolet divergence is proportional to the volume of the compactification torus, it disappears in the zero-volume limit, and the exact (non-perturbative) R^4 term of ten-dimensional type IIB [8] is reproduced by finite terms alone.

The expression for the modular function $f(\rho, \bar{\rho})$ in the IIB theory will be motivated in section 2.1 by matching the known R^4 terms that arise at tree level and one loop in string perturbation theory together with the expected structure of multiply-charged D-instanton contributions. A single D-instanton of charge $N = \hat{m}n$ can be identified, following T-duality, with configurations in the IIA theory in which the world-line of a charge-n D-particle winds $\hat{m}$ times around a compact dimension. Precisely the same expression will be obtained in section 2.2 from the low momentum limit of the one-loop four graviton scattering amplitude in eleven-dimensional supergravity compactified on T^2. I will also make some comments concerning the relation of the higher-order terms in the momentum expansion of the one-loop supergravity amplitude with string loop diagrams (related comments are made in [9]).

The R^4 term is related by supersymmetry to a large number of other terms. In the language of the type IIB theory these terms include a sixteen-fermion term, $f_{16}(\rho, \bar{\rho})\lambda^{16}$, where λ is the complex spin-$\frac{1}{2}$ chiral fermion [10]. This is the analogue of the 't Hooft fermion vertex in a conventional Yang–Mills instanton background. These interaction terms can be expressed as integrals over half the ten-dimensional on-shell superspace (there is no ten-dimensional off-shell superspace formalism). Such protected terms again only receive perturbative contributions at tree-level

[1]The notation R^4 represents a particular contraction of four Riemann curvature tensors that will be reviewed below.

and at one loop in string theory and appear to be determined entirely by one-loop diagrams in eleven-dimensional supergravity. A brief overview of such terms is given in section 2.3.

The expressions for the instanton contributions in the expansions of these interaction terms defines a measure on the space of charge-N D-instantons. In [11] it was argued that this should equal the partition function of the zero-dimensional Yang–Mills model, which is an integral over bosonic and fermionic matrices in the Lie algebra of $SU(N)$ [12]. This, in turn, is related to the bulk term in the Witten index for charge-N D-particles [13, 14]. This circle of arguments which is presented in detail in [11] will be briefly reviewed at the end of section 2.3.

It would be very interesting to determine how much of the structure of the string perturbation expansion can be determined from eleven-dimensional supergravity beyond these protected terms. For example, the R^4 term is the lowest-order term in a momentum expansion of the exact four-graviton scattering amplitude in type IIB string theory. There are very likely to be terms of higher order in momenta that are also protected by supersymmetry. To what extent can the momentum expansion of the effective action be determined by perturbation theory in eleven dimensions? A more limited question is to what extent the momentum expansion of the string tree-level amplitude can be reproduced by eleven-dimensional supergravity perturbation theory? This depends on the systematics of the multi-loop diagrams in eleven-dimensional supergravity as will be described in section 3. These diagrams have finite pieces that depend on the moduli of the compact dimensions. New primitive divergences arise corresponding to terms which have derivatives acting on R^4, denoted symbolically as $\partial^{2n} R^4$. It would be interesting to see whether these might be determined by the same kind of arguments that led at one loop to the determination of the R^4 term. However, the arguments that I present beyond one loop are based on dimensional analysis rather than explicit evaluation of the rather complicated Feynman diagrams. In this way it is easy to see how specific multi-loop diagrams could give specific finite terms in the M-theory action that correspond to terms that arise from string perturbation theory. However, it seems likely that the values of the counterterms are fixed unambiguously only at low orders in the momentum expansion where they are protected by supersymmetry.

2. Higher-order terms in M-theory

2.1. $SL(2,Z)$-INVARIANT R^4 TERMS IN TYPE IIB

The leading perturbative contributions to the four graviton scattering amplitude in type IIB superstring theory are of the form (in string frame)

$$S_{R^4}^{pert} = \int \sqrt{g^B} \left(\zeta(3) \rho_2^2 + \frac{\pi^2}{3} \right) t^8 t^8 R^4, \tag{2.1}$$

where $g^B_{\mu\nu}$ is the IIB string-frame metric and

$$t_8 t_8 R^4 \equiv t^{\mu_1 \dots \mu_8} t_{\nu_1 \dots \nu_8} R^{\nu_1 \nu_2}_{\mu_1 \mu_2} \cdots R^{\nu_7 \nu_8}_{\mu_7 \mu_8}, \tag{2.2}$$

where the rank-eight tensor, t^8, is defined in [15]. The first term in (2.1) is the tree-level contribution [16, 17] that is of order α'^3 relative to the leading Einstein–Hilbert term. The second term is the one-loop contribution [18] and has no dependence on the dilaton in the string frame. Non-perturbative contributions to the $t_8 t_8 R^4$ term also arise from single D-instantons with charge N [8] which give an infinite series of non-perturbative contributions that have the form

$$S_{R^4}^{nonpert} = \sum_{N>0} c_N(\rho_2)(e^{2\pi i N\rho} + e^{-2\pi i N\bar\rho}) t_8 t_8 R^4. \tag{2.3}$$

One way of counting these D-instantons makes use of T-duality between the IIB and the IIA theories in nine dimensions. From the IIA point of view a D-instanton is associated with the world-line of a D-particle of charge n and mass $e^{-\phi^A} n$ (where ϕ^A is the IIA dilaton) winding around the compact ninth dimension of radius r_A. The euclidean action of this configuration is $S = 2\pi\hat m n(C^{(1)} + i r_A e^{\phi^A})$ where $C^{(1)}$ is the IIA one-form and $\hat m$ is the winding number of the world-line. After T-duality this leads to the expression $S = 2\pi N\rho$ for the D-instanton action where $N = \hat m n$. Although the coefficients c_N are probably very hard to determine directly they are fixed by the requirement that the total action should be invariant under $SL(2,Z)$ transformations. This means that it must have the form,

$$S_{R^4} \equiv S_{R^4}^{pert} + S_{R^4}^{nonpert} = \int \sqrt{g^B} \rho_2^{1/2} f(\rho,\bar\rho) t_8 t_8 R^4 \, d^{10}x, \tag{2.4}$$

where $f(\rho,\bar\rho)$ is a modular function since R is invariant under $SL(2,Z)$ transformations in the Einstein frame.

The precise coefficients of the known perturbative contributions, S^{pert}, together with the general form of the instanton corrections motivates the suggestion [8] that f is given by

$$f(\rho,\bar\rho) = \pi^{-2} \sum_{(\hat l_1,\hat l_2)\neq(0,0)} \frac{\rho_2^{3/2}}{|\hat l_1 + \rho \hat l_2|^3} = \pi^{-2}\zeta(3) E_{\frac{3}{2}}(\rho,\bar\rho), \tag{2.5}$$

where the nonholomorphic Eisenstein series E_s is defined by [19]

$$\begin{aligned} \zeta(2s)E_s(\rho,\bar\rho) &= \sum_{(\hat l_1,\hat l_2)\neq(0,0)} \frac{\rho_2^s}{|\hat l_1 + \rho \hat l_2|^{2s}} \\ &= \zeta(2s)\rho_2^s + \frac{\sqrt{\pi}\,\Gamma(s-\frac{1}{2})\zeta(2s-1)}{\Gamma(s)} \rho_2^{1-s} + \mathcal{R}_s. \end{aligned} \tag{2.6}$$

The last equality is a large-ρ_2 expansion and $\mathcal{R}_s$ indicates a specific sum of exponentially decreasing terms. In the special case $s = 3/2$ this sum is given by

$$\mathcal{R}_{\frac{3}{2}} = \sum_{N>0} \left(\sum_{N|\hat m} \frac{1}{\hat m^2} \right) N^{\frac{1}{2}} \left(e^{2\pi i N\rho} + e^{-2\pi i N\bar\rho}\right) \times \sum_{k=0}^{\infty} (2\pi N\rho_2)^{-k} \frac{\Gamma(\frac{1}{2}-k)}{\Gamma(-\frac{1}{2}-k)}, \tag{2.7}$$

where $\sum_{N|\hat{m}}$ indicates a sum over the divisors of N. There are only two power-behaved terms in the expansion (2.6) and they correspond precisely to the known tree-level and one-loop terms in the R^4 effective action of the IIB theory, while the series of exponentials in (2.7) correspond to D-instanton corrections with the expected instanton number $N = \hat{m}n$. This lends weight to the suggestion that (2.5) is the exact result, in which case there should be a perturbative non-renormalization theorem [8] that forbids corrections to the R^4 term beyond one loop. Recently there has been a certain amount of evidence for the validity of such a theorem [20, 21] (although an apparent contradiction in the literature [22] deserves closer analysis). The measure factor $\sum_{N|n} \frac{1}{n^2}$ in the instanton sum (2.7) can be related [11] to the expression for the Witten index of relevance in the analysis of D-particle threshold bound states [13][14] as will be described at the end of section 2.3.

2.2. ONE LOOP IN ELEVEN DIMENSIONS

The leading low energy behaviour of the one-loop four-graviton amplitude of eleven-dimensional supergravity compactified on a torus with radii R_{10} and R_{11} was considered in [7]. Here we will consider the complete momentum dependence of the same amplitude (a similar argument was also given in [9]) and expand the expression in a power series in the Mandelstam invariants, S, T and U,

$$S = -(k_1 + k_2)^2, \qquad T = -(k_1 + k_4)^2, \qquad U = -(k_1 + k_3)^2, \tag{2.8}$$

so that S+T+U =0.

The terms of interest arise from the sum of all one-loop diagrams with four external gravitons and with the graviton, gravitino or antisymmetric three-form potential circulating around the loop. This sum is most succinctly calculated in a first-quantized light-cone gauge formalism in which the amplitude is described as a trace over the states of an eleven-dimensional super-particle circulating around the loop and coupled to the four external gravitons by vertex operators (as in [7]). The result is given by

$$A_4 = \frac{1}{\kappa_{11}^2} \tilde{K}[I(S,T) + I(S,U) + I(U,T)], \tag{2.9}$$

where $\tilde{K}$ is the linearized approximation to R^4 (which is eighth order in momenta and symmetric under the interchange of any pair of gravitons) and the function $I(S,T)$ has the form of a Feynman integral for a massless scalar field theory. It is given by

$$I(S,T) = \frac{1}{\pi^{5/2}\mathcal{V}_2} \int \prod_{r=1}^{4} d\tau_r \int d^9q \sum_{\{l_1,l_2\}} e^{-G^{IJ} l_I l_J \tau - \sum_{r=1}^{4} p_r^2 \tau_r}, \tag{2.10}$$

where $\tau = \sum_{r=1}^{4} \tau_r$ and q_i $(i = 1, \cdots, 9)$ is the nine-dimensional loop momentum transverse to T^2. The parameters τ_i label the relative positions of the four vertices. The sum is over the Kaluza–Klein momenta (l_1 and l_2) in the two compact dimensions $(I,J=1,2)$ and the momenta in the legs of the loop are given by

$$p_r = q + \sum_{s=1}^{r} k_s, \tag{2.11}$$

where the external momenta, k_r^μ satisfy $k_r^2 = 0$ and $\sum_{r=1}^4 k_r = 0$. The particular kinematic configuration has been chosen in which the external momenta have zero components in the directions of the torus, i.e. $k_r^I = 0$. The eleven-dimensional coupling constant, κ_{11}, has dimension, $(length)^{9/2}$ and will be set equal to 1 in most of the following. The inverse metric on T^2 is defined by

$$G^{IJ} l_I l_J = \frac{1}{\mathcal{V}_2 \Omega_2} |l_1 + \Omega l_2|^2, \tag{2.12}$$

where $\mathcal{V}_2$ is the volume of the torus with complex structure, $\Omega = \Omega_1 + i\Omega_2$ (where Ω_1 is an angular parameter and $\Omega_2 = R_{10}/R_{11}$). Setting all the $k_r = 0$ in (2.10) gives the lowest-order result of [7]. The full amplitude (2.9) gives rise to terms in the M-theory effective action compactified on T^2 of the form

$$S_4 = \frac{1}{\kappa_{11}^2} \int \sqrt{G^{(9)}} \mathcal{V}_2 h(\mathcal{V}_2, \Omega; \partial^2) R^4 d^9x. \tag{2.13}$$

The function h is a modular function of Ω and its argument ∂^2 symbolically indicates derivatives acting on the fields in R^4, corresponding to the dependence of A_4 on the momenta k_r.

After completing a square in the exponent of (2.10) and then performing the shifted loop integral the expression becomes

$$I(S,T) = \frac{\pi^2}{\mathcal{V}_2} \int \prod_{r=1}^{4} d\tau_r \tau^{-9/2} \sum_{\{l_1,l_2\}} e^{-G^{IJ} l_I l_J \tau + \frac{1}{\tau}(S\tau_1\tau_3 + T\tau_2\tau_4)}. \tag{2.14}$$

This integral is to be evaluated in the region $S,T<0$ where it converges and then analytically continued to the physical region.

The momentum-independent terms in (2.14) can be isolated by writing

$$I(S,T) = I_0 + I'(S,T), \tag{2.15}$$

where

$$I_0 \equiv I(0,0) = \frac{\pi^{3/2}}{\mathcal{V}_2} \int_0^\infty d\tau \tau^{-3/2} \sum_{\{l_1,l_2\}} e^{-\pi G^{IJ} l_I l_J \tau}, \tag{2.16}$$

which is the expression considered in [7]. It has a divergence in the limit $\tau = 0$. A double Poisson resummation reexpresses I_0 as a sum over $\hat{l}_1$ and $\hat{l}_2$ that may be identified with the winding numbers of the euclidean world-line of the supergraviton around the directions R_{11} and R_9 of the torus, respectively. The result is

$$\begin{aligned} I_0 &= \pi^{3/2} \int_0^\infty d\tau \tau^{-5/2} \sum_{\{\hat{l}_1,\hat{l}_2\}} e^{-\pi G_{IJ}\hat{l}_I\hat{l}_J\frac{1}{\tau}} \\ &= \pi^{3/2} \int_0^\infty d\hat{\tau}\hat{\tau}^{\frac{1}{2}} \sum_{\{\hat{l}_1,\hat{l}_2\}} e^{-\pi G_{IJ}\hat{l}_I\hat{l}_J\hat{\tau}}, \end{aligned} \tag{2.17}$$

where $\hat{\tau} = 1/\tau$. This isolates the divergence in the zero winding term ($\hat{l}_1 = \hat{l}_2 = 0$). This is presumably regularized by a microscopic theory, such as the matrix model [23], but its regularized value is also determined uniquely by requiring that I_0 reproduce the IIA and IIB string theory R^4 terms in nine dimensions. Since the loop diverges as Λ^3, where Λ is a momentum cut-off, the regularized value of this term has the form $c\kappa_{11}^{-2/9}$ where c is a dimensionless constant. The remaining terms in (2.17) depend on the volume and complex structure of T^2 and are finite. The $\hat{\tau}$ integral is trivial for these terms and (2.17) can be written as

$$I_0 = c\kappa_{11}^{-2/9} + \mathcal{V}_2^{-3/2}\zeta(3)E_{\frac{3}{2}}(\Omega,\bar{\Omega}). \tag{2.18}$$

Substituting into (2.15) and (2.9) leads to a R^4 contribution to the M-theory effective action that can be expressed in terms of the IIB theory compactified on a circle.[2]

In the limit $\mathcal{V}_2 \to 0$ the radius, r_B, of the IIB circle becomes infinite and the second term in (2.18) dominates, leading to an expression for the action for the decompactified IIB theory which coincides with (2.4) (with f defined by (2.5)). The IIB string tree-level contributions arise from terms in (2.17) with $\hat{l}_1 \neq 0$ and $\hat{l}_2 = 0$ so the loop has non-zero winding only in the eleventh dimension. The string one-loop and D-instanton terms in (2.4) can be extracted from the $\hat{l}_2 \equiv \hat{m} \neq 0$ terms in (2.17) by performing a Poisson summation that takes the winding numbers $\hat{l}_1$ into Kaluza–Klein charges n and identifying $N = \hat{m}n$. The $n = 0$ term gives rise to the term in (2.4) that is independent of the dilaton (the one-loop contribution) while the charge-N D-instanton contributions in (2.7) come from the $N \neq 0$ contributions.

Although the zero winding number term with coefficient c does not contribute in the $\mathcal{V}_2 \to 0$ limit it does contribute to the finite-$\mathcal{V}_2$ amplitude. By transforming to the type IIA coordinates it is easy to see that c is the coefficient of the one-loop contribution to the R^4 term in the IIA theory in ten dimensions and takes precisely the same value as the one-loop term in S^{pert}, thereby ensuring the T-duality of the type IIA and IIB theories [7].

There is an intriguing analogy between this calculation and the calculation of the exact prepotential of four-dimensional $N = 2$ supersymmetric Yang–Mills theory from a one-loop amplitude in five dimensions [24].

[2]This makes use of the usual relations between the parameters of M-theory on T^2 and the type II string theories on S^1: $r^A = (r^B)^{-1} = R_{10}(R_{11})^{\frac{1}{2}}$, $e^{\phi^B} = (r^A)^{-1}e^{\phi^A} = R_{11}/R_{10}$, where r^A is the radius of the IIA circle in string units and ϕ^A and ϕ^B are the dilatons of the two type II theories that are related by T-duality.

The momentum-dependent terms in $I(S,T)$ in (2.15) are contained in $I'(S,T)$. This is evaluated by separating the $l_1 = l_2 = 0$ term from the rest,

$$I'(S,T) = I^0(S,T) + \sum_{n=2}^{\infty} I_n(S,T). \tag{2.19}$$

The term with zero Kaluza–Klein momenta, I^0, gives a contribution to the amplitude that is non-analytic in the Mandelstam invariants while I_n is a homogeneous polynomial in S and T of degree n. After translating to string coordinates the term I^0 can be identified with a corresponding one-loop term in string perturbation theory. The polynomials $I_n(S,T)$, $I_n(T,U)$ and $I_n(U,S)$ can be identified with terms that should arise from multi-loop amplitudes in the supergravity theory. More details are given in the Appendix which overlaps with [9].

Compactification of the supergravity loop amplitude to eight dimensions on T^3 raises new issues associated with the presence of new instantons that are identified with wrappings of the M-theory membrane world-volume around T^3. The complete expression for the R^4 term is now obtained by combining the one-loop calculation with the constraints of U-duality [25, 7, 26]. Compactification to lower dimensions raises yet more issues analogous to those that arise in the compactification of the matrix model.

2.3. SIXTEEN-FERMION AND RELATED TERMS

The $f(\rho,\bar{\rho})t_8t_8R^4$ term is one of a large number of terms of the same dimension that are related to each other by supersymmetry. Prominent among these is the analogue of the 't Hooft multi-fermion vertex of the IIB theory,

$$S_{\lambda^{16}} = \int \sqrt{g^B}\rho_2^{1/2} f_{16}(\rho,\bar{\rho})\lambda^{16} d^{10}x, \tag{2.20}$$

where λ is the complex chiral spin-$\frac{1}{2}$ fermion. This field transforms with weight 3/2 under the $U(1)$ denominator in the coset $SL(2,Z)/U(1)$ [27, 28, 29] so that a general $SL(2,Z)$ transformation acts as

$$\lambda \to \left(\frac{c\bar{\rho}+d}{c\rho+d}\right)^{\frac{3}{4}} \lambda. \tag{2.21}$$

As a consequence of $SL(2,Z)$ invariance of the action it follows that the coefficient f_{16} in (2.20) is a non-holomorphic modular form of weight $(12,-12)$ (where the notation indicates the holomorphic and anti-holomorphic weights, respectively).

The expression for f_{16} can also be obtained by a one-loop calculation in eleven-dimensional supergravity compactified on T^2. This time the process has sixteen external gravitini in polarization states that correspond to λ in the IIB description. With this choice of external states the loop diagram vanishes in the limit of eleven non-compact dimensions and is finite for generic T^2. It was evaluated in a recent paper [10] and the result is

$$f_{16} = \frac{1}{\pi^2}\rho_2^{3/2}\Gamma(27/2) \sum_{(\hat{l}_1,\hat{l}_2)\neq(0,0)} \frac{(\hat{l}_1+\hat{l}_2\bar{\rho})^{24}}{|\hat{l}_1+\hat{l}_2\rho|^{27}}, \tag{2.22}$$

where $\hat{l}_1$ and $\hat{l}_2$ are the winding numbers of the world-line around the cycles of the torus. This expression has the large ρ_2 expansion,

$$\begin{aligned}\rho_2^{1/2} f_{16} &= \pi^{-2}\Gamma(27/2)\zeta(3)(\rho_2)^2 + \pi^{-2}\Gamma(23/2)\zeta(2) \\ &+ 2^{24}\pi^{23/2}\sum_{N>0}\left(\sum_{N|\hat{m}}\frac{1}{\hat{m}^2}\right)(N\rho_2)^{25/2}e^{2\pi i N\rho}(1+O(\rho_2^{-1})),\end{aligned} \tag{2.23}$$

which again indicates the presence of string perturbation theory contributions at tree-level and one-loop together with an infinite number of D-instanton contributions (the anti D-instantons enter with coefficients that are of higher order in ρ_2^{-1}). There are presently no sixteen-fermion perturbative string calculations in the literature with which to compare the two leading terms.

In [10] it was also shown that the functions f_{16} and f are related by

$$f_{16}(\rho,\bar{\rho}) = \rho_2^{12}\mathcal{D}^{12}f(\rho,\bar{\rho}). \tag{2.24}$$

The covariant derivative $\mathcal{D}$ is defined by

$$F_{d+2,\bar{d}} = \mathcal{D}_d F_{d,\bar{d}} = i\left(\frac{\partial}{\partial\rho}+\frac{d}{(\rho-\bar{\rho})}\right)F_{d,\bar{d}}, \tag{2.25}$$

where $F_{d,d'}$ is a modular form of holomorphic weight d and anti-holomorphic weight d' and

$$\mathcal{D}^k F_{d,\bar{d}} \equiv \mathcal{D}_{d+2(k-1)}\mathcal{D}_{d+2(k-2)}\cdots\mathcal{D}_d F_{d,\bar{d}}. \tag{2.26}$$

The relationship (2.24) between f and f_{16} is consistent with the consequences of linearized supersymmetry. The physical on-shell fields of linearized IIB supergravity are contained in a superfield $\Phi(x,\theta)$ (where θ is a complex chiral Grassmann spinor of $SO(9,1)$) satisfying the holomorphic constraint $D^*\Phi = 0$ and the on-shell condition $D^4\Phi = D^{*4}\Phi^*$ where D and D^* are supercovariant derivatives [29] This field has an expansion in powers of θ,

$$\Phi = \rho_0 + a - \frac{2i}{g}\bar{\theta}\lambda - \cdots - \frac{i}{48g}\bar{\theta}\Gamma^{\mu\nu\eta}\theta\bar{\theta}\Gamma_\eta{}^{\sigma\tau}\theta R_{\mu\nu\sigma\tau} + \cdots, \tag{2.27}$$

where $\rho_0 = \rho - a$ is a constant background that defines the coupling constant in the linearized theory, R is the linearized approximation to the curvature and $\cdots$ indicates the other terms in the expansion. The matrices Γ^μ are $SO(9,1)$ gamma matrices and $\bar{\theta} = \Gamma^0\theta$ (with no complex conjugation). The linearized supersymmetric on-shell action has the form

$$S' = \mathrm{Re}\int d^{10}x d^{16}\theta\, g^4 F[\Phi], \tag{2.28}$$

and the various component interactions are contained in the θ^{16} term in the expansion of $F[\Phi]$. In addition to the R^4 and λ^{16} terms there are many others – some of these are singlets under the $U(1)$ (and arose at one string loop in [30]) while others carry a net even number of units of $U(1)$ charge [10] (some of these terms have also been discussed in [31]).

Although the fully nonlinear theory cannot easily be expressed in terms of Φ the leading behaviour of the charge-N instanton term is obtained by substituting $F = N^{-7/2} e^{2\pi i N\Phi} \sum_{N|n} \frac{1}{\hat{m}^2}$ into (2.28). In this manner the leading instanton contributions to the R^4, λ^{16} and related terms are all contained in a simple superspace expression.

The expression for the measure on single charge-N D-instantons, $\sum_{N|n} \frac{1}{\hat{m}^2}$, can be read off from the expressions for the instanton contributions to the various processes described above. This can then be compared with the expression that should follow from the matrix description of D-instantons. According to [12] this is determined by the partition function for $U(N)$ ten-dimensional supersymmetric Yang–Mills theory reduced to zero dimensions, in which case the configuration space consists of the ten $N \times N$ bosonic matrices, A_μ, and sixteen $N \times N$ fermionic matrices, ψ^a. The partition function is given by

$$Z_{U(N)} = \int d^{10}y d^{16}\epsilon Z_{SU(N)}, \tag{2.29}$$

where the center of mass super-translation zero modes have been separated from the measure factor,

$$Z_{SU(N)} = \int_{SU(N)} DAD\psi \exp(-S(A,\psi)). \tag{2.30}$$

This integral, which enters into the matrix model of [34], has not been explicitly evaluated apart from the $N = 2$ case considered in [13][14] (which gave $Z_{SU(2)} = 5/4$). However, equating it with the measure for N coincident D-instantons obtained from the $t_8 t_8 R^4$ and related terms strongly suggests that

$$Z_{SU(N)} = \sum_{N|\hat{m}} \frac{1}{\hat{m}^2}, \tag{2.31}$$

for all N [11]. It would obviously be of interest to check this by explicit evaluation of the finite-dimensional integral (2.30).

3. Tree-level string theory and multi-loop supergravity.

We have seen that the lowest-dimensional terms in the string theory effective action correspond to terms that arise at one loop in eleven-dimensional supergravity. We would now like to see whether higher-order string-theory terms can be obtained from multi-loop amplitudes in eleven dimensions. To be explicit we will compare the expansion of the ten-dimensional IIA string tree amplitude with the compactification of multi-loop diagrams of eleven-dimensional supergravity on a circle of radius R_{11}. The calculation of multi-loop diagrams is obviously very difficult but we will be able to infer certain systematic properties from dimensional considerations.

First, consider the tree amplitude for four-graviton scattering in either of the type II superstring theories which is given by

$$\begin{aligned} A_4^{tree} &= e^{-2\phi}\frac{\tilde{K}}{stu}\frac{\Gamma(1-\alpha' s)\Gamma(1-\alpha' t)\Gamma(1-\alpha' u)}{\Gamma(1+\alpha' s)\Gamma(1+\alpha' t)\Gamma(1+\alpha' u)} \\ &= e^{-2\phi}\frac{\tilde{K}}{stu}\exp\left(\sum_{n=1}^{\infty}\frac{2\zeta(2n+1)}{2n+1}(s^{2n+1}+t^{2n+1}+u^{2n+1})\right), \end{aligned} \tag{3.1}$$

where the string-frame Mandelstam invariants are related to those in the M-theory coordinates by

$$s=\frac{S}{R_{11}}, \qquad t=\frac{T}{R_{11}}, \qquad u=\frac{U}{R_{11}}. \tag{3.2}$$

It is important that every term in the exponent can be expressed as a polynomial in s and t multiplied by stu, as can be seen from the identity,

$$s^{2n+1}+t^{2n+1}+u^{2n+1} = stu\left[\sum_{r=1}^{n}\sum_{q=0}^{2n-2r}\frac{(2n+1)!}{r!(2n+1-r)!}(-1)^q s^{2n-1-r-q}t^{r+q-1}\right], \tag{3.3}$$

where $n \geq 1$. This means that the massless poles only contribute to the first term in the expansion of the exponential.

When expressed in terms of the Mandelstam invariants in the M-theory metric the expression (3.1) has the low-energy expansion,

$$\begin{aligned} A_4^{tree} \sim \tilde{K}\Bigg(&\frac{1}{STU}+\frac{2\zeta(3)}{R_{11}^3}+\frac{2\zeta(5)}{R_{11}^5}(S^2+ST+T^2)+\frac{2\zeta(3)^2}{R_{11}^6}STU \\ &+\frac{2\zeta(7)}{R_{11}^7}(S^4+2S^3T+3S^2T^2+2ST^3+T^4)+\frac{2\zeta(3)\zeta(5)}{R_{11}^8}STU(S^2+ST+T^2) \\ &+\frac{2\zeta(9)}{R_{11}^9}(S^6+4S^5T+\cdots+T^6)+\frac{4}{3R_{11}^9}\zeta(3)^3S^2T^2U^2+\cdots\Bigg). \end{aligned} \tag{3.4}$$

The first term in the expansion combines with the kinematic factor, $\tilde{K}$, to give the tree-level amplitude that is described by the Einstein–Hilbert action of eleven-dimensional supergravity compactified on a circle of radius $R_{11} = e^{2\phi/3}$. The subsequent term, with coefficient $\zeta(3)$ is the term considered earlier that comes from the R^4 term in the effective action. We saw that this term is reproduced by the one-loop diagram of eleven-dimensional supergravity compactified on a circle. The higher-order terms in (3.4) come from terms in the effective action with derivatives acting on R^4. The question is whether the whole series might be reproduced by summing loop diagrams of eleven-dimensional supergravity compactified on a circle.

We will restrict ourselves here to considering some simple multi-loop contributions that should reproduce the first few terms in the low-energy expansion of

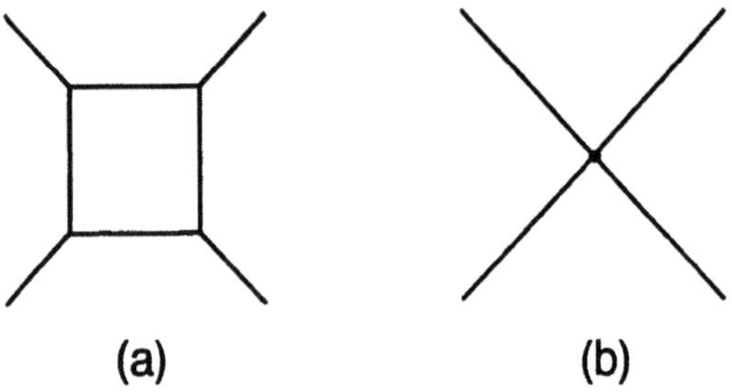

Figure 1. (a) The finite part of the one-loop diagram contributing to the R^4 term of compactified eleven-dimensional supergravity comes from the non-zero windings of the circulating particles. (b) The primitively divergent R^4 term that arises from the regularized zero winding sector is regularized to a specific finite value consistent with supersymmetry and various duality symmetries.

the string tree-level amplitude, (3.4). We have already seen that the constant term proportional to $\zeta(3)(R_{11})^{-3}$ comes from the finite part of the one-loop diagram of figure 1(a). Figure 1(b) represents the counterterm for the primitive divergence of the one-loop diagram that is regularized to a finite value by the considerations outlined earlier.

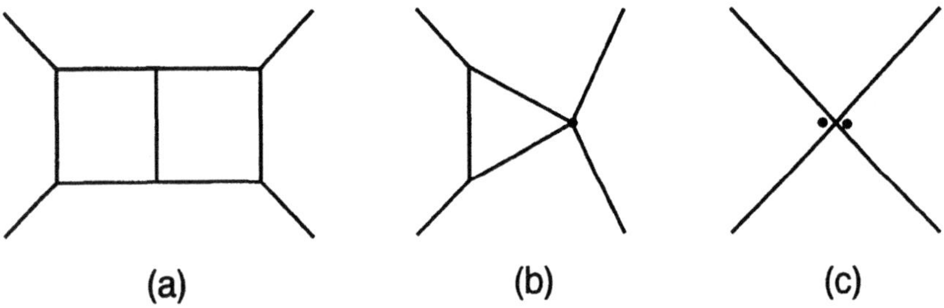

Figure 2. (a) A representation of the non-zero windings of the particles circulating in a two-loop contribution to four-graviton scattering. The dimension, $(momentum)^{20}$, of this finite contribution could give terms of the form s^3 in the ten-dimensional string tree-level amplitude. (b) A contribution in which one loop is replaced by the R^4 counterterm. has dimension $(momentum)^{17}$ and could contribute terms of the form s^2 to the string tree-level amplitude. (c) A counterterm for the two-loop primitive divergence.

Just as with the one-loop diagrams, there are finite contributions from multi-loop diagrams which come from non-zero windings of the internal loops around the compact dimension(s). At each order there is also a new primitive divergence. For example, there are many two-loop diagrams that must be summed to give the complete amplitude. These have superficial degrees of divergence Λ^{20}, where Λ is a momentum space cut-off and is $O(\kappa_{11})^4$. However, we are discussing terms in the amplitude proportional to derivatives acting on R^4 so there are at least eight powers of the external momenta, reducing the naive divergence to Λ^{12}, or less (depending on the number of derivatives). This leads to new (two-loop) primitive divergences indicated by figure 2(c), where the two dots indicate the fact that this is a combination of two-loop counterterms. More generally, at n loops there are primitive divergences of the form $\Lambda^{9n-6}R^4$, where the powers of the cut-off Λ may be substituted by powers of the external momenta. The ultra-violet divergences coming from the zero winding-number sector give rise to counterterms that are independent of R_{11}.

However, just as with the one-loop diagrams there will be finite terms arising from the effects of the internal propagators winding around the compact direction(s). These are collectively represented, in the case of the two-loop diagrams, by the ladder diagram in figure 2(a) which has dimension $(momentum)^{20}$. After accounting for the eight powers of the external momenta in the overall R^4, twelve powers of momenta remain that must be replaced by appropriate powers of the dimensional parameters, R_{11} and S, T, U. This diagram may give a contribution to the tree-level string amplitude which is proportional to $(R_{11})^{-3}$. In such a contribution there are nine powers of momentum left to be soaked up by a combination involving Mandelstam invariants. In the analogous expressions derived from string theory, such as (3.4), each Mandelstam invariant comes along with one power of R_{11}^{-1}. Thus, the nine powers of momenta should be associated with a linear combination of S^3/R_{11}^3, T^3/R_{11}^3, S^2T/R_{11}^3 and T^2S/R_{11}^3. We would like to identify this with the term in (3.4) that is cubic in the invariants. At n loops the Feynman diagrams analogous to figure 2(a) can give finite contributions of the form s^{3n-3} in the expansion of the string tree amplitude (3.4).

In addition to these finite contributions there are also many other loop diagrams in which the counterterms are inserted as vertices. In fact, the diagram with the smallest primitive degree of divergence (other than the one-loop diagrams of figure 1) is the one shown in figure 2(b), in which the vertex indicated by a dot is the R^4 counterterm that was required to make sense of the one-loop diagram. Since all the particles of the supermultiplet circulate around the loop the supersymmetric partners of the R^4 vertices are also involved. These couple the two external gravitons to two $C^{(3)}$'s, or to two gravitini. Figure 2(b) is meant to represent all diagrams of this order containing a single one-loop counterterm. These have dimension $(momentum)^{17}$ so that after allowing for the eight powers of external momenta in R^4 there are nine powers of momentum left to account for. These diagrams have not been explicitly evaluated but if it is assumed that they contribute to the string tree-level process there must be an overall factor of $e^{-2\phi^A} = R_{11}^{-3}$, absorbing three of the nine momentum powers. Then dimensional counting implies that the the only possible finite function is *quadratic* in the invariants, $s, t, u \sim S/R_{11}, T/R_{11}, U/R_{11}$. This quadratic term ought to be precisely the first momentum-dependent correction to the R^4 term deduced by expanding the tree diagram. It should therefore turn out that an explicit calculation of figure 2(b) gives the finite contribution,

$$\frac{2\zeta(5)}{R_{11}^3}\tilde{K}\left(\frac{S^2+ST+T^2}{R_{11}^2}\right), \tag{3.5}$$

where $\tilde{K}$ is the linearized approximation to $t_8t_8R^4$ and is eighth-order in the external momenta. Although the diagram in figure 2(b) has not been explicitly evaluated it is tempting to imagine that it has the form $\tilde{K}I(S,T,U)$, where I is the expression for the loop amplitude of the same geometry in scalar field theory. This has a coefficient that includes a factor of $\zeta(5)$ just as there was a factor of $\zeta(3)$ in the finite part of the box diagram of figure 1(a). Verifying that the precise coefficient is the one in (3.5) would lend support to the suggestion that quantum corrections to compactified eleven-dimensional supergravity coincide with the string tree-level expressions at this order. In a similar manner the term with coefficient $\zeta(7)$ in (3.4) could arise from the diagam with two R^4 vertices and two propagators. However, this exhausts the contributions from one-loop diagrams with R^4 vertices and it is not at all clear how the terms of higher order in momenta in (3.4) would emerge in any systematic manner.

Of course, such diagrams also contribute to string loop effects. For example, there are the generalizations of the non-analytic one-loop massless threshold term, (A.8). In either of the type II string theories these terms can be written symbolically as

$$\int \sqrt{g_E^{(10)}}\, s_E^n f_n(t_E/s_E) R^4 d^{10}x = \int \sqrt{g^{(10)}}\, e^{\frac{n-1}{2}\phi^A} s^n f_n(t/s) R^4 d^{10}x, \qquad (3.6)$$

where g_E is the type II string metric in the Einstein frame. The function $f_n(t/s)$ involves up to n powers of t/s as well as factors of $\log(t/(s+t))$ and $\log(s/(s+t))$. The $n=1$ case is given implicitly by (A.8). Since these terms arise in string perturbation theory, which is an expansion in $e^{2\phi}$, it must be that $n = 4m+1$, where m is an integer. In the Einstein frame these terms are independent of the dilaton so they are inert under the action of $SL(2,Z)$ in the IIB theory. They are therefore not constrained by S-duality. In the IIA theory such terms are directly related to analogous thresholds in genus-$(m+1)$ multi-loop eleven-dimensional supergravity amplitudes. These Feynman diagrams are dimensionally of the form $(\text{momentum})^{9m+3}R^4$. After compactification their contribution to the genus-$(m+1)$ string amplitude has a factor of $e^{2m\phi^A} = R_{11}^{3m}$ from powers of the coupling. The total momentum dimension of this contribution is then written as $R_{11}^{3m}(\text{momentum})^{12m+3}R^4$, which must therefore be made up of terms of the form $e^{2m\phi^A}s^{4m+1}f_n(t/s)R^4$ in the string coordinates, using the fact that $s = S/R_{11}$. This agrees in structure with the terms in (3.6) if the function f contains the appropriate logarithmic terms.

There are many complications in understanding in detail the systematics of the correspondence between the higher-loop supergravity diagrams and string diagrams. Whereas the R^4 and related terms of the same dimension are integrals over half the superspace, terms with more derivatives are integrals over a higher fraction of the superspace. Each power of momentum is equivalent to two powers of θ so that terms with less than eight powers of momentum acting on R^4 should be protected and may be determined in this manner. This would include the terms up to the $\zeta(3)^2$ term in (3.4) and possibly also the $\zeta(7)$ term. Whether it is possible to go beyond this and relate terms in string perturbation theory at higher order in the momentum expansion to eleven-dimensional supergravity is much less obvious.

A. Momentum dependence of one-loop amplitude

The momentum dependence of the four-graviton amplitude is contained in the functions $I_n(S,T)$ $(n = 2, \cdots, \infty)$ in (2.19). This will be a homogeneous polynomial of degree n. There is no $n = 1$ term after adding the contributions of $I(S,U)$ and $I(T,U)$ since there is no nonvanishing linear symmetric combination. The term with zero Kaluza–Klein momenta is given by

$$I^0(S,T) = \frac{1}{\mathcal{V}_2}\int \prod_{r=1}^{4} d\tau_r (\tau)^{-9/2}\left(e^{(S\tau_1\tau_3+T\tau_2\tau_4)/\tau} - 1\right), \tag{A.1}$$

where the -1 accounts for the fact that the constant terms are contained in I_0. The variables τ_r may be redefined by

$$\omega_1 = \frac{\tau_1}{\tau}, \quad \omega_2 = \frac{1}{\tau}(\tau_1+\tau_2), \omega_3 = \frac{1}{\tau}(\tau_1+\tau_2+\tau_3), \quad \tau = \sum_{r=1}^{4}\tau_r, \tag{A.2}$$

which are to be integrated over the region $0 \leq \omega_1 \leq \omega_2 \leq \omega_3 \leq 1$ so that

$$I^0(S,T) = \frac{1}{\mathcal{V}_2}\int_0^\infty d\tau \int \prod_{r=1}^{3} d\omega_r (\tau)^{-3/2}\left(e^{-Q(S,T;\omega_r)\tau} - 1\right), \tag{A.3}$$

where

$$Q(S,T;\omega_r) = -S\omega_1(\omega_3-\omega_2) - T(\omega_2-\omega_1)(1-\omega_3). \tag{A.4}$$

The integral in (A.3) is simply evaluated by noting that $\partial I^0/\partial Q = \sqrt{\pi}\int\prod d\omega_r Q^{-\frac{1}{2}}$ so that

$$I^0(S,T) \equiv 2\sqrt{\pi}\mathcal{S}^{\frac{1}{2}} = 2\sqrt{\pi}\int\prod d\omega_r(-Q(S,T;\omega_r))^{\frac{1}{2}}, \tag{A.5}$$

where, more generally, $\mathcal{S}^n$ is defined by

$$\mathcal{S}^n = \int\prod d\omega_r(-Q(S,T;\omega_r))^n. \tag{A.6}$$

Similarly, $\mathcal{T}^n$ and $\mathcal{U}^n$ will be defined by cyclically permuting S, T and U in the function Q.

This non-analytic term (A.5) in nine-dimensional M-theory translates into a similar term in either nine-dimensional string theory (as was also noticed in [9]) making use of the familiar relation, by $g^{(9)} = R_{11}G^{(9)}$, between the nine-dimensional type II metric, $g^{(9)}_{\mu\nu}$, and the M-theory metric, $G^{(9)}$, compactified on T^2.

Using the relation between the M-theory and string theory Mandelstam invariants (3.2) it follows that

$$2\sqrt{\pi}\int\sqrt{G^{(9)}}R^4(\mathcal{S}^{\frac{1}{2}}+\mathcal{T}^{\frac{1}{2}}+\mathcal{U}^{\frac{1}{2}})d^9x = 2\sqrt{\pi}\int\sqrt{g^{(9)}}R^4(\mathcal{S}_s^{\frac{1}{2}}+\mathcal{T}_s^{\frac{1}{2}}+\mathcal{U}_s^{\frac{1}{2}})d^9x, \tag{A.7}$$

where the expressions $\mathcal{S}_s, \mathcal{T}_s$ and $\mathcal{U}_s$ are defined in terms of the Mandelstam invariants of string theory. The analogous expression in the limit of decompactification, $R_{10} \to \infty$, is of the form,

$$\int \sqrt{G^{(10)}} R^4 (\mathcal{S} \ln \mathcal{S} + \mathcal{T} \ln \mathcal{T} + \mathcal{U} \ln \mathcal{U}) d^{10}x$$
$$= \int \sqrt{g^{A(10)}} R^4 (\mathcal{S}_s \ln \mathcal{S}_s + \mathcal{T}_s \ln \mathcal{T}_s + \mathcal{U}_s \ln \mathcal{U}_s) d^{10}x. \quad \text{(A.8)}$$

The scale of the logarithms cancels out after using the condition $s + t + u = 0$. Both (A.7) and (A.8) have imaginary parts corresponding to the massless normal thresholds determined by unitarity. However, the real parts, which might have given rise to arbitrary constants, are here fixed to precise values.

The terms $I_n(S,T)$ in (2.19) are given by by expanding the exponential in (2.14) giving,

$$\begin{aligned} I_n(S,T) &= \int_0^\infty \frac{d\tau}{\tau^{\frac{3}{2}-n}} \int \prod_{r=1}^{3} d\omega_r \sum_{(l_1,l_2)\neq(0,0)} e^{-G^{IJ} l_I l_J \tau} \frac{(-Q(S,T;\omega_r))^n}{n!} \\ &= \Gamma(n-\frac{1}{2})\zeta(n-\frac{1}{2}) E_{n-\frac{1}{2}}(\Omega,\bar{\Omega}) \int \prod_{r=1}^{3} d\omega_r \frac{(-Q(S,T;\omega_r))^n}{n!}. \quad \text{(A.9)} \end{aligned}$$

Putting all the terms together the complete expression for the R^4 term in the one-loop effective action for eleven-dimensional supergravity on T^2 is given by (2.13) with the function h defined by the amplitude A_4 in (2.9) where

$$\begin{aligned} & I(S,T) + I(T,U) + I(U,S) = \\ & I_0 + 2\sqrt{\pi}\mathcal{W}^{\frac{1}{2}} + \sum_{n=2}^{\infty} \frac{\mathcal{W}^n \mathcal{V}_2^{n-\frac{3}{2}}}{n!} \left[\Gamma(n-\frac{1}{2})\zeta(2n-1) \left(\frac{R_{10}}{R_{11}}\right)^{n-\frac{1}{2}} \right. \\ & \left. + \sqrt{\pi}\Gamma(n-1)\zeta(2n-2) \left(\frac{R_{10}}{R_{11}}\right)^{\frac{3}{2}-n} \right] + \text{non-perturbative terms,} \quad \text{(A.10)} \end{aligned}$$

and

$$\mathcal{W}^n = \mathcal{S}^n + \mathcal{T}^n + \mathcal{U}^n. \quad \text{(A.11)}$$

The S, T or U-dependent terms in (A.10) exponentiate in the form $e^{S\mathcal{V}_2}$ which vanishes in the infinite volume limit, $\mathcal{V}_2 \to \infty$.

It follows from (A.10) that the expression for the nine-dimensional effective action arising from the one-loop supergravity amplitude can be written in IIA coordinates as

$$\int \sqrt{g^{A(9)}} r^A R^4 \left[I_0 + \frac{1}{r_A} \mathcal{W}_s^{\frac{1}{2}} + \sum_{n=2}^{\infty} \left(\frac{1}{r_A^2} \frac{r_A^{2n} \mathcal{W}_s^n}{n!} + e^{-2\phi^A} \frac{e^{2n\phi^A} \mathcal{W}_s^n}{n!} \right) \right] d^9x$$
$$+\text{non-perturbative terms,} \quad \text{(A.12)}$$

where

$$\mathcal{W}_s^n = \mathcal{S}_s^n + \mathcal{T}_s^n + \mathcal{U}_s^n, \tag{A.13}$$

and I_0 can be reexpressed in type IIA parameters, r_A and ϕ^A. The first sum in round parentheses gives terms of the form $e^{r_A^2 s}$ which are exponentially suppressed in the ten-dimensional limit, $r_A \to \infty$. The ten-dimensional limit in the IIA theory is therefore given by a power series in $e^{2\phi^A} s$ or, for constant dilaton, by a series of powers of $g_A^2 s$ where $g_A = e^{\phi^A}$. Each power, n, should presumably be identified with a contribution from a string loop amplitude of genus $n+1$. This suggests that the leading low-momentum term at genus n of the IIA theory should be of order $(g_A^2 k^2)^{n-1}$, where $g_A = e^{\phi^A}$ for constant ϕ^A (and k^2 represents some combination of s, t and u).

In the IIB coordinates the action becomes

$$\sum_n \int \sqrt{g^{B(9)}} r^B R^4 \left[I_0 + \frac{1}{r^B} \mathcal{W}_s^{\frac{1}{2}} + \sum_{n=2}^{\infty} \left(\frac{1}{n!} \frac{\mathcal{W}_s^n}{r_B^{2n}} + \frac{e^{-2\phi^B}}{n!} \frac{\mathcal{W}_s^n e^{2n\phi^B}}{r_B^{2n}} \right) \right] d^9x$$
$$+\text{non} - \text{perturbative terms.} \tag{A.14}$$

All the momentum-dependent terms in this expression vanish in the limit $r_B \to \infty$ with fixed coupling e^{ϕ^B}. This means that these terms do not contribute to the loop amplitudes of the ten-dimensional type IIB theory. However, the four-graviton loop amplitudes are identical in the IIA and IIB theories as a consequence of T-duality. This means that there must be terms in an expansion in powers of $e^{2\phi^B} s = g_B^2 s$ that have another eleven-dimensional origin. For example, such terms can arise from eleven-dimensional multi-loop diagrams compactified on a torus.

References

1. E. Cremmer, B. Julia and J. Scherk, *Supergravity theory in eleven dimensions*, Phys. Lett. **76B** (1978) 409.
2. M.J. Duff, J.T. Liu and R. Minasian, *Eleven-dimensional origin of string-string duality: a one loop test*, hep-th/9506126, Nucl. Phys. **452B** (1995)261.
3. C. Vafa and E. Witten, *A one loop test of string duality*, hep-th/9505053, Nucl. Phys. **B447** (1995)261.
4. Blum and J. Harvey, *Anomaly inflow for gauge defects*, hep-th/9310035; Nucl. Phys. **B416** (1994) 119.
5. P. Aspinwall, *Some Relationships Between Dualities in String Theory*, in Proceedings of 'S-duality and mirror symmetry', Trieste 1995, hep-th/9508154, Nucl. Phys. Proc. **46** (1996) 30.
6. J.H. Schwarz, *Lectures on Superstring and M-theory dualities*, hep-th/9607201;J.H. Schwarz, *An Sl(2, Z) multiplet of type IIb superstrings*, hep-th/9508143 Phys. Lett. **360B** (1995) 13.
7. M.B. Green, M. Gutperle and P. Vanhove, *One loop in eleven dimensions*, hep-th/9706175, Phys. Lett. **409B** (1997)177.
8. M.B. Green and M. Gutperle, *Effects of D-instantons*, hep-th/9701093, Nucl. Phys. **B498** (1997)195.
9. J.G. Russo and A.A. Tseytlin, *One loop four graviton amplitude in eleven-dimensional supergravity* hep-th/9707134.
10. M.B. Green, M. Gutperle and H. Kwon, *Sixteen-fermion and related terms in M theory on T^2*, hep-th/9710151.

11. M.B. Green and Michael Gutperle, *D-particle bound states and the D-instanton measure* hep-th/9711107.
12. E. Witten, *Bound States of Strings and p-branes*, Nucl. Phys. **B460** (1996) 335.
13. S. Sethi and M. Stern, *D-brane bound states redux*, hep-th/9705046.
14. P. Yi, *Witten index and threshold bound states of D-branes*, hep-th/9704098.
15. M.B. Green and J.H. Schwarz, *Supersymmetric dual string theory (III). Loops and renormalization*, Nucl. Phys. **198B** (1982) 441.
16. M.T. Grisaru, A.E.M Van de Ven and D. Zanon, *Two-dimensional supersymmetric sigma models on Ricci flat Kähler manifolds are not finite*, Nucl. Phys. **B277** (1986) 388; *Four loop divergences for the N=1 supersymmetric nonlinear sigma model in two-dimensions*, Nucl. Phys. **B277** (1986) 409.
17. D.J. Gross and E. Witten, *Superstring modifications of Einstein's equations*, Nucl. Phys. **B277** (1986) 1.
18. M.B. Green and J.H. Schwarz, *Supersymmetrical string theories*, Phys. Lett. **109B** (1982) 444.
19. A. Terras, *Harmonic Analysis on Symmetric Spaces and Applications*, vol. I, Springer–Verlag (1985).
20. I. Antoniadis, B. Pioline and T.R. Taylor, *Calculable* $e^{1/\lambda}$ *effects*, hep-th/9707222.
21. N. Berkovits, *Construction of* R^4 *terms in N=2 D=8 superspace*, hep-th/9709116.
22. R. Jengo and C.-J. Zhu, *Two loop computation of the four particle amplitude in heterotic string theory*, Phys. Lett. **212B** (1988) 313.
23. T. Banks, W. Fischler, S.H. Shenker and L. Susskind, *M-Theory as a matrix model: a conjecture*, hep-th/9610043, Phys. Rev. **D55** (1997) 5112.
24. N. Nekrasov and A. Lawrence, *Instanton sums and five dimensional gauge theories*, hep-th/9706025.
25. M.B. Green and P. Vanhove, *D-instantons, strings and M-theory*, hep-th/9704145, Phys. Lett. **408B** (1997)122.
26. E. Kiritsis and B. Pioline, *On* R^4 *threshold corrections in IIB string theory and (p,q) string instantons*, hep-th/9707018
27. J.H. Schwarz and P.C. West, *Symmetries and transformations of chiral N=2 D=10 supergravity*, Phys. Lett. **126B** (1983) 301.
28. J.H. Schwarz, *Covariant field equations of chiral N=2 D=10 supergravity*, Nucl. Phys. **B226** (1983) 269.
29. P.S. Howe and P.C. West, *The complete N=2 D=10 supergravity*, Nucl. Phys. **B238** (1984) 181.
30. D.J. Gross and J.H. Sloan, *The quartic effective action for the heterotic string*, Nucl. Phys. **291B** (1987) 41.
31. A. Kehagias and H. Partouche, *The exact quartic effective action for the type IIB superstring*, hep-th/9710023.
32. E. Kiritsis and B. Pioline, *On* R^4 *threshold corrections in IIB string theory and (p,q) string instantons*, hep-th/9707018;
A. Strominger, *Loop corrections to the universal hypermultiplet*, hep-th/9706195.
I. Antoniadis, S. Ferrara, R. Minasian and K.S. Narain, R^4 *couplings in M and type II theories*, hep-th/9707013.
33. C. Bachas, C. Fabre, E. Kiritsis, N.A. Obers and P. Vanhove, *Heterotic/type I duality and D-brane instantons*, hep-th/9707126.
E. Kiritsis and N.A. Obers, *Heterotic/type-I duality in* $D < 10$ *dimensions, threshold corrections and D-instantons*, hep-th/9709058.
34. N. Ishibashi, H. Kawai and Y. Kitazawa and A. Tsuchiya, *A large N reduced model as superstring*, hep-th/9612115; Nucl.Phys.B498:467-491,1997.
35. C. Bachas, *Heterotic versus type I*, Talk at STRINGS'97 (Amsterdam, June 16-21 1997) and HEP-97 (Jerusalem, August 19-26 1997), hep-th/9710102.

BLACK HOLES AND D-BRANES

JUAN M. MALDACENA
Lyman Lab., Harvard University,
Cambridge, MA 02138 USA

1. Introduction

Under a wide variety of conditions General Relativity predicts that singularities will develop [1]. The cosmic censorship hypothesis states that under generic physical situations leading to gravitational collapse the resulting singularities will be covered by an event horizon [2]. This conjecture has not been yet proved but there exists great evidence that it is correct [3]. The area of the horizon is an interesting quantity since it always increases upon classical evolution [4], this looks very similar to the second law of thermodynamics. The analogy became more precise when Hawking showed [5] that quantum mechanics implies that black holes emit thermal radiation with a temperature obeying the first law of thermodynamics $dM = T_H dS$, where the entropy is $S = \frac{A_H}{4G_N\hbar}$ [6], M is the black hole mass and A_H is the horizon area (from now on we set $\hbar = 1$ but keep $G_N \neq 1$). The area increase law becomes the second law of thermodynamics. If one includes Hawking radiation, the black hole mass decreases and so does the area of the horizon, but the total entropy, defined as $S = A_H/4G_N + S_{rad}$, increases. For any physical system we expect that the entropy is the logarithm of the number of states with given macroscopic properties, like the mass, charge, angular momentum, etc. It has been a longstanding puzzle to find the degrees of freedom that give rise to this entropy. It seems clear that some quantum gravity will be necessary to describe the microstates. String theory [7] is a theory of quantum gravity so one would naturally expect that it should give an answer to this question. But string theory is defined perturbatively and black holes involve strong interactions due to their large mass. Only when some non-perturbative tools became available [8, 9] could precise calculations be made [10]. There are, however, rough counting arguments that produce the right scaling for the entropy using just string perturbation theory [11, 12, 13]. Our focus will be to explain the calculations that produce precise results, which are valid presently only for a subset of all possible black hole configurations. For other reviews on this subject see [14, 15, 16]. Recently there has been some studies of Schwarzschild black holes using Matrix theory [17] and agreement up to numerical coefficients was found for the entropy and other quantities [18].

L. Baulieu et al. (eds.), Strings, Branes and Dualities, 219–240.

We are going to be treating charged black holes. The cosmic censorship hypothesis gives a bound for the mass of a black hole in terms of its charge $M \geq Q$ (in appropriate units). The black hole with $M = Q$ is called extremal. We will study extremal and near extremal ($M - Q \ll Q$) black holes and we focus on black holes in five spacetime dimensions. In four spacetime dimensions the discussion is similar. In section 2 we write down the classical supergravity solutions that we are going to describe. In section 3 we present the D-brane description of extremal black holes in five dimensions. In section 4 we study near extremal black holes, their entropy and their decay rates and we compare them to the semiclassical results. In section 5 we compute the greybody factors for emission of massless scalars.

2. Classical solutions

In this section we describe in some detail the classical black holes that we will study. We consider the maximally supersymmetric supergravity theory but, in fact, some of the solutions are also solutions of the plain Einstein-Maxwell Lagrangian. This theory contains gravity, many U(1) fields, several scalar fields and many fermion fields, we can view it as just gravity plus some matter fields (this is an oversimplification because the gravitino is not an ordinary matter field). We consider the supergravity theory in 4+1 spacetime dimensions. The discussion in 3+1 spacetime dimensions is similar, but just slightly more complicated. The full quantum theory is type IIB string theory compactified on T^5 which reduces at low energies to the maximally supersymmetric supergravity theory in 4+1 dimensions. It has 32 supersymmetry generators and it is the dimensional reduction of ten dimensional type IIB supergravity. The theory contains 27 abelian gauge fields. The full string theory contains charged objects that couple to each of these gauge fields. These objects are: 5 Kaluza Klein momenta, 5 string winding directions, 5 D-string winding directions, 10 possible D-3-brane wrapping modes, a solitonic NS fivebrane, a D-fivebrane. All these charges are interchanged by U-duality transformations [22] and they are all quantized. Therefore, we measure charges in integer multiples of the elementary units. All these objects are viewed as point particles by the 4+1 dimensional observer.

A charged black hole solution is characterized in principle by 27 charges. Fortunately, using the U-duality symmetries it is possible to transform any solution into a solution carrying three charges. Since each abelian vector field is associated to an object in the full string theory that is charged under it, we can specify the charge of the configuration by naming the object and the magnitude of the charge, which will be an integer. we consider a black hole solution which has the same charge as Q_5 D-fivebranes wrapped on T^5, Q_1 D-1-branes wrapped on an S_1 (we choose it as the direction $\hat{9}$), and momentum $P = N/R$ also along the direction of the D-string (direction $\hat{9}$). When we mention D-branes in this context of classical solutions we only refer to the charge that the solution is carrying, there are no explicit D-branes in the sense of [9, 8] anywhere in spacetime. We choose this set of charges because the string theory description is simpler. Other black holes are related to this by U-duality transformations.

We start by presenting the ten dimensional solution [25] (in string metric)

$$e^{-2(\phi-\phi_\infty)}=\left(1+\frac{r_0^2\sinh^2\gamma}{r^2}\right)\left(1+\frac{r_0^2\sinh^2\alpha}{r^2}\right)^{-1}, \tag{2.1}$$

$$\begin{aligned} ds^2_{str} &= \left(1+\tfrac{r_0^2\sinh^2\alpha}{r^2}\right)^{-1/2}\left(1+\tfrac{r_0^2\sinh^2\gamma}{r^2}\right)^{-1/2}\times \\ &\Big[-dt^2+dx_9^2+\tfrac{r_0^2}{r^2}(\cosh\sigma dt+\sinh\sigma dx_9)^2 \\ &+\left(1+\tfrac{r_0^2\sinh^2\alpha}{r^2}\right)(dx_5^2+\ldots+dx_8^2)\Big] \\ &+\left(1+\tfrac{r_0^2\sinh^2\alpha}{r^2}\right)^{1/2}\left(1+\tfrac{r_0^2\sinh^2\gamma}{r^2}\right)^{1/2}\times \\ &\left[\left(1-\tfrac{r_0^2}{r^2}\right)^{-1}dr^2+r^2d\Omega_3^2\right]. \end{aligned} \tag{2.2}$$

Also some components of the Ramond-Ramond three-form field strength $H'_{\mu\nu\rho}$ are nonzero since the solution carries D1-brane and D5-brane charge. This solution is parameterized by the four independent quantities α,γ,σ,r_0. There are two extra parameters which enter through the charge quantization conditions which are the radius of the 9th dimension R_9 and the product of the radii in the other four compact directions $V\equiv R_5R_6R_7R_8$. The three charges are

$$\begin{aligned} Q_1 &= \tfrac{V}{4\pi^2 g}\int_{S^3}e^{2\phi_6}*H'=\tfrac{Vr_0^2}{2g}\sinh 2\alpha, \\ Q_5 &= \tfrac{1}{4\pi^2 g}\int_{S^3}H'=\tfrac{r_0^2}{2g}\sinh 2\gamma, \\ N &= \tfrac{R^2Vr_0^2}{2g^2}\sinh 2\sigma, \end{aligned} \tag{2.3}$$

where $*$ is the Hodge dual in the six dimensions $x^0,..,x^5$. For simplicity we set from now on $\alpha'=1$. All charges are normalized to be integers. We have chosen a convention such that $g\to 1/g$ under S-duality. Further explanations on the charge quantization conditions can be found in [8, 26].

Reducing (2.2) to five dimensions using the standard dimensional reduction procedure [27], the solution takes the simple and symmetric form:

$$ds_5^2=-\lambda^{-2/3}hdt^2+\lambda^{1/3}\left(\frac{dr^2}{h}+r^2d\Omega_3^2\right), \tag{2.4}$$

where

$$\begin{aligned} \lambda &= \left(1+\tfrac{r_1^2}{r^2}\right)\left(1+\tfrac{r_5^2}{r^2}\right)\left(1+\tfrac{r_n^2}{r^2}\right), \\ h &= 1-\tfrac{r_0^2}{r^2} \end{aligned} \tag{2.5}$$

$$r_1^2=r_0^2\sinh^2\alpha\,,\quad r_5^2=r_0^2\sinh^2\gamma\,,\quad r_n^2=r_0^2\sinh^2\sigma \tag{2.6}$$

This is just the five-dimensional Schwarzschild metric with the time and space components rescaled by different powers of λ. The event horizon is at $r=r_0$. Several thermodynamic quantities can be associated to this solution. They can be computed in either the ten dimensional or five dimensional metrics and yield the same answer. For example, the ADM energy is

$$M=\frac{RVr_0^2}{2g^2}(\cosh 2\alpha+\cosh 2\gamma+\cosh 2\sigma). \tag{2.7}$$

The Bekenstein-Hawking entropy is

$$\begin{aligned} S = & \frac{A_{10}}{4G_N^{10}} = \frac{A_5}{4G_N^5} \\ = & \frac{2\pi R V r_0^3}{g^2} \cosh\alpha \cosh\gamma \cosh\sigma. \end{aligned} \tag{2.8}$$

where A is the area of the horizon and we have used that the Newton constant is $G_N^{10} = 8\pi^6 g^2$. The Hawking temperature is

$$T = \frac{1}{2\pi r_0 \cosh\alpha \cosh\gamma \cosh\sigma}. \tag{2.9}$$

The extremal limit corresponds to $r_0 \to 0$, $\alpha, \gamma, \sigma \to \infty$ keeping the charges (2.3) finite. In that limit the entropy (2.8) becomes [23, 24, 15]

$$S = 2\pi\sqrt{NQ_1Q_5} \tag{2.10}$$

and the temperature vanishes. Note that the extremal entropy is independent of any continuous parameters [28, 29]. The extremal black hole backgrounds preserve some space time supersymmetries and therefore they are BPS states. In this case the cosmic censorship bound becomes identical to the supersymmetry BPS bound [30].

The near extremal limit corresponds to r_0 small and α, γ, σ large. The relative values of α, γ, σ are related to the total contribution of the different charges to the mass (2.7). The near extremal black holes that are easiest to analyze in terms of D-branes are those where $\sigma \ll \alpha, \gamma$, or $r_0, r_n \ll r_1, r_5$, which means that the contribution to the mass (2.7) due to the D-branes is much bigger than the contribution due to the momentum excitations. This limit is called "dilute gas" [31] [32]. In this limit, the mass and entropy of the near extremal black hole become

$$M = \frac{Q_5 R V}{g} + \frac{Q_1 R}{g} + \frac{R V r_0^2}{2g^2} \cosh 2\sigma \ , \tag{2.11}$$

$$S = 2\pi \frac{R\sqrt{V} r_0}{g} \sqrt{Q_1 Q_5} \cosh\sigma \ . \tag{2.12}$$

It is interesting to note [19, 20] that the near horizon geometry of black holes in the dilute gas limit becomes some quotient of $AdS_3 \times S^3$ where the cosmological constant and the radius of S^3 in Planck units are $\Lambda^{-1} \sim R^2 \sim \sqrt{Q_1 Q_5}$. These identifications are precisely the ones that lead to the Bañados-Teitelboim-Zanelli [21] black hole in 2+1 dimensions.

Note that the five dimensional Reissner-Nordstrøm solution corresponds to the case of $\alpha = \gamma = \sigma$ which is not included in the dilute gas limit.

All these black hole solutions will be well defined if curvatures are everywhere much smaller than α', since otherwise α' corrections to the low energy action become important. This generically implies that the sizes of the black hole should obey $r_1, r_5, r_n \gg 1$ (remember that $\alpha' = 1$). The precise condition will be a little more complicated if some scale is very different from the others. If the compactification sizes are of the order of α' this implies that

$$gQ_1 \gg 1, \quad gQ_5 \gg 1, \quad g^2N \gg 1 \tag{2.13}$$

Note that we cannot enforce (2.13) by setting $Q = 1$ and making $g \gg 1$ since the condition (2.13) was derived in weakly coupled string theory, if $g \gg 1$ the corresponding bound comes from light D-strings and implies again that $Q \gg 1$. Actually, in the S-dual picture we obtain a black hole carrying NS fivebrane charge. In that case the black hole condition is just $Q \gg 1$. Therefore black holes always involve large values of the charges.

3. D-brane description of extremal 5d black holes

We continue with type IIB string theory on $T^5 = T^4 \times S^1$. We consider a configuration of Q_5 D-fivebranes wrapping the whole T^5, Q_1 D-strings wrapping the S^1 and momentum N/R_9 along the S^1, choosing this S^1 to be in the direction $\hat{9}$. All charges N, Q_1, Q_5 are integers. For a review of D-branes see [33]. We take the coupling constant g to be small and the radius R_9 to be large. The total mass of the system is

$$M = \frac{Q_5 RV}{g} + \frac{Q_1 R}{g} + \frac{N}{R} \tag{3.1}$$

and it saturates the corresponding BPS bound.

We will calculate the entropy of this state in perturbative string theory. This calculation was first done by Strominger and Vafa [10]. We will describe three equivalent ways of doing the calculation, they are related by duality. We present it in three ways to emphasize the different aspects. All three use properties of D-branes.

3.1. DESCRIPTION IN TERMS OF OPEN STRINGS

This picture was proposed in [36, 37, 26]. Since extremal D-branes are boost invariant along the directions parallel to the branes they cannot carry momentum along S_1 by just moving rigidly. Our first task will be to identify the D-brane excitations that carry the momentum. The BPS mass formula for the whole system implies that these excitations have to be massless and moving along the S^1 since the excitation energy, defined as the total mass of the system minus the mass of the onebranes and fivebranes, is equal to the momentum. If any excitation fails to be massless it would contribute more to the energy than to the momentum and the BPS mass formula would be violated. Excitations of the branes are described by open strings. There are many types of open strings to consider: those that go from one 1-brane to another 1-brane, which we denote as (1,1) strings, as well as the corresponding (5,5), (1,5) and (5,1) strings (the last two being different because the strings are oriented). Each of these types of strings have some massless and some massive modes, we will concentrate on the massless modes. The fact that the (1,5) strings have massless modes is non-trivial and it is basically due to the fact that D1-branes and D5-branes preserve some common supersymmetries. It will be important that the (1,5) strings have 4 bosonic and 4 fermionic massless modes. Actually, in our case we have $4Q_1Q_5$ massless modes since the strings can go from any D1-brane to any D5-brane.

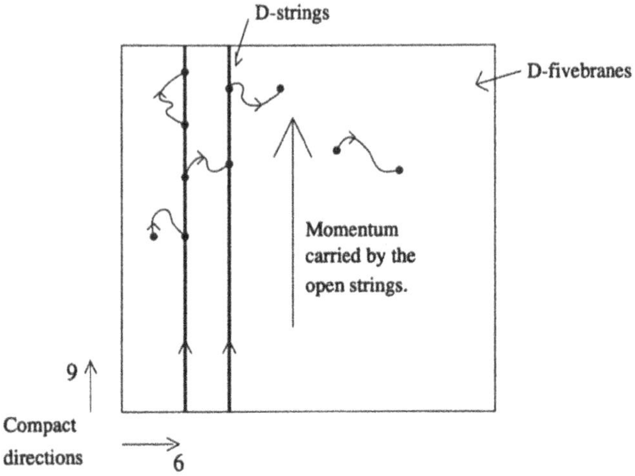

Figure 1. Configuration of intersecting D-branes open strings go between different branes

The (1,1), (5,5) and (1,5) strings interact among themselves. We are interested in the low energy limit of these interactions. This corresponds to the field theory limit of the system of branes. If we take the size of the S_1 along $\hat{9}$ to be very large this will be a 1+1 dimensional gauge theory. The Lagrangian is then determined by supersymmetry and gauge invariance. The (1,1) and (5,5) strings are $U(Q_1)$ and $U(Q_5)$ gauge bosons respectively and the (1,5) strings (and (5,1) strings) are in the fundamental (antifundamental) of $U(Q_1) \times U(Q_5)$[1]. This Lagrangian was studied by Douglas [34]. It turns out that after these interactions are taken into account only $4Q_1Q_5$ truly massless degrees of freedom remain. In gauge theory terms, one is interested in the Higgs branch of the theory which is $4Q_1Q_5$-dimensional. We will spare the details which can be found in [26] (page 52), and conclude that the number of massless states is $4Q_1Q_5$ (with the same number of bosonic and fermionic states since the theory on the branes is supersymmetric). These massless degrees of freedom are described by a two dimensional conformal field theory [10], which accounts for the low energy excitations of the D-brane system. In fact, it is a (4,4) superconformal field theory, i.e. it has four left moving and four right moving supersymmetry generators. The rotational symmetry $SO(4)_{1234} \sim SU(2)_L \times SU(2)_R$ of the four spatial extended dimensions, that are transverse to

[1]The gauge theory is in reality a bit more complicated since we are on a torus and one has to include the multiple "images" of the D1-branes. Notice however that most of the extra states are highly massive, they are only necessary to ensure the right periodicity conditions on the torus.

the black hole, acts on this superconformal field theory as the $SU(2)_L \times SU(2)_R$ R-symmetries of the N=4 supersymmetry algebra in two dimensions. Notice that the chirality in space becomes correlated with the chirality of the 1+1 dimensional theory. So that $SU(2)_L$ in spacetime acts on the leftmovers and $SU(2)_R$ on the right movers [35] .

The BPS state that we are interested in has only left moving excitations so the rightmovers are in their ground state $E_R = 0$, The state counting is the same as that of a one dimensional gas of left moving particles with $N_{B,F} = 4Q_1Q_5$ bosonic and fermionic species with energy $E = E_L = N/R_9$ on a compact one dimensional space of length $L = 2\pi R_9$. The standard entropy formula gives [10] [36]

$$S_e = \sqrt{\pi(2N_B + N_F)EL/6} = 2\pi\sqrt{Q_1Q_5N} \ , \tag{3.2}$$

in perfect agreement with (2.10), including the numerical coefficient. It might seem surprising that a system could have entropy at zero temperature, but this is a common phenomenon. Consider for example a gas of massless particles in a box with periodic boundary conditions constrained to have a fixed amount of momentum, at $T = 0$ the entropy remains nonzero. This reason is *exactly* the same reason that black hole entropy is nonzero at $T = 0$.

In our previous argument we implicitly took the D-strings and the fivebranes to be singly wound since we were assuming that the excitations carried momentum quantized in units of $1/R$. For large N, $N \gg Q_1Q_5$, the entropy (3.2) is the same no matter how the branes are wound. However for $N \sim Q_1Q_5$ the winding starts to matter. The reason is that in order for the asymptotic entropy formula to be correct for low N we need to have enough states with small energies [37]. Let us study the effect of different wrappings. We first simplify the problem and consider a set of Q_1 1-branes wrapped on S^1, ignoring for the time being, the 5-branes. We may distinguish the various ways the branes interconnect. For example, they may connect up so as to form one long brane of total length $R' = RQ_1$. At the opposite extreme they might form Q_1 disconnected loops. The spectra of open strings is different in each case. For the latter case the open strings behave like Q_1 species of 1 dimensional particles, each with energy spectrum given by integer multiples of $1/R$. In the former case they behave more like a single species of 1 dimensional particle living on a space of length Q_1R. The result [38] is a spectrum of single particle energies given by integer multiples of $\frac{1}{Q_1R}$. In other words the system simulates a spectrum of fractional charges. For consistency the total charge must add up to an integer multiple of $1/R$ but it can do so by adding up fractional charges.

Now let us return to the case of both 1 and 5 branes. By suppressing reference to the four compact directions orthogonal to x^9 we may think of the 5 branes as another kind of 1 brane wrapped on S^1. The 5-branes may also be connected to form a single multiply wound brane or several singly wound branes. Let us consider the spectrum of (1,5) type strings (strings which connect a 1-brane to a five-brane) when both the 1 and 5 branes each form a single long brane. The 1-brane has total length Q_1R and the 5-brane has length Q_5R. A given open string can be indexed by a pair of indices $[i, \bar{j}]$ labeling which loop of 1-brane and 5-brane it ends on. As a simple example choose $Q_1 = 2$ and $Q_5 = 3$. Now start with the $[1,1]$ string which connects the first loop of 1-brane to the first loop of 5-brane. Let us transport this string around the S^1. When it comes back to the starting point it is a $[2,2]$ string. Transport it again and it becomes a $[1,3]$ string. It must be cycled 6 times before returning to the $[1,1]$ configuration. It follows that such a string has a spectrum of a single species living on a circle of size $6R$. More generally, if Q_1 and Q_5 are relatively prime the system simulates a single species on a circle of size Q_1Q_5R. If the $Q's$ are not relatively prime the situation is slightly more complicated but the result is the same. A more detailed picture of how this happens is presented in [39].

We can easily see that this way of wrapping the branes gives the correct value for the extremal entropy. As above, the open strings have 4 bosonic and 4 fermionic degrees of freedom and carry total momentum N/R. This time the quantization length is $R' = Q_1Q_5R$ and the momentum is quantized in units of $(Q_1Q_5R)^{-1}$. Thus instead of being at level N the system is at level $N' = NQ_1Q_5$. In place of the original Q_1Q_5 species we now have a single species. The result is

$$S = 2\pi\sqrt{N'} = 2\pi\sqrt{NQ_1Q_5} \tag{3.3}$$

So we have a long effective string that is moving along the fivebrane. In the extremal case, this effective string picture follows precisely from an analysis of the moduli spaces of BPS states [40]. What one actually has is a sum over multiple string states which one can call "second quantized" strings on a fivebrane [41]. The state in which they are all connected into a single long string is the one having most entropy.

3.2. DESCRIPTION IN TERMS OF INSTANTONS

We now present the same calculation but in a picture where we start with just D-fivebranes and we build up the charges as excitations of the D-fivebranes. We start with Q_5 D-fivebranes. The low energy theory on the fivebranes is an U(Q_5) supersymmetric Yang Mills with 16 supersymmetries (same amount as N=4 in d=4). This theory contains BPS string solitons which are constructed as follows: if the fivebranes are along the directions 56789, take an instanton configuration that involves the directions 5678 and the gauge fields along those directions. The corresponding field configuration could be localized in the directions 5678 but will be extended along the direction 9, so it is a string soliton. Notice that, even though we call this solution an "instanton" (in the sense that the Yang-Mills fields are self dual solutions of a YM theory in four Euclidean dimensions (5678)), the physical interpretation is that we have a string "soliton" which exists for all times. It turns out that each instanton that lives on the fivebrane world volume carries one unit of D-string charge [34] due to a Chern Simons coupling on the fivebrane of the form

$$\int d^{1+5}x \; B_2^{RR} \wedge TrF \wedge F \tag{3.4}$$

since $TrF \wedge F$ will be proportional to the instanton number. We are interested in the case that the instanton number is Q_1. So D-strings dissolve into instantons when they get into fivebranes. In fact, giving an expectation value to the (1,5) strings corresponds to giving a size to the instanton [34, 42]. This description makes sense, in principle, only when the mass of the fivebranes is much bigger than the mass of the D-1-branes in (3.1), since otherwise the fivebranes would contain so much energy that they would no longer be described by the low energy YM theory. This instanton configuration is characterized by $4Q_1Q_5$ continuous parameters which specify the instanton positions on the branes as well as their relative orientation inside the $U(Q_5)$ gauge group, the space of these parameters is called "moduli space". When we put some momentum along the direction $\hat{9}$, this momentum can be carried by small oscillations of the instanton configuration. We denote the instanton parameters by ξ^a, $a = 1,..,4Q_1Q_5$. They can be slowly varying functions $\xi^a(t - x^9)$ representing traveling waves moving along the instantons.

These are small oscillations in the parameters specifying the instanton configuration, i.e. oscillations in moduli space. Each bosonic mode has a fermionic superpartner and together they form a (4,4) superconformal field theory with central charge $c = 6Q_1Q_5$. In principle we can have a nontrivial metric on the moduli space of instantons. This metric will be hyperkähler, by supersymmetry. This implies that we indeed have a conformal field theory and furthermore, the central charge is given in terms of the dimension of the hyperkähler space without further corrections. A configuration carrying momentum N corresponds to states in the SCFT with $L_0 = N$ and $\bar{L}_0 = 0$, the entropy of such states can be calculated using the CFT formula $d(N) \sim e^{2\pi\sqrt{Nc/6}}$ for the degeneracy of states at level N. This yields (3.2) again. The moduli space of instantons is, topologically, a symmetric product: $\mathcal{M} = (T^4)^{Q_1Q_5}/S(Q_1Q_5)$ [43, 44]. This is the target space of the SCFT, in other words: $\xi^a(t, x^9)$ defines a map from $R \times S_1$ to $\mathcal{M}$. Since we have twisted sectors there are low lying modes with energies of the order of $1/RQ_1Q_5$ which give rise to the long effective string picture described above. This picture where we start with only one kind of branes is the one that we would naturally use [45, 46] if we are working in the M(atrix) theory of [17].

We should also address the question of whether the system really forms a bound state, this is a question on the behavior of the zero modes on the moduli space. The analysis of [10] shows that they indeed form a bound state. Note that also for entropic reasons the state would stay together for a long time.

It is interesting that when the momentum is not uniformly distributed along the string (the $\hat{9}$ direction) the D-brane calculation still agrees with the corresponding black hole result [48].

3.3. A MORE GEOMETRIC PICTURE

We now consider a T-dual configuration which has two sets of D3-branes intersecting along a line. We have Q_3 D3-branes oriented along the directions 569 and Q_3' oriented along 789. We also have, as above, momentum N along the direction $\hat{9}$. This configuration is related by T-duality along 56 to the one considered above. Along the directions 5678 the configuration looks like two sets of two dimensional planes intersecting at points. We imagine the planes to be separated from each other. According to what we saw above, we expect that the degrees of freedom come from 33' strings. Exciting these massless strings actually corresponds to deforming the two intersecting planes into a smooth Riemann surface. In order to state this more precisely, let us consider a pair of intersecting D3- and D3'-branes. We will consider what happens locally at the intersection point. Let us define $z = X^5 + iX^6$ and $w = X^7 + iX^8$. We concentrate on what is happening in the directions 5678, the configuration will be translationally invariant (for the moment) along the direction $\hat{9}$. The two intersecting branes are described by the equation $zw = 0$, in other words, the brane is sitting at the points (z, w) which satisfy the above equation, which are, of course the plane $w = 0$ (one D3-brane) and $z = 0$ (the other D3'-brane). Exciting the 33' strings corresponds to deforming the shape of the threebranes so that they obey the equation $zw = c$ where the constant c is related to the expectation value of the 33' massless strings. We now have a single smooth D3-brane which is wrapping once around each of the planes 56 and 78. A nonzero value for c smoothes out or "blows up" the singular intersection of the D3-branes. The three-branes will preserve some supersymmetries as long as the equation characterizing their shape is holomorphic [47]. Since we have Q_3Q_3' intersection points we can deform each of them in this way. We end up with a single smooth D3 brane that is wrapping Q_3 times the plane 56 and Q_3' times the plane 78. This gives $2Q_3Q_3'$ real parameters characterizing the shape of the surface. There are other $2Q_3Q_3'$ parameters coming from Wilson lines of the U(1) gauge field that lives on this single D3 brane. If we look at each intersection point locally, once the brane is obeying the equation $zw = c \neq 0$ we see that we can have a gauge field $A_z = b/z$. Note that the degrees of freedom come mostly from deforming these intersection points, the global positions of the $Q_3 + Q_3'$ branes is a subleading effect in Q. Actually, the counting of degrees of freedom can be done precisely [43, 44] and the result is that we have precisely $4Q_3Q_3'$ bosonic degrees of freedom. By supersymmetry we have an equal number of fermionic degrees of freedom. They arise from the fermion that lives on the D3-brane. Since the D3-brane has a complicated shape, this fermion field will have $4Q_3Q_3'$ zero modes.

Now the momentum along $\hat{9}$ will be carried by oscillations of these massless degrees of freedom and the entropy is calculated as above, once we know the number of massless degrees of freedom. Notice that in this picture we are implicitly assuming that the radii of the compact dimensions are all very big.

Notice that this picture relies only on the low energy properties of a *single* curved D3 brane. It only uses that a single D3 brane is described by a vector supermultiplet. We are not using Polchinski's characterization of D-branes in terms of open strings [8]. In retrospect the black hole counting could have been done in this fashion before the discovery of the open-string description of D-branes. A similar counting argument can be done for four dimensional black holes in which only low energy properties of M-theory fivebranes are used [49].

3.4. JUSTIFICATION OF THE BPS COUNTING

In performing these calculations we have assumed that the coupling was weak. One might naively think that the only condition is $g \ll 1$. However, since there is a large number of branes there could be large $N(= Q)$ effects which grow as gQ. At low energies these are just the large N effects of the Yang-Mills theory (note that the YM coupling is $g_{YM} = \sqrt{g}$). In string theory they correspond to the possibility of inserting a hole on the worldsheet, each hole has a power of g and a factor of Q coming from the trace over Chan Paton factors. So the above calculations are correct when

$$gQ \ll 1 \tag{3.5}$$

On the other hand the classical black hole solution is well defined when $gQ \gg 1$ (2.13). The reason that we expect agreement (and *find* it) is that we are counting the number of BPS bound states and this number is not expected to change as we vary the coupling constant. Notice that indeed the extremal black hole entropy (2.10) is independent of the coupling and all other continuous parameters. Similarly, we assumed that the compactification radii were large. The same argument shows that the number of BPS states does not change when we change the radii.

4. Near Extremal black holes

We now turn to a discussion of near extremal five-dimensional black holes (2.4) (2.11)(2.12). For the reasons that we have just discussed we might naively not expect agreement in this case. However, in the dilute gas regime, we are very close to a configuration of extremal D1 and D5 branes and supersymmetry non-renormalization arguments do indeed help us [50, 51] and explain the agreement that we are going to find. We are going to consider a weakly coupled system of D-branes but we will always restrict to the low energy approximation. We will see that for the near-extremal case the agreement between the two approaches is just as impressive as in the extremal case. The D-brane model is a low energy approximation to the full quantum dynamics of black holes. The energy should be low compared to the scale set by the gravitational size of the black hole r_s defined as the radius at which the redshift of a static observer becomes of order one, $r_s^2 \sim gQ \sim r_1^2, r_5^2$. The condition on the energy becomes $\omega r_s \ll 1$, where ω represents the typical energy of the brane excitations, as well as the Hawking temperature of the system [50]. We will not go into the details of the justification of this extrapolation which can be found in [50].

In the previous section we saw that we always end up with a conformal field theory in 1+1 dimensions at low energies. The different counting arguments presented in the last section boiled down to different ways of computing the central charge of this conformal field theory. We start with a system of 1D-branes and 5-D branes as before and we add some extra energy and momentum to the system. This energy excites the massless left and right moving modes of the instanton configuration. In the instanton picture we say that we are creating left and right moving excitations on the moduli space, $\xi^I(t, x^9)$. Using the black hole formulas

(2.3),(2.11) we calculate the energies of the left and right movers

$$M - \frac{Q_5 RV}{g} - \frac{Q_1 R}{g} = \frac{RVr_0^2}{2g}\cosh 2\sigma = \frac{N_L + N_R}{R} , \tag{4.1}$$

$$N = \frac{R^2 V r_0^2}{2g^2}\sinh 2\sigma = N_L - N_R . \tag{4.2}$$

In this fashion we can calculate $N_{L,R}$ in terms of the black hole parameters. The entropy calculation proceeds as in the extremal case. We work in the multiply wound picture with 4 bosons and 4 fermions with effective $N'_{L,R} = Q_1 Q_5 N_{L,R}$. We find that

$$S = 2\pi(\sqrt{N'_L} + \sqrt{N'_R}) = 2\pi\sqrt{Q_1 Q_5 N_L} + 2\pi\sqrt{Q_1 Q_5 N_R} . \tag{4.3}$$

We see that this result agrees with the near extremal entropy in the dilute gas limit (2.12) once we use (4.1) (4.2). This is the simplest case, if we want to consider more general near extremal black holes, including Reissner-Nordstrøm, one has to include other excitations besides the right movers [36, 52] and the arguments are not so well justified.

These non-BPS states will decay. The simplest decay process is a collision of a right moving excitation with a left moving one to give a closed string mode that leaves the brane. We will calculate the emission rate for uncharged particles. The basic process is a right moving mode with momentum $p_9 = n/R_9 Q_1 Q_5$ colliding with a left moving one of momentum $p_9 = -n/R_9 Q_1 Q_5$ to give a closed string mode of energy $\omega = 2n/RQ_1 Q_5$. Notice that we are considering the branes to be multiply wound since that is the configuration that had the highest entropy. If the momenta are not exactly opposite the outgoing string carries some momentum in the 9th direction and we get a charged particle from the five dimensional point of view. Notice that the momentum in the 9th direction of the outgoing particle has to be quantized in units of $1/R_9$, only particles on the branes can have fractional momenta!. This means that outgoing charged particles have a very large mass, and that they are thermally suppressed when R_9 is small. All charged particles have masses of at least the compactification scale.

In other words, we have a very long effective string winding around the compact direction $\hat{9}$, it can oscillate along the other 4 compact dimensions (5678) and it emits gravitational quadrupole radiation. The graviton is polarized along the compact directions and is a scalar from the point of view of the five dimensional observer.

We will calculate the rate for this process according to the usual rules of relativistic quantum mechanics and show that the radiation has a thermal spectrum if we do not know the initial microscopic state of the black hole.

The state of the D-branes is specified by giving the left and right moving occupation numbers of each of the bosonic and fermionic oscillators. In fact, the near extremal D-branes live in a subsector of the total Hilbert space that is isomorphic to the Hilbert space of a 1+1 dimensional CFT. The initial state $|\Psi_i\rangle$ can emit a closed string mode and become $|\Psi_f\rangle$. The rate, averaged over initial states and

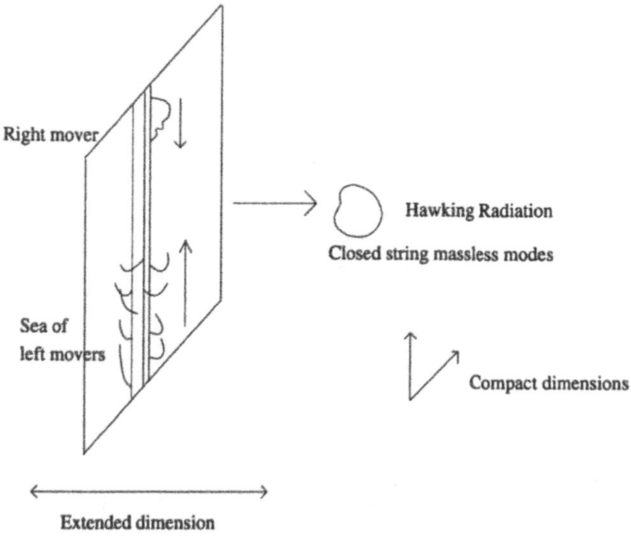

Figure 2. D-brane picture of the Hawking radiation emission process.

summed over final states (as one would do for calculating the decay rate of an unpolarized atom) is

$$d\Gamma \sim \frac{d^4k}{\omega} \frac{1}{p_0^R p_0^L V R} \delta(\omega - (p_0^R + p_0^L)) \sum_{i,f} |\langle \Psi_f | H_{int} | \Psi_i \rangle|^2 \tag{4.4}$$

We have included the factor due to the compactified volume RV. The relevant string amplitude for this process is given by a correlation function on the disc with two insertions on the boundary, corresponding to the two open string states and an insertion in the interior, corresponding to the closed string state. We consider the case when the outgoing closed string is a spin zero boson in five dimensions, so that it corresponds to the dilaton, the internal metric, internal $B_{\mu\nu}$ fields, or internal components of RR gauge fields. This disc amplitude, call it $\mathcal{A}$, is proportional to the string coupling constant g and to ω^2 [53]. The reason for this last fact is that it has to vanish when we go to zero momentum, otherwise it would indicate that there is a mass term for the open strings (since one can vary the vacuum expectation value of the corresponding closed string fields continuously). In conclusion, up to numerical factors,

$$\mathcal{A} \sim g\omega^2. \tag{4.5}$$

Note that performing the average over initial and summing over final states will just produce a factor of the form $\rho_L(n)\rho_R(n)$ with

$$\rho_R(n) = \frac{1}{N_i} \sum_i \langle \Psi_i | a_n^{R\dagger} a_n^R | \Psi_i \rangle \tag{4.6}$$

where N_i is the total number of initial states and a_n^R is the creation operator for one of the 4 bosonic open string states. The factor $\rho_L(n)$ is similar. Since we are just averaging over all possible initial states with given value of N_R, this corresponds to taking the expectation value of $a_n^\dagger a_n$ in the microcanonical ensemble with total energy $E_R = N_R/R_9 = N_R'/R_9 Q_1 Q_5$ of a one dimensional gas. Because N_R' is large compared to one, we can calculate (4.6) in the canonical ensemble. The occupation number is then

$$\rho_R(\omega) = \frac{e^{\frac{-\omega}{2T_R}}}{1 - e^{\frac{-\omega}{2T_R}}} .$$

We can read off the "right moving" temperature

$$T_R = \frac{1}{\pi}\frac{1}{R}\sqrt{\frac{N_R}{Q_1 Q_5}} . \tag{4.7}$$

There is a similar factor for the left movers ρ_L with a similar looking "temperature"

$$T_L = \frac{1}{\pi}\frac{1}{R}\sqrt{\frac{N_L}{Q_1 Q_5}} . \tag{4.8}$$

In fact it would be more accurate to say that there is only one physical temperature of the gas, which agrees with the Hawking temperature of the corresponding black hole,

$$\frac{1}{T_H} = \frac{1}{2}(\frac{1}{T_L} + \frac{1}{T_R}) \tag{4.9}$$

and that $T_{L,R}^{-1} = T_H^{-1}(1 \pm \mu)$ are some natural combination of the temperature and the chemical potential, which gives the gas some net momentum. Using the values for $N_{L,R}$ from (4.1) (4.2) we find

$$T_L = \frac{1}{\pi}\frac{r_0 e^{\sigma}}{2 r_1 r_5} , \qquad T_R = \frac{1}{\pi}\frac{r_0 e^{-\sigma}}{2 r_1 r_5} . \tag{4.10}$$

The expression for the rate is, up to a numerical constant,

$$d\Gamma \sim \frac{d^4 k}{\omega}\frac{1}{p_0^R p_0^L R V}|\mathcal{A}|^2 Q_1 Q_5 R \rho_R(\omega)\rho_L(\omega) \tag{4.11}$$

where $\mathcal{A}$ is the disc diagram result. The factor $Q_1 Q_5 R$ is a volume factor, which arises from the delta function of momenta in (4.4) $\sum_n \delta(\omega - 2n/R Q_1 Q_5) \sim R Q_1 Q_5$. The final expression for the rate is, using (4.5) in (4.11),

$$d\Gamma = \frac{\pi^3 g^2}{V} Q_1 Q_5 \omega \frac{1}{e^{\frac{\omega}{2T_L}} - 1}\frac{1}{e^{\frac{\omega}{2T_R}} - 1}\frac{d^4 k}{(2\pi)^4} \tag{4.12}$$

We have not shown here how to calculate the precise numerical constant in front of (4.12) , this precise calculation was done in [54], and we refer the reader to it for the details.

If we are considering a black hole which is very close to extremality with nonzero momentum $N \sim N_L \gg N_R$ then we find from (4.8)(4.7) that $T_L \gg T_R$. Examining the expression for the rate (4.12) we see that the typical emitted energies are of the order of T_R. Therefore, we can approximate the left moving thermal factor by

$$\rho_L \sim \frac{2T_L}{\omega} \tag{4.13}$$

and replacing it in (4.12) we find

$$d\Gamma = \frac{2\pi^2 g^2}{RV}\sqrt{Q_1 Q_5 N}\frac{1}{e^{\frac{\omega}{2T_R}}-1}\frac{d^4k}{(2\pi)^4} = A_H \frac{1}{e^{\frac{\omega}{2T_R}}-1}\frac{d^4k}{(2\pi)^4} \tag{4.14}$$

where A_H is the area of the horizon. We conclude that the emission is thermal, with a physical Hawking temperature

$$T_H = 2T_R \tag{4.15}$$

which exactly matches the classical result (2.9). The area appeared correctly in (4.12) [55] . Actually, the coupling constant coming from the string amplitude $\mathcal{A}$ is hidden in the expression for the area (area $= 4G_N^5 S$). The overall coefficient in (4.14) matches precisely with the semiclassical result [54].

Notice that if we were emitting a spacetime fermion then the left moving mode could be a boson and the right moving mode a fermion, this produces the correct thermal factor for a spacetime fermion. The opposite possibility gives a much lower rate, since we do not have the enhancement due to the large ρ_L (4.13).

When separation from extremality is very small, then the number of right movers is small and the statistical arguments used to derive (4.12) fail. Semiclassically this should happen when the temperature is so low that the emission of one quantum at temperature T causes the temperature to change by an amount of order T. This means that the specific heat is of order one. This happens when the mass difference from extremality is [56]

$$\delta M_{min} = M - M_{BPS} \sim \frac{G_N^5}{r_e^4} \tag{4.16}$$

for a Reissner-Nordstrøm black hole, with r_e being the Schwarzschild radius of the solution. The D-brane approach suggests the existence of a mass gap

$$\delta M_{min} \sim \frac{2}{Q_1 Q_5 R} \tag{4.17}$$

which using (2.3) scales like (4.16). This is an *extremely* small energy for a macroscopic extremal black hole. In fact, it is of the order of the kinetic energy that the black hole would have, due to the uncertainty principle, if we want to measure its position with an accuracy of the order of its typical gravitational radius r_s: $\delta M \sim (\Delta p)^2/M$ with $\Delta p \sim 1/r_s$.

Now we calculate the entropy of a rotating black hole in five dimensions [35, 57]. The angular momentum is characterized by the eigenvalues on two orthogonal two-planes, J_1, J_2, for example J_1 corresponds to rotations of the 12 plane and J_2 to rotations of the 34 plane. In terms of the J_3 eigenvalues J_L, J_R of the $SU(2)_L \times SU(2)_R \sim SO(4)$ decomposition of the spatial rotation group we find

$$J_1 = J_L + J_R \qquad J_2 = J_L - J_R \tag{4.18}$$

As we mentioned above J_R, J_L are carried by right and left movers respectively. They are eigenvalues of U(1) currents, are part of the $SU(2)_L \times SU(2)_R$ R-symmetries of the 1+1 dimensional conformal field theory. States carrying U(1) eigenvalue J have conformal weight bigger than $\Delta = 6J^2/c$ where $c = 6Q_1Q_5$ is the central charge. The state with minimum conformal weight is $e^{iJ\phi}|0\rangle$, where $j = \frac{c}{12}\partial\phi$ is the U(1) current. Therefore, in the total left moving energy N_L there is an amount $\Delta = J_L^2/Q_1Q_5$ which we are not free to distribute. It is fixed by the condition that the system has angular momentum J_L, so the effective number of left movers that we are free to vary is $\tilde{N}_L = N_L - J_L^2/Q_1Q_5$. The same is true for the right movers, so that the entropy becomes

$$\begin{aligned} S = & \; 2\pi\sqrt{Q_1Q_5}(\sqrt{\tilde{N}_L} + \sqrt{\tilde{N}_R}) = \\ = & \; 2\pi\sqrt{N_LQ_1Q_5 - J_L^2} + 2\pi\sqrt{N_RQ_1Q_5 - J_R^2} \end{aligned} \tag{4.19}$$

which agrees with the classical entropy formula of a rotating black hole in the dilute gas regime [57]. Actually, in five dimensions we can have rotating BPS black holes by setting $N_R = J_R = 0$, this implies $J_1 = J_2$. Again the corresponding formula agrees with the classical entropy formula but the restriction to the dilute gas regime is no longer necessary since the computation is protected by supersymmetry.

5. Greybody factors

The D-brane emission rate into massless scalars is given by (4.12). More precisely that is the emission into minimally coupled scalars, scalars that in the supergravity theory are not coupled to the vector fields that are excited in the black hole background. In other words, the action for these scalars is

$$S \sim \int d^5x \; \sqrt{g} g^{\mu\nu} \partial_\mu\phi\partial_\nu\phi \tag{5.1}$$

where g is the Einstein metric in five spacetime dimensions. According to the semiclassical analysis the emission rate should be

$$d\Gamma = \sigma(\omega, r_0, \alpha, \gamma, \sigma)\frac{1}{e^{\frac{\omega}{T_H}} - 1} \tag{5.2}$$

where $\sigma(\omega, r_0, \alpha, \gamma, \sigma)$ is the absorption cross section of the black hole which is a function of the various parameters specifying the black hole solution (2.4). In the usual Schwarzschild black hole case the only scale in the solution would be the Schwarzschild radius r_s. This emission rate (5.2) has the same form for any body emitting thermal radiation. The absorption cross section comes in because of detailed balance: in order for the body to be in equilibrium with a bath of radiation it has to absorb as much as it emits. The prefactor in (5.2) is usually called greybody factor, since it is what makes bodies grey instead of black. At first sight, the semiclassical rate (5.2) does not seem to be in agreement with the D-brane rate (4.12) since one has two exponential factors and the the other seems to have only one. In order to see whether they really agree we should calculate the greybody factor. It turns out that the greybody factor is precisely such that these two calculations to agree. We now describe this calculation.

We consider the scattering of scalars from a five dimensional black hole in the dilute gas limit $r_0, r_n \ll r_1, r_5$. We also restrict to low energies satisfying $\omega \ll 1/r_1, 1/r_5$ but there is no restriction on $\frac{\omega r_0}{r_1 r_5}$ or $\frac{\omega r_n}{r_1 r_5}$, in other words, no restriction on ω/T_L, ω/T_R. We follow the notation of [32], where further details of the geometry may also be found. The wave equation in this background becomes

$$\frac{h}{r^3}\frac{d}{dr}r^3 h\frac{d\phi}{dr} + \omega^2\lambda\phi = 0, \tag{5.3}$$

where λ, h are defined in (2.5).

We divide space into a far region $r \gg r_1, r_5$ and a near region $r \ll 1/\omega$ and we will match the solutions in the overlapping region. In the far region, the equation is solved by the Bessel functions

$$\phi = \frac{1}{\rho}\left[\alpha J_\nu(\rho) + \beta J_{-\nu}(\rho)\right]. \tag{5.4}$$

with $\rho = \omega r$ and $\nu^2 = 1 - \epsilon$, where $\epsilon = \omega^2(r_1^2 + r_5^2)$ is very small and we keep it to simplify the form of the intermediate equations but will disappear from the final answer. From the large ρ behavior the incoming flux is found to be

$$f_{in} = Im(\phi^* r^3 \partial_r \phi) = \frac{1}{2\pi\omega^2}|\alpha e^{i\nu\pi/2} + \beta e^{-i\nu\pi/2}|^2. \tag{5.5}$$

On the other hand, the small ρ behavior of the far region solution is

$$\begin{aligned}\phi = \ & \tfrac{1}{\rho}\Big[\alpha(\tfrac{\rho}{2})^\nu(\tfrac{1}{\Gamma(\nu+1)} - O(\rho^2)) \\ & +\beta(\tfrac{\rho}{2})^{-\nu}(\tfrac{1}{\Gamma(-\nu+1)} - O(\rho^2))\Big].\end{aligned} \tag{5.6}$$

Now we turn to the solution in the near region $r \ll 1/\omega$. Defining $v = r_0^2/r^2$, the near region wave equation is

$$(1-v)^2\frac{d^2\phi}{dv^2} - (1-v)\frac{d\phi}{dv} + \left(C + \frac{D}{v} + \frac{E}{v^2}\right)\phi = 0 \tag{5.7}$$

where

$$C = \left(\frac{\omega r_n r_1 r_5}{2r_0^2}\right)^2, \quad D = \frac{\omega^2 r_1^2 r_5^2}{4r_0^2} + \frac{\nu^2 - 1}{4}, \quad E = -\frac{\nu^2-1}{4}. \tag{5.8}$$

Defining

$$\phi = v^{-(\nu-1)/2}(1-v)^{-i\frac{\omega}{4\pi T_H}} AF \tag{5.9}$$

with A a constant, we find that the solution to (5.7) with only ingoing flux at the horizon is given by (5.9) with the hypergeometric function

$$\begin{aligned} F = \ & F(a,b,c;1-v) \\ a \ & = -\nu/2 + 1/2 + i\tfrac{\omega}{4\pi T_L} \\ b \ & = -\nu/2 + 1/2 + i\tfrac{\omega}{4\pi T_R} \\ c \ & = 1 + i\tfrac{\omega}{2\pi T_H} \end{aligned} \tag{5.10}$$

The behavior for small v can be calculated by expressing the hypergeometric function (5.10), which depends on $1-v$, in terms of hypergeometric functions depending on v and then expanding in v. Matching this with (5.6) we find

$$\alpha/2 = A\left[\frac{\Gamma(1+i\frac{\omega}{2\pi T_H})}{\Gamma(1+i\frac{\omega}{4\pi T_L})\Gamma(1+i\frac{\omega}{4\pi T_R})}\right], \qquad \beta \ll \alpha \tag{5.11}$$

The absorbed flux is

$$f_{abs} = Im(\phi^* h r^3 \partial_r \phi) = \frac{\omega r_0^2}{2\pi T_H}|A|^2. \tag{5.12}$$

The absorption cross section for the radial problem is given by the ratio of the two fluxes (5.12) (5.5). The plane wave cross section is obtained by multiplying by $4\pi/\omega^3$

$$\sigma_{\rm abs} = \pi^3 r_1^2 r_5^2 \omega \frac{e^{\frac{\omega}{T_H}} - 1}{\left(e^{\frac{\omega}{2T_L}} - 1\right)\left(e^{\frac{\omega}{2T_R}} - 1\right)} \tag{5.13}$$

where the exponential terms come from the gamma functions in (5.11). We see that it has precisely the right form to make the D-brane result (4.12) agree with the semiclassical calculation (5.2).

Notice that we are doing two very different calculations. On one hand we are considering a quantum field propagating on a fixed classical supergravity background and on the other we are doing a computation in flat space with some D-branes at the origin. This might seem surprising at first but it is not once we realize that the low energy condition $\omega r_g \ll 1$ implies that the black hole looks like a point for such long wavelengths. Actually the most surprising and interesting fact is that in the D-brane calculation we are computing a *thermal* rate, agreement is found only after we average over the states of the D-branes. This suggests that the classical gravity approximation is a *thermodynamic* approximation to quantum gravity.

These greybody factor calculations have been generalized to various cases. One possible generalization is to consider the emission of scalars that are not minimally coupled, in some cases precise agreement is found [58] [59]. From the D-brane point of view the difference between these scalars and the one that we have been considering is in the conformal weight of the operator on the effective SCFT that they couple to. The minimally coupled scalars couple to operators of dimension (1,1) (like $\partial X \bar{\partial} X$) while the scalars in [58] couple to operators of conformal weights (2,2) or (1,2) in [59].There is still some puzzling disagreement for the case of operators of weight (3,1) [60], which hopefully will be resolved soon!. Greybody factors for fermions and vectors were calculated and found to agree with the D-brane model [64]. Another generalization is to consider the emission of higher partial waves [61, 63, 62]. These calculations of greybody factors shows that some of the features of the near extremal geometry are encoded in the dynamics of the 1+1 dimensional gas (or the CFT). Since the wavelengths of the particles we scatter are much bigger than the size of the black holes it is hard to get precise information about the metric. A more direct way to obtain information about the metric is by using D-brane probes [65]. In that approach one starts from D-branes in flat space and by integrating out the massive stretched open strings one obtains an action for the probe D-brane that is at some distance from the rest of the branes. This action is then interpreted as the action of a D-brane in the presence of some classical supergravity background. This works to one loop [65, 66] but the status of the higher loop contributions is unclear.

6. Final remarks

We have seen how string theory, using D-branes provides a microscopic understanding of the black hole entropy. It is possible to study extremal and near extremal black holes in some particular limit ("dilute gas"). The description not only accounts for the entropy but also captures the full low energy dynamics of the black hole, such as low energy absorption and emission cross sections. Hawking radiation coming from these near extremal black holes can always be described within the low energy approximation. Since the D-brane model is unitary we can conclude that the low energy dynamics of these black holes is unitary. One might wonder whether there are higher energy processes that violate unitarity. Well, it could certainly be the case. Notice, however, that the information loss paradox for these black holes could be put just in terms of low energy processes: sending in some low energy waves and watching the black hole evaporate, this is a process that according to the semiclassical arguments would produce a large amount of entropy [70]. The D-brane model, on the other hand predicts that very little entropy (if any) is created. We say very little because it is possible that the higher energy processes might be not unitary. Some reasons why the traditional semiclassical arguments would not apply for this case were explored in [71].

The greybody factor calculation shows that many features of the geometry close to the horizon (close relative to the size of the black hole, but far compared to the Planck length) are reproduced by the gas of left and right movers. Complicated S-matrix elements for absorption and emission are reproduced. It is important that in order to reproduce them we have to do an average over the black hole microstates. This is saying that the classical gravity approximation itself is, in some sense, a thermodynamic approximation. Ideas along this line have been envisioned in [72]. This is, in my opinion, the tip of the iceberg in a new understanding of quantum gravity and the relation of gauge theories to gravitation.

In 2+1 dimensions the entropy was explained in terms of boundary degrees of freedom of a Chern Simons theory by Carlip [73], since 2+1 gravity can be shown to be equivalent to a Chern Simons theory. It would be interesting to understand the relation between that picture and the string theory picture.

There is clear relation of this description of black holes and Matrix theory [17]. In the D-brane description of black holes the spacetime physics close to the horizon is described in terms of some gauge theory living on D-branes. In Matrix theory the whole spacetime theory arises from the gauge theory living on branes. The Matrix description of black holes is just the D-brane description reinterpreting the D5-brane charge as momentum [46, 45].

The four dimensional black holes have a similar description [67, 68, 69, 49], and there is a similar "dilute gas" limit. It was actually shown in [74] that there is another dilute gas limit for the near extremal Kerr black hole in five dimensions which also has greybody factors suggesting a description in terms of a 1+1 dimensional gas.

Acknowledgements

I thank C. Callan, G. Horowitz, D. Lowe, A. Strominger, L. Susskind and A. Peet, for many fruitful collaborations on this subject.

References

1. S. Hawking and G. Ellis, *The Large Scale Structure of Spacetime*, Cambridge Univ. Press, 1973.
2. R. Penrose, *Phys. Rev. Lett.* **14** (1965) 57. R. Penrose, "Singularities and Time Asymmetry" and R. Geroch and G. Horowitz "Global Structure of Spacetimes", both in *General Relativity, an Einstein Centenary Survey*, ed. S. Hawking and W. Israel (Cambridge University Press)(1979)
3. R. Wald, Talk at the American Physical Society meeting, Washington, April 1996.
4. S. Hawking, *Phys. Rev. Lett.* **26** (1971)1344.
5. S.W. Hawking, *Comm. Math. Phys.* **43** (1975) 199.
6. J. Bekenstein, *Phys. Rev.* **D7** (1973) 2333; *Phys. Rev.* **D 9** (1974) 3292; S. W. Hawking, *Phys. Rev.* **D13** (1976) 191.
7. For some reviews see: M. Green, J. Schwarz, and E. Witten, "Superstring Theory," two volumes (Cambridge University Press, 1987); A.M. Polyakov, "Gauge Fields and Strings," (Harwood, 1987); M. Kaku, "Introduction to Superstrings," (Springer-Verlag, 1988) ; D. Lust and S. Theisen, "Lectures on String Theory," (Springer-Verlag, 1989).
8. J. Polchinski, *Phys. Rev. Lett.* 75 (1995) 4724
9. hep-th/9510017;J. Dai, R. Leigh and J. Polchinski, *Mod. Phys. Lett.* **A 4** (1989) 2073; P. Horava, *Phys. Lett.* **B231** (1989) 251.
10. A. Strominger and C. Vafa, *Phys. Lett.* **B379** (1996) 99, hep-th/9601029.
11. L. Susskind, RU-93-44, hep-th/9309145
12. A. Sen, *Mod. Phys. Lett.* **A10** (1995) 2081, hep-th/9504147.
13. G. Horowitz and J. Polchinski, hep-th/9612146.
14. G. Horowitz, gr-qc/9704072; gr-qc/9604051.
15. D. Youm, hep-th/9710046, to appear in *Phys. Rep.*.
16. A. Peet, hep-th/9712253.
17. T. Banks, W. Fischler, S. H. Shenker, L. Susskind, *Phys.Rev.* **D55** (1997) 5112, hep-th/9610043.
18. T. Banks, W. Fischler, I.R. Klebanov and L. Susskind *Phys. Rev. Lett.* 80 (1998) 226, hep-th/9709091; hep-th/9711005; I. Klebanov and L. Susskind,

hep-th/9709108;G. Horowitz and E. Martinec, hep-th/9710217; M. Li hep-th/9710226; S. Das, S. Mathur, S.Kalyana Rama and P. Ramadevi, hep-th/9711003; H. Liu and A.A. Tseytlin, hep-th/9712063; T. Banks, W. Fischler and I.R. Klebanov, hep-th/9712236; F. Englert and E. Rabinovici, hep-th/9801048; M. Li and E. Martinec hep-th/9801070.

19. S. Hyun, hep-th/9704005.
20. H. Boonstra, B. Peeters and K. Skenderis, *Phys. Lett.* **B411** (1997) 59, hep-th/9706192.
21. M. Banados, C. Teitelboim and J. Zanelli *Phys. Rev. Lett.* **69** (1992) 1849, hep-th/9204099.
22. C. Hull and P. Townsend, *Nucl. Phys.* **B438** (1995) 109, hep-th/9410167.
23. A. Tseytlin, *Mod. Phys. Lett.* **A11** (1996) 689, hep-th/9601177.
24. M. Cvetic, Review talk, hep-th/9701152 and references therein.
25. G. Horowitz, J. Maldacena and A. Strominger, *Phys. Lett.* **B383** (1996) 151, hep-th/9603109.
26. J. Maldacena, Ph.D. Thesis, Princeton University, hep-th/9607235
27. J. Maharana and J. Schwarz, Nucl.Phys. **B390** (1993) 3; A. Sen, *Nucl. Phys.* **D404** (1993) 109.
28. S. Ferrara, R. Kallosh and A. Strominger, *Phys. Rev.* **D 52** (1995) 5412, hep-th/9508072; M. Cvetic and D. Youm, *Phys. Rev.* **D53** (1996) 584, hep-th/9507090; G. Gibbons and P. Townsend, *Phys. Rev. Lett.* **71** (1993) 3754.
29. F. Larsen and F. Wilczek, *Phys. Lett.* **B375** (1996) 37, hep-th/9511064.
30. R. Kallosh, A. Linde, T. Ortín, A. Peet, A. Van Proeyen, *Phys. Rev.* **D46** (1992) 5278, hep-th/9205027.
31. G. Horowitz and A. Strominger, *Phys. Rev. Lett.* **77** (1996) 2368, hep-th/9602051.
32. J. Maldacena and A. Strominger, *Phys. Rev.* **D55**(1997)861, hep-th/9609026.
33. J. Polchinski, S. Chaudhuri and C. Johnson, *Notes on D-Branes*, hep-th/9602052; J. Polchinski, *Tasi Lectures on D-branes*, hep-th/9611050.
34. M. Douglas, hep-th/9512077.
35. J. Breckenridge, R. Myers, A. Peet and C. Vafa, Phys. Lett. **B391** (1997) 93, hep-th/9602065.
36. C. Callan and J. Maldacena, *Nucl. Phys.* **B472** (1996) 591, hep-th/9602043.
37. J. Maldacena and L. Susskind, *Nucl. Phys.* **B475** (1996) 679, hep-th/9604042.
38. S.R. Das and S.D. Mathur, Phys. Lett. **B375** (1996) 103, hep-th/9601152.
39. S. Hassan and S. Wadia, Phys. Lett. **B402** (1997) 43, hep-th/9703163; hep-th/9712213.
40. R. Dijkgraaf, G. Moore, E. Verlinde and H. Verlinde, Commun. Math. Phys. **185** (1997) 197, hep-th/9608096.
41. R. Dijkgraaf, E. Verlinde and H. Verlinde, *Nucl. Phys.* **B486** (1996) 77, hep-th/9603126; *Nucl. Phys.* **B486** (1996) 89, hep-th/9604055; *Nucl. Phys.* **B484** (1996) 543, hep-th/9607026.
42. E. Witten, *Nucl. Phys.* **460** (1996) 541, hep-th/9511030.
43. C. Vafa, *Nucl. Phys.* **B463** (1996) 415, hep-th/9511088; Nucl .Phys. **B463** (1996) 435, hep-th/9512078
44. M. Bershadsky, C. Vafa and V. Sadov, *Nucl. Phys.* **B463** (1996)398,hep-th/9510225.
45. M. Li and E. Martinec, Class. Quant. Grav. **14** (1997) 3187, hep-th/9703211; hep-th/9704134.
46. R. Dijkgraaf, E. Verlinde and H. Verlinde, *Nucl. Phys.* **B506** (1997) 121, hep-th/9704018.
47. K. Becker, M. Becker and A. Strominger, *Nucl. Phys.* **B456** (1995) 130, hep-th/9507158.
48. G. Horowitz and D. Marolf, Phys .Rev. **D55** (1997) 835, hep-th/9605224; *Phys. Rev.* **D55** (1997) 846, hep-th/9606113.
49. J. Maldacena, A. Strominger and E. Witten, hep-th/9711053, to appear in JHEP.

50. J. Maldacena, *Phys. Rev.* **D55** (1997) 7645, hep-th/9611125.
51. S. Das, *Phys. Rev.* **D56** (1997) 3582, hep-th/9703146.
52. J. Maldacena, *Nucl. Phys.* **B477** (1996) 168, hep-th/9605016; I. Klebanov and A. Tseytlin, *Nucl. Phys.* **B475** (1996) 164, hep-th/9604166; V. Balasubramanian and F. Larsen, *Nucl. Phys.* **B478** (1996) 199, hep-th/9604189; A. Hanany and I. Klebanov, *Nucl. Phys.* **B482** (1996) 105, hep-th/9606136; I. Klebanov and A. Tseytlin, *Nucl. Phys.* **B479** (1996) 319, hep-th/9607107
53. A. Hashimoto and I. Klebanov, Phys. Lett. **B381** (1996) 437, hep-th/9604065; M. Garousi and R. Myers, Nucl .Phys. **B475** (1996) 193, hep-th/9603194.
54. S. Das and S. Mathur *Nucl. Phys.* **B478** (1996) 561, hep-th/9606185; *Nucl. Phys.* **B482** (1996) 153, hep-th/9607149.
55. A. Dhar, G. Mandal and S. Wadia Phys. Lett. **B388** (1996) 51, hep-th/9605234.
56. J. Preskill, P. Schwarz, A. Shapere, S. Trivedi and F. Wilczek, Mod. Phys. Lett.**A 6** (1991) 2353; C. Holzhey and F. Wilczek, *Nucl. Phys.* **B380** (1992) 447, hep-th/9202014; P. Kraus and F. Wilczek, *Nucl. Phys.* **B433** (1995) 403.
57. J. Breckenridge, D. Lowe, R. Myers, A. Peet, A. Strominger and C. Vafa, Phys. Lett. **B381** (1996) 423, hep-th/9603078.
58. C. Callan, S. Gubser, I. Klebanov, A. Tseytlin, *Nucl. Phys.* **B489** 65, hep-th/9610172; I. Klebanov and M. Krasnitz, *Phys. Rev.* **D55** (1996) 3250, hep-th/9612051.
59. I. Klebanov, A. Rajaraman, A. Tseytlin, *Nucl. Phys.* **B503** (1997) 157, hep-th/9704112; H.W. Lee, Y.S. Myung and J. Y. Kim, hep-th/9708099.
60. M. Krasnitz, I. Klebanov, *Phys. Rev.* **D56** (1997) 2173, hep-th/9703216.
61. J. Maldacena and A. Strominger, *Phys. Rev.* **D56** (1997) 4975, hep-th/9702015.
62. S. Mathur, hep-th/9704156.
63. S. Gubser, *Phys. Rev.* **D56** (1997) 4984, hep-th/9704195.
64. S. Gubser, *Phys. Rev.* **D56** (1997) 7854, hep-th/9706100; M. Cvetic and F. Larsen, hep-th/9712118; K. Hosomichi, hep-th/9711072.
65. M. Douglas, J. Polchinski and A. Strominger, hep-th/9703031.
66. J. Maldacena, hep-th/9705053.
67. J. Maldacena and A. Strominger, *Phys. Rev. Lett.* **77** (1996) 428, hep-th/9603061; C. Johnson, R. Kuhri and R. Myers, Phys. Lett. **B378** (1996) 78, hep-th/9603062.
68. G. Horowitz, D. Lowe and J. Maldacena, *Phys. Rev. Lett.* **77** (1996) 430, hep-th/9603195.
69. S. Gubser and I. Klebanov, *Phys. Rev. Lett.* **77** (1996) 4491, hep-th/9609076.
70. S. Hawking, *Phys. Rev.* **D14** (1976) 2460.
71. T. Jacobson, hep-th/9705017.
72. T. Jacobson, Phys .Rev. Lett. **75** (1995) 1260, gr-qc/9504004.
73. S. Carlip, *Phys. Rev.* **D51** (1995) 632, gr-qc/9409052.
74. M. Cvetic and F. Larsen, *Nucl. Phys.* **B506** (1997) 107, hep-th/9706071; *Phys. Rev.* **D56** (1997) 4994, hep-th/9705192.

M-THEORY AND $N = 2$ STRINGS

EMIL MARTINEC
Enrico Fermi Institute and Dept. of Physics
5640 South Ellis Ave., Chicago, IL 60637-1433

1. Introduction

A few years ago, if asked to describe string theory, the average practitioner would have classified its different manifestations according to their various worldsheet gauge principles. On the 1+1 dimensional worldsheet, there can be (p,q) supersymmetries that square to translations along the (left,right)-handed light cone; one says that the worldsheet has (p,q) gauged supersymmetry. The bosonic string has no supersymmetry; $p = q = 0$. The supersymmetric string theories have, say, $q = 1$. Thus type IIA/B string theory has (1,1) supersymmetry. The type I/IA strings are the orbifold of these by worldsheet parity, and the heterotic strings are in the class (0,1). Remarkably, we now understand that all the supersymmetric string theories – type IIA/B, type I, and heterotic – appear to describe asymptotic expansions of a single nonperturbative master theory: M-theory. This theory has many miraculous duality properties that are now only beginning to be unravelled; other lecturers at this school will review the current state of affairs. In these lectures, I will give an overview of a relatively unexplored corner of string theory, namely the N=2 strings [1, 2, 3] (more specifically strings with (2,2) or (2,1) gauged worldsheet supersymmetry).

The driving force behind the recent unification has been the recognition of the fundamental importance of the spacetime (as opposed to worldsheet) supersymmetry algebra. The small (BPS) representations of the supersymmetry algebra form a quasi-topological sector of the theory. By tracking this BPS spectrum across moduli space, one can deduce the interconnections of the various string limits. The issue we wish to address is the role of the heterotic (2,1) string, which is also a stringy realization of spacetime supersymmetry and therefore ought to play a role somehow. As we will see [4, 5], many of the basic objects of M-theory are realized in the (2,1) string. Self-duality and integrability are further features of the (2,1) string, arising from the chiral sector with N=2 worldsheet supersymmetry.

We will see that the chiral critical dimensions of the (2,1) string are $d = 4$ (2 space, 2 time) for the N=2 sector, and $d = 12$ (10 space, 2 time) for the N=1 sector. To see the relation to M-theory, consider standard heterotic target

L. Baulieu et al. (eds.), Strings, Branes and Dualities, 241–265.

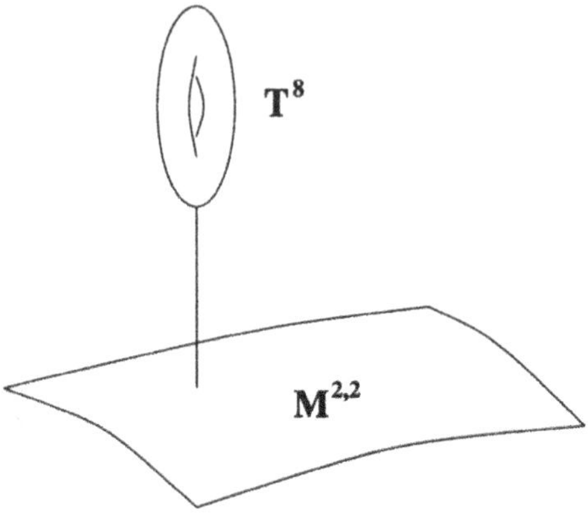

Figure 1. (2,1) heterotic geometry.

space geometry. Here this is a 2+2 dimensional base manifold for the dimensions common to both chiralities, with the additional left-movers parametrizing an eight-dimensional torus of stringy dimensions fibered over it (see figure 1).

The geometrical fields are the graviton h, antisymmetric tensor field b, gravitino χ, and the gauge connection ε on the fiber. These fields are further restricted by the extra constraints of N=2 local supersymmetry on the worldsheet, giving rise to prepotentials

$$\begin{aligned} h_{\mu\nu} + b_{\mu\nu} &\rightarrow I_\mu^\lambda \partial_\lambda a_\nu \\ \varepsilon_\mu^a &\rightarrow I_\mu^\lambda \partial_\lambda \varphi^a \\ \chi_\mu^\alpha &\rightarrow I_\mu^\lambda \partial_\lambda \psi^\alpha \,. \end{aligned} \tag{1.1}$$

Here I_μ^λ is an almost complex structure on the base space, which is required by the local N=2 worldsheet supersymmetry. Under linearized gauge transformations, these restricted fields transform as

$$\begin{aligned} \delta(h+b) = \partial\xi &\rightarrow \delta a \sim \xi \\ \delta\varepsilon = \partial\Lambda &\rightarrow \delta\varphi \sim \Lambda \\ \delta\chi = \partial\eta &\rightarrow \delta\psi \sim \eta \,. \end{aligned} \tag{1.2}$$

In other words, the remnant fields are Nambu-Goldstone fields of spontaneously broken symmetries (spacetime antisymmetric tensor field gauge transformations, translations, supersymmetries) on a D-brane. More precisely, one has in a complex coordinate basis $\varepsilon = i(\partial - \bar\partial)\varphi + (\partial + \bar\partial)\theta = \partial(\theta + i\varphi) + \bar\partial(\theta - i\varphi)$, where the (real) gauge symmetry is $\delta\theta = \Lambda$. However, one can go to a holomorphic basis by complexifying the gauge group G; then φ is a coordinate on $G_{\mathbf{C}}/G$ whose dynamics is determined by holomorphic gauge symmetry much as in the 2d WZW model [6, 7, 8]. This virtually guarantees us a connection to brane physics, since brane dynamics is almost by definition given by the nonlinear Lagrangian of the

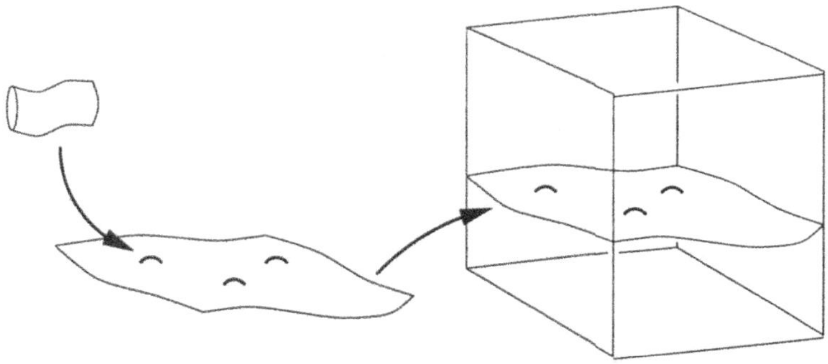

Figure 2. Chain of brane embeddings implied by (2,1) string states.

spacetime symmetries broken by the brane. A picture of this aspect of (2,1) string theory is shown in figure 2. The (2,1) string worldsheet maps into the worldvolume of a D-brane, which is itself embedded in spacetime. The quanta of this brane are the (2,1) strings themselves. Since the transverse fluctuations of the brane can be traced to those of the fiber connection $\varepsilon \sim \partial\varphi$, while the longitudinal directions are those of the base space, we see that spacetime itself is the total space of the heterotic geometry of figure 1.

The dimension of the brane is determined by the number of independent momentum components in string vertex operators $\mathcal{O}_\alpha e^{ik\cdot x}$. When the (2,1) string target space is $\mathbb{R}^{2,2} \times T^8$, this kinematics is 1+1 or 2+1 dimensional; when the spatial dimensions are further compactified, the kinematics is more or less 9+1 dimensional – the target is a kind of ninebrane.

As mentioned above, an additional geometric structure is self-duality. The almost complex structure I_μ^λ is one of a triplet of such structures preserved by the target space geometry (these almost complex structures are not integrable, since $(\partial - \bar{\partial})I = db = torsion$). Thus one can bring to bear the machinery of twistor theory to characterize the classical solution space. One of the characteristic properties of self-dual gravity is the symmetry group of area preserving diffeomorphisms $SDiff_2 \sim SU(\infty)$, suggesting a connection to matrix theory [9].

There are indeed a number of intriguing analogies among the matrix model of M-theory, (2,1) strings, and other matrix models:

1. The (BFSS) matrix model of M-theory [9] realizes the membrane as a collective phenomenon of D0-brane supergravitons, i.e. as a state in the large N collective field theory. In a sense, the BFSS matrix model gives a map from a *single* noncommutative torus into 'spacetime'. In the graviton limit, the matrices approximately commute; in the membrane limit, the commutators are large. Similarly, the (2,1) string describes a map from a *single* brane into spacetime.
2. Other examples where large N collective field theory generates string theory as an asymptotic expansion around a particular master field include:

   (a) The 1+1d noncritical string based on the original matrix model of [10].

(b) 2d Yang-Mills [11].

(c) 2+1d SU(k) Chern-Simons theory as $k, N \to \infty$ [12].

(d) 2+2d self-dual gravity [13],[1]. This connection will be outlined in section 2 below.

3. (2,1) string perturbation theory is an expansion around the infinite tension limit; $T_{\rm brane} \sim g_{\rm str}^{-2}$. In other words, it is an expansion in low energy relative to the brane tension. This is similar to 1+1d noncritical string theory, where the effective expansion parameter is the energy of excitations relative to some scale (in that case the cosmological constant μ of Liouville theory, or equivalently the fermi energy of the matrix partons).

Thus one sees common themes cropping up in diverse settings.

The focus of these lectures will be the structure of N=2 strings, and their possible relation to matrix dynamics. We begin in section 2 with the somewhat simpler (2,2) string, in order to introduce the novel features of local N=2 worldsheet supersymmetry and its associated self-duality structure. Included are an overview of the relation between the self-dual gravity of the (2,2) string and $SDiff_2$ dynamics; as well as an aside on the D-brane spectrum of (2,2) strings, which leads to yet another interesting connection to matrix theory. Section 3 introduces (2,1) strings – their worldsheet gauge algebra, spectrum, and simple compactifications. The target space effective action for (2,1) strings is derived in section 4; parallels with matrix theory are explored in section 5.

2. (2,2) Strings

The gauged N=2 supersymmetry on the string worldsheet consists of currents T (the stress tensor), $G^{\pm}$ (the two supercurrents), and J (a U(1) R-symmetry current). Their algebra is schematically

$$\begin{aligned} [T, G^{\pm}] &\sim \tfrac{3}{2} G^{\pm} \\ [T, J] &\sim J \\ [G^{+}, G^{-}] &\sim T + J \\ [J, G^{\pm}] &\sim \pm G^{\pm} \ . \end{aligned} \tag{2.1}$$

In the conformal gauge, the contributions to the Virasoro central charge from the Faddeev-Popov ghosts is $-26 + 2 \cdot 11 - 2 = -6$, so that the critical dimension is $d_{\rm crit} = 2/3\ c_{\rm matter} = 4$. A free field representation of the above currents is

$$\begin{aligned} T &= -\tfrac{1}{2} \partial X \cdot \partial X - \psi \partial \psi \\ J &= I_{\mu\nu} \psi^{\mu} \psi^{\nu} \qquad , \qquad \mu = 0, ..., 3 \\ G^{\pm} &= (\eta_{\mu\nu} \pm I_{\mu\nu}) \psi^{\mu} \partial X^{\nu} = G_{\rm N=1} \pm G_2 \ . \end{aligned} \tag{2.2}$$

Here, $I_{\mu\nu}$ is a self-dual tensor which acts as a complex structure. Preserving this structure under analytic continuation requires the signature to be Euclidean (4+0) or ultrahyperbolic (2+2). Even though the indefinite signature case has more than one negative metric string coordinate, there are no negative metric states. Each gauge invariance removes two fields' worth of degrees of freedom; in fact, the two bosonic (T, J) and two fermionic $(G^{\pm})$ gauge symmetries kill *all* the oscillator modes of the string.

One of the powerful consequences of this fact is the triviality of the string S-matrix. On the one hand, there can be no Regge behavior in scattering amplitudes, since the sequence of poles in the S-matrix are associated with physical oscillator excitations of the string, of which there are none. On the other hand, the covariant formalism generates a Koba-Nielsen (integral) representation for the S-matrix amplitude, which exhibits such poles. The tension between these two properties is resolved by the vanishing of any amplitude beyond the three-point function (the three-point function is protected because it does not involve Koba-Nielsen integrals). The single nontrivial S-matrix element is[1]

$$\langle V(1)V(2)V(3)\rangle = (k_1 \cdot I \cdot k_2)_\ell (k_1 \cdot I \cdot k_2)_r \ . \tag{2.3}$$

Since the only center of mass of the string can be excited, it is possible to find a representative of the string vertex operator which is a simple exponential

$$V = \bar{\Sigma}_{gh} \Sigma_{gh}\ e^{ik\cdot x} \tag{2.4}$$

apart from ghost (measure) factors Σ_{gh} (see Appendix A for a brief summary). A BRST-equivalent representative (or 'picture') of the vertex represents it as an integral over (2,2) superspace,

$$V = \int d^2\theta d^2\bar{\theta}\ e^{ik\cdot X} = \int d\theta d\bar{\theta}|_{\mathrm{N=1}} (k^\mu I_{\mu\nu} \bar{D}_{\mathrm{N=1}} X^\nu)(k^\lambda I_{\lambda\sigma} D_{\mathrm{N=1}} X^\sigma) e^{ik\cdot X} \ , \tag{2.5}$$

i.e. from the N=1 point of view the single physical state is a graviton whose only physical polarization is a fluctuation in the Kahler potential – $g_{i\bar{j}} = \partial_i \partial_{\bar{j}} K$ in complex coordinates.

Another important feature of the (2,2) string is the absence of target space supersymmetry. Consider a Wilson line of the U(1) R-current, $exp[i\alpha \int^{z} J]$. Parallel transport of a complex fermion ψ around the point z picks up a phase $exp[i\alpha]$; however J is gauged, and so the phase cannot have physical significance. In particular, there is no physical distinction between periodic (NS) and antiperiodic (R) boundary conditions, hence no target space fermions and no target space supersymmetry.

The effective action which generates the S-matrix (2.3) is

$$S_{\mathrm{eff}} = \frac{1}{g_{\mathrm{str}}^2} \int d^4x [\tfrac{1}{2}\partial K \cdot \bar{\partial} K + \tfrac{1}{3} K \partial\bar{\partial} K \wedge \partial\bar{\partial} K] \ , \tag{2.6}$$

from which one obtains the equation of motion

$$I \wedge \partial\bar{\partial} K + \partial\bar{\partial} K \wedge \partial\bar{\partial} K = 0 \ , \tag{2.7}$$

[1]Similar to the 1+1d noncritical string, where the 'bulk' S-matrix is trivial, the effective action is cubic (see below).

known as the Plebanski equation. This is the equation that governs the dynamics of self-dual gravity. It is a straighforward exercise in 2+2 kinematics to show that the quartic S-matrix computed from the action (2.6) vanishes [1].

An alternate route to the Plebanski equation proceeds from the generalized beta-function of the background sigma model for string propagation. In (2,2) supersymmetry, there are two kinds of scalar superfields: chiral and twisted chiral [14]. The simplest situation has the sigma model background described entirely in terms of chiral superfields; then the target space holonomy lies in U(d). The sigma model beta function equations are the conditions for SU(d) holonomy

$$R_{\mu\nu}[\Gamma] = 2\nabla_\mu\nabla_\nu\Phi \tag{2.8}$$

($\Gamma^\mu_{\nu\lambda} = \{{}^{\;\mu}_{\nu\lambda}\} - \frac{1}{2}H^\mu_{\nu\lambda}$ is the connection with torsion H). These equations may be integrated in complex coordinates to yield

$$\mathrm{logdet}[g_{i\bar{j}}] = 2\Phi + f(x) + \bar{f}(\bar{x}) \ . \tag{2.9}$$

A further analysis of the conditions for (2,2) supersymmetry in the sigma model coupled to worldsheet gravity [15] shows that the dilaton Φ is locally the real part of a holomorphic function; thus *locally* one can choose coordinates such that the dynamical equation is

$$\det[g_{i\bar{j}}] = 1 \ . \tag{2.10}$$

Some solutions of these equations have been described in [15]. The single physical degree of freedom of the N=2 string is the center of mass mode $K(x,\bar{x})$ describing fluctuations of the Kähler potential $g_{i\bar{j}} = I_{i\bar{j}} + \partial_i\partial_{\bar{j}}K$, where I is the background Kähler form. Expanding (2.10) in this way yields the Plebanski equation (2.7). Note that although K contains the field-theoretic degrees of freedom of the string, there may be additional moduli in the global modes of the metric, antisymmetric tensor, and dilaton – for instance the action (2.6) on K3 depends implicitly on the full 80 moduli of the conformal field theory, as well as the string coupling $\kappa = e^{-2\Phi}$. These are analogues of the special states of 2d noncritical string theory [16] that exist only for discrete momenta.

Backgrounds involving one chiral and one twisted chiral superfield must have two commuting complex structures [14] and are thus essentially trivial [17] – the quasi-Kähler potential $\tilde{K}$ describing the background geometry satisfies a free field equation. When spacetime has a translational Killing vector field with compact orbits, these backgrounds are related to the above self-dual geometry by T-duality [14, 15]. Thus one might call the theory discribed by (2.6) the N=2B theory, and the theory described by the free quasi-Kähler potential $\tilde{K}$ the N=2A string.

There are also interesting solutions which fall outside the class described by constant dilaton, for example the NS instanton [18],[15]

$$\begin{aligned} e^{2\Phi} &= e^{2\Phi_0} + \frac{n\alpha'}{r^2} \\ H_{\mu\nu\lambda} &= -\epsilon_{\mu\nu\lambda}{}^{\rho}\nabla_\rho\Phi \\ G_{\mu\nu} &= e^{2\Phi}\delta_{\mu\nu} \ , \end{aligned} \tag{2.11}$$

which is the counterpart of the NS five-brane in the usual 10d theory (an instanton is the magnetic dual to a string in 4d). In particular, this object is not that same as the D-instanton encountered below.

The Ricci-flat four dimensional geometry required by the N=2 string admits a hyperKahler structure, and the metric is self-dual. It is easy to see that the N=2 local supersymmetry generates N=4 global supersymmetry in the critical dimension $d_{\rm crit} = 4$. The canonical normalization of the U(1) R-current

$$J(z)J(w) \sim \frac{1}{(z-w)^2} \tag{2.12}$$

implies that the exponential fields $J^{\pm}(z) = exp[\pm i\sqrt{2}\int^z J]$, together with $J(z)$, form an SU(2) current algebra; furthermore, $J^{\mp}G^{\pm} \sim (z-w)^{-1}\tilde{G}^{\pm}$ are two additional supersymmetry currents. These enlarge the N=2 currents (2.1) to the N=4 current multiplet. Since any solution to the background field equations of motion preserves an SU(2) symmetry in the tangent space (in Euclidean signature; SL(2,R) in ultrahyperbolic signature), the target space is hyperKahler. There is a two-sphere's worth of choices of which U(1) inside this global SU(2) is the gauged local R-symmetry. Parametrizing this choice is the *twistor variable* $\zeta \in \mathbb{C}\mathrm{P}^1 = SU(2)/U(1)$ of self-dual geometry. For more on the relation of twistor geometry to the (2,2) string, see for instance [1].

This self-duality structure leads to an infinite number of conservation laws, and it is likely that this integrability underlies the triviality of the S-matrix seen above. Another way to exhibit these conservation laws reveals a connection to the area-preserving diffeomorphism group $SDiff_2$ [13]. Consider the zero-curvature condition in two dimensions

$$[\partial_t + \lambda A_t, \partial_x + \lambda A_x] = 0 \ , \tag{2.13}$$

where λ is an arbitrary (spectral) parameter, and A_α ($\alpha = 1,2$) is an $SDiff_2$-valued connection. One can represent $SDiff_2$ as the algebra of canonical transformations of a one-dimensional phase space parametrized by coordinates (p,q). Then (2.13) at order λ implies

$$\begin{aligned} A_t(x,t;p,q) &= -(\partial_t\partial_p\Omega)\partial_q + (\partial_t\partial_q\Omega)\partial_p \\ A_x(x,t;p,q) &= -(\partial_x\partial_p\Omega)\partial_q + (\partial_x\partial_q\Omega)\partial_p \ ; \end{aligned} \tag{2.14}$$

the $o(\lambda^2)$ equations then read (in an appropriate choice of coordinates)

$$(\partial_t\partial_p\Omega)(\partial_x\partial_q\Omega) - (\partial_t\partial_q\Omega)(\partial_x\partial_p\Omega) = 1 \ , \tag{2.15}$$

i.e. the Plebanski equation. The symmetry group of (2.13) is the loop group of $SDiff_2$; the self-dual gravity equations are essentially the same as the two-dimensional chiral model equations for the group $SDiff_2$, where (x,t,p,q) parametrizes the four-dimensional spacetime.

2.1. OPEN STRINGS, D-BRANES, AND MATRIX THEORY

The connection of the target space dynamics of the (2,2) string to $SDiff_2$ leads to an intriguing toy version of the circle of ideas underlying matrix theory. N=2 strings have very little dynamics; the N=(2,2) string has no physical transverse excitations. The center-of-mass mode of the closed string describes fluctuations of the Kähler potential of the self-dual metric; the corresponding mode of the open string describes fluctuations of a self-dual Yang-Mills field. The open string S-matrix is again essentially trivial, vanishing beyond the three-point function. In such a simplified dynamical setting, one might expect to be able to make exact statements about N=2 string D-brane dynamics as well, perhaps even nonperturbatively. The D-branes should strongly affect the strong-coupling behavior of the theory, since they become light in this regime.

In a sense, in N=2 strings self-duality plays a role similar to the BPS condition. There is no BPS structure per se, due to the absence of spacetime supersymmetry; the almost topological nature of the dynamics makes up for this, however. This is because spacetime supersymmetry of the usual superstring is intimately related to the spectral flow that analytically relates the NS and R sectors in worldsheet N=2 supersymmetry [19]. In the N=2 string, this spectral flow is gauged, so in a sense one has *only* the 'BPS' sector[2]. The spectral flow also generates the global SU(2) symmetry whose preservation is tantamount to the hyperKähler condition on the target space, and hence self-duality.

The fascinating proposal of matrix theory [9] casts the full light-cone gauge dynamics of M-theory in terms of the quantum mechanics of the BPS sector of zero-branes. Only the unexcited open strings stretching between the zero-branes are kept, yet one recaptures a remarkable amount of the complete theory in this way. One might hope that N=2 string D-branes serve as a toy model of this dynamics, since N=2 local supersymmetry permits *only* such states in the physical state space of open strings – all open string oscillations are unphysical.

Appendix B is devoted to a cursory explanation of the open string sector of (2,2) strings. The upshot is that there are D-branes, just as in the (1,1) and (0,0) strings. The latter lead to conventional (super)Yang-Mills dynamics on the brane, dimensionally reduced from 10 or 26 dimensions. Thus it is not surprising that, since the (2,2) open string sector describes four dimensional self-dual gauge theory, the D-branes are governed by the dimensional reduction of the self-duality equation:

[2]More precisely, the sort of BPS states that preserve the self-dual supercharge, i.e. Q_α and not $Q_{\dot\alpha}$

worldvolume dimension	integrable system	equation
4	SDYM	$F + \tilde{F} = 0$
3	Bogomolny	$F_{ij} + \epsilon_{ijk} D_k \Phi = 0$
2	Hitchin	$F + [X, \bar{X}] = 0$
1	Nahm	$D_t X^i + \epsilon^i{}_{jk}[X^j, X^k] = 0$
0	ADHM	$[X^\mu, X^\nu] + \epsilon^{\mu\nu}{}_{\lambda\sigma}[X^\lambda, X^\sigma] = 0$

We have denoted the branes by their total worldvolume dimensions rather than the space dimension, since that notation is a bit ambiguous in signature 2+2. Curiously, for worldvolume dimension equal to two (D-strings), the equation is very similar to (2.13). In fact one can approximate matrix theory rather closely, in the following sense. Consider the limit of N coincident D-strings, so that the $[X, \bar{X}]$ term in the Hitchin equation drops out. then the equation governing the D-string is a zero-curvature condition for its two-dimensional gauge field. In the limit $N \to \infty$, one has the equations of motion of the two dimensional chiral model with gauge group SU(∞) $\sim$ $SDiff_2$; this can be shown to be equivalent to the self-duality equations (c.f. the last of refs. [13]). Thus one has a situation where one can recover the low-energy theory corresponding to the N=2 string from the large-N limit of one of its D-branes. The principal difference is the lack of a connection to the infinite momentum frame, and the corresponding appearance of an extra dimension at strong coupling. Also, the transverse space to the D-branes does not appear to be related to the extra dimensions arising from the matrix phase space; one starts from a two-dimensional object in a four-dimensional space whose transverse position is nondynamical, and grows a different two dimensions by taking the large N limit. It may be that the appropriate starting point is a six-dimensional theory [20]. Perhaps the (0+2)-brane is close to an analogue of the D-particle in the construction of [9] since it spans both time directions of 2+2 spacetime. It would indeed be intriguing if one could realize self-dual gravity in terms of D-brane 'constituents'; perhaps this would provide an interpretation of the 'entropy' of the nuts and bolts of compactified self-dual solutions [21].

Independent of matrix theoretic ideas, the D-branes are the light objects of the theory at strong N=2 string coupling; they, together with the NS instanton (2.11), should dominate the nonperturbative behavior of self-dual gravity.

The N=2 string D-brane system is also a convenient laboratory for the exploration of the Nahm transform; T-duality relates the different branes upon, for instance, compactification on T^4 (mirror symmetry plays an analogous role on K3). Another intriguing point is that the moduli space of k instantons in SU(N) gauge theory has been proposed as the bosonic part of the configuration space of matrix noncritical strings [22]. N=2 strings generate exactly this moduli space as the space of physical deformations, with the D-instantons regulating the singularity at zero instanton scale size; and manifestly incorporate T-duality. Thus there might be a close relationship to the dynamics of this so-far mysterious 6d noncritical string theory to quantum mechanics on the moduli space of N=2 strings.

3. (2,1) Strings

Heterotic (2,1) strings [2] combine the self-dual, integrable structure of (2,2) strings with the spacetime supersymmetry present in strings with N=1 gauged worldsheet supersymmetry. A free field representation for the N=1 currents is

$$\begin{aligned} T &= -\tfrac{1}{2}\partial x \cdot \partial x - \psi \cdot \partial \psi - \tfrac{1}{2}\partial y \cdot \partial y - \lambda \cdot \partial \lambda \\ G &= \psi \cdot \partial x + \lambda \cdot \partial y \ . \end{aligned} \tag{3.1}$$

Here y^a, $a = 1, \ldots, 8$ is a chiral boson and λ^a a Majorana-Weyl fermion.[3] Using these currents as gauge constraints in conformal gauge is not sufficient to remove both of the two timelike oscillator mode towers of x and ψ; one needs another bosonic and fermionic gauge current. The simplest choice [2] is a U(1) supercurrent

$$\begin{aligned} J &= v \cdot \partial x + v_{\rm int} \cdot \partial y \\ \Psi &= v \cdot \psi + v_{\rm int} \cdot \lambda \ . \end{aligned} \tag{3.2}$$

In order to gauge this current, it must be anomaly free, which forces $\mathbf{v} = (v, v_{\rm int})$ to be a null vector: $v^2 + v_{\rm int}^2 = 0$. The ghosts for J, Ψ have $c = -3$; therefore, they increase the critical dimension by two, so $d_{\rm crit} = 12 = 10 + 2$. One appears to have a 'superstring in d=12', but not really; the additional gauge constraints project momenta and polarizations to be orthogonal to $\mathbf{v}$. If $x \in \mathbb{R}^{2,2}$, then the (y, λ) system must form a $c = 12$ holomorphic superconformal field theory (just as the internal sector of the (1,0) string in $\mathbb{R}^{9,1}$ must form a $c = 16$ conformal field theory). The unique choice preserving spacetime supersymmetry, modular invariance, etc., is for y to live on the Cartan torus of E_8.

Massless vertex operators are BRST equivalent to

$$\begin{aligned} V_{NS}^{grav} &= (\bar{\Sigma}_{gh}^{(N=2)})(\Sigma_{NS}^{(N=1)}\, \xi_\mu(k)\psi^\mu)e^{ik\cdot x} \\ V_{NS}^{gauge} &= (\bar{\Sigma}_{gh}^{(N=2)})(\Sigma_{NS}^{(N=1)}\, \xi_a(k)\lambda^a)e^{ik\cdot x} \\ V_R &= (\bar{\Sigma}_{gh}^{(N=2)})(\Sigma_{R}^{(N=1)}\, u^\alpha(k)S_\alpha)e^{ik\cdot x} \ . \end{aligned} \tag{3.3}$$

As in the (2,2) case, an equivalent representative of the graviton is

$$V = \int d\theta d^2\bar{\theta}\; e^{ik\cdot X} = \int d\theta d\bar{\theta}|_{\rm N=1}\; I_\mu^\lambda k_\lambda \xi_\nu (\bar{D}_{\rm N=1} X^\mu D X^\nu) \exp[ik_\mu X^\mu + ik_a Y^a] \ , \tag{3.4}$$

so that the graviton/antisymmetric tensor has a special polarization $h_{\mu\nu} + b_{\mu\nu} \sim I_\mu^\lambda \partial_\lambda \xi_\nu$, verifying the claim (1.1) of the introduction. Similarly, the gauge field $\varepsilon_\mu^a \sim I_\mu^\lambda \partial_\lambda \varphi^a$ and gravitino $\chi_\mu^\alpha \sim I_\mu^\lambda \partial_\lambda \psi^\alpha$. If the target space is $\mathbb{R}^{2,2} \times T_{\rm int}^8$, only the states (3.3) satisfy level matching

$$\begin{aligned} \bar{L}_0 &= k_\mu k^\mu = 0 \\ L_0 &= k_M k^M + N_\ell = 0 \end{aligned} \tag{3.5}$$

[3]Our conventions are as follows: the N=1 chiral sector will be the left-moving oscillations; the N=2 chiral sector will be right-movers. The indices are $\mu = 0, 1, 2, 3$ for $x^\mu \in \mathbb{R}^{2,2}$; $a = 1, \ldots, 8$ for $y^a \in T^8$; often we will use a combined index $M = (\mu, a) = 0, \ldots, 11$ when dealing with purely left-moving quantities.

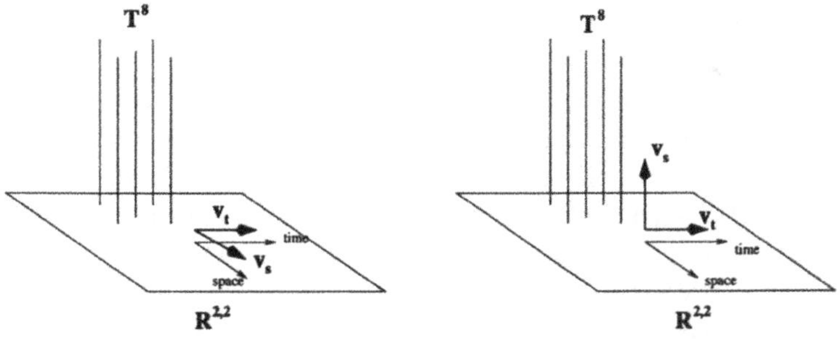

(a) (b)

Figure 3. Choices of null constraint on the left-movers.

In this case the only physical states are at the massless level, $N_\ell = 0$, much like the (2,2) string; with no momentum in the internal T^8 directions, $k_a = 0$.

The Virasoro/null current constraints impose restrictions on the polarizations and momenta:

$$\begin{array}{llll} \text{Virasoro} & k_M \xi^M = 0 & \xi \sim \xi + \alpha k & \text{NS} \\ & \not{k} u = 0 & & \text{R} \\ \text{Null current} & \mathbf{v}_M \xi^M = 0 & \xi \sim \xi + \alpha \mathbf{v} & \text{NS} \\ & \not{\mathbf{v}} u = 0 & & \text{R}\,. \end{array}$$

These constraints reduce the 12 polarizations of the NS vector to 8 transverse, and the 32 components of the Majorana-Weyl R sector spinor to 8 physical states, as expected. The spacetime interpretation of these physical states depends on the specific orientation of the null vector $\mathbf{v}$, see figure 3.

In figure 3a, the null vector is oriented entirely within the $\mathbb{R}^{2,2}$ base space. The kinematics consists of 1+1 dimensional momenta k (recall that the level matching constraint eliminates any momentum components in the E_8 directions), with the physical polarizations consisting of a 1+1 gauge field a_μ, 8 scalars φ^a, and 8 fermions ψ^α. The spectrum is that of the type IIB D-string! On the other hand, if the null vector has its time component in $\mathbb{R}^{2,2}$ and its space component in T^8 (see figure 3b), then the kinematics is 2+1 dimensional, and the physical polarizations are the 2+1 gauge potential a_μ, 7 scalars φ^a, and 8 fermions ψ^α. The spectrum in this case is that of the type IIA D2-brane.

One always has sixteen supersymmetries

$$Q_\alpha = \oint \Sigma_{gh,R}^{(N=1)} S_\alpha \quad , \qquad \not{\mathbf{v}} Q = 0 \,, \tag{3.6}$$

just as a type II D-brane breaking half the supersymmetries. The algebra of these supercharges is (on-shell, since all considerations are modulo BRST equivalence of the (2,1) string)

$$\{Q_\alpha, Q_\beta\} = (\gamma^{MN})_{\alpha\beta} P_M \mathbf{v}_N \ . \tag{3.7}$$

If our interpretation is correct, one would expect there to be an additional sixteen supersymmetries which are spontaneously broken by the brane. These supersymmetries would be nonlinearly realized in the worldvolume theory, and hence not visible in the single-string Hilbert space. One might see them in a careful study of the vertex algebra. In any event, the (2,1) string would appear to describe D1- or D2-branes in static gauge, stretched across the noncompact spatial directions of $\mathbb{R}^{2,2}$. The transverse directions appear to be compactified on a torus, probably the E_8 Cartan torus.

3.1. TOROIDAL COMPACTIFICATION OF (2,1) STRINGS

Qualitatively new features arise upon further compactification of the (2,1) string, since the level-matching constraints (3.5) prove to be much less restrictive. In fact, the physical spectrum consists of a stringy tower of 'Dabholkar-Harvey' states – ground states on the right-moving N=2 side, oscillator excitation at arbitrary level on the left-moving N=1 side, with momentum and winding to compensate. Note that since it is the worldsheet chirality that carries spacetime supersymmetry that must be excited, states with $N_\ell \neq 0$ break target space supersymmetry; i.e. these states are not BPS-saturated.

Consider then the further compactification of the spatial x^2, x^3 coordinates of $\mathbb{R}^{2,2}$, so that the target is $\mathbb{R}^2(\text{time}) \times T^2 \times T^8_{\text{int}}$. More generally, one could consider an arbitrary spatial torus corresponding to a point in the Narain moduli space $\mathcal{N}^{10,2}$, but a product torus will suffice for illustrative purposes. The (2,1) string will in general have both momentum and winding

$$p^i_{\ell,r} = \frac{n^i}{R_i} \pm \frac{m^i R_i}{2} \quad , \qquad i = 2,3 \ . \tag{3.8}$$

Reexamining the level-matching constraints

$$\begin{aligned} 0 &= -p_0^2 - p_1^2 + p_{2,r}^2 + p_{3,r}^2 \\ 0 &= -p_0^2 - p_1^2 + p_{2,\ell}^2 + p_{3,\ell}^2 + \vec{p}_{T^8}^{\,2} + 2N_\ell \ , \end{aligned} \tag{3.9}$$

one sees that many more states are now available – one need only satisfy the mass shell condition, the level-matching constraint, and the null constraint (as well as the highest weight condition under the current algebra):

$$\begin{aligned} 0 &= -p_0^2 - p_1^2 + \left(\frac{n^i}{R_i}\right)^2 + \left(\tfrac{1}{2} m^i R_i\right)^2 + \tfrac{1}{2}\vec{p}_{T^8}^{\,2} + N_\ell \\ 0 &= 2m^i n^i + \vec{p}_{T^8}^{\,2} + 2N_\ell \\ 0 &= \mathbf{v} \cdot p_\ell \ . \end{aligned} \tag{3.10}$$

Note that the constraints now allow the internal momenta $\vec{p}_{T^8}$ to be arbitrarily excited, so that the kinematics is more or less ten dimensional. These states are

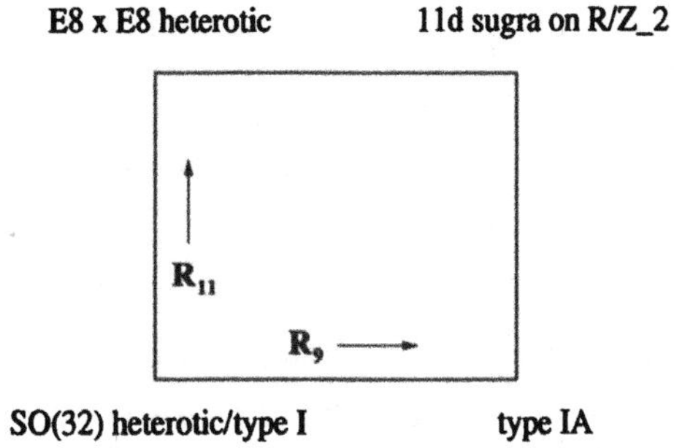

Figure 4. Moduli space of $S^1 \times S^1/\mathbf{Z}_2$ compactifications of M-theory.

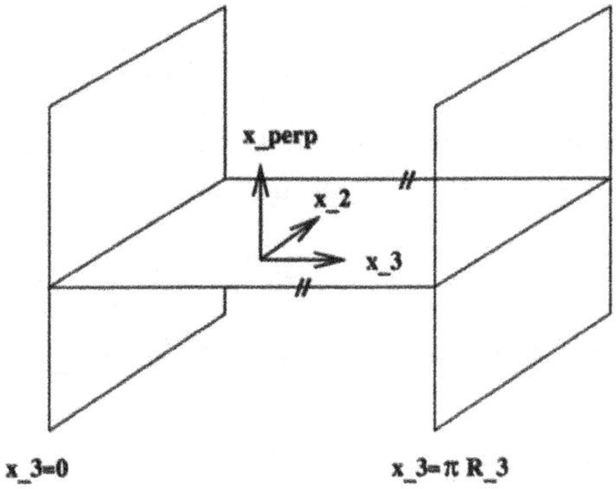

Figure 5. The heterotic membrane geometry generated by the (2,1) string.

similar to the perturbative BPS states of the type II superstring; for instance, the level density grows exponentially. Even with this change of circumstances, one can check that the four-point function continues to vanish, indicating that the triviality of the S-matrix continues to hold. T-duality of the (2,1) string seems to imply that there is a minimum 'size' to the target circle(s) over which the target D-brane is stretched: $R_{\rm min} \sim \ell_{\rm str}^{(2,1)}$.

3.2. HETEROTIC/TYPE I CONSTRUCTION

It is now well-understood that the heterotic and type IA string theories are particular asymptotic limits in the moduli space of $\mathbf{Z}_2$ orientifolds of M-theory, see figure 4. The simplest situation has 16 ninebranes at each of the two orientifold planes to cancel anomalies; the low-energy dynamics consists of 11d supergravity in the bulk, and 10d SYM on these walls. Since the (2,1) string seems only to describe a brane in spacetime, to realize these vacua as a (2,1) string background one might look for a wrapped membrane/string that sees this structure, see figure 5. In other words, one looks to describe a cylindrical, open membrane stretched between the orientifold planes $x_3 = 0, \pi R_3$. Its boundaries should have 1+1 dimensional fields describing rank eight current algebra, such that one obtains a heterotic string when $R_2 \gg R_3$; on the other hand, when $R_2 \ll R_3$, these fields will be frozen, leaving the finite number of Chan-Paton labels of type IA theory. The orientifold symmetry acts as $x_3 \to -x_3$, $A_{MNP} \to -A_{MNP}$. The sign flip of the three-form gauge field indicates orientation reversal of the M-theory membrane worldvolume. Since one appears to be in static gauge, the orbifold $x_3 \to -x_3$ of the (2,1) string will accomplish both the orbifold of spacetime and the orientation reversal of the brane worldvolume. One simply wants to find a consistent supersymmetric orbifold of the (2,1) string with this operation as part of the orbifold group. A straightforward chain of logic leads to a unique answer satisfying the orbifold level-matching constraints, preserving half the sixteen spacetime supersymmetries, and treating all the internal coordinates y on the same footing. Right-moving worldsheet supersymmetry is preserved only if, in addition to $x_3 \to -x_3$, one simultaneously flips $\bar\psi_3 \to -\bar\psi_3$. Now consider the right-moving U(1) R-current, e.g. $\bar J = \bar\psi^0\bar\psi^3 + \bar\psi^1\bar\psi^2$; there are two choices for the $\mathbf{Z}_2$ to have a well-defined action on the gauge algebra: (1) $\bar\psi_0 \to -\bar\psi_0$, which one can show does not lead to a spacetime supersymmetric solution; and (2) $\bar\psi_1 \to -\bar\psi_1$, which gives the twisted N=2 algebra [23]

$$\bar J \to -\bar J \quad , \qquad \bar G^+ \leftrightarrow \bar G^- \; . \tag{3.11}$$

Proceeding along these lines, one finds a unique $\mathbf{Z}_2$ satisfying the above requirements:

$$\begin{aligned}
(x_1, x_3; y_1, .., y_8) &\to -(x_1, x_3; y_1, .., y_8)\\
(\psi_1, \psi_3; \lambda_1, ..., \lambda_8) &\to -(\psi_1, \psi_3; \lambda_1, ..., \lambda_8)\\
(\bar\psi_1, \bar\psi_3) &\to -(\bar\psi_1, \bar\psi_3) \quad ,
\end{aligned}$$

with all other coordinates invariant.

Vertex operators describe physical fluctuations of the target brane, and have the form $\mathcal{O}\cos[k_3 x^3]$ if the polarization operator $\mathcal{O}$ is $\mathbf{Z}_2$ even, and $\mathcal{O}\sin[k_3 x^3]$ if $\mathcal{O}$ is $\mathbf{Z}_2$ odd. At the massless level, one finds Neumann boundary conditions ($\cos[k_3 x^3]$) for polarizations along $(0, 2; 4, ..., 11)$, and Dirichlet boundary conditions ($\sin[k_3 x^3]$) for polarizations along $(1, 3)$. Thus, in space there are nine Neumann and one Dirichlet boundary condition; in time, one Dirichlet and one Neumann. Since we wish to avoid thinking about what a Dirichlet boundary condition means in physical time, we place the time component of the vector $\mathbf{v}$ (defining the null projection J) in the x_1 direction; thus its space component must also be one of the $\mathbf{Z}_2$ twisted directions $(2, 4, ..., 11)$. Choosing the latter in one of the internal directions, one finds now an open membrane stretched between eight-branes in a type IIA description; that is, when the target (x_2, x_3) torus orbifold is a large cylinder, in the low-energy limit one has a 2+1 target space gauge field coupled to 7 scalars with Neumann boundary conditions. This is precisely the bulk dynamics of an open membrane stretched between D8-branes. One can lift to this effective theory to eleven dimensions by dualizing the vector to another scalar (which one easily checks also has Neumann boundary conditions) to find an open membrane stretched between orientifold ninebranes.

The boundary dynamics comes from the twisted sectors of the (2,1) string orbifold, as these states are pinned to the fixed points $x_3 = 0, \pi R_3$ which are the orientifold planes. Since the orbifold is an asymmetric one, the number of states at each of the two fixed points $x_3 = 0, \pi R_3$ is the square root of the number of fixed points of the internal $T^8/\mathbf{Z}_2$ orbifold, $\sqrt{2^8} = 16$. Consistency of the operator algebra shows that these states appear in the Ramond sector, and an analysis of the BRST constraints shows that they are chiral in the target space. Thus there are sixteen massless fermion fields living on each boundary of the open target membrane, which describe a rank eight current algebra, and are inert under spacetime supersymmetry. Half the supersymmetry charges (3.6) are broken by the orbifold. Note that this is exactly the spectrum found on the D8-brane boundaries of the matrix theory description of the heterotic string, a fact that will be important for us below.

There is a very similar orbifold [5], also breaking half the supersymmetry, obtained if one relaxes the condition that all internal coordinates are treated identically. Twisting half rather than all the y^a, λ^a

$$\begin{aligned}(x_1, x_3; y_1, .., y_4) &\to -(x_1, x_3; y_1, .., y_4)\\ (\psi_1, \psi_3; \lambda_1, ..., \lambda_4) &\to -(\psi_1, \psi_3; \lambda_1, ..., \lambda_4)\\ (\bar\psi_1, \bar\psi_3) &\to -(\bar\psi_1, \bar\psi_3) \ ,\end{aligned}$$

is also a consistent asymmetric orbifold. The twisted sector ground states transform as a hypermultiplet under the eight remaining supersymmetries.

Finally, it should be noted that we have only described the low-energy spectrum; there is again a stringy tower of left-moving states just as in the untwisted, toroidally compactified (2,1) string.

4. S-matrix, effective action and geometry of (2,1) strings

The tree-level, three-point S-matrix factorizes between left- and right-movers, hence we can immediately deduce the answer by combining (2,2) string result (2.3) with that of the superstring; for three massless metric perturbations, the result is

$$\langle V_\xi(1) V_\xi(2) V_\xi(3)\rangle = (k_1 \cdot I \cdot k_3)_r (\xi_1 \cdot \xi_2 \, k_1 \cdot \xi_3)_\ell + \text{cyclic} \ , \tag{4.1}$$

leading to the cubic effective Lagrangian

$$\mathcal{L}^{(3)}_{\text{eff}} = \tfrac{1}{2} F_{\mu\nu} F_{\nu\lambda} F_{\lambda\sigma} F_{\sigma\mu} - \tfrac{1}{8} (F_{\alpha\beta} F_{\beta\alpha})(F_{\mu\nu} F_{\nu\mu}) \tag{4.2}$$

with $F_{\mu\nu} = \partial_{(\mu}\xi_{\nu)}$. Curiously, this is a term in the expansion of the Born-Infeld Lagrangian around the background $F_{\mu\nu} = I_{\mu\nu}$. One can incorporate the effects of the internal scalars φ simply by letting the polarization indices in (4.2) run over 10d (or 12d, since we are imposing the null constraints by hand), while keeping the kinematics 1+1 or 2+1 dimensional. Regarding the fermions, their S-matrix is

$$\langle V_\psi(1) V_\xi(2) V_\psi(3) + V_\psi(1) V_\varphi(2) V_\psi(3)\rangle = (k_1 \cdot I \cdot k_3)_r [\bar u_1 (\xi^\mu \Gamma_\mu + \zeta^a \Gamma_a) u_3]_\ell \ , \tag{4.3}$$

where Γ_M are twelve dimensional gamma matrices. One obtains the effective Lagrangian

$$\mathcal{L}^{(3)}_{\rm eff} = (\bar\psi\Gamma^\mu\partial_\nu\psi)I^{\nu\lambda}F_{\lambda\mu} + (\bar\psi\Gamma^a\partial_\nu\psi)I^{\nu\lambda}\partial_\lambda\varphi^a \ . \tag{4.4}$$

These are the only nonzero S-matrices[4] in $\mathbb{R}^{2,2} \times T^8$, but this does not mean that we have found the full effective action. It happens that the iteration of these cubic field-theoretic vertices generates a four-point S-matrix, such that kinematics in $\mathbb{R}^{2,2}$ allows the pole terms to cancel among s, t, and u channels, leaving an explicit four-point contact term. For the full S-matrix to vanish, one must add a cancelling quartic term in the effective action. Continued iteration yields terms to all orders in small fluctuations. This situation contrasts with the (2,2) string, where the symmetry is powerful enough to cause the potential four-point contact term to vanish, and allows the effective action to cubic order to in fact be the exact theory.

The simplest way to extract the full answer is to realize that the iteration of the cubic terms will generate contact terms of the form $k^n(\delta\phi)^n$ at $n^{\rm th}$ order in fluctuations, while fluctuations in the geometrical fields h, b, ε are all of the form $\partial(\delta\phi)$ due to the (2,1) supersymmetry constraints. Hence the effective action found from the beta function will come entirely from one loop (higher loops will come with extra factors of $\alpha' k^2$, and hence more powers of momenta than small fluctuations). The beta function equations (2.8) can be integrated once to give

$$\Gamma_\mu \equiv \Gamma^\rho_{\nu\mu} I^\nu_\rho = 0 \tag{4.5}$$

with the effective action

$$S^{\rm grav}_{\rm eff} = \frac{1}{g^2_{\rm str}} \int d^4x \, \det{}^{\frac12}[\eta_{i\bar\jmath} + F_{i\bar\jmath}] \ , \tag{4.6}$$

at least for the gravity sector. The gauge fields may be added via an analysis of sigma model anomalies [24],[8]; the result is

$$S_{\rm eff} = \frac{1}{g^2_{\rm str}} \int d^4x \, \det{}^{\frac12}[\eta_{i\bar\jmath} + F_{i\bar\jmath} + \alpha' \partial_i\varphi^a\partial_{\bar\jmath}\varphi^a] \ . \tag{4.7}$$

It has been checked [24] that this action reproduces all cubic and induced quartic vertices involving only bosons, and that the S-matrix vanishes at quartic order. The action is remarkably similar to the static gauge effective action for D-branes. There are two significant differences, however: (1) For target space $\mathbb{R}^{2,2} \times T^8$, it is exact to all orders in $\alpha' = \ell^2_{\rm str}$, presumably related to the fact that there are no oscillator excitations of the underlying (2,1) string; and (2) the dynamics is integrable. The vanishing of the S-matrix in the fully compactified case indicates that this integrability continues to hold even when the kinematics is essentially ten-dimensional. It would be very interesting to work out the generalization of (4.7) in this case. It would also be interesting to explore the possibility of nontrivial solutions along the lines of (2.11).

[4]Hence the analogy with the 1+1d noncritical string made in the introduction, wherein the 'bulk' S-matrix is also trivial.

The terms in the effective action involving fermions should in principle be determined by supersymmetry. However, it is difficult to compute quantities involving nontrivial Ramond vertex backgrounds (there is no sigma-model approach). A better understanding of the spacetime supersymmetry current algebra and its anomaly structure would be helpful, since it was essentially this structure which enabled us to determine the dependence of the effective action on the scalars φ^a. As mentioned in the introduction, the physical fields on the target brane are all Nambu-Goldstone modes of the spacetime symmetries spontaneously broken by the brane, and therefore the effective action is expected to be largely determined by the various broken and unbroken symmetries. Because the bosonic structure is so similar to the Dirac-Born-Infeld/Nambu-Goto action, it is natural to guess that the full answer including fermions has a structure similar to that of Dp-branes in static gauge

$$\begin{aligned} S^{(p)} &= \int d^{p+1}\sigma \det{}^{\frac{1}{2}}\Big[\eta_{\mu\nu} + F_{\mu\nu} + \partial_\mu\phi^a\partial_\nu\phi^a \\ &\qquad -2\bar\psi(\Gamma_\mu + \Gamma^a\partial_\mu\phi^a)\partial_\nu\phi + (\bar\psi\Gamma^a\partial_\mu\psi)(\bar\psi\Gamma^a\partial_\nu\psi)\Big] . \end{aligned} \tag{4.8}$$

The sorts of terms appearing in the expansion of this action are compatible with with the cubic S-matrix (4.4), but the full structure is far from understood. This fermionic completion of of the action (4.7) represents a rather nontrivial coupling of SDYM to self-dual gravity with torsion, with an additional fermionic symmetry whose geometry ought to be quite intriguing.

5. Connections between (2,1) strings and matrix theory

In this final section of these lectures, I would like to present some evidence for the idea that (2,1) strings are closely related to the formulation of M-theory on T^9 as a matrix model.

A striking feature of supergravity U-duality is the appearance of exceptional groups in less than six noncompact spacetime dimensions. Naive extrapolation leads to a duality group $E_{9(9)}(\mathbf{Z})$ in two dimensions [25]. However, the solutions to low-energy supergravity which exhibit the various BPS charges permuted under duality, become more and more singular the lower the dimension. For instance, in 2+1 dimensions, n fundamental strings wound around a given cycle of the internal T^7 has a dilaton background of the form [26]

$$e^{-\Phi} = e^{-\Phi_0} - \sum_{i=1}^{n} 8G_N\ell_{\rm str}^{-2} ln|\vec r - \vec r_i| \; ; \tag{5.1}$$

one can introduce an arbitrary number of such sources, but the dynamics becomes strongly coupled near the core of each one. Presumably the correct solution receives modifications due to the nonperturbative states that become light there (c.f. [27]). This is supported by the observation that, in the IIB theory, one can wrap a D7-brane around the T^7; these are in the same U-duality multiplet as the fundamental string, yet they have a rather different classical solution – the core is nonsingular (from the viewpoint of F-theory), but one cannot introduce more than 24 sources. In 1+1 dimensions, the situation is even worse [28]; any source generates a response in the dilaton of the form

$$e^{-\Phi(x^+)} = e^{-\Phi_0} - \int\int_{-\infty}^{x^+} T_{++} \ ; \tag{5.2}$$

positive stress-energy of the source forces a singularity in the dilaton at finite x^+, so again information of a more nonperturbative nature is needed. Finally, if one considers the velocity-dependent forces between these objects in low dimensions, eventually they become confining even between the BPS states. For instance, in matrix theory the velocity dependent force between two matrix partons on $\mathbb{R}^{10-d,1} \times T^d$ is v^4/r^{7-d}, becoming confining below 3+1 dimensions.

It is not clear what to make of these observations. Two possibilities are

- Supergravity makes sense as a low-energy theory, but that the branes that have been our guide to defining matrix theory do not make sense – or at least cannot be present in the arbitrary numbers needed to define a large-N continuum limit.[5]
- Matrix theory exists in all dimensions; however, the confining potential between matrix partons means that there will be no moduli space for the gauge dynamics defining matrix theory in low dimensions, therefore no low-energy approximation that one could call spacetime, therefore no supergravity.

The connection between (2,1) strings and brane dynamics cries out for some connection to M-theory. Could (2,1) strings describe the 'low-dimensional phase' of matrix M-theory conjectured above? There are a number of reasons to think so:

1. At least eight coordinates are compact on the scale $\ell_{\rm str}^{(2,1)}$.
2. Massless physical states of the fully compactified (2,1) string have interactions similar to what one would expect from SYM with gauge group $SDiff_2$.[6]
3. The $\mathbf{Z}_2$ orbifold that gives the heterotic string acts as expected in matrix theory [29].
4. In the fully compactified (2,1) string, the decompactification limits are analogous to the limits which yield matrix IIA/B strings.

Consider again the vertex operators of the fully compactified (2,1) string

$$\begin{aligned} V_{NS} &= (\text{ghosts})\big(\xi_M(p_\ell)\psi^M\big)\exp[ip_\ell\cdot x_\ell + ip_r\cdot x_r] \equiv \mathcal{A}_M(x_\ell,x_r) \\ V_R &= (\text{ghosts})\big(u_A(p_\ell)S^A\big)\exp[ip_\ell\cdot x_\ell + ip_r\cdot x_r] \equiv \lambda_A(x_\ell,x_r)\ . \end{aligned} \tag{5.3}$$

The three-point function of, for instance, the bosons (4.1) generates the cubic coupling

$$\mathcal{L}^{(3)} = g_{\rm str}\, \partial_\ell^M \mathcal{A}^N \{\mathcal{A}_M, \mathcal{A}_N\}\ , \tag{5.4}$$

[5]Also the branes have a strong back-reaction on spacetime, perhaps precluding the existence of the null Killing vector needed to define the light cone gauge.

[6]In matrix theory on T^9, the theory cannot quite be SYM even at low energies, due to anomalies.

where $\{F, G\} = (\partial F/\partial x_r^\mu) I^{\mu\nu} (\partial G/\partial x_r^\nu)$ is a Poisson bracket induced by the complex structure of the right-moving N=2 supersymmetry. This cubic vertex has the same structure as Yang-Mills with a gauge group of symplectic diffeomorphisms. In this interpretation, we wish to regard x_ℓ as transverse 'spacetime' coordinates, and x_r as phase space coordinates on the Lie algebra of $SDiff_2$ à la matrix theory. In fact, the left- and right-moving degrees of freedom are not independent; they are coupled through the level-matching constraints (3.9). Moreover, the lattice of momenta $\Gamma^{10,2}$ is generically not factorized as $\Gamma_\ell^{10} \times \Gamma_r^2$. Nevertheless, there may be a sense in which a structure of $SDiff_2$ gauge theory is present off-shell [30]. It is possible that this gluing of gauge and spacetime fluctuations is an artifact of the expansion about a particular classical solution, much as a monopole ties rotations in space and isospin.

To see the action of $SDiff_2$, consider the right-moving chiral part of the vertex operators, taking sets which are mutually local in their OPE's; this means

$$p_i \;, \quad i = 1, 2, 3 \; : \qquad p_i^2 = \sum_{i=1}^{3} p_i = 0 \; . \tag{5.5}$$

The generic construction of such sets is as follows: Take two orthogonal null vectors $n_{(1)}, n_{(2)}$ (i.e. $n_{(1)}^2 = n_{(2)}^2 = n_{(1)} \cdot n_{(2)} = 0$, and $n_{(1)} \neq \alpha n_{(2)}$). This is possible in signature 2+2. The $n_{(a)}$ span a *self-dual null plane* $\mathcal{N}$. Now focus on momenta $p = an_{(1)} + bn_{(2)}$ which are arbitrary linear combinations of the two basis vectors. The chiral vertex operators $J_p = \oint dz \int d^2\bar{\theta} \exp[ip \cdot x_r]$ satisfy the algebra

$$[J_p, J_q] = p \cdot I \cdot q \; J_{p+q} \tag{5.6}$$

modulo BRST equivalence. Momenta $p \notin \mathcal{N}$ do not have mutually local OPE's, and hence no well-defined algebra structure. These on-shell $p_r \in \mathcal{N}$ are completely determined by their spatial components; the null condition forces the length of the spacelike part to equal that of the timelike part, so that the triple $(p, q, -p-q)$ form congruent triangles in the temporal x_0-x_1 and spatial x_2-x_3 planes (an overall relative orientation of these two triangles is determined by the choice[7] of $\mathcal{N}$). The upshot is that the algebra (5.6) is an algebra of two-dimensional and not four-dimensional symplectic diffeomorphisms, i.e. $SDiff_2 \sim \mathrm{SU}(\infty)$.

Thus the structure of the fully compactified (2,1) string strongly resembles that of matrix theory on T^9, although there are profound differences due to the constraints relating left- and right-moving coordinates (x_ℓ, x_r). Further support for this idea is provided by the fact that the decompactification limit of the (2,1) string is described by a D-brane style Lagrangian for either strings or membranes (depending on the orientation of the null vector $\mathbf{v}$); in fact it is precisely the D-string action if the null vector $\mathbf{v} \in \mathbb{R}^{2,2}$.[8] D-string dynamics is precisely what one finds in matrix theory if one shrinks a compactified circle to a radius $R_i \ll \ell_{\rm pl}$;

[7]This choice is not unique; in fact the set of self-dual null planes is parametrized by a coordinate $\zeta \in \mathbb{RP}^2$ which is the twistor parameter.

[8]The left-right constraints are most powerful here, since they restrict all momenta to $p_\ell = p_r$. It would seem then that this limit describes only a single D-string and not

then the dual circle on which the SYM dynamics takes place lives on a circle of radius $\Sigma \sim \ell_{\rm pl}^3/R_i R$, so that the IR dynamics becomes 1+1d SYM in the limit [22].

The heterotic/type I construction in section 3.2 follows precisely the pattern one would expect of matrix theory on T^9. In matrix theory, the heterotic/type I theory on T^d arises from a SYM orientifold on $\tilde{T}^d \times \tilde{S}^1$; the orientifold $\mathbf{Z}_2$ acts simultaneously on the torus and acts as the involution that sends $SU(N) \rightarrow SO(N)$. One must also add by hand 32 fermionic matter multiplets in order to cancel gauge anomalies; these represent the couplings of the zero-brane/supergravitons to the nine-branes of the background spacetime. All of this structure has a direct parallel in the (2,1) string. The orbifold group of section 3.2 acts on the left-moving 'spacetime' coordinates x_ℓ by reflecting nine spatial coordinates (of which one is removed by null projection) while leaving one untouched – in other words, as though the parameter space were $\tilde{T}^8/\mathbf{Z}_2 \times \tilde{S}^1$. Simultaneously, it acts to flip the right-moving U(1) R-current $\bar{J} \rightarrow -\bar{J}$. Thus the $SDiff_2$ algebra (5.6) will undergo a $\mathbf{Z}_2$ quotient which one can perhaps think of as $SO(\infty)$. Finally, one finds the requisite 32 fermionic states in the twisted sector (and here they are not simply put in by hand). Decompactification of a circle again resembles the construction of the matrix heterotic string [29], although again the left-right gluing hides the gauge group structure in this limit.

The alternate orbifold presented at the end of section 3.2 seems to be a $\tilde{T}^4/\mathbf{Z}_2 \times \tilde{T}^5$ orbifold, but rather than sixteen hypermultiplets appearing in the twisted sector (as expected in matrix theory on this space), we find only one.

At this point several remarks are in order. First, the (2,1) string appears to lack the parameters needed to explore the full 128-parameter moduli space of matrix theory on T^9. Eight circles are fixed to the scale $\ell_{\rm str}^{(2,1)}$, so it is not clear exactly how to connect the construction to the rest of M-theory. This aim might be furthered by understanding how to construct U-duality multiplets in the context of (2,1) string theory. The perturbative string states will be the momentum states of the SYM theory under the present interpretation (which are longitudinally wrapped membranes in matrix theory). The duality group is probably $E_{9(9)}(\mathbf{Z})$ [25] and does not act entirely within the perturbative theory.[9] In fact, one expects [32] states whose masses scale as arbitrarily high powers of $1/g_{\rm YM}^2$. Since $g_{\rm YM} \propto g_{\rm str}$, we are not going to find these easily.

A. Technology of local N=2 worldsheet supersymmetry

The ghost operators in superstring vertices provide the appropriate superconformally covariant geometrical measure for their integration over the worldsheet (for details, see [33, 34]). The superghosts have a number of different 'pictures', or

$N \rightarrow \infty$ of them. It may be that this freezing of U(N) down to U(1) is an artifact of the string perturbation expansion, which takes place about a particular classical solution; perhaps this symmetry is broken in the particular vacuum seen by the (2,1) string.

[9]This corrects an erroneous interpretation in [31], where an attempt was made to connect the duality group to the symmetries of the internal E_8 torus.

BRST representatives, because the splitting of the supersymmetry ghosts into creation and annihilation operators is ambiguous. For the N=1 superconformal algebra, there is a single pair of such ghosts β, γ; for N=2 one has two – $\beta_\pm$, $\gamma_\pm$ for the two supersymmetries $G_\pm$. The ghost current may be bosonized as[10] $\partial\phi_\pm = \beta_\pm\gamma_\pm$. The exponentials $exp[\alpha\phi_\pm]$ interpolate between the various vacua; $\phi_\pm$ are free fields.

In the NS sector there are two canonical pictures for a given chiral sector of the usual N=1 superstring; the N=2 string doubles this. Consider a superfield, with expansion in components

$$\mathcal{O} = \mathcal{O}^{(0)} + \bar\theta_+\mathcal{O}^{(1)}_- + \bar\theta_-\mathcal{O}^{(1)}_+ + \bar\theta_+\bar\theta_-\mathcal{O}^{(2)} \tag{A.1}$$

under the right-moving N=2 supersymmetry. For example, the exponential $exp[ik\cdot X]$ expands as

$$\begin{aligned} e^{ik\cdot X} &= e^{ik\cdot x}\big[1 + i\bar\theta_+\, k\cdot(\mathbf{1}+I)\cdot\bar\psi + i\bar\theta_-\, k\cdot(\mathbf{1}-I)\cdot\bar\psi \qquad \text{(A.2)}\\ &\quad +\bar\theta_+\bar\theta_-\big(k\cdot I\cdot\bar\partial x - (k\cdot\bar\psi)(k\cdot I\cdot\bar\psi)\big)\big] \;. \qquad \text{(A.3)} \end{aligned}$$

BRST invariant vertex operators are then

$$\begin{aligned} V^{(0)} &= e^{-\bar\phi_+ - \bar\phi_-}e^{ik\cdot x} \\ V^{(1,\pm)} &= e^{-\bar\phi_\pm}[k\cdot(\mathbf{1}\mp I)\cdot\bar\psi]e^{ik\cdot x} \\ V^{(2)} &= \big[k\cdot I\cdot\bar\partial x - (k\cdot\bar\psi)(k\cdot I\cdot\bar\psi)\big]e^{ik\cdot x} \;. \end{aligned} \tag{A.4}$$

Any combination of these with ($\pm$) ghost charges adding up to -2 (to cancel the background charge on the sphere) will yield the same on-shell S-matrix amplitude (2.3). The N=1 structure is as usual [33] in terms of the bosonization $\partial\phi = \beta\gamma$, with additional ghost factors in the Ramond sector from the spin field for the supercurrent Ψ, equation (3.2). These are again constructed in the standard way: the corresponding spin 1/2 ghosts $\tilde\beta$, $\tilde\gamma$ are bosonized via $\partial\rho = \tilde\beta\tilde\gamma$, and the standard picture for the Ramond sector vertices involves a factor $\Sigma^{(N=1)}_{gh,R} = e^{-\frac{1}{2}\phi+\frac{1}{2}\rho}$. The NS vertex operators in equations (2.4),(3.3) have been written in the zero picture for both left- and right-movers.

B. (2,2) open strings and D-branes

The open string sector describes self-dual Yang-Mills, in a formulation due to Yang [1],[35]. The equations of motion are

$$\begin{aligned} F^{(2,0)} &= F^{(0,2)} = 0 \\ k\wedge F &= 0 \end{aligned} \tag{B.1}$$

The single physical string mode is again the center-of-mass, which is an adjoint scalar φ related to the gauge field by

$$A = (De^{-\varphi})e^{\varphi} \quad , \qquad \bar A = (\bar De^{\varphi})e^{-\varphi} \; , \tag{B.2}$$

[10] For the twisted N=2 algebra, it is more convenient to bosonize the twist eigenstates $\beta_+\gamma_+ \pm \beta_-\gamma_-$ rather than the charge eigenstates.

with D a background covariant derivative. This ansatz automatically solves the (2,0) and (0,2) components of (B.1), and converts the (1,1) part into a wave equation for φ. In the modern interpretation of gauge charge, the Chan-Paton indices are carried by branes filling spacetime, whose number may be determined by closed string tadpole cancellation (of course, such restrictions are not active at open string tree level, where we can formally consider any classical Chan-Paton group). A calculation of the tadpole [35] indicates a gauge group G=SO(2); however, it should be noted that there are many massless tadpoles in N=2 string loop amplitudes, whose interpretation is currently unclear, rendering the determination of G somewhat ambiguous.

B.1. BOUNDARY CONDITIONS AND D-BRANES

The open string boundary conditions of N=2 superconformal field theory have been investigated by [36]. They fall into two classes:

$$\begin{array}{lll} \mathrm{A-type:} & J_\ell = -J_r \ , & G_\ell^+ = \pm G_r^- \\ \mathrm{B-type:} & J_\ell = +J_r \ , & G_\ell^+ = \pm G_r^+ \ , \end{array} \tag{B.3}$$

where d is the complex dimension of the target space and $p+1$ is the brane dimension[11]. These two types of boundary condition are related to the two types of N=2 superfield [14]: If a chiral field obeys type B boundary conditions, then its T-dual is a twisted chiral superfield satisfying type A boundary conditions, and vice-versa. Also, for a single free superfield, type B boundary conditions arise when the scalar components are either both Neumann or both Dirichlet, whereas type A has one Dirichlet and one Neumann boundary condition.

The dynamics of type II branes in ten dimensions is, at long wavelengths, the dimensional reduction of the Yang-Mills theory governing 10d open strings, the dynamics of the Dirichlet coordinates being "frozen" on the brane. Correspondingly, one expects the dynamics of N=2 string D-branes to be governed by the dimensional reduction of the self-dual equations (B.1). For instance, the D-string equations are the Hitchin equations

$$\bar{D}_A X = D_A \bar{X} = 0 \quad , \qquad F_A + [X, \bar{X}] = 0 \ . \tag{B.4}$$

Here the dynamical field is a 2d scalar φ in the adjoint of U(N_2) determining the reduced gauge field (B.2). The covariantly constant (and therefore nondynamical) Higgs field X describes the transverse positions of the N_2 D-strings. When N_2 D-strings coincide, these equations become the 2d chiral model equations. Similarly, one can show that the D-particle is described by the Nahm equations

$$DX^i = \epsilon^{ijk}[X^j, X^k] \ , \tag{B.5}$$

[11]Unfortunately, this convention is not particularly well-adapted to N=2 strings with signature 2+2; nevertheless we will adhere to it, using the notation (s+t)-brane for a brane with s space and t time dimensions when necessary.

and the D2-brane by the Bogomolny equations (for recent related work, see [37])

$$F_{ij} = \epsilon_{ijk} D_k X \ . \tag{B.6}$$

There are other possible reductions, for instance imposing Dirichlet boundary conditions along a null direction, or wrapping branes around homology cycles of nontrivial self-dual four-manifolds such as K3; we will not discuss them here. Note that the equations (B.4)-(B.6) are all in the family of Uhlenbeck-Yau type equations describing BPS configurations of D-branes (c.f. [38]), another indication of the resemblance of the dynamics to the BPS sector of superstrings.

The D-instanton is also of some interest, in that its amplitudes may serve to define an off-shell continuation of N=2 string dynamics, and thus an off-shell quantum theory of self-dual gravity and Yang-Mills. These amplitudes are in some respects similar to those of macroscopic loops in noncritical string theory [39].

One may also consider composite systems of p- and p'-branes. For example D-instantons in the open string theory are composites of a (0+0)-brane and a (4+0)-brane which regularize an abelian instanton [12]. The 0-4 strings should regulate the moduli space of ADHM instantons (although this needs to be checked).

B.2. COUPLING D-BRANES TO SELF-DUAL GRAVITY

The SDYM open string dynamics couples to the self-dual gravity of N=2 closed strings, since as usual open string loops factorize on closed string exchange. Moreover, performing a Kähler gauge transformation $\delta K = \Lambda(X) + \Lambda^*(X^*)$ on the worldsheet action

$$S = \int_\Sigma d^2z d^2\theta d^2\bar{\theta}\, K(X, X^*) + \oint_{\partial\Sigma} ds d^2\theta\, \varphi(X, X^*) \ , \tag{B.7}$$

one sees that Λ can be pushed onto gauge transformations of the U(1) part of the prepotential φ which encodes the self-dual U(N)=SU(N)×U(1) open string gauge background (B.2).

Marcus has shown [35] that the leading correction to the closed string equation of motion (2.10) is

$$\det[g_{i\bar{j}}] = 1 - \frac{2\kappa^2}{g^2} \mathrm{Tr}[F_{i\bar{j}} F^{i\bar{j}}] + O(g^4) \ , \tag{B.8}$$

where g and κ are the open and closed string couplings, related by $\kappa \sim \sqrt{\hbar} g^2$. Using the open string equation of motion $k \wedge F = 0$, one may rewrite this in a way that manifests the above gauge symmetry:

$$\tfrac{1}{N} \mathrm{Tr}[(k + gN^{\frac{1}{2}} F) \wedge (k + gN^{\frac{1}{2}} F)] = \Omega \wedge \Omega^* \tag{B.9}$$

[12] Assuming the tadpole cancellation giving gauge group SO(2); otherwise – for instance at open string tree level – it may be consistent to consider multiple (4+0)-branes and thus nonabelian (e.g. SU(N_{4+0}) self-dual YM gauge group. Then the D-instantons carry fundamental representation gauge quantum numbers.

where Ω is the holomorphic two-form.

It seems, however, that D-branes will not act as linearized sources for static gravitational and/or axion-dilaton fields. A static string solution, as for instance that of the fundamental string [26] is not self-dual (as one may see for instance by the set of supersymmetries it preserves).

The D-branes of ordinary 10d string theory carry charges under the antisymmetric tensor fields of the R-R sector. In the N=2 string these fields are gauge equivalent to NS-NS fields by spectral flow in the N=2 U(1); effectively, there are no R-R gauge fields, and the D-branes serve as sources for the NS sector fields (as one sees by inserting closed string vertex operators on the disk). There appear to be no gauge charges carried by N=2 string D-branes, apart from those of the self-dual gravitational field.

Acknowledgements

It is my great pleasure to thank Martin O'Loughlin and especially David Kutasov for the intense collaboration which led to the ideas presented here. Thanks also to Tom Banks, Jeff Harvey, Chris Hull, Greg Moore, and Stephen Shenker for fruitful discussions. The author is supported in part by DOE grant DE-FG02-90ER-40560.

References

1. H. Ooguri and C. Vafa, *Nucl. Phys.* **B361** (1991) 469.
2. H. Ooguri and C. Vafa, *Nucl. Phys.* **B367** (1991) 83.
3. For a review, see for instance N. Marcus, talk at the Rome String Theory Workshop (1992); hep-th/9211059.
4. D. Kutasov and E. Martinec, hep-th/9602049, *Nucl. Phys.* **B477** (1996) 652.
5. D. Kutasov, E. Martinec and M. O'Loughlin, hep-th/9603116, *Nucl. Phys.* **B477** (1996) 675.
6. V. Nair and J. Schiff, *Nucl. Phys.* **B371** (1992) 329; *Phys. Lett.* **246** (1990) 423.
7. A. Losev, G. Moore, N. Nekrasov, and S. Shatashvili, hep-th/9509151, *Nucl. Phys. Proc. Suppl.* **46** (1996) 130; hep-th/9606082, *Nucl. Phys.* **B484** (1997) 196.
8. C. Hull, hep-th/9702067.
9. T. Banks, W. Fischler, S. Shenker and L. Susskind, hep-th/9610043; *Phys. Rev.* **55** (1997) 5112.
10. E. Brezin, C. Itzykson, G. Parisi, and J.-B. Zuber, *Comm. Math. Phys.* **59** (1978) 35.
11. D. Gross and W. Taylor, hep-th/9301068, *Nucl. Phys.* **B400** (1993) 181.
12. M. Douglas, hep-th/9403119.
13. Q.-H. Park, *Phys. Lett.* **238** (1990) 287
    R.S. Ward, *Class. Quant. Grav.* **7** (1990) L217
    V. Husain, *Class. Quant. Grav.* **11** (1994) 927.
14. S.J. Gates, C. Hull, and M. Rocek, *Nucl. Phys.* **B248** (1984) 157.
15. E. Kiritsis, C. Kounnas, and D. Lust, hep-th/9308124; *Int. J. Mod. Phys.* **9** (1994) 1361.
16. A. Polyakov, *Mod. Phys. Lett.* **A6** (1991) 3272.
17. C. Hull, hep-th/9606190; *Phys. Lett.* **387** (1996) 497.
18. S.-J. Rey, talk given at Workshop on Superstrings and Particle Theory, Tuscaloosa, Alabama, 1989; L. Clavelli and B. Harms (eds.), World Scientific; A. Strominger, *Nucl. Phys.* **B343** (1990) 167; C. Callan, J. Harvey, and A. Strominger, *Nucl. Phys.*

B359 (1991) 611.
19. W. Boucher, D. Friedan, A. Kent, S.H. Shenker, and E. Witten, unpublished; T. Banks, L. Dixon, D. Friedan, and E. Martinec, *Nucl. Phys.* **B299** (1986) 71..
20. J. Plebanski and M. Przanowski, `hep-th/9605233`; Phys. Lett. **A212** (1996) 22.
21. G.W. Gibbons and S.W. Hawking, *Comm. Math. Phys.* **66** (1979) 291.
22. L. Motl, `hep-th/9701025`; T. Banks and N. Seiberg, `hep-th/9702187`; R. Dijkgraaf, E. Verlinde, and H. Verlinde, `hep-th/9703030`.
23. W. Boucher, D. Friedan, and A. Kent, *Phys. Lett.* **172** (1986) 316; A. Schwimmer and N. Seiberg, *Phys. Lett.* **184** (1987) 191; S. Ketov, O. Lechtenfeld, and A. Parkes, `hep-th/9312150`, *Phys. Rev.* **51** (1995) 2872; J. Bischoff, S. Ketov, and O. Lechtenfeld, `hep-th/9406101`, *Nucl. Phys.* **B438** (1995) 373.
24. D. Kutasov and E. Martinec, `hep-th/9612102`.
25. C. Hull and P. Townsend, `hep-th/9410167`; *Nucl. Phys.* **B438** (1995) 109.
26. A. Dabholkar, G. Gibbons, J. Harvey, and F. Ruiz-Ruiz, *Nucl. Phys.* **B340** (1990) 33.
27. R. Gregory, J. Harvey, and G. Moore, `hep-th/9708086`.
28. T. Banks and L. Susskind, `hep-th/9511193`; *Phys. Rev.* **55** (1996) 1677.
29. U. Danielsson and G. Ferretti, `hep-th/9610082`; S. Kachru and E. Silverstein, 9612162; L. Motl, `hep-th/9612198`; N. Kim and S.-J. Rey, `hep-th/9701139`; T. Banks and L. Motl, `hep-th/9703218`; D. Lowe, `hep-th/9704041`; S.-J. Rey, `hep-th/9704158`; P. Hořava, `hep-th/9705055`.
30. A. Giveon and A. Shapere, `hep-th/9203008`; *Nucl. Phys.* **B386** (1992) 43.
31. E. Martinec, `hep-th/9706194`.
32. S. Elitzur, A. Giveon, D. Kutasov, and E. Rabinovici, `hep-th/9707217`.
33. D. Friedan, E. Martinec, and S.H. Shenker, *Nucl. Phys.* **B271** (1986) 93.
34. S. Giddings and E. Martinec, *Nucl. Phys.* **B278** (1986) 91.
35. N. Marcus, `hep-th/9207024`; *Nucl. Phys.* **B387** (1992) 263.
36. H. Ooguri, Y. Oz, and Z. Yin `hep-th/9606112`; *Nucl. Phys.* **B477** (1996) 407.
37. K. Lee and P. Yi, `hep-th/9702107`.
38. J. Harvey and G. Moore, `hep-th/9609017`.
39. T. Banks, M. Douglas, N. Seiberg, and S. Shenker, *Phys. Lett.* **238** (1990) 279; for a review, see e.g. P. Ginsparg and G. Moore, `hep-th/9304011`.

BRANES WITHIN BRANES†

MICHAEL R. DOUGLAS
Dept. of Physics and Astronomy
Rutgers University
Piscataway, NJ 08855-0849
and
I.H.E.S.
Bures-sur-Yvette, France 91440

1. Introduction

Superstring theory has a rich spectrum of solitonic states, and over the last years much has been learned about their important roles in the theory, in strong-weak coupling duality and in resolving singularities [1]. The most important are the BPS states (those in reduced multiplets of supersymmetry) as one can make exact statements about them even at strong coupling, and thus dualities can predict their existence. Many dualities exchange elementary string states with states having Ramond-Ramond gauge charge, which strongly suggests that the two should be considered equally fundamental.

In the remarkable work [3] Polchinski showed that the Dirichlet branes of the Type I and II superstrings are RR charged BPS states. A Dirichlet (D) brane [2] is simply an allowed endpoint for open strings and these objects are much more tractable than previously studied string solitons. At low energies, a D-brane is described by an effective world-volume theory, essentially dimensionally reduced supersymmetric Yang-Mills theory. [4, 11] The difficulty of the non-renormalizability of p-brane theories for $p > 1$ is resolved: these theories are the low-energy limits of UV-finite open string theories. (Clearly it will be very interesting to elucidate this point. An old proposal of Green (for $p = 1$) was in this spirit. [5])

Many of the objects required by string duality have now been identified as D-branes, such as the heterotic string soliton [17]. In another remarkable recent work [19] the 5-brane of type I superstring theory was studied by Witten, who proposed that it was the zero size limit of a gauge instanton. The instanton 5-branes had been studied as field theory solitons and as conformal field theories (in the dual heterotic theory). [9] which by the anomalous Bianchi identity $dH = \mathrm{tr}\ R^2 - \mathrm{tr}\ F^2$ is an

†This work appeared originally as `hep-th/9512077`.

L. Baulieu et al. (eds.), Strings, Branes and Dualities, 267–275.

object with RR charge. The complete solution has metric and dilaton dependence as well, and in the limit of small instanton scale size a long tube develops with large string coupling. Witten's treatment provides a simple description of the resulting physics: the zero size limit is an $Sp(1)$ gauge theory in the world-volume.

This world-volume theory can have non-perturbative physics of its own, and in this note we study the properties of its instanton solution, a 1-brane confined to the 5-brane, and show that it can be described as a bound state of the Dirichlet 1-brane with the 5-brane. We also mention type II generalizations to $p-4$ branes within p-branes.

2. RR fields in superstring theory

We begin by reviewing the RR fields of superstring theory and their generalized Green-Schwarz couplings. The massless RR gauge fields are naturally combined into a bispinor potential $C_{\alpha\tilde{\beta}}(x)$ and field strength $H_{\alpha\tilde{\beta}}$ in the $(16+16')\otimes(16+16')$ Dirac spinor representations of the algebra of fermion zero modes ψ_0^μ and $\tilde{\psi}_0^\mu$. This is equivalent to a sum of differential forms, e.g. $C = \sum C^{(p+1)}_{\mu_1\ldots\mu_{p+1}}\Gamma^{(p+1)}_{\mu_1\ldots\mu_{p+1}}$, with $\delta_{\alpha\tilde{\beta}}$ as the 0-form and using [6, 7] $b^{*\mu} \equiv dx^\mu \equiv \frac{1}{2}(\psi_0^\mu + i\tilde{\psi}_0^\mu)$ and $b^\mu \equiv *dx^\mu * \equiv \frac{1}{2}(\psi_0^\mu - i\tilde{\psi}_0^\mu)$. The on-shell conditions for the RR vertex operator $\hat{H}_{\alpha\tilde{\beta}}S^\alpha\tilde{S}^{\tilde{\beta}}\ \mathrm{exp}ikX$ are separate Dirac equations $b^{*\mu}\partial_\mu\hat{H} = b^\mu\partial_\mu\hat{H} = 0$, whose sum and difference are $d\hat{H} = *d*\hat{H} = 0$. Thus $\hat{H}$ is an RR field strength and the free equation of motion and Bianchi identity appear on the same footing. The GSO projections $(-1)^{F+\tilde{F}}H = (-1)^{\tilde{F}}H = H$ will restrict H to contain only even (for IIa) or odd (I and IIb) forms, and impose $\hat{H} = *\hat{H}$. In type I theory we also impose $\Omega\hat{H} = \hat{H}$, where the twist operator Ω acts as $H \to i^{r-3}H$, keeping only $H^{(3)}$ and its dual $H^{(7)}$.

The tension and RR charge of a D-brane can be determined from one-point functions on a disk with boundary on the D-brane. By usual string coupling counting these would be $O(1/\lambda)$, correct for the tension but not the charge. This is because the conserved RR charges are not surface integrals of $\hat{H}$. A simple way to see this is to add a linear dilaton background $D \propto X$. [3] The world-sheet supercurrents become $G = \psi\cdot\partial X + (\partial D)\cdot\partial\psi$ and the Bianchi identity becomes

$$d\hat{H} - \partial_\mu D dx^\mu \hat{H} = e^{-D}de^{-D}\hat{H} = 0. \tag{2.1}$$

(We also incorporated the usual $e^{-2D} = 1/\lambda^2_{str}$ for an inverse propagator. This is a bit cavalier but will suffice for our purpose; see [13] for a better treatment.) Thus the RR charges are surface integrals of $H \equiv e^{-D}\hat{H} \equiv dC$.

A D-brane acts as an electric $p+1$-form source in (2.1),

$$dH = \frac{\mu_p}{\alpha} * J^{(p+1)} \tag{2.2}$$

where $J^{(p+1)}$ is the $p+1$-form tangent to the brane. μ_p is the charge and the RR action is $\sum_{p\leq 3}\frac{1}{2}\alpha\int H^{p+2} * H^{p+2}$. (The common origin of the RR fields makes such a convention natural. We will take $\alpha' = 1$, $\alpha = 1/2\pi$, and $\mu_p = (2\pi)^{3-p}$).

This was inferred from the closed string exchange in [3]; it can also be computed along the lines of the open string tadpole computation in [7, 8]. The result in the massless sector can be derived from the result $\delta_{\alpha\tilde{\beta}}$ for a free boundary ($p = 9$) by changing to Dirichlet boundary conditions in the orthogonal directions, which takes $\tilde{\psi}^\mu \to -\tilde{\psi}^\mu$. The allowed values of p in a particular string theory are those for which the appropriate form exists, so in type I theory 1, 5 and 9-branes exist.

The result (2.2) was generalized to a constant background gauge field by Li [12], using the open string results of [8, 7]. The zero momentum limit in the massless sector is particularly simple and can be derived simply by letting the boundary action for the gauge field $\exp \int d\sigma\ F_{\mu\nu}\psi^\mu\psi^\nu$ act on the boundary state. Since the gauge field vertex operators have μ, ν tangent to the brane they are just $F_{\mu\nu}b^{*\mu}b^{*\nu} = F_{\mu\nu}dx^\mu dx^\nu$ and the result is

$$dH = \frac{\mu_p}{\alpha}\mathrm{Str}\ e^F * J^{p+1}. \tag{2.3}$$

As argued in [8] this generalizes to the non-abelian case with $F = dA + A^2$. This implies a world-volume action

$$S = \int T_p e^{-D}\sqrt{\det(h+F)} + \mu_p\ C \wedge \mathrm{Str}\ e^F. \tag{2.4}$$

The 9-brane result includes both the $dH = \mathrm{tr}\ F^2$ term in the anomalous Bianchi identity and the $C^{(2)}\mathrm{tr}\ F^4$ coupling required by the Green-Schwarz mechanism. The generalization shows the close connection of these with the field theory gauge anomaly.

Using T-duality, it is just as easy to put in the dependence on the transverse modes X^i of a general p-brane:

$$dH = \frac{\mu_p}{\alpha}\mathrm{Str}\ e^{F+dX^i b_i} * J^{p+1}. \tag{2.5}$$

One role of this term is to implement Lorentz covariance: if we take $X^i = R^i_\alpha x^\alpha$, we get a rotated or boosted D-brane, and the boundary state is proportional to $\prod d(X^i + R^i_\alpha x^\alpha)$. In the spirit of Bachas [16], we could also argue in the other direction: the generalized Green-Schwarz coupling (and by extension the field theoretic gauge anomaly) follows simply from T-duality and Lorentz invariance!

In type II string theories, there is a $U(1)$ factor in the D-brane gauge symmetry, which will mix with the NS 2-form tensor to preserve its gauge invariance $\delta B = -d\Lambda$, modifying (2.3) as $F \to F - B$. [22, 11] The $B^k C$ couplings produced by this should be those predicted by Morrison from considerations of mirror symmetry. [20]

Another test of (2.3) is that localized gauge field solutions in a D-brane should be sources with correctly quantized charge. Now terms in the expansion of Str $\exp iF/2\pi$ are the Chern characters (which appear in the index theorem) which will indeed integrate to integers on localized gauge fields such as instantons. The 2π's needed are reproduced by the dependence of the μ_p on p.

There is a subtlety in that we must insist on the word 'localized' in the above statement. For example, the dyonic 1-branes of IIb theory are sources of C^0 with charge λm, which is not quantized in the unit set by the D-instanton charge. Now there is no paradox yet as the Dirac argument [21] does not apply to a zero-form potential, but clearly the same phenomenon could make higher branes non-quantized sources of the higher potentials. As a test case, consider a flux tube, say $F_{01} \neq 0$ in a 5-brane, which we take to satisfy Gauss' law by simply requiring $\partial_0 F_{01} = \partial_1 F_{01} = 0$. This will be a source of C^4, but only components with an index in the 01 dimensions, for which the Dirac argument again does not go through.

Because of (2.3), an instanton in a p-brane will be an magnetic source of $H^{(12-p)}$ (or electric source of $H^{(p-2)}$). This generalizes the result of [9], for the solitonic 5-brane constructed from a type I instanton ($p = 9$), to general p. It motivates a generalization of Witten's proposal: the zero scale size limit of the instanton in a p-brane will be a Dirichlet $p-4$ brane, bound to the p-brane. We will study this in detail for $p = 5$: the instanton in Witten's 5-brane, which on general grounds is a 1-brane, will turn out in the zero size limit to become a bound state of the Dirichlet 1-brane with the Dirichlet 5-brane.

3. Branes in type I theory

This brings us to the structure of the 1 and 5-branes of type I theory, analyzed by Polchinski and Witten. [17, 19] First consider the Dirichlet 1-brane, which will have electric $C^{(2)}$ charge. This and its string tension $T \sim 1/\lambda$ make it a candidate for the string soliton required by type I-heterotic duality. The first point to check is its supersymmetry, which is half that of the type I theory: in other words $16 = 8^{+} + 8'^{-}$ under $SO(8) \times SO(1,1)$ of which the first part is unbroken, so the world-sheet theory will have a chiral $(0,8)$ supersymmetry.

To make a more detailed comparison of the world-sheet theories, we must realize that there are two important modifications from the SYM theory (2.4). First, we have an additional projection on the world-sheet theory, $\Omega = 1$. With a single 1-brane this produces $SO(1)$ gauge theory, which has no massless gauge sector. On the other hand, Ω acts with opposite sign on the vertex operators $\partial_\tau X$ and $\partial_\sigma X$, so the transverse coordinates X survive the projection. The Ramond sector contains their superpartners ψ, a spinor 8 of $SO(8)$. These are chiral on the world-sheet, both because the supersymmetry is chiral, but as can also be seen by considering the action of Ω.

Second, we have an additional sector of open strings (the 'DN sector') with mixed boundary conditions, one on the 1-brane and one free (i.e. on a 9-brane). A free superfield with mixed Dirichlet-Neumann boundary conditions has moding shifted by $1/2$ and ground state energy shifted by $1/8$, so the NS sector has ground state energy $-1/2+8/8 > 0$ and no massless states, while the R sector is a $SO(1,1)$ spinor λ^I in the 32 of $SO(32)$, chiral after the GSO projection. We thus reproduce exactly the zero modes of the light-cone heterotic string.

We turn to Dirichlet 5-branes. One such object can be obtained by T-duality from the 9-brane, but Witten's 5-brane [19] is different: it has $Sp(1)$ Chan-Paton factors. In other words, the open strings ending on the 5-brane have a two-valued index A, and the twist Ω acts on this index as $T : v_A \rightarrow \epsilon_{AA'} v_{A'}$. The opposite sign for the displacement vertex operator makes it a two-index antisymmetric tensor of $Sp(1)$, i.e. a scalar, so this also describes a single object, now with charge 2. This charge is consistent with the normalizations in (2.3) if we remember that a minimally embedded $SO(N)$ instanton has charge $1/8\pi^2 \int \text{tr } F^2 = 2$.

The type I supersymmetry now decomposes under $SO(5,1) \times SO(4)$ as $16 = (4,2)+(4',2')$. Taking the first half as unbroken, we get a single Weyl $d = 6$ spinor, so the world-volume theory will contain $N = 1$, $d = 6$ $Sp(1)$ SYM theory and a singlet hypermultiplet X. The Ω projection acts on the gauginos as $\Gamma^1\Gamma^2\Gamma^3\Gamma^4 T$, correlating their symmetry as $Sp(1)$ tensors with their $d = 6$ chirality, as required by supersymmetry.

The DN sector is a $(2,32)$ of $Sp(1) \times SO(32)$. The NS sector has ground state energy $-1/2 + 4/8 = 0$ so contributes six dimensional scalars, which evidently must be in hypermultiplets. Indeed the fermion zero modes ψ_0^i make these states a spinor $(1,2)$ of $SO(4)$ (after GSO). Since the two ends of the string are different, the Ω projection relates these fields with their charge conjugates. It includes an ϵ_{AB} on the Chan-Paton factors, and an $\epsilon_{\alpha\beta}$ to undo the $SO(4)$ charge conjugation, leaving 'half-hypermultiplets' whose scalars satisfy

$$H^*_{\alpha AI} = \epsilon_{AB}\epsilon_{\alpha\beta}H_{\beta BI}. \tag{3.1}$$

The R sector is a chiral $d = 6$ spinor with $\chi^{AI*} = C\chi^{BI}\epsilon_{AB}$.

The scalars $\phi_{\alpha AI}$ provide flat directions which break $SO(32)$. Indeed Witten showed that this moduli space coincides precisely with the ADHM description of instanton moduli space. It is the space of solutions of the D-flatness conditions (equivalent to the D and F conditions of the more familiar $N = 2$, $d = 4$ case)

$$0 = \sum_I H^*_{\alpha AI}\sigma^i_{\alpha\beta}t^j_{AB}H_{\beta BI}. \tag{3.2}$$

A way to see the content of these conditions is to write the ϕ as $SO(4)$ vectors $H_{\alpha AI} = \sigma^\mu_{\alpha A}H^\mu_I$. The reality condition (3.1) then just states that the H^μ_I are real. Since the 9 D-flatness conditions form an $SO(4)$-covariant set, they must be expressable as $H^\mu \cdot H^\nu = c\delta^{\mu\nu}$, and the four vectors $H^\mu/\sqrt{c}$ form an orthonormal frame. Thus the moduli space is a Grassmannian $SO(32)/SO(28) \times SO(4)$ times a $\mathbf{R}^4$ including scale size and an overall $SU(2)$ gauge rotation. By considering k coincident 5-branes, the construction generalizes to k-instanton moduli space. The effective theory is $Sp(k)$ gauge theory with matter $X_{[AB]}$ and $H_{\alpha AI}$.

4. Branes within branes

We now turn to our main example, and the question: Now that we know a 5-brane has $Sp(1)$ gauge symmetry, what are the properties of a instanton solution of this gauge theory? It will be a 1-brane embedded in the 5-brane, and it will break half of the $d = 6$ supersymmetry. This decomposes under $SO(4) \times SO(1,1)$ as the complex spinors $2^+ + 2'^-$ and we see that the world-sheet theory will have $(0,4)$ supersymmetry. Its tension is proportional to the instanton action in the 5-brane, of order $1/\lambda$. Finally, from (2.3), we expect it to have the electric $H^{(3)}$ charge of a Dirichlet 1-brane.

Before exploring the possibility that this is a D-brane, let us work out the field content of the solitonic 1-brane of the 5-brane theory. Each zero mode of the theory dimensionally reduced to $N = 2$, $d = 4$ will provide a world-sheet field. There is the usual 5-dimensional moduli space of $Sp(1)$ instantons, position and a scale size, giving transverse fluctuations and an additional scalar, and we include global $Sp(1)$ gauge rotations as well for 8 bosonic zero modes. The gauginos have 8 zero modes which are partners under the broken supersymmetry of these bosonic modes. Finally, the fermions in the matter hypermultiplets will provide 32 zero modes (these are half-flavors of $SU(2)$ in $d = 4$), while there are no covariantly constant scalars in the instanton background.

Both the charge and the counting of zero modes tell us that if this 1-brane is a D-brane, it must be a bound state between the 5-brane and the 1-brane we already discussed! Let us consider a state containing a 1-brane embedded in the 5-brane. The space-time symmetry is now $SO(4)_I \times SO(4)_E \times SO(1,1)$ with the factor 'I' inside the 5-brane and the factor 'E' external to both. The 5-brane supersymmetry will decompose as $(2,2)^+ + (2',2)^-$, with the first factor unbroken by the one-brane. Besides the fields (X, ψ) and λ, there is an open string sector DN' with strings stretching from the 1-brane to the 5-brane, producing fields in the 2 of $Sp(1)$. The NS sector again has ground state energy zero and will provide world-sheet scalars Y in the $(2,1)$ of $SO(4)_I$. The R sector has fermion zero modes both in the world-sheet dimensions and in the 4 dimensions transverse to both 1-brane and 5-brane. Thus it provides fermions ρ in the $(1,2)^+$ and $\bar{\rho}$ in the $(1,2')^-$.

The proposal is that an instanton in the 5-brane is just this object with the fields Y turned on, and that doing this will make the components of (X, ψ) perpendicular to the 5-brane massive, binding it to the 5-brane. The 32 zero modes are present, while the transverse displacements will be a subset of (X, ψ) and the gaugino zero modes are ρ. Now the DN' states have masses $m^2 \sim (X_1 - X_5)^2$ (the displacement of the 1-brane perpendicular to the 5-brane), so there must be a potential term $Y^2(X_1 - X_5)^2$ as well as mass terms $\rho\bar{\rho}(X_1 - X_5)$. At $X_1 = X_5$ the Y's become massless, and assuming there is no potential for Y (which we will argue below), we can move out on the branch $Y \neq 0$, giving a mass to the transverse components of $X_1 - X_5$. We also need their fermionic partners to become massive, so there must also be couplings $\psi\bar{\rho}Y$. In fact all of these couplings are related by $(0,4)$ supersymmetry.

By bringing k 1-branes and N 5-branes together, we get an $SO(k) \times Sp(N)$ gauge theory, whose vacua should parameterize the moduli space of k instantons in $Sp(N)$ gauge theory. The matter content is a symmetric tensor $X_{(ij)}$ of $SO(k)$, their left-moving partners $\psi_{(ij)}$ in the $(2,2) + (2',2')$, right-moving antisymmetric $\bar{\psi}_{[ij]}$ in the $(2,2') + (2',2)$ and λ_{iI} in the $(k,1,32)$, scalars Y in the $(k,2N,1)$ and the 2 of $SO(4)_I$, and complex fermions in the $(k,2N,1)$ and $(1,2)^+ + (1,2')^-$. Turning on enough Y to bind all of the 1-branes, the moduli space will have the correct dimension $4kN + 2k(k+1) - 2k(k-1) = 4k(N+1)$, if the potential is the same one we would have in $d = 6$, $\mathcal{N} = 1$ $SO(k)$ gauge theory, a point we will argue shortly. Leaving out the λ's (which are obvious zero modes of χ) there are $4k(k+1) + 4kN$ left-moving and $4k(k-1) + 4kN$ right-moving fermions, out of which we need $4k(N+1)$ left movers to stay massless, $4k^2$ left and right to pair up, and the remaining $4k(N-1)$ right movers must correspond to additional zero modes in a k-instanton background of the 5-brane theory. Indeed this is the right number to come from the antisymmetric tensor.

We needed the potential for Y to be that of a $d=6$, $\mathcal{N}=1$ supersymmetric gauge theory, but the 1-brane is a $d=2$, $(0,4)$ theory. We argue as follows: the DN vertex operators for Y are twist fields in the four dimensions transverse to the 1-brane and internal to the 5-brane, and trivial in the other six dimensions. In fact they are exactly the same DN vertex operators as in Witten's 5-brane, there twist operators in the four dimensions transverse to the 5-brane and trivial in the other dimensions. Thus their correlation functions must be identical, and the potential will be equal to that of the $d=6$, $\mathcal{N}=1$ theory.

Rather than make this more explicit here, we argue that a related $d=6$, $\mathcal{N}=1$ Lagrangian will also describe the instanton moduli space. First, it is worth noting that the whole set up can be considerably generalized in the type II theories, which contain all even (IIa) or odd (IIb) p-branes. By bringing k branes together we get a world-volume theory with $U(k)$ symmetry, and now (2.3) tells us that an instanton in this theory has the right charge to be a Dirichlet $p-4$-brane. This suggests that the dimension-dependent details in the above discussions must not be essential to the basic phenomenon, that a D-brane can be equivalent to an instanton.

This also suggests that we do the comparison of the moduli space with the ADHM moduli space in the highest dimension possible, which will give the strongest constraints on the world-volume theory. Let us consider the type I $Sp(N)$ and $U(N)$ theories. Although choices other than $SO(32)$ are not completely sensible string theories, the problems come in at one open-string loop (anomalies) or equivalently closed string tree level (tadpoles). The classical open string sector will be a sensible 6-dimensional supersymmetric effective theory, which we can use to determine the moduli spaces for the 5-branes. This moduli space will then generalize to gauge fields in any p-brane – the idea is that no matter what dimension brane we consider, the essential structure of the theory is its content in the four dimensions transverse to the branes. Using this, we find that the $Sp(N)$ k-instanton moduli space will be reproduced by an $SO(k)$ gauge theory containing $(k,2N)$ half-hypermultiplets and a symmetric tensor of $SO(k)$, and the $U(N)$ moduli space will be reproduced by a $U(k)$ gauge theory containing hypermultiplets in the (k,N), adjoint and singlet. Indeed all this fits with the description of the ADHM moduli space given in section 4 of [23].

Other details of our explicit example will clearly generalize. For example, by adding additional branes, new open string sectors will produce a fairly general set of fields on the original p-brane, and it must be that the zero modes of these fields in the instanton background are reproduced by the open strings to the lower dimensional brane. The discussion of the transverse moduli should also be simpler in a more general context. We plan to give a more general treatment of these points elsewhere.

Let us mention another aspect of the equivalence between gauge soliton and D-brane, for concreteness using Witten's original example. [1] Since we have two

[1]This idea was developed in discussion with T. Banks, M. Berkooz and N. Seiberg.

descriptions of the same object, the two sources

$$dH^{(3)} = -\frac{1}{8\pi^2}\text{tr } F^2 - 2 * J^6 \tag{4.1}$$

are really indistinguishable in string theory. In ordinary gauge theory, gauge fields with different instanton number are topologically distinct, so asserting that the limit $F = 0$ is connected to an instanton implies a sort of 'topology change' which deserves better understanding. Consider an S^3 surrounding the instanton (at large distance), and consider the enclosed magnetic charge $Q = \int H$. One contribution to this is the integral of the Chern-Simons form, which is not gauge invariant. It measures the instanton number if we define the gauge field using a single patch, but by a large gauge transformation on the sphere (singular in the interior) we can change it by an integer. Evidently this large gauge transformation in string theory must create a B field on the sphere to keep Q constant, one which could be produced by a D-brane source in the interior. Perhaps this can be extended to all of space-time, and we can think of the equivalence between instanton and D-brane as implemented by a large gauge transformation.

Another speculation which this suggests is that in a higher-dimensional formulation ("M-theory"), the instanton in $d = 10$ (e.g. in the boundary theory of [18]) might be a bound state of a fundamental 5-brane with the boundary.

In conclusion, the equivalence of instantons (and possibly other gauge'q topological solitons) with D-branes is a very general feature of superstring theory, which surely is an important clue to its underlying geometry.

Acknowledgements

I would like to thank T. Banks, M. Berkooz, R. Leigh, J. Polchinski and N. Seiberg for enlightening discussions.

This work was supported in part by DOE grant DE-FG05-90ER40559, NSF PHY-9157016 and the A. P. Sloan Foundation.

References

1. J. H. Schwarz and A. Sen, Phys. Lett. **B312** (1993) 105;
   C. Hull and P. Townsend, Nucl. Phys. **B438** (1995) 109;
   E. Witten, Nucl. Phys. **B443** (1995) 85;
   A. Strominger, Nucl. Phys. **B451** (1995) 96.
2. J. Dai, R. G. Leigh and J. Polchinski, Mod. Phys. Lett. **A4** (1989) 2073.
3. J. Polchinski, "Dirichlet branes and Ramond-Ramond charges," hep-th/9510017.
4. R. G. Leigh, Mod. Phys. Lett **A4** (1989) 2767.
5. M. B. Green, Nucl. Phys. **B293** (1987) 593-611.
6. M. B. Green, J. Schwarz and E. Witten, Superstring Theory vol. II, section 14.3.4, pp.387-390.
7. J. Polchinski and Y. Cai, Nucl. Phys. **B296** (1988) 91.
8. C. G. Callan, C. Lovelace, C. R. Nappi and S. A. Yost, Nucl. Phys. **B308** (1988) 221.
9. C. Callan, J. Harvey and A. Strominger, Nucl. Phys. **B367** (1991) 60.
10. E. Witten, "Sigma Models and the ADHM Construction of Instantons," hep-th/9410052.

11. E. Witten, "Bound States of Strings and p-branes," hep-th/9510135.
12. M. Li, "Boundary States of D-Branes and Dy-Strings," hep-th/9510161.
13. M. Li, "Dirichlet Boundary State in Linear Dilaton Background," hep-th/9512042.
14. A. Sen, "U-duality and Intersecting D-branes," hep-th/9511026.
15. C. G. Callan and I. R. Klebanov, "D-Brane Boundary State Dynamics," hep-th/9511173.
16. C. Bachas, "D-brane Dynamics," hep-th/9511043
17. J. Polchinski and E. Witten, "Evidence for Heterotic-Type I String Duality," hep-th/9510169.
18. P. Horava and E. Witten, "Heterotic and Type I String Dynamics from Eleven Dimensions," hep-th/9510209.
19. E. Witten, "Small Instantons in String Theory," hep-th/9511030.
20. D. Morrison, "Mirror Symmetry and the Type II String," hep-th/9512016.
21. R. Nepomechie, Phys. Rev. **D31** (1985) 1921;
C. Teitelboim, Phys. Lett. **B167** (1986) 63, 69.
22. E. Cremmer and J. Scherk, Nucl. Phys. **B72** (1974) 117.
23. N. H. Christ, E. J. Weinberg and N. K. Stanton, Phys. Rev. **D18** (1978) 2013-2025.

REVIEW OF MATRIX THEORY

D. BIGATTI[†] AND L. SUSSKIND
Stanford University

Lecture zero: Introduction

Matrix theory [1] is a nonperturbative theory of fundamental processes which evolved out of the older perturbative string theory. There are two well-known formulations of string theory, one covariant and one in the so-called light cone frame [2]. Each has its advantages. In the covariant theory, relativistic invariance is manifest, a euclidean continuation exists and the analytic properties of the S matrix are apparent. This makes it relatively easy to derive properties like CPT and crossing symmetry. What is less clear is that the theory satisfies all the rules of conventional unitary quantum mechanics. In the light cone formulation [3], the theory is manifestly hamiltonian quantum mechanics, with a distinct non-relativistic flavor. There exists a space of states and operators which is completely gauge invariant and there are no ghosts or unphysical degreees of freedom. Among other things, this makes it easy to identify the spectrum of states. Matrix theory is also formulated in the light cone frame and enjoys the same advantages and disadvantages as LCF string theory. Unlike perturbative string theory, matrix theory is capable of addressing many of the nonperturbative questions raised by string theory and quantum gravity.

The need for a nonperturbative theory became very apparent a few years ago from two lines of research: black holes and duality. In the case of black holes, it was clear to a few of us that explaining the Bekenstein Hawking entropy would require a microscopic description involving degrees of freedom more fundamental than those that are available in ordinary general relativity [4]. Furthermore, although string theory offered some insight, the weakly coupled theory could never completely solve this problem since black holes are fundamentally non perturbative objects. In addition, the so-called information paradox of Hawking suggested that there is something very naive and wrong with our usual ideas of locality that we inherited from quantum field theory. As we shall see, matrix theory does not seem to be

[†]On leave of absence from Università di Genova, Italy.

L. Baulieu et al. (eds.), Strings, Branes and Dualities, 277–318.

a space-time quantum field theory even approximately, except in the low energy perturbative regime.

The need for a more powerful approach was made even clearer when it was discovered that the various different string theories are really part of a vast "web" of solutions of a single 11-dimensional theory called M-theory [5]. The web is parametrized by the "moduli" describing the many different compactifications. The perturbative string theories describe remote corners of the moduli space, but most of the space is beyond their reach. Each string corner has a spectrum of light weakly coupled states such as IIa, IIb, I or heterotic strings. Perhaps the right way to say it is that all of the objects are present in all of moduli space, but in each corner some particular set becomes light and weakly coupled. However, perturbation theory in any of the corners can not detect the other corners and the web that connects them.

It is evident that any fundamental degrees of freedom which can describe all these stringy excitations as well as 11-dimensional gravitons, membranes, 5-branes, D-branes and black holes must be very special and unusual. An even more surprising indication that the underlying degrees of freedom are unlike anything previously encountered is their ability to grow new dimensions of space. The simplest example is the compactification of the 11-dimensional M-theory on a two torus [6]. This obviously leads to a theory with 9 noncompact dimensions. Now let the torus shrink to zero size. The surprise is that in this limit a new 10th noncompact dimension appears. Nothing like this was ever imagined in older ideas based on quantizing the classical gravitational field.

Thus it is very satisfying that a simple system of degrees of freedom which does all of the above has been identified. In the following set of lectures we will describe the "matrix theory" of this system of protean degrees of freedom and how it accomplishes the necessary feats. The lectures are self contained and do not require an extensive knowledge of string theory. However they far from exhaust the important and interesting things that have been learned in the last year.

We would like to warn the reader that here and there an attempt has been made to keep track of numerical factors, but they should not be expected to be completely consistent.

Lecture 1: The light cone frame

In this lecture the light cone method will be reviewed [3] . The dimension of space-time is D. The space-time dimensions are labeled $X^{\mu} = (t, z, X^1, \cdots, X^{D-2})$. One of the spatial coordinates z has been singled out. It is called the longitudinal direction and the space $(X^1, \cdots, X^{D-2})$ is called the transverse plane. Now introduce two light-like coordinated $X^{\pm}$ to replace (t, z) :

$$X^{\pm} = \frac{t \pm z}{\sqrt{2}} \tag{1.1}$$

In the light cone frame X^+ is used as a time variable and the space $X^+ = 0$ is the surface of initial conditions. A quantum state in the Schrödinger picture describes data on a surface $X^+ =$ constant. A Hamiltonian is used to propagate the quantum state from one value of X^+ to another:

$$|X_2^+\rangle = e^{iH(X_2^+ - X_1^+)}|X_1^+\rangle \tag{1.2}$$

Let us consider the form of H for an arbitrary system with center of mass energy M. The system need not be a single particle. Let us also introduce conjugate momenta $P_\pm = \frac{(P_t \pm P_z)}{\sqrt{2}}$. The "on shell" condition is

$$P_\mu P^\mu = M^2 \tag{1.3}$$

or

$$2P_+P_- - P_i^2 = M^2 \tag{1.4}$$

Now let us define P_+ to be the hamiltonian H. We find

$$H = \frac{P_i^2}{2P_-} + \frac{M^2}{2P_-} \tag{1.5}$$

Eq. (1.5) has a nonrelativistic look to it. Let's compare it with the general expression for energy in a Galilean invariant theory in $D-2$ spatial dimensions.

Let μ and p be the total mass and momentum of the system and let the Galilean invariant internal energy be called $U/2\mu$. Then the total energy is

$$H = \frac{p^2}{2\mu} + \frac{U}{2\mu} \tag{1.6}$$

Thus, if we identify the nonrelativistic mass of a system with its P_- and U with the invariant M^2, the formulas are the same. Is this a coincidence or there is an underlying reason for this? The answer is that there is a subgroup of the Poincaré group, relevant for light cone physics, which is isomorphic to the Galilean group.

Let us review the Galilean group. Classically it is generated by the following:

p	spatial translations
h	time translations
l_{ij}	angular momentum in the ij plane
$\mu x^i_{C.M.}$	Galilean boosts.

To see that $\mu x^i_{C.M.}$ generates boosts consider the action of $exp(iV \cdot \mu x^i_{C.M.})$ on p. From the fact that $x_{C.M.}$ and p are conjugate variables

$$e^{[iV \cdot \mu x^i_{C.M.}]}\, p\, e^{-[iV \cdot \mu x^i_{C.M.}]} = p + \mu V \tag{1.7}$$

which has the form of a boost with velocity V.

One more generator belongs to the Galilean group. The commutation relation

$$[\mu x^i_{C.M.}\,,\, p_i] = i\mu \tag{1.8}$$

indicates that the mass μ must be included to close the algebra. The mass commutes with the other generators, so it is a central charge. We leave it as an exercise for the reader to compute the remaining commutation relations.

Now consider the Poincaré generators of D dimensional space-time:

P_i	transverse translations
$P_+ = H$	X^+ translations
P_-	X^- translations
L_{ij}	transverse rotations
L_{iz}	rotations in (X^i, z) plane
K_{0z}	Lorentz boosts along z
K_{0i}	Lorentz boosts along X^i

Now define the following correspondence:

$$\begin{aligned} P_i &\longleftrightarrow p_i \\ P_+ = H &\longleftrightarrow h \\ P_- &\longleftrightarrow \mu \\ L_{ij} &\longleftrightarrow l_{ij} \\ B_i = \tfrac{K_{0i}+L_{0i}}{\sqrt{2}} &\longleftrightarrow \mu x^i_{C.M.} \end{aligned}$$

Using the standard Poincaré commutation relations one finds an exact isomorphism between the Galilean group and the Poincaré subgroup generated by P_i, H, P_-, L_i and B_i. Thus it follows that relativistic physics in the light cone frame is Galilean invariant and must have all the properties which follow from this symmetry. For example the Hamiltonian must have the form

$$H = \frac{P_i^2}{2P_-} + E_{\text{internal}} \tag{1.9}$$

where E_{internal} is Galilean invariant.

The Poincaré generator K_{0z} is not part of the Galilean subgroup, but it gives important information about E_{internal}. Using the commutation relations

$$[K_{0z}, H] = H \tag{1.10}$$

$$[K_{0z}, P_-] = -P_- \tag{1.11}$$

we see that the product HP_- is boost invariant. It follows that E_{internal} has the form $M^2/(2P_-)$ with M^2 being invariant under Galilean boosts and Lorentz transformations. In other words H must scale like $1/P_-$ under a rescaling of all P_- in the system. It is interesting to think of rescaling the P_- axis as a kind of scale transformation. The invariance of physics under longitudinal boosts is understood as the existence of a renormalization group fixed point [7] .

Let us now consider the formulation of quantum field theory in the light cone frame. Let ϕ be a scalar field with action

$$A = \int d^D X \left\{ \frac{\partial_\mu \phi \partial^\mu \phi}{2} - \frac{m^2 \phi^2}{2} - \lambda \phi^3 \right\} \tag{1.12}$$

In terms of light cone coordinates this becomes

$$A = \int dX^{+} \mathcal{L} \tag{1.13}$$

$$\mathcal{L} = \int dX^{-} dX^{i} \left\{ \dot{\phi}\, \partial_{-}\phi - \frac{(\partial_i \phi)^2}{2} - m^2\phi^2 - \lambda\phi^3 \right\} \tag{1.14}$$

where the notation $\dot{\phi} = \partial_{+}\phi$ has been used. We can now identify the canonical momentum π conjugate to ϕ

$$\pi = \frac{\delta \mathcal{L}}{\delta \dot{\phi}} = \partial_{-}\phi \tag{1.15}$$

from which we deduce the equal time commutation relations

$$[\phi(X^i, X^-), \phi'(Y^i, Y^-)] = i\,\delta(X^i - Y^i)\,\delta(X^- - Y^-) \tag{1.16}$$

We may Fourier transform the field ϕ with respect to X^-:

$$\phi(X^i, X^-) = \int_0^{\infty} dk_- \, \frac{\phi(X^i, k_-)}{\sqrt{2\pi|k_-|}} e^{ik_- X^-} +$$

$$+ \frac{\phi^*(X^i, k_-)}{\sqrt{2\pi|k_-|}} e^{-ik_- X^-} \tag{1.17}$$

The Fourier coefficients ϕ, ϕ^* have the non relativistic commutation relations

$$[\phi(X^i, k_-), \phi(Y^i, l_-)] = 0 \tag{1.18}$$

$$[\phi^*(X^i, k_-), \phi^*(Y^i, l_-)] = 0 \tag{1.19}$$

$$[\phi(X^i, k_-), \phi^*(Y^i, l_-)] = \delta(X^i - Y^i)\,\delta(k_- - l_-) \tag{1.20}$$

The Hamiltonian following from (1.14) has the form

$$H = \int dk_- dX^i \frac{\nabla\phi^* \cdot \nabla\phi + m^2\phi^*\phi}{2k_-} + H_{\text{interaction}} \tag{1.21}$$

The first term in H may be compared with the hamiltonian of a system of free particles in nonrelativistic physics. Let $\psi_k(X)$ be the second quantized Schrödinger field for the kth type of particle. Then

$$H_{\text{N.R.}} = \sum_k \frac{\nabla\psi_k^\dagger(x)\, \nabla\psi_k(x)}{2\mu_k} + \frac{U_k}{2\mu_k}\psi^\dagger\psi \tag{1.22}$$

Evidently, the Hamiltonian in eq. (1.21) has the nonrelativistic form, except that the discrete sum over particle type is replaced by an integral over k_-.

A very important point to notice is that the k_- integration runs only over non negative k_-. There are no quanta with $k_- < 0$. This of course is analogous to the positivity of mass in non relativistic quantum mechanics.

The interaction term $H_{\text{interaction}}$ has the form

$$H_{\text{interaction}} \propto \int dk_- \, dl_- \, \frac{\phi^\dagger(k_-)}{\sqrt{|k_-|}} \frac{\phi^\dagger(l_-)}{\sqrt{|l_-|}} \frac{\phi(k_- + l_-)}{\sqrt{|k_- + l_-|}} + \text{c. c.} \tag{1.23}$$

The important thing to note is that the value of k_- is conserved by the interaction. This, together with the positivity of k_-, insures that there are no terms like $\phi^\dagger\phi^\dagger\phi^\dagger$ which create quanta from the Fock space vacuum. For this reason the Fock space vacuum is the true vacuum in the Fock space quantization.

Perturbative processes induced by $H_{\text{interaction}}$ are generated by vertices which allow one particle to split into two, or the reverse, conserving k_-.

1.1. DLCQ

Matrix theory is based on a form of light cone quantization called "discrete light cone quantization" [8]. To define DLCQ, the light like coordinate X^- is compactified to a circle of circumference $2\pi R$. The effect of this compactification is to discretize the spectrum of P_-:

$$P_- = \frac{N}{R} \tag{1.24}$$

where N is a non negative integer. Since P_- is conserved, the system splits up into an infinite number of superselection sectors characterized by N.

Equations (1.17), (1.18), (1.19), (1.20) are replaced by

$$\phi = \phi_0 + \sum_{n=1}^{\infty} \frac{\phi_n(X^i)}{\sqrt{2\pi n}} e^{in\frac{X^-}{R}} + \text{c. c.} \tag{1.25}$$

$$[\phi_n^\dagger, \phi_m^\dagger] = [\phi_n, \phi_m] = 0 \tag{1.26}$$

$$[\phi_n^\dagger(X^i), \phi_m(Y^i)] = \delta_{nm}\delta(X^i - Y^i) \tag{1.27}$$

where ϕ_0 is the mode of ϕ with $k_- = 0$. The Hamiltonian becomes a discrete series of terms

$$H = R\left\{\sum_{n=1}^{\infty} \frac{\nabla\phi_n^* \cdot \nabla\phi_n + m^2}{2n} + \right.$$

$$\left. + \, const. \sum_{n,m=0}^{\infty} \frac{\phi^\dagger(n, X^i)}{\sqrt{n}} \frac{\phi^\dagger(m, X^i)}{\sqrt{m}} \frac{\phi(n+m, X^i)}{\sqrt{n+m}} \right\} + \text{c. c.} \tag{1.28}$$

The "zero mode" ϕ_0 is non dynamical and can be integrated out, giving rise to new terms in H. These new terms conserve P_- and preserve the Galilean symmetry (provided that N is conserved). Other than that, they may be of arbitrary complexity. For example $\phi^\dagger\phi\phi\phi$, $\phi^\dagger\phi^\dagger\phi\phi$, ... terms may be induced.

Quantum mechanics within a given N sector is much simpler than in the uncompactified theory. For example, in the sector $N = 1$ nothing interesting can happen. The spectrum is a single particle which can not split into constituents. For $N = 2$ the Hilbert space is a sum of one particle states with two units of P_- and two particle states, each particle carrying $N = 1$. The only processes which can occur are splitting of the 2-unit particle into two one-unit particles and scattering of the 2-unit particles. However, as N grows the number of allowed processes grows.

Physical applications require that the limit $N \to \infty$ be taken. To see this, we need only note that the quantum number N is given by

$$N = P_- R \tag{1.29}$$

so that if we fix the physical component of momentum and let the radius R tend to infinity, N also becomes infinite.

Any attempt to use DLCQ as a numerical approximation scheme should begin with an estimate of how large N needs to be in order to achieve a given degree of accuracy for a given problem [9]. A rough estimate can be obtained from geometrical considerations. Consider a system with a given mass M whose largest spatial dimension is of order ρ. Assume the system is at rest in the transverse plane. That is,

$$P_i = 0 \tag{1.30}$$

Let us also boost the system along the z axis until it is at rest with $P_z = 0$. Its momentum vector is purely timelike with $P_- = P_+ = M$. In this frame the longitudinal size Δz of the system is no larger than ρ. Obviously for DLCQ to give a good approximation the size of the system should not exceed the compactification scale R. Thus, the condition for a good approximation is

$$\rho \lesssim R \tag{1.31}$$

Multiplying by $P_- = M$ gives

$$M \lesssim RP_- \tag{1.32}$$

But RP_- is N, so that the condition is

$$N \gtrsim M\rho \tag{1.33}$$

Later we will use this condition in studying black holes.

1.2. ANOTHER VIEW OF DLCQ

The light cone frame can be characterized by its metric

$$ds^2 = dX^+ dX^- - (dX^i)^2 \tag{1.34}$$

In order to resolve some of the ambiguities inherent in lightlike compactification, it is useful to introduce a frame in which the metric has the "regularized" form [10]

$$ds^2 = d\tilde{X}^+ d\tilde{X}^- - \epsilon^2 (d\tilde{X}^-)^2 - (dX^i)^2 \tag{1.35}$$

The limit $\epsilon \to 0$ defines the light cone frame, but now the direction $\tilde{X}^-$ is a true spacelike direction. Compactification of $\tilde{X}^-$ on a spacelike circle involves only standard procedures. Note, however, that if X^- is periodic with radius R the proper size of the spacelike circle is

$$R_c = \epsilon R \tag{1.36}$$

Evidently DLCQ may be interpreted as the limit of spacelike compactification in which the compactification size shrinks to zero. However that is not all there is to it. Consider the sector of DLCQ with $P_- = N/R$. This may now be interpreted in the spacelike compactified theory as the sector with spacelike momentum

$$\tilde{P}_- = \frac{N}{R_c} = \frac{1}{\epsilon}\frac{N}{R} \tag{1.37}$$

In other words DLCQ in sector N is obtained by the following procedure:

1. compactify on a spacelike circle of size R_c;
2. consider the sector with N units of quantized spacelike momentum;
3. holding N fixed let $R_c \to 0$; in this case the spacelike $\tilde{P}$ tends to ∞.
4. Now boost the system back to the original frame in which the compactification radius is R. The boost factor or time dilation factor is R_c/R and therefore goes to infinity. The time dilation factor also implies a rescaling of the hamiltonian by a factor R/R_c.

Thinking about DLCQ from this point of view illuminates the problems associated with the zero modes ϕ_0. The zero mode sector of a compactified theory defines a field theory in a lower dimension but with a coupling which blows up as $R_c \to \infty$. Thus integrating the zero modes out of a field theory is nontrivial and involves the solution to a strong coupling field theory which can not be obtained by perturbative means [11].

One may wonder how it is possible to retrieve all of the physics of the uncompactified theory from the theory on the vanishingly small circle of size R_c. The answer is Lorentz contraction. Suppose we focus on an object of size b and mass M. Let us boost this object so that its momentum in the z direction is N/R_c. Its longitudinal size is reduced to $b' = b\frac{MR_c}{N}$. Evidentally as N gets large the system under consideration becomes much smaller than R_c. In the frame of the moving system the compactification radius becomes so big that it has no effect of the system. Thus we see that the system compactified on R_c contains all the information about the uncompactified theory.

1.3. SUPERGALILEAN SYMMETRY

For supersymmetric theories, the Galilean symmetry can be augmented with anticommuting spinorial supercharges to form the supergalilean group. Let us focus on the 11-dimensional case which will be of particular interest in the following chapters. We begin with the rotation group $O(9)$ representing rotations in the 9-dimensional transverse space. The spinor representation of $O(9)$ is real and 16 dimensional. Spinors are labeled θ_α with $\alpha = (1, \ldots, 16)$. The Dirac matrices $\gamma^i_{\alpha,\beta}$ are real.

The superalgebra of 11 dimensional supergravity has 32 real supercharges which can be separated into two sets of 16, namely Q_α and q_α. The Q_α anticommute to give the light cone Hamiltonian

$$[Q_\alpha, Q_\beta]_+ = \delta_{\alpha\beta} H = \delta_{\alpha\beta} P_+ \tag{1.38}$$

Similarly, the q_α close on P_-:

$$[q_\alpha, q_\beta]_+ = \delta_{\alpha\beta} P_- \tag{1.39}$$

and the mixed anticommutation relations are

$$[Q_\alpha, q_\beta]_+ = \gamma^i_{\alpha\beta} P_i \tag{1.40}$$

A simple one-particle representation of the supergalilean group may be constructed as follows. The particle states are labeled by P_-, P_i and 16 anticommuting coordinates θ_α. The supercharges are

$$Q_\alpha = \frac{P_i \gamma^i_{\alpha\beta} \theta_\beta}{\sqrt{2P_-}} \tag{1.41}$$

$$q_\alpha = \theta_\alpha \sqrt{P_-} \tag{1.42}$$

The Hamiltonian is easily seen to be

$$H = \frac{P_i^2}{2P_-} \tag{1.43}$$

Lecture 2: Matrix theory

2.1. TYPE IIA STRING THEORY AND M THEORY

One of the big string theory surprises of the last few years was the discovery that string theory implies the existence of an 11-dimensional theory which itself is not a string theory. This theory, called M-theory [5], is the limit of 10-dimensional IIA string theory in which the string coupling g_{st} tends to infinity .

There were a number of hints that an 11-D theory might underlie IIA string theory. The first was the existence of the dilaton field in the low energy action. Typically when a gravitational theory is compactified (let us call the compact dimension z), the component of the metric g_{zz} behaves like a scalar field in the lower dimensional theory. Furthermore, it enters the action of a lower dimensional theory in the same way that the dilaton does. This suggests that the dilaton is really the local compactification radius of the z direction. The second piece of evidence was the existence of an abelian gauge field in the 10-dimensional IIA theory. Generally speaking, all other gauge fields that occur in string theory can be thought of as Kaluza-Klein gauge fields, either before or after some kind of duality transformation. The so-called RR gauge boson of type IIA theory was not obviously a K-K object. However, with the interpretation of IIA as the K-K compactification of M-theory, the R-R gauge field found its natural interpretation in terms of the $g_{\mu z}$ components of the 11-D metric.

What was missing from perturbative string theory were the charged sources of the R-R field. In Kaluza Klein theory these sources are the quantized z-components of momenta of particles moving in 11 dimensions. However the KK charges were soon located in the form of Polchinski's D-branes [12]. In particular, the only objects carrying the appropriate charge are the D0-branes, particle-like objects on which strings may end. One of the main things we will need to know about these objects is their mass. Let us begin with a massless 11-dimensional particle with an unit of Kaluza-Klein momentum. The mass of this particle in the lower dimensional theory is given by

$$M_0 = \frac{1}{R_C} \tag{2.1}$$

where R_C is the compactification radius of the z direction; in other words it is the background value of the dilaton field.

On the other hand, D0-branes have a mass which can be computed in string theory and is given by [12]

$$M_0 = \frac{1}{g_{st} l_{st}} \tag{2.2}$$

where l_{st} is the basic length scale in string theory.

Evidently, the KK interpretation requires

$$R_C = g_{st} l_{st} \tag{2.3}$$

This explains why the KK charges do not appear in perturbative string theory. Their masses diverge as $g_{st} \to 0$ and they decouple from perturbation theory. In this limit the compactification scale $R_C \to 0$ and becomes invisible.

On the other hand, the limit $g_{st} \to \infty$ is characterized by very light D0-branes and by the compactification radius becoming infinite. This limit is called M-theory.

What do we know about M-theory? First of all it is a 11-D gravitational theory with a Planck length l_{11}. The gravitational coupling in 11-D is

$$G_{11} = l_{11}^9 \tag{2.4}$$

Furthermore, since IIA string theory is supersymmetric with 32 real supercharges, so must be M-theory. This fits conveniently with the known properties of 11-D supergravity. Thus we assume that low energy M-theory is governed by 11-D supergravity.

Besides the gravitational field $g_{\mu\nu}$, 11-D supergravity has another bosonic field, a 3-form $A_{\mu\nu\rho}$ and a fermionic gravitino ψ. In addition to the quanta of these fields, M-theory is postulated to have two other forms of matter. Both are associated with the form field $A_{\mu\nu\rho}$. Just as a one-form (a gauge field) couples to the world lines of particles, a 3-form can couple to the 3-dimensional world volume swept out by a membrane or 2-brane. Since in M-theory the only natural length scale is the 11-D Planck length, the energy of the membrane is governed by a tension (energy/area) of order

$$T_2 = \frac{1}{l_{11}^3} \tag{2.5}$$

The other form of matter is related to $A_{\mu\nu\rho}$ by an analogue of electric/magnetic duality. The second form of matter is 5-branes which sweep out 6-dimensional world volumes.

It is now possible to understand the origin of strings in M-theory when one of the space coordinates z is compactified. Consider a membrane with the topology of a torus. Its spatial volume is parametrized by two angles, θ_1 and θ_2. Let us suppose it is embedded in 11-D so that one cycle, say θ_1, is wrapped around the compact coordinate z:

$$\theta_1 = \frac{z}{R_C} \tag{2.6}$$

The other spatial directions $X_1, \ldots, X_9$ are functions of θ_2.

Such a configuration defines a string in the 9 noncompact space dimensions

$$X^i = X^i(\theta_2) \qquad i = 1, \ldots, 9 \tag{2.7}$$

The tension of the string is easily computed since the energy per unit area of the membrane is

$$T_2 = \frac{1}{l_{11}^3} \tag{2.8}$$

The energy per unit length of the string is

$$T_1 = \frac{R}{l_{11}^3} \tag{2.9}$$

On the other hand, the string tension and string length scale are related by

$$T_1 = \frac{1}{l_s^2} \tag{2.10}$$

Thus

$$l_s^2 = \frac{l_{11}^3}{R_C} \tag{2.11}$$

Combining this with equation (2.3) we find

$$g_{st}^2 = \frac{R_C^2}{l_{st}^2} = \frac{R_C^3}{l_{11}^3} \tag{2.12}$$

Equations (2.11) and (2.12) define the connection between string theoretic and M-theoretic quantities.

2.2. DLCQ OF M-THEORY

Let us now construct the DLCQ of M-theory. According to sect. 1.2 we begin by compactifying M-theory on a space-like circle of radius R_c. But we have just seen that this is equivalent to type IIA string theory on a circle of size

$$R_c^3 = g_{st}^2 l_{11}^3 \tag{2.13}$$

Next we focus on the sector with spacelike momentum N/R_c. In IIA language this means the sector with N units of D0-brane charge. Finally, holding N fixed we pass to the limit $R_c \to 0$ or equivalently $g_{st} \to 0$. In other words we pass to the limit of infinitely weakly coupled string theory. One subtle point is that in passing to the limit we do not hold the string scale fixed. We are interested in phenomena on the scale of l_{11} so it is appropriate to keep this scale fixed. The string scale goes to infinity according to eq. (2.11).

Note that even in 11 dimensional Planck units the (10 dimensional) masses of the D0-branes tend to infinity as $R_c \to 0$. This means that the D0-branes become nonrelativistic in the 10-D sense. It is therefore natural that they are described by Galilean quantum mechanics in 9 spatial dimensions. This Galilean symmetry is of course exactly the Galilean symmetry of DLCQ.

Thus we come to the remarkable conclusion that M-theory is equivalent to the $N \to \infty$ limit of the nonrelativistic quantum mechanics of N D0-branes in weakly coupled limit of IIA string theory.

2.3. D0-BRANE QUANTUM MECHANICS

Fortunately the quantum mechanics of N nonrelativistic D0-branes has been worked out in [13], [14].

Let us begin with the theory of a single D0-brane. The particle is characterized by a transverse location X^i and a 10-D mass $\frac{1}{R_c}$. The D0-brane Lagrangian is

$$\mathcal{L} = \frac{\epsilon}{2R_c}(\dot{X}^i)^2 \tag{2.14}$$

where the factor ϵ represents the time dilation factor from 10-D time to 11-D X^+ described in sect. 2.

The Hamiltonian of the system satisfies

$$H = \frac{1}{2\epsilon}R_c P_i^2 = \frac{1}{2}RP_i^2 \tag{2.15}$$

In order to make the system supersymmetric we add 16 fermionic zero mode coordinates θ as in sect. 2. The D0-brane then has the spectrum and quantum numbers of 11-D supergraviton.

Next, let us consider N D0-branes. We begin with N positions and fermionic partners X_a^i, θ_a $(a = 1, \ldots N)$. The obvious action is

$$\mathcal{L} = \sum_a \left[\frac{1}{2R_c} \dot{X}_a^2 + \dot{\theta}_{\dot{a}} \theta_a \right] \tag{2.16}$$

To describe the interactions between D0-branes, we recall that in IIA string theory a D0-brane is the Dirichlet boundary where a string may end. Thus we may consider strings which connect the D0-branes (a, b). Since the string tension is given by $\frac{R_c}{l_{11}^3}$, the minimum energy of such a string is

$$E(a,b) = \frac{|X_a - X_b|\, R_c}{l_{11}^3} \tag{2.17}$$

Consider an excited string state with additional energy

$$\Delta E \sim \frac{1}{l_s} = \left(\frac{R_c}{l_{11}^3} \right)^{1/2} \tag{2.18}$$

Note that in the limit $R_c \to 0$ the excitation energy of a string becomes infinitely bigger than the minimum excitation energy $E(a, b)$. Thus one may expect that in the limit $R_c \to 0$ only the lightest strings connecting (a, b) are relevant.

In string theory, strings are similar to quanta of a field that can be created and annihilated. It is therefore plausible to describe these lightest states by field operators. A systematic analysis shows that the strings can be either bosons or fermions and that the bosonic strings carry a polarization index i which transforms as a vector under 9-dim. rotations. Similarly, the fermions form spinors. Accordingly we label the string fields X_{ab}^i, θ_{ab}.

If we relabel the coordinate locations and the thetas by

$$X_a \to X_{aa} \tag{2.19}$$

$$\theta_a \to \theta_{aa} \tag{2.20}$$

we see that the degrees of freedom can be assembled into $N \times N$ matrix degrees of freedom that include positions (diagonal elements) and stretched strings as well as their fermionic superpartners.

The action for the matrix degrees of freedom can be derived from IIA string theory [14]. In this lecture we will not carry out the derivation but merely write down the result:

$$\mathcal{L} = Tr \left[\frac{1}{2R_c} \dot{X}^2 + \frac{R_c}{4\, l_{11}^6} [X^i, X^j]^2 + \text{fermions} \right] \tag{2.21}$$

The action (2.21) has a symmetry that might appear accidental from the incomplete discussion in this paper. The symmetry is $U(N)$ where the action of the symmetry group is

$$X \to U^{\dagger} X U \tag{2.22}$$

$$\theta \to U^{\dagger} \theta U \tag{2.23}$$

This symmetry has as its origin the underlying gauge symmetry of string theory. It is the same symmetry that leads to gauge simmetries for open string theories and it should be treated as a gauge symmetry. To make the gauge invariance manifest we may introduce a vector potential A_0 in the adjoint of $U(N)$. The time derivatives in the action can be replaced by covariant derivatives:

$$\frac{dx}{dt} \to \frac{dx}{dt} + [A_0, X] \tag{2.24}$$

The only effect of including A_0 into the action is to introduce a Gauss law constraint which requires the generators of $U(N)$ to vanish on physical states. In other words only the states which are invariant under the $U(N)$ symmetry are allowed in the physical spectrum.

The Hamiltonian resulting from eq. (2.21) is obtained straightforwardly. The momentum conjugate to X^i_{ab} is

$$P_i^{ab} = \frac{1}{R_c} \dot{X}^i_{ab} \tag{2.25}$$

The Hamiltonian (after rescaling by the time dilation factor R/R_c) is

$$H = R\,Tr\left[\frac{1}{2}P_i^2 - \frac{1}{4\,l_{11}^6}[X^i, X^j]^2 + fermions\right] \tag{2.26}$$

Thus we arrive at the precise form of the matrix theory conjecture:

1. Whatever M-theory is, it may be formulated in a world with a compact lightlike direction and quantized according to the rules of DLCQ;
2. In the sector of the theory with $P_- = \frac{N}{R}$ the theory is exactly described by the hamiltonian (2.26) describing supersymmetric quantum mechanics with 16 real supersymmetries and $U(N)$ symmetry.

The supergalilean symmetry of eq. (2.26) required by the DLCQ interpretation is easy to prove. For example a Galilean boost acts on Π according to

$$\Pi \to \Pi + \frac{1}{R} v \otimes I \tag{2.27}$$

where v is a parameter and I is the $N \times N$ identity matrix. Note also that the quartic potential is invariant under translations

$$X^i \to X^i + a^i \otimes I \tag{2.28}$$

2.4. THE SUPERGRAVITON

M-theory must contain in its spectrum the massless graviton and its various superpartners. Let us consider how the spectrum of single and multiple supergravitons arise. We begin with $N = 1$ in which case the Hamiltonian is

$$H = \frac{R}{2} P_i^2 \tag{2.29}$$

This is the correct expression in DLCQ for any massless particle with $P_- = 1/R$. In this case the θ do not appear in H but they do provide the spin degrees of freedom. The 16 θ can be grouped into 8 complex fermionic creation operators and their conjugates. Therefore the algebra of the θ can be realized on a space of dimensionality 2^8. The 256 states consist of 128 bosons and 128 fermions. It is found that the bosons include 44 gravitons which transform under $O(9)$ as a symmetric traceless tensor and 84 objects transforming as an antisymmetric tensor with three indices. This of course is exactly the bosonic content of the 11-D supergraviton. The 128 fermions are the gravitinos.

Now let us consider the general case of arbitrary N. To do this we divide the degrees of freedom into $U(1)$ and $SU(N)$ degrees of freedom by separating off the trace. This is identical to a separation into center of mass and relative motions.

$$X_{c.m.} = \frac{1}{N} TrX \tag{2.30}$$

$$\theta_{c.m.} = \frac{1}{\sqrt{N}} Tr\theta \tag{2.31}$$

$$P_{c.m.} = TrP \tag{2.32}$$

$$X - X_{c.m.} = X_{rel} \tag{2.33}$$

$$\theta - \frac{1}{\sqrt{N}} \theta_{c.m.} = \theta_{rel} \tag{2.34}$$

$$P - \frac{P_{c.m.}}{N} = P_{rel} \tag{2.35}$$

From the form of H we see that the c. m. degrees of freedom are completely decoupled from the relative variables. The hamiltonian may be written as

$$H = \frac{R P_{c.m.}^2}{2N} + H_{rel} \tag{2.36}$$

where H_{rel} is the Hamiltonian with $SU(N)$ symmetry for the relative variables.

Let us suppose that H_{rel} has a normalizable supersymmetric ground state. Supersymmetry requires the eigenvalue of H_{rel} to vanish. Thus for the ground state of the relative motion the Hamiltonian reduces to

$$H = \frac{RP^2_{c.m.}}{2N} = \frac{P^2_{c.m.}}{2P_-} \tag{2.37}$$

which again has the right form for a massless particle. Since $\theta_{c.m.}$ does not enter in H it again provides the spin degeneracy of the supergraviton.

The existence of a normalizable ground state for the c. m. variables is not at all trivial. To see why, let us consider the potential energy $-Tr\ R[X^i, X^j]^2$. This term provides a potential which confines the $X_{c.m.}$ to the region near the origin. However there are "flat directions" along which they can escape. For example consider matrices X^i which all commute. To understand the meaning of such configurations we note that the mutual commutation implies that the X can be simultaneously diagonalized. If we do so, then the configuration can be labeled by N 9-vectors X^i_a representing the location of the N D0-branes. In other words the flat directions could allow the D0-branes to leak away from the bound state. As we shall argue, the flat directions are not lifted by quantum corrections, so the likelyhood remains that the bound configurations representing a supergraviton of N units of P_- may be unstable.

Thus, it is very significant that the existence of a normalizable ground state for the relative variables has recently been proved by Sethi and Stern and by Yi. The theorem is surprisingly difficult to prove and in fact depends crucially on the very special properties of the system, including its 16 supersimmetries.

Having proved the existence of single supergraviton states, we now turn to multigraviton configurations. In particular, the theory must admit free multigraviton configurations in the limit of large separation. Thus let us try to construct a two supergraviton state. To do this we temporarily restrict attention to $N \times N$ matrices X, θ which have the form of $N_1 \times N_1 + N_2 \times N_2$ block diagonal matrices. Let us call the elements in the upper block X^i_{ab}, θ_{ab} and the ones in the lower block Y^i_{ab}, $\phi_{\alpha\beta}$. The off diagonal blocks are temporarily frozen to zero. If we now substitute such matrices into the action (eq. (2.21)) we find that the system separates into two uncoupled matrix models, one with $N_1 \times N_1$ matrices and the other with $N_2 \times N_2$ matrices. Quantizing this system naturally provides a pair of supergravitons with momenta N_1/R and N_2/R.

Freezing the off diagonal elements is, of course, not a legitimate thing to do. But let us assume that the center of masses of the two gravitons are very distant:

$$|\frac{1}{N_1}TrX - \frac{1}{N_2}TrY| = |X_{c.m.} - Y_{c.m.}| \gg l_{11} \tag{2.38}$$

Now consider the equation of motion for an off diagonal element which we call w. The action for w in the background of X and Y has the form

$$\mathcal{L}_z = \frac{1}{2R}\dot{w}^2 - \frac{R|X_Y|^2}{l^6_{11}}w^2 \tag{2.39}$$

where the second term arises from the commutator term in (2.21). The off diagonal elements behave like oscillators with frequency

$$\omega = \frac{R|X-Y|}{l_{11}^3} \tag{2.40}$$

Note that this is exactly the energy of a string stretched from X to Y. This was to be expected since the off diagonal elements of the matrices correspond to string field operators.

The important thing to note about eq. (2.40) is that as $|X-Y|$ grows the frequency of the mode w tends to infinity. One can therefore expect that the off diagonal elements w are naturally frozen into their ground states. This is correct but it is not quite legitimate, in general, to ignore the effects of the zero point motion of w. The total ground state energy of the oscillators is of order

$$N_1 N_2 \omega = \frac{N_1 N_2 R\,|X-Y|}{l_{11}^3} \tag{2.41}$$

and would give rise to a confining potential between gravitons. This would be completely unacceptable. Here is where supersymmetry comes to the rescue. The point is that each boson fluctuation is accompanied by a fermionic partner whose zero point energy exactly cancels the bosonic term. This is an exact consequence of supersymmetry.

There is a very close analogy with the Higgs effect. The X correspond to scalar fields in the adjoint of $U(N)$. The background

$$X = \begin{pmatrix} X_{c.m.} & 0 \\ 0 & Y_{c.m.} \end{pmatrix} \tag{2.42}$$

breaks the $U(N)$ symmetry to $U(N_1) \times U(N_2)$ and the off diagonal elements are analogous to W-bosons which become massive as a consequence of the symmetry breaking. The vanishing of the zero point energy is identical to the vanishing of the ground state energy density in super-Higgs theories.

Let us next consider the velocity dependence of the forces between supergravitons [14] . The simplest way to think about these interactions is to continue the analogy with the Higgs effect. The effective action for slowly varying Higgs field can be computed by integrating out the massive W fields. For example, the effective action for $X_{c.m.} - Y_{c.m.} \equiv r$ can be computed to quadratic order in the velocities by computing Feynman diagrams with the external lines representing the fields $\dot{X}_{c.m.}$ and $\dot{Y}_{c.m.}$. The diagrams are completely analogous to vacuum polarization diagrams which renormalize the kinetic terms in QFT. Fortunately the calculation is not necessary. The high degree of supersymmetry insures that the result exactly vanishes.

The first non-vanishing interaction arises at the quartic level in velocities. It has been exactly computed to the one and two loop order of the matrix quantum mechanics (0+1 field theory). Here we will only sketch the method.

From the matrix theory lagrangian we can read off the Feynman rules for the theory. For example the propagator for the massive w is

$$\Delta_F = \frac{R}{\omega^2 + m^2} \tag{2.43}$$

$$m \sim \frac{Rr}{l_{11}^3} \tag{2.44}$$

and the quartic bosonic vertex is

$$vertex = \frac{R}{l_{11}^6} \tag{2.45}$$

Consider the one-loop diagram with four external X lines at zero frequency (Figure 1). It has the form

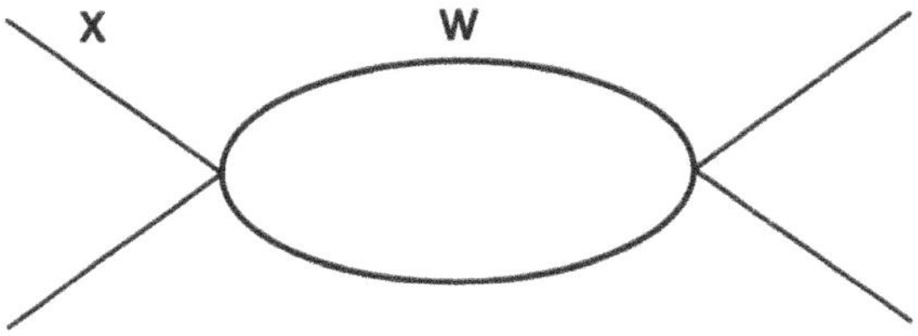

Figure 1.

$$\int \frac{d\omega}{(\omega^2 + m^2)^2} \frac{R^4}{(l_{11})^{12}} N_1 N_2 \tag{2.46}$$

The factor $N_1 N_2$ is just because there are $N_1 N_2$ off diagonal W-bosons to integrate out. Such diagrams each contribute to the effective potential to order X^4, but, as we have argued, they all add up to zero as a consequence of supersymmetry. However we are not interested in the static potential, but rather in terms involving velocities. One finds that the high degree of supersymmetry implies that the first nonvanishing contributions are at the level of 4 powers of velocity. Roughly speaking, this introduces two more propagators, so that the effective action will have the form

$$S_{eff} \sim |\dot{r}|^4 \int \frac{d\omega}{(\omega^2 + m^2)^4} \frac{N_1 N_2 R^4}{l_{11}^2} \sim \frac{1}{m^7} \dot{r}^4 \frac{N_1 N_2 R^4}{l_{11}^3} \tag{2.47}$$

Now, using

$$m = \frac{rR}{l_{11}^3} \tag{2.48}$$

we find

$$S_{eff} \sim \left[\frac{l_{11}^9}{r^7 R^3} N_1 N_2\right] |\dot{r}|^4 = \frac{N_1 N_2 G_N}{r^7 R^3} |\dot{X} - \dot{Y}|^4 \tag{2.49}$$

where G_N is the 11-dim. Newton constant.

The point of this effective action is that it can be used to compute the scattering of two supergravitons for large impact parameter. The simplest way to do this is to convert S_{eff} to an interaction Hamiltonian. From the free term in the action

$$S_{free} = \frac{N_1}{R}\frac{\dot{X}^2}{2} + \frac{N_2}{R}\frac{\dot{Y}^2}{2} \tag{2.50}$$

we identify the velocities in terms of 9-dimensional momenta

$$\dot{X} = \frac{P_1}{N_1} R \tag{2.51}$$

$$\dot{Y} = \frac{P_2}{N_2} R \tag{2.52}$$

$$|\dot{X} - \dot{Y}|^4 = (\frac{P_1}{N_1} - \frac{P_2}{N_2})^4 R^4 \tag{2.53}$$

The interaction Hamiltonian is obtained from Legendre transforming the Lagrangian. It has the form

$$H_{eff} = R\left[\frac{P_1^2}{2N} + \frac{P_2^2}{2N} + \frac{N_1 N_2 G_N}{r^7}\left(\frac{P_1}{N_1} - \frac{P_2}{N_2}\right)^4\right] \tag{2.54}$$

or, using $P_- = \frac{N}{R}$,

$$H_{eff} = R\left[\frac{P_1^2}{2P_-^{(1)}} + \frac{P_2^2}{2P_-^{(2)}} + \frac{a}{R^2}\frac{P_-^{(1)} P_-^{(2)} G_N}{r^7}\left(\frac{P_1}{P_-^{(1)}} - \frac{P_2}{P_-^{(2)}}\right)^4\right] \tag{2.55}$$

where a is a numerical constant which has been computed [14].

The interaction term can now be treated in Born approximation giving rise to a scattering amplitude at large impact parameter. The resulting scattering amplitude can be compared with a corresponding scattering amplitude obtained from tree diagrams in supergravity. In order to make this comparison, we must restrict the process to the kinematic situation in which the momentum transfer is purely transverse.

Processes in which P_- is exchanged correspond in matrix theory to events in which D0-branes are transfered from one cluster (N_1 branes) to the other (N_2). The process computed in this section does not involve exchange of N and so it must be compared with supergravity amplitudes with no P_- exchange. With this restriction, the supergravity scattering amplitude arising from single graviton scattering is found to exactly agree with the result obtained from (2.55).

Originally, it was thought that this should generalize to a great number of processes that

- involve such large impact parameters that only tree diagrams should contribute;
- involve no exchange of P_-.

For example, one might attempt a similar calculation for three body scattering in which all three supergravitons are far from another in tranverse space. The supergravity calculation involves the diagrams of the form shown in figure 2.

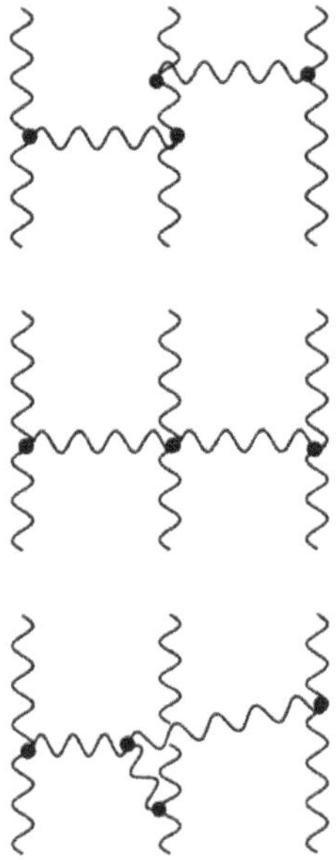

Figure 2.

If we allow all distances between gravitons to be proportional to r and velocities to v the amplitude scales like

$$G_N \frac{v^6}{r^{14}} \tag{2.56}$$

It is not difficult in matrix theory to find a contribution to the effective action which has this form. It arises from a two loop matrix theory diagram. However when calculated in detail the supergravity and matrix theory computations disagree [16].

Although this disagreement is disappointing, it does not necessarily threaten the validity of matrix theory. Let us first consider the situation from the supergravity side. According to sect. 1.2, DLCQ it may be thought of as compactification on a vanishingly small spacelike circle with the longitudinal momentum kept fixed in units of $\frac{1}{R_C}$. As we discussed in lecture 1, in QFT there are zero modes which carry $\tilde{P}_- = 0$ which must be integrated out. In the limit $R_C \to 0$ these zero modes become strongly coupling with coupling of order $\frac{1}{R_C}$. There is no obvious reason why the diagrams involving non linear processes in the zero mode sector should adequately represent the correct physics [11].

In the present situation, we are considering the compactification of 11-D sugra down to 10-D with a Newton constant of order

$$G_{10} \sim \frac{l_{11}^9}{R_C} \tag{2.57}$$

Now, for given R_C, a length scale exists which defines the region for which perturbation theory in G_{10} is valid. Thus for distance satisfying

$$r > G_{10}^{1/8} = \frac{l_{11}^{9/8}}{R_C^{1/8}} \tag{2.58}$$

the tree approximation for the zero modes should be adequate. However, the length scales which interest us in M-theory are fixed in units of l_{11}. Thus as $R_C \to 0$ the tree approximation of sugra tells us nothing about the DLCQ of M-theory. Possible exceptions to this would be special amplitudes which supersymmetry protects. Thus we might conjecture that the two body amplitude in order $\dot{X}^4$ is protected but that the three body amplitude in order $\dot{X}^6$ is not. In fact there is evidence that this is the case.

Now consider the matrix theory side. In principle, the amplitudes for both two and three body scattering are subject to matrix theory loop corrections. The form of these corrections is illuminating. As an example consider the loop corrections to the $\frac{\dot{X}^4}{r^7}$ two body force. Simple power counting shows that the leading corrections are a power series in $\frac{l_{11}^3 N}{r^3}$, assuming N_1 and $N_2 \sim N$.

Thus, the one loop amplitude could be corrected by a function

$$F\left(\frac{Nl_{11}^3}{r^3}\right) = 1 + c_2\frac{Nl_{11}^3}{r^3} + c_3\left(\frac{Nl_{11}^3}{r^3}\right)^2 + \dots \tag{2.59}$$

Now, for fixed N as $r \to \infty$ the higher corrections vanish. But this is not the limit that interests us. Instead we should fix $\frac{r}{l_{11}}$ and let $N \to \infty$. Therefore the correct physics is determined by the behaviour of F as its argument tends to infinity, not to zero.

On the other hand we have seen that the one loop amplitude exactly agrees with expected supergravity results. This suggests that a supersymmetric nonrenormalization theorem may be at work. Thus far no general theorem has been derived, but the coefficient C_2 has been computed [17] and, encouragingly, the result vanishes.

In the case of the three body amplitude studied in [16] similar corrections to (2.59) are expected. However in this case there is unlikely to be a nonrenormalization theorem. The amplitude is almost certainly not constrained by supersymmetry since we already know that two different supersymmetric theories give different results, namely tree diagram supergravity and one loop matrix theory. Thus, we believe that the correct conjecture is not that matrix theory (at finite N) and tree diagram sugra should agree, but rather

1. when they agree, a nonrenormalization theorem will be discovered;
2. when they do not, higher loop diagrams will not vanish.

This could be confirmed by a three loop computation of the $\dot{X}^6$ effective action.

Lecture 3: Objects and dualities in Matrix theory

3.1. 2-BRANES AND 5-BRANES

In the last lecture we demonstated how the Fock space of supergravitons arises out of matrix degrees of freedom. M-theory has a wide variety of other objects, including membranes, 5-branes and black holes. In addition when the theory is compactified it should have strings and D-branes. In the first half of this lecture we will develop the theory of membranes and describe what is known about 5-branes. Then we will move on to the theory of compactification. This will allow us to explore the origins of string theory dualities in Matrix Theory.

There are various membrane configurations that can occur in uncompactified M-theory. Large but finite membranes with spherical, toroidal or higher genus topology are possible. These membranes are unstable, as they will eventually collapse into a black hole. These computations are of course very complicated, even classically, and have never actually been carried out, but it's clear what's going to happen: the "surface tension" of the membrane is a force which acts always in the same direction, so the brane is going to shrink till it approaches its Schwartzschild radius. We can however take very big (but not infinite) membranes for which the decay time is very long and which are so large that a semiclassical description is appropriate. We may also consider stable infinite membranes which extend over an entire plane in the 10 dimensional space. Finally, if the theory is compactified, there are stable finite membranes wrapped on two cycles of the compactification manifold. In this lecture we consider only the first two situations.

A membrane is endowed with a world-volume in the same sense in which a string has a world sheet. The world volume is a three dimensional space consisting of time and a two dimensional space locally parametrized by two coordinates. In this lecture we will consider only the toroidal topology although generalization to other topologies is not difficult.

The geometry of the world space of the membrane in Matrix Theory is not conventional. It is a kind of space that mathematicians call a non-commuting geometry [18]. Such geometries generalize classical geometry much the same way as quantum mechanics generalizes the phase space of classical mechanics. Begin by defining two angular coordinates parametrizing the world space we call p, q. However, the coordinates are not ordinary variables. They are noncommuting with commutation relations

$$[p, q] = \frac{2\pi i}{N} \tag{3.1}$$

The non-commuting torus geometry allows us to construct a useful representation for all $N \times N$ matrices. To this aim, a more convenient labeling is

$$U = e^{ip} \tag{3.2}$$

$$V = e^{iq} \tag{3.3}$$

so that

$$UV = e^{\frac{2\pi i}{N}} VU \tag{3.4}$$

We can now represent any $N \times N$ matrix in terms of the U, V, just by slightly generalizing the Fourier sum expansion:

$$Z = \sum_{n,m=1}^{N} Z_{nm}(U)^n(V)^m \tag{3.5}$$

The representation (3.5) defines a function of two periodic variables for every $N \times N$ matrix. As long as N is finite we have operator ordering ambiguities, but when N is very large the description becomes semiclassical. In fact from (3.1) we see that $1/N$ plays the role of Planck's constant and we encounter the same pattern of the $\hbar \to 0$ limit as in ordinary quantum mechanics. In these lectures we will not consider the issue of convergence to the classical limit. However the reader should be aware that the convergence is subtle and not all configurations have a good classical description even as $N \to \infty$.

The mathematical correspondence between matrices and functions of p, q includes rules for replacing the trace operation and the commutator.

$$Tr\, Z \to N \int dp\, dq Z(p,q) \tag{3.6}$$

$$[Z,W] \to \frac{i}{N}\left[\frac{\partial Z}{\partial q}\frac{\partial W}{\partial p} - \frac{\partial Z}{\partial p}\frac{\partial W}{\partial q}\right] = \frac{i}{N}\{Z,W\}_{\text{Poisson bracket}} \tag{3.7}$$

Using these semi-classical rules we can now rewrite the matrix Lagrangian

$$\mathcal{L} = Tr\left\{\frac{1}{2R}\dot{X}^i\dot{X}^i + \frac{R}{4\, l_{11}^6}[X^i,X^j]^2 + fermions\right\} \tag{3.8}$$

in the large N limit, inserting the above replacements:

$$H = \frac{R}{N}\int dp\, dq\left\{\frac{1}{2}\left(\Pi_i(p,q)\right)^2 - \frac{1}{4\, l_{11}^6}\{X^i,X^j\}^2_{\text{Poisson bracket}} + fermions\right\} \tag{3.9}$$

This Hamiltonian obviously describes a toroidal membrane whose embedding in transverse space is given by $X(p,q)$. In fact it is precisely the Hamiltonian of a supermembrane in light cone frame [22].

The first thing to note about eq. (3.9) is the proportionality to $\frac{R}{N} = \frac{1}{P_-}$; the energy excitations go as $\frac{1}{P_-}$, which is the right kind of behaviour for a localized object in the light cone frame. In this way we have tested, even if only semiclassically, something which was not obvious; from now on, we can treat the p and q as "true" coordinates and assume we have recovered the previously studied two-brane.

We now turn to the infinite 2-brane. The rigorous way to do this is to first compactify the theory on a 2-torus of size $L \times L$ and consider a membrane wrapped around the torus. The limit $L \to \infty$ defines the infinite membrane. However, since we have not yet studied compactification we will adopt an informal less rigorous approach.

To this aim let's introduce an IR cutoff L. We then define

$$X^1 = \frac{pL}{2\pi} \tag{3.10}$$

$$X^2 = \frac{qL}{2\pi} \tag{3.11}$$

The area of the membrane in classical physics would be obtained from the integral of the Poisson bracket of X^1, X^2. The correspondence with matrices gives

$$\frac{1}{N} Tr[X^1, X^2] \sim L^2 \equiv Area\, of\, the\, I.R.\, (cut\, off)\, 2 - brane \tag{3.12}$$

Let us now turn to another example [20]. This time we are going to consider a new kind of object. So far the objects we have studied can be thought of as longitudinally localized "lumps" which look like blurred world lines. These objects Lorentz contract as N increases. They are characterized by having energy which decreases like $\frac{1}{P_-} = \frac{R}{N}$. Now we are going to consider a brane which is wrapped around the light like direction X^-. Consider a p-brane with tension T with one of its dimensions wrapped around X^- and the other $p-1$ dimensions stretched out over a volume L^{p-1}. The energy of such an object is $E = TL^{p-1}R$. Note that it scales like R but does not contain the factor $1/N$.

Consider the following configuration. Let $N = n^2$. Define p and q to be $n \times n$ matrices which satisfy $[q,p] = \frac{2\pi i}{n}$. Let

$$X^1 = p \otimes I \cdot L \tag{3.13}$$

$$X^2 = q \otimes I \cdot L \tag{3.14}$$

$$X^3 = I \otimes p \cdot L \tag{3.15}$$

$$X^4 = I \otimes q \cdot L \tag{3.16}$$

The matrix Hamiltonian for such static configurations becomes

$$E = \frac{R}{4\,l_{11}^6}\,Tr\left\{[X^i, X^j]^2\right\} = \frac{RL^4}{4\,l_{11}^6}\,\frac{1}{n^2}\,Tr(I \otimes I) = \frac{RL^4}{4\,l_{11}^6} \tag{3.17}$$

That is

$$E \sim \frac{L^4 R}{l_{11}^6} \tag{3.18}$$

Notice that energy goes as R, not as $\frac{R}{N}$ as for "lumplike objects". Thus the object is a wrapped brane. Evidently it is a wrapped 5-brane of some sort since its energy scales like 5 powers of a length. The 5-brane discussed in this section is not a "pure" 5-brane. It also has 2-brane charge [20]. It is a 5-brane with "dissolved" 2-branes oriented in the 1,2 and the 3,4 planes. However it is the simplest 5-brane configuration to describe in terms of Matrix Theory.

Presumably the theory also contains various kinds of 5-branes not wrapped on X^- but these are more difficult to describe.

3.2. COMPACTIFICATION

To go beyond, and in particular to test dualities, we have to compactify the theory. To avoid possible confusion, we want to point out that up to now we have a theory with a strange (lightlike) compactification along X^-, that is, in the longitudinal plane. We will undo this in the end, but in any case the X^- compactification has nothing to do with true (spacelike) compactification. This we will choose to do in the transverse space. After such compactification of a single direction matrix theory becomes IIA string theory in light cone frame and we have all of the KK modes discussion of par. 2.1.

So, let's proceed to compactify along a (matrix!) direction. Usually the compactification is carried out along the 11th direction; we are going, instead, to do it along X^9.

There is a recipe for compactifying a matrix direction [23]. Consider infinite matrices of the form (where $X^i(n)$ are $N \times N$ matrices)

$$X^i = \begin{pmatrix} X^i(1) & X^i(2) & X^i(3) & \cdots & \cdots \\ \cdots & X^i(1) & X^i(2) & X^i(3) & \ddots \\ \vdots & . & \ddots & X^i(2) & \ddots \\ . & . & \cdots & X^i(1) & \ddots \end{pmatrix} \qquad i \neq 9 \tag{3.19}$$

$$X^9 = \begin{pmatrix} X^9(1) - 2L & X^9(2) & X^9(3) & \cdots \\ \cdots & X^9(1) - L & X^9(2) & \ddots \\ \vdots & . & X^9(1) & \ddots \\ . & . & . & X^9(1) + L \\ . & . & \cdots & \ddots \end{pmatrix} \tag{3.20}$$

that is, the directions are $N \times N$ block matrices with suitable shifts of L, $2L$, etc. on the diagonal for the compact direction. (This is the usual notion of compactification: we can imagine to put side by side various copies of the compactified space, which are identical but shifted by the compactification length L).

We can write $\mathcal{L}$ in terms of suitable "matrix fields". We have a matrix representation

$$\begin{pmatrix} -2L & 0 & 0 & 0 & 0 \\ 0 & L & 0 & 0 & 0 \\ 0 & 0 & 2L & 0 & 0 \\ 0 & 0 & 0 & 3L & 0 \\ 0 & 0 & 0 & 0 & \ddots \end{pmatrix} = -iL\frac{\partial}{\partial\sigma} \tag{3.21}$$

$$\begin{pmatrix} 0 & 1 & 0 & 0 & \cdots \\ 0 & 0 & 1 & 0 & \cdots \\ 0 & 0 & 0 & 1 & \cdots \\ 0 & 0 & 0 & 0 & \cdots \\ \cdots & \cdots & \cdots & \cdots & \cdots \\ 1 & 0 & 0 & 0 & \cdots \end{pmatrix} = e^{i\sigma} \tag{3.22}$$

$$\begin{pmatrix} 0 & \cdots & 1 & 0 & \cdots \\ 0 & 0 & \cdots & 1 & \cdots \\ 0 & 0 & 0 & \cdots & \cdots \\ \cdots & \cdots & \cdots & \cdots & \cdots \end{pmatrix} = e^{in\sigma} \tag{3.23}$$

$$X^9 = i\frac{\partial}{\partial\sigma} + \sum X^9(n)e^{in\sigma} = i\frac{\partial}{\partial\sigma} + A(\sigma) \tag{3.24}$$

$$X^i = \phi^i(\sigma) \qquad i \neq 9 \tag{3.25}$$

Let us work out the terms of the Lagrangian:

$$\textit{Kinetic terms} \Longrightarrow \frac{1}{R}\int d\sigma\left(\frac{\dot{A}^2}{2} + \frac{\dot{\phi}^2}{2}\right) \tag{3.26}$$

$$[X^i, X^j]^2\,, \quad i \neq j \neq 9 \Longrightarrow R\int d\sigma[\phi^i(\sigma, \phi^j(\sigma)] \tag{3.27}$$

$$[X^i, X^9]^2 \Longrightarrow R\int d\sigma\left(\frac{\partial\phi^i}{\partial\sigma} + i[A, \phi^i]\right)^2 = R\int d\sigma|D_\sigma\phi|^2 \tag{3.28}$$

Similar considerations apply to fermions.

This is a super Yang Mills theory; in this case, it is endowed with 16 susy and lives in 1+1 dimensions, over a (dual) 1-torus. After suitable rescalings of fields and of σ the Lagrangian appears as (ϕ is the X field after rescaling)

$$\mathcal{L} = \frac{1}{g_{YM}^2} \int_0^{\Sigma} \left(\frac{\dot{A}^2}{2} + \frac{\dot{\phi}^2}{2} - \frac{(D\phi)^2}{2} + fermions \right) \tag{3.29}$$

where Σ is the compactified radius of the field theory and g_{YM} its coupling constant. We have thus three quantities, L, Σ and g_{YM}, of which only L was known before; actually, the theory before had really just one parameter. Thus we will have to relate to L the other two.

Let us, first of all, summarize our notations. The rules on a d-torus are basically the same. M theory lives on a torus (chosen to be transverse) of sides L_i and volume $V = \prod_i L_i$; it has only one physical parameter, in addition to L_i: the eleven dimension Planck scale l_{11}. However, it contains another (unphysical) length scale, the radius of lightlike compactification R. The super Yang Mills theory lives on a dual torus of sides Σ_i and volume $V_D = \prod_i \Sigma_i$. To describe the YM parameters in terms of L_i and l_{11}, let's remember again that $\dot{\phi}$ and $\dot{A}$ are the velocities (derivatives of X) respectively in the noncompactified and compactified direction, and accordingly the electric energy is the kinetic energy for the compact directions. The electric term of YM theory is

$$L \sim \frac{1}{g_{YM}^2} \int d^d x \frac{\dot{A}_i^2}{2} \sim \frac{1}{g_{YM}^2} V_D \frac{\dot{A}_i^2}{2} \tag{3.30}$$

$$H = \frac{g_{YM}^2}{2V_D} \Pi_{A_i}^2 = \frac{g_{YM}^2}{2V_D} \Sigma_i^2 n^2 \tag{3.31}$$

where

$$\Pi_{A_i} = \frac{\partial \mathcal{L}}{\partial \dot{A}} \tag{3.32}$$

Eq. (3.31) and in particular the interpretation of $A_i \Sigma_i$ as an angle and the quantization rule $\Pi_{A_i} = \sum_i n$ arise from the Wilson loop behaviour of a electromagnetic-like theory on a circle, that is, after one direction has been compactified. In principle one needs the theory to be abelian, but if this is not the case it is enough to refer to the $U(1)$ contained in the gauge group. The $U(1)$ we are referring to corresponds to separating off one zero-brane.

M theory gives for the kinetic energy the expressions

$$kinetic\ energy = \frac{p_i^2}{2p_-} = \frac{n^2 R}{2L_i^2} \tag{3.33}$$

where $P_- = \frac{1}{R}$, $P_i = \frac{n}{L_i}$.

Comparison between (3.31) and (3.33) gives

$$g_{YM}^2 = \frac{V_D R}{L_i^2 \Sigma_i^2} \tag{3.34}$$

Another equation for Σ can be obtained. We remember that in the SYM picture a KK quantum of momentum N/σ_i has energy

$$E = \frac{n}{\Sigma_i} \tag{3.35}$$

The correspondent matrix theory energy is the one of the winding modes

$$E = \frac{nL_i R}{l_{11}^3} \tag{3.36}$$

Equating these we get

$$L_i \Sigma_i = l_{11}^3 \frac{1}{R} \tag{3.37}$$

and

$$V_D = \frac{l_{11}^{3d}}{R^d} \frac{1}{V} \tag{3.38}$$

Inserting (3.37) and (3.38) in (3.34) gives

$$g_{YM}^2 = \frac{l_{11}^{(3d-6)} R^{3-d}}{V} \tag{3.39}$$

We can also introduce a dimensionless coupling

$$\tilde{g}^2 = \frac{l_{11}^3}{L^3} \equiv g_{YM}^2 \Sigma^{d-3} \tag{3.40}$$

whose expression does not depend on d.

3.3. T-DUALITY

We would now like to see that matrix theory embodies the expected dualities of M theory and string theory [24], [25], [26]. Let us act with T-duality on the strings. We know that the result is

$$L_i \longrightarrow \frac{l_s^2}{L_i} \equiv \tilde{L}_i \tag{3.41}$$

for all the L_i over which we dualize (a subset of the total). We know also that T-dualizing over an odd number of L_i interchanges IIA and IIB string theories, while doing it over an even number of L_i sends them respectively in themselves.

To study the even case we must have at least L_i, i=1, 2, 3 (which means a string theory living in 8 dimensions); we choose L_1 to be a "dilaton direction" and will dualize L_2, L_3. We have

$$\tilde{L}_2 = \frac{l_s^2}{L_2} \tag{3.42}$$

$$\tilde{L}_3 = \frac{l_s^2}{L_3} \tag{3.43}$$

and

$$\tilde{l}_s = l_s \tag{3.44}$$

$$\tilde{G}_N^{(8)} = G_N^{(8)} \tag{3.45}$$

One has also

$$l_s^2 = \frac{l_{11}^3}{L_1} \tag{3.46}$$

where l_s is the characteristic string length if we think of a string as limit of a wrapped 2-brane:

$$l_s^2 = \frac{l_{11}^3}{R} \tag{3.47}$$

The effect of T-duality is thus, explicitly in terms of L_i, $i = 1, ...3$ and l_{11},

$$\tilde{L}_1 = \frac{l_{11}^3}{L_2 L_3} \tag{3.48}$$

$$\tilde{L}_2 = \frac{l_{11}^3}{L_2 L_1} \tag{3.49}$$

$$\tilde{L}_3 = \frac{l_{11}^3}{L_1 L_3} \tag{3.50}$$

$$\tilde{l}_{11}^3 = \frac{l_{11}^6}{L_1 L_2 L_3} \tag{3.51}$$

Using (3.39) for the dual torus, it turns out that

$$\tilde{g}_{YM}^2 g_{YM}^2 = (2\pi)^2 \tag{3.52}$$

which has the form of electric/magnetic duality. This SYM theory is actually believed to enjoy exact Montonen-Olive duality. It should be noted that the argument presented in this lecture is the first demonstration of T-duality which does not depend on ordinary string perturbation theory.

We will come back to this topic in the next section where we will discuss how a new (and unexpected) spatial direction can arise in the theory if we shrink a 3-torus to a 2-torus and, more generally, what is the relation between coordinates and fluxes (which gives to space a very special and unusual role). For the moment, let us summarize that (by means of the identification between electric and magnetic fields and velocities in noncompact and compact directions) the electric/magnetic duality, which exchanges quanta of electric flux and of magnetic flux, is translated, in the matrix theory picture, as a T-duality which interchanges momentum in the compact direction with the quanta (corresponding to topological quantum numbers) representing winding of strings or wrapping of membranes.

Lecture 4: The emergence of space in Matrix theory

The role of space in matrix theory is very unusual if compared to what happens in a field theory. Spatial directions materialize in several new and surprising ways, some of which we will now list.

- The usual transverse coordinates X arise as the moduli space of a super Yang Mills theory.
- The longitudinal direction X^- is described as the conjugate to P_- which is proportional to the N of $U(N)$ super YM theory.
- New unexpected directions can emerge as in the case of compactification on a 2-torus. As the 2-torus shrinks to zero a new 10th direction opens up; it is connected to the Yang Mills magnetic flux.

Our guide to understanding of the above properties will be the symmetries, and in particular rotation invariance; the unexpected features will, in the following, appear as unexpected extensions of these symmetries.

We have discussed, in the previous chapter, the relation between electric/magnetic duality in the super Yang-Mills gauge theory and T-duality on the matrix theory side. In that case we considered a theory on the T^3 torus; now we move to a 2-torus. For simplicity we work only with a rectangular torus of sides L_1, L_2. Now consider the limit in which the face of the torus is "shrunk":

$$L_1 L_2 \to 0 \ , \qquad \frac{L_1}{L_2} \ fixed \tag{4.1}$$

This causes on the YM side the blowing up of the dual torus. As $L_1 L_2$ is shrunk, the cost of a unit of magnetic flux goes down (the magnetic field is inversely proportional to the area and the energy stored in it goes as $B^2\cdot$ area). The energy turns out to be proportional to the area and to l_{11}^{-3}:

$$E = \frac{L_1 L_2}{l_{11}^3} \tag{4.2}$$

It is evident that this configuration describes a membrane of tension $1/l_{11}^3$ wrapped on the 2-torus. The energy cost decreases, leading to an accumulation of a vast number of very low energy states. As we will see, these states can be identified with a new spatial dimension, Y, decompactifying as $L_1L_2 \to 0$ [6]. The new direction is identified through its conjugate momentum which is obviously proportional to the two brane wrapping number or equivalently the yang mills magnetic flux which we call n.

Let's compute the compactification radius L_Y of the hypothetical new direction. The energy of a KK mode along the Y direction is

$$E = \frac{1}{L_Y} \tag{4.3}$$

Equating (4.2) and (4.3) gives the volume of the 3-torus (Y,1,2):

$$L_Y L_1 L_2 = l_{11}^3 \tag{4.4}$$

For L_1, L_2 finite L_Y is also finite, but when $L_1L_2 \to 0$, $L_Y \to \infty$ and the transverse space is generated by one more vector: it is the span of $\{X^3, \ldots, X^9, Y\}$.

How do we know that the correct interpretation of the low energy states is in terms of a new direction of space? It would obviously be sufficient to prove the existence of a symmetry which rotates Y into one of the noncompact directions $(X^3, \ldots, X^9)$. In order to demonstrate such an invariance we will temporarily compactify X^3 so that its radius is identical to L_Y. The theory now becomes a 3+1 SYM theory which was discussed in the previous lecture. The choice $L_3 = L_Y$ gives [27]

$$L_1 L_2 L_3 = l_{11}^3 \tag{4.5}$$

and using (3.39) we compute the Yang-Mills coupling

$$g_{YM}^2 = 2\pi \tag{4.6}$$

Thus the YM theory is at its self dual point at which it enjoys the symmetry

$$E_i \to B_{ij}\epsilon_{ijk} \tag{4.7}$$

Now the electric field E_i is the conjugate to A_i, which also means that it is momentum conjugate to X^3. Using

$$P_3 = E_3 \tag{4.8}$$

$$P_Y = B_{12}\epsilon_{123} \tag{4.9}$$

we see that there is a discrete symmetry $(P_3 \leftrightarrow P_Y)$.

We have then a set of spatial directions

$$(Y; X^3, X^4, \ldots, X^9) \tag{4.10}$$

which enjoys the (discrete) exchange symmetry between Y and X^3. On the other hand, in the limit $L_1L_2 \to 0$, L_3 becomes infinite and the $O(7)$ invariance is restored. The only way the theory can have $O(7)$ invariance and $Y \leftrightarrow X^3$ invariance is for it to become fully $O(8)$ invariant. Thus, if we let L_Y and L_3 go to infinity, we'll regain $O(8)$, the complete continuous rotational symmetry.

There is another way to see the $O(8)$ invariance in the limit $L_1, L_2 \to 0$. Returning to the 2+1 dimensional theory, eq. (3.39) gives

$$g_{YM}^2 = \frac{(l_{11})^{3d-6}}{V} R^{3-d} \tag{4.11}$$

and for its dimensionless analogue

$$\tilde{g}_{YM} = \frac{l_{11}^3}{(L_1L_2)^{3/2}} \tag{4.12}$$

Since $\tilde{g}_{YM}$ is dimensionless it makes sense to discuss if it is large or small; let's notice, then, that it goes to infinity as the torus shrinks. If we prefer we can state that, keeping fixed the dimensional coupling constant, the YM torus becomes larger and larger. In the limit we are driven to a strongly coupled scale invariant fixed point of a SCFT, whose behaviour has been classified. In the present case, it is known that the relevant fixed point has $O(8)$ invariance [29].

4.1. SUPERGRAVITON SCATTERING WITH EXCHANGE OF P_Y

In this section we will discuss supergraviton scattering with exchange of momentum in the Y direction [28] (but no exchange of P_-). Before doing that, we suppose that the two graviton scattering involves no transfer either of longitudinal or Y momentum (for uncompactified theory in 11 dim.)

$$\Delta P_Y = \Delta P_- = 0 \tag{4.13}$$

Ti is not difficult to generalize the argument of section 2.4 to the case of scattering with vanishing longitudinal and Y momentum. The only difficulty is that the momentum integral in eq. (2.47) is now replaced by an integration over d^3k, since the matrix theory is now a 2+1 dimensional quantum field theory. We obtain

$$\dot{X}^4 \int \frac{d^3k}{k^8} \sim \frac{\dot{X}^4}{\rho^5} \tag{4.14}$$

where ρ is distance in the 7 dimensional space or impact parameter. This again agrees with the supergravity tree diagram analysis.

We are now going to study processes where $\Delta P_Y \neq 0$. We are going to argue that the exchange of momentum is actually exchange of magnetic flux and that the whole process is an instantonic one [15]; it corresponds actually to the creation of a Polyakov instanton.

Let us consider two supergravitons of longitudinal momentum $\frac{N}{2}$ each:

$$\begin{pmatrix} (X^I) & \cdot \\ \cdot & (X^{II}) \end{pmatrix} \tag{4.15}$$

X^I, X^{II} are matrices $\frac{N}{2} \times \frac{N}{2}$; $X^I - X^{II}$ is very large. The $U(N)$ group breaks to a $U(\frac{N}{2}) \times U(\frac{N}{2})$.

If we work in transverse center of mass coordinates, (4.15) becomes

$$\begin{pmatrix} (X) & \cdot \\ \cdot & (-X) \end{pmatrix} \tag{4.16}$$

Now, the two $U(1)$s in $U(\frac{N}{2}) \times U(\frac{N}{2})$ are the center of mass position of the super-gravitons and the two abelian magnetic fluxes n_1, n_2 are the Y momenta.

Let us now consider a process in which a single unit of magnetic flux is exchanged:

$$n_1 \to n_1 + 1 \tag{4.17}$$

$$n_2 \to n_2 - 1 \tag{4.18}$$

Another way to describe this process is that a unit of magnetic flux is created in a $U(1)$ contained in a $SU(2)$ subgroup of $U(N)$. Such a process in which an $SU(2)$ magnetic flux changes is actually an instanton process in 2+1 dimensions.

Thus the process we are considering is an instanton in an $SU(2)$ theory spontaneously broken to $U(1)$. The Higgs expectation value is proportional to the separation of the gravitons with the precise connection given by

$$\langle \phi \rangle = g_{YM}^2 \rho / L_Y \tag{4.19}$$

The instanton action is

$$S_{inst} = \frac{2\pi}{g_{YM}^2} \langle \phi \rangle = 2\pi \frac{\rho L_1 L_2}{l_{11}^3} = \frac{2\pi\rho}{L_Y} \tag{4.20}$$

where $\langle \phi \rangle$ is the v. e. v. of the Higgs field and, if we remember that the moduli space is the X space, is not only a v. e. v. but also a distance; that is, ρ is to be interpreted as the distance between the branes.

The amplitude for the exchange of one unit of momentum in the Y direction in the regime $\rho \gg L_Y$ is proportional to

$$\frac{v^4 e^{-2\pi\rho/L_Y}}{\rho^3} \tag{4.21}$$

The exponential suggests that this is the formula for an exchange process and this is indeed the case: the mass for the Kaluza-Klein first mode is

$$M_{KK} = \frac{2\pi}{L_Y} \tag{4.22}$$

and the corresponding amplitude is straightforwardly proportional to

$$e^{-(M_{KK}\rho)} = e^{-\frac{2\pi\rho}{L_Y}} \tag{4.23}$$

Thus we see how exchange processes involving Y momentum are generated by nonperturbative instanton processes [32].

In these lectures we have considered matrix theory compactified on 2 and 3-tori. The case on the 1-torus is also well studied. As we might expect, in the limit of small compactification radius the theory on a 1-torus becomes IIA string theory in the light cone frame. This has been beautifully demonstrated in the papers by Motl, Banks and Seiberg and Dijkgraaf, Verlinde and Verlinde. Unfortunately there is not enough time in this lectures to review their extremely interesting work.

The theory compactified on T^d with $d > 3$ is much more problematic. In this case the relevant SYM theories are non renormalizable and therefore require new short distance data to define them. Interesting proposals for the theory on 4,5 and 6 tori exist in the literature, but they are beyond the scope of this elementary review.

Finally, the case of T^d which would define a world with 3+1 noncompact dimensions appears to require an entirely new approach.

Lecture 5: Black holes in matrix theory

5.1. CHOOSING N

Our focus in these lectures has been on identifying and studying all the various objects that matrix theory needs to describe. We have seen how supergravitons, membranes and 5-branes arise. In addition, when compactified on a small S^1, weakly coupled IIA string theory emerges [30]. This leaves only black holes.

From the traditional relativists' point of view, black holes are extremely mysterious objects. They are described by unique classical solutions of Einstein's equations. All perturbations quickly die away leaving a featureless "bald" black hole with "no hair". On the other hand Bekenstein and Hawking have given persuasive arguments that black holes possess thermodynamic entropy and temperature which point to the existence of a hidden microstructure. In particular, entropy generally represents the counting of hidden microstates which are invisible in a coarse grained description. In this lecture we will address the problem of black hole entropy in matrix theory.

Most string theory studies of black hole entropy have concentrated on BPS black holes and their low lying excitations. Supersymmetry allows an extraordinary degree of mathematical control over these objects and quite precise state counting can be done. The results are in remarkable agreement with the Bekenstein Hawking formula. Much of this work can probably be translated into matrix theory terms [33] but in this lecture we will concentrate on another class of black holes -the Schwartzschild black holes. These objects are as far as possible from BPS supersymmetric states and are therefore uncontrolled by the constraints of supersymmetry. We will concentrate on a particular special case, which in many respects is the simplest, Schwartzschild black holes in $7 + 1$ dimensions, obtained from matrix theory compactified on a 3-torus.

An ultimate exact treatment of objects in matrix theory requires a passage to the infinite N limit. Unfortunately this limit is extremely difficult. The limit is in many respect similar to taking the continuum limit of a lattice gauge theory. In the limit of vanishingly small lattice spacing, $a \to 0$, the number of degrees of freedom used to describe a given hadronic system diverges. Obviously there is no real need to describe a system by so many d. o. f.: a coarse grained description in which the lattice spacing a is a few times smaller than the probed wavelengths is sufficient to capture the relevant properties. In this sense we may introduce the idea of a maximum lattice spacing a_{max} which will allow an accurate computation of the properties of the system. The value of a_{max} depends on the system or process. For example, if we are interested in inelastic electroproduction, the value of a_{max} must decrease as the momentum transfer increases.

Very similar remarks apply to the choice of N. The $N \to \infty$ limit introduces far more d. o. f. than are actually needed to study any given system. Therefore our first task in applying matrix theory as a quantitative tool is to choose a value of N which is large enough to capture the physics of a given black hole. Thus let us define $N_{min}(S)$ to be the minimum value of N which will allow a given degree of accuracy in calculating properties of a black hole of entropy S.

The value of N_{min} may be obtained from a simple spacetime argument in light-like compactification of M-theory. The basic requirement of an accurate description of a given system is that the compactification radius R be large enough to easily contain the system without distortion. Thus consider a black hole of mass M and Schwartzschild radius R_s. Let us choose the black hole to have momentum

$$P_+ = P_- \sim M \tag{5.1}$$

$$P_i = 0 \tag{5.2}$$

In other words, the black hole is at rest. In this frame, the condition for the black hole to fit safely in the compact longitudinal space is

$$R > R_s \tag{5.3}$$

Now, multiplying both sides by P_-, we find

$$RP_- > R_s M \tag{5.4}$$

But the product RP_- is the integer N, so we find

$$N_{min} \sim R_s M \tag{5.5}$$

Another view of the same condition is given in [31], [9].

5.2. PROPERTIES OF SCHWARTZSCHILD BLACK HOLES

Let us now consider the properties of black holes that we wish to reproduce. All formulas will be up to numerical constants.

If M-theory is compactified on a d-torus it becomes a $D = 11 - d$ dimensional theory with Newton constant

$$G_D = \frac{G_{11}}{L^d} = \frac{l_{11}^9}{L^d} \tag{5.6}$$

A Schwartzschild black hole of mass M has a radius

$$R_s \sim M^{\left(\frac{1}{D-3}\right)} G_D^{\left(\frac{1}{D-3}\right)} \tag{5.7}$$

According to Bekenstein and Hawking the entropy of such a black hole is

$$S = \frac{Area}{4G_D} \tag{5.8}$$

where Area refers to the $D-2$ dimensional hypervolume of the horizon:

$$Area \sim R_s^{D-2} \tag{5.9}$$

Thus

$$S \sim \frac{1}{G_D}(MG_D)^{\frac{D-2}{D-3}} \sim M^{\frac{D-2}{D-3}} G_D^{\frac{1}{D-3}} \tag{5.10}$$

Now consider the value of $N_{min}(S)$ given in eq. (5.5):

$$N_{min}(S) = MR_s = M(MG_D)^{\frac{1}{D-3}} = S \tag{5.11}$$

We see that the value of N_{min} in every dimension is proportional to the entropy of the black hole.

In what follows, we will see that the thermodynamic properties of super Yang Mills theory can be estimated by standard arguments only if $S \lesssim N$. Thus we are caught between conflicting requirements. For $N \gg S$ we don't have tools to compute. For $N \ll S$ the black hole will not fit into the compact geomrtry. Therefore we are forced to study the black hole using $N = N_{min} = S$.

5.3. SUPER YANG MILLS THERMODYNAMICS

As we have seen in lecture 2, matrix theory compactified on a d-torus is described by $d+1$ super Yang Mills theory with 16 real supercharges. For $d = 3$ we are dealing with a very well known and special quantum field theory. In the standard 3+1 dimensional terminology it is $U(N)$ Yang Mills theory with 4 supersymmetries and with all fields in the adjoint repersentation. This theory is very special. In addition to having electric/magnetic duality, it enjoys another property which makes it especially easy to analyze, namely it is exactly scale invariant.

Let us begin by considering it in the thermodynamic limit. The theory is characterized by a "moduli" space defined by the expectation values of the scalar fields ϕ. Since the ϕ also represents the positions of the original D0-branes in the non compact directions, we choose them at the origin. This represents the fact that we are considering a single compact object -the black hole- and not several disconnected pieces.

Now consider the equation of state of the system, defined by giving the entropy S as a function of temperature. Since entropy is extensive, it is proportional to the volume Σ^3 of the dual torus. Furthermore, the scale invariance insures that S has the form

$$S = constant\ T^3 \Sigma^3 \tag{5.12}$$

The constant appearing in this equation counts the number of degrees of freedom. Here is what we know about it. For vanishing coupling constant, the theory is described by free quanta in the adjoint of $U(N)$. This means that the number of degrees of freedom is $\sim N^2$. Thus for small g_{YM}

$$S = N^2 T^3 \Sigma^3 \tag{5.13}$$

Furthermore, for very large g_{YM}, the strong/weak duality again requires the same equation of state. Although it has not been proved, we will assume that eq. (5.13) is roughly correct for all g_{YM}.

Finally we may use the standard thermodynamic relation

$$dE = TdS \tag{5.14}$$

to obtain the energy of the system:

$$E \sim N^2 T^4 \Sigma^3 \tag{5.15}$$

We will be interested in relating the entropy and mass of the black hole. Thus let us eliminate the temperature from eq. (5.13) and (5.15). We find

$$S = N^2 \Sigma^3 \left(\frac{E}{N^2 \Sigma^3} \right)^{3/4} \tag{5.16}$$

5.4. BLACK HOLE THERMODYNAMICS

Now the energy of the quantum field theory is identified with the light cone energy of the system of D0-branes forming the black hole. That is

$$E = \frac{M^2}{2P_-} \approx \frac{M^2}{N} R \tag{5.17}$$

Plugging (5.17) into (5.16) gives

$$S = N^2 \Sigma^3 \left(\frac{M^2 R}{N^3 \Sigma^3} \right)^{3/4} \tag{5.18}$$

Using eq. (3.37) we obtain

$$S = \frac{1}{N^{1/4}} M^{3/2} \left(\frac{l_{11}^9}{L^3} \right)^{1/4} \tag{5.19}$$

As we shall see, this formula only makes sense when $N \lesssim S$.

For $N \gg S$ we need more powerful methods to compute the equation of state. But, as we have seen, $N \sim S$ is the minimal acceptable value for making reliable estimates of black hole properties. Thus we evaluate eq. (5.19) at $N = S$ to obtain

$$S = M^{6/5} G_8^{1/5} \tag{5.20}$$

This is precisely the correct form for the black hole entropy in terms of the mass. In order to appreciate the significance of this formula let us consider the most general behaviour of S consistent with dimensional analysis. Since S is dimensionless and G_8 has dimensions (length)^6 we have

$$S = M^a G^b l_{11}^{(a-6b)} \tag{5.21}$$

There are two undetermined exponents in eq. (5.21). Matrix theory gets them both correct!

The next thing we would like to understand is the temperature of the black hole. The standard Hawking temperature is given by

$$T_H = \frac{1}{R_s} \tag{5.22}$$

This is the temperature in the rest frame. In the light cone frame the temperature is red shifted by a boost factor $\frac{M}{P_-} = \frac{RM}{N}$. Thus the light cone temperature is

$$T_{l.c.} = \frac{R}{R_s}\frac{M}{N} \tag{5.23}$$

On the other hand, from (5.13) we see that at $N = S$ the Yang Mills temperature is

$$T = \left(\frac{1}{N}\right)^{1/3}\frac{1}{\sigma} = \frac{RL}{l_{11}^3}\frac{1}{N^{1/3}} \tag{5.24}$$

By identifying N with S and using the $R_s^{D-3} = GM$ relation we find that the temperatures in (5.23) and (5.24) agree. In other words, the Yang Mills temperature T is the Hawking temperature boosted to the light cone frame.

One may wonder why we can not simply use (5.13) and (5.15) for $N \ggg S$. If we did so, then eq. (5.19) would indicate that black hole entropy would depend on N and not just on the mass and Newton constant. We can construct an equation of state which does reproduce the correct black hole entropy for all N. Using eq. (5.20) and (5.17) we obtain

$$S = E^{3/5}\left(\frac{N}{R}\right)^{5/3} G_8^{1/5} \tag{5.25}$$

Using $dE = TdS$ we compute the temperature

$$T = S^{2/3}\frac{R}{N}(G_8)^{-1/3} = S^{2/3}\frac{R}{N}\frac{L}{l_{11}^3} = \frac{S^{2/3}}{N}\frac{1}{\Sigma} \tag{5.26}$$

or

$$S = (T\Sigma N)^{3/2} \tag{5.27}$$

This equation of state agreees with (5.13) at $S = N$, but also guarantees correct black hole thermodynamics for all N. The important feature of the equation of state (5.27) is that the entropy is not extensive. We will return to this shortly.

A consistent picture can now be formulated. At high temperatures for which $S \gg N$ the equation of state is given by (5.13). The entropy is extensive here because the typical wavelenght of a quantum is much smaller than Σ, the size of the dual torus. As the temperature decreases (at fixed N), eq. (5.13) continues to hold until we come to the point

$$S = N \tag{5.28}$$

$$T = \frac{1}{\Sigma N^{1/3}} \tag{5.29}$$

At this point the system begins to behave like a black hole and, as we have seen, agreement with black hole thermodynamics results. However, at this point, a transition occurs in the behaviour of $S(T)$. Although we don't know how to derive the very low temperature behaviour in eq. (5.27), we will see good reason to expect a breakdown of (5.13) at the transition point.

5.5. LOW TEMPERATURES

Ordinarily, extensivity of the equation of state for a field theory will break down when the temperature becomes so low that the typical wavelength of a quantum becomes of the same order of the size of the system. Thus, the equation of state for a free scalar field will hold down to temperature

$$T_{min} \approx \frac{1}{\Sigma} \tag{5.30}$$

Eq. (5.28), (5.29) suggest that the bulk equation of state continues to much lower temperature as N becomes large.

Similar behaviour has previously been seen in applying super Yang Mills theory to black hole problems [34].

To illustrate the reason for the equation of state to be continued to such low temperature we will consider the example of 1+1 dimensional super Yang Mills theory. In addition to the usual local fields, the degrees of freedom include global "Wilson loops" degrees of freedom. These Wilson loops are unitary $N \times N$ matrices which determine how the fields transform when they are transported once around Σ. Thus if the Wilson loop satisfies

$$W = 1 \tag{5.31}$$

all fields are periodic. If on the other hand W is not unity, the adjoint fields transform as

$$\phi(\sigma + \Sigma) = W^{\dagger}\phi(\sigma)W \tag{5.32}$$

Now consider the case where W has the form of a shift matrix

$$W = \begin{pmatrix} 0 & 1 & 0 & 0 & \cdots \\ 0 & 0 & 1 & 0 & \cdots \\ 0 & 0 & 0 & 1 & \cdots \\ 0 & 0 & 0 & 0 & \cdots \\ \cdots & \cdots & \cdots & \cdots & \cdots \\ 1 & 0 & 0 & 0 & \cdots \end{pmatrix} \tag{5.33}$$

The fields satisfy

$$\phi_{a,b}(\sigma + \Sigma) = \phi_{a+1,b+1}(\sigma) \tag{5.34}$$

The periodicity of the field ϕ is significantly modified. In fact the field only returns to its original value after cycling around the σ axis N times. The result of this is that the system can support waves with effective wavelength

$$\lambda = \Sigma N \tag{5.35}$$

and the bulk equation of state continues to temperature of order

$$T_{min} \sim \frac{1}{\Sigma N} \tag{5.36}$$

In effect, the system behaves as if it lived on a circle of length N times larger than its real length.

In the 3+1 dimensional case of present interest it can be shown that Wilson loop configurations exist for which the effective volume of the dual torus is N times bigger. In other words the dual torus behaves as if it had radii $\Sigma' = \Sigma N^{1/3}$. Thus it is completely natural that the bulk thermodynamics continues down to $\Sigma T = \frac{1}{N^{1/3}}$. This represents a remarkable confirmation on the Yang Mills side that a transition will happen just when the system fits the compact X^- axis.

(5.37)

Conclusions

In these lectures we have seen the remarkable ways in which the various objects of M-theory and string theory arise out of the underlying degrees of freedom of matrix theory. The list includes supergravitons, membranes, 5-branes, strings, D-branes and black holes.

Given this ability to arrange themselves into exactly the correct objects, it seems very likely that matrix theory correctly captures the nonperturbative physics of string theory. There are however many unresolved questions. First of all, compactification on 4, 5 and 6 tori involves non-renormalizable field theories. Even worse, the 7-torus introduces major difficulties of principle. The problems of probing Lorentz invariance or the existence of the large N limit are unsolved. Perhaps most important is that we have no clear general understanding of the connection of Matrix theory and classical general relativity. Circumstancial evidence exists that the low energy theory involves Einstein gravity but no clear and comprehensive derivation exists at present. Hopefully the situation will soon change.

Acknowledgements

This work was supported in part by NSF grant No. PHY-9219345. The authors would like to thank the Institute for Advanced Study for hospitality and support during preparation of the article.

References

1. M Theory As A Matrix Model: A Conjecture; T. Banks, W. Fischler, S. H. Shenker, L. Susskind, Phys. Rev. **D55** (1997) 5112-5128
2. Green Schwarz and Witten, Superstring theory, Cambridge University Press 1987
3. The Parton Picture Of Elementary Particles, Kogut and Susskind, Phys. Rept. **8** (1973) 75
4. On The Quantum Structure Of A Black Hole, G. 't Hooft, Nucl.Phys. **B256** (1985) 727
   Black hole evaporation without information loss, C. R. Stephens, G. 't Hooft and B. F. Whiting, Class. Quant. Grav. **11** (1994) 621
   `hep-th/9306069`, The stretched horizon and black hole complementarity, L. Susskind, L. Thorlacius, and J. Uglum, Phys. Rev. **D48** (1993) 3743-3761
5. `hep-th/9506126`, Eleven Dimensional Origin of String/String Duality: A One Loop Test, M. J. Duff, J. T. Liu, R. Minasian, Nucl.Phys. **B452** (1995) 261
   `hep-th/9410167`, Unity of Superstring Dualities, C. M. Hull, P. K. Townsend, Nucl.Phys. **B438** (1995) 109
   `hep-th/9503124`, String Theory Dynamics In Various Dimensions, E. Witten, Nucl.Phys. **B443** (1995) 85
6. `hep-th/9508154`, Some Relationships Between Dualities in String Theory, Paul S. Aspinwall, Nucl. Phys. Proc. Suppl. **46** (1996) 30-38
   `hep-th/9510086`, The Power of M Theory, John H. Schwarz, Phys.Lett. **B367** (1996) 97
7. `hep-th/9409089`, The World as a Hologram, L. Susskind, J. Math. Phys. **36** (1995) 6377
8. `hep-th/9704080`, Another Conjecture about M(atrix) Theory, Leonard Susskind
9. `hep-th/9711063`, A note on discrete light cone quantization, Daniela Bigatti, Leonard Susskind, submitted for publication in Physics Letters B
10. `hep-th/9710009`, Why is the Matrix Model Correct?, Nathan Seiberg, Phys. Rev. Lett. **79** (1997) 3577-3580
    `hep-th/9709220`, D0 Branes on T^n and Matrix Theory, Ashoke Sen
11. `hep-th/9711037`, Compactification in the Lightlike Limit, Simeon Hellerman, Joseph Polchinski
12. `hep-th/9510017`, Joseph Polchinski, Phys. Rev. Lett. **75** (1995) 4724
    `hep-th/9602052`, Notes on D-Branes, Joseph Polchinski, Shyamoli Chaudhuri, Clifford V. Johnson
13. `hep-th/9603081`, D-particle Dynamics and Bound States, Ulf H. Danielsson, Gabriele Ferretti, Bo Sundborg, Int. J. Mod. Phys. **A11** (1996) 5463-5478
    `hep-th/9603127`, A Comment on Zero-brane Quantum Mechanics, Daniel Kabat, Philippe Pouliot, Phys. Rev. Lett. **77** (1996) 1004-1007
14. `hep-th/9511043`, D-brane dynamics, C. Bachas, Phys. Lett. **B374** (1996) 37-42

15. hep-th/9704029, Membrane Scattering with M-Momentum Transfer, Joseph Polchinski, Philippe Pouliot, Phys. Rev. **D56** (1997) 6601-6606
16. hep-th/9710174, Multigraviton Scattering in the Matrix Model, M. Dine, A. Rajaraman
17. hep-th/9705091, A Two-Loop Test of M(atrix) Theory, Katrin Becker, Melanie Becker, to appear in Nuclear Physics B
18. Alain Connes, Non commutative geometry, Academic Press 1994
19. hep-th/9610236, Five-branes in M(atrix) Theory, Micha Berkooz, Michael R. Douglas, Phys. Lett. **B395** (1997) 196-202
20. hep-th/9612157, Branes from Matrices, Tom Banks, Nathan Seiberg, Stephen Shenker, Nucl. Phys. **B490** (1997) 91-106
21. hep-th/9701025, Proposals on nonperturbative superstring interactions, Lubos Motl
hep-th/9702187, Strings from Matrices, Tom Banks, Nathan Seiberg, Nucl. Phys. **B497** (1997) 41-55
hep-th/9703030, Matrix String Theory, R. Dijkgraaf, E. Verlinde, H. Verlinde, Nucl. Phys. **B500** (1997) 43-61
22. B. de Wit, J. Hoppe and H. Nicolai, Nucl. Phys. **B 305** (1988)
23. hep-th/9611042, D-brane field theory on compact spaces, Washington Taylor, Phys. Lett. **B394** (1997) 283-287
24. hep-th/9611164, T Duality in M(atrix) Theory and S Duality in Field Theory, Leonard Susskind
25. hep-th/9611202, Branes, Fluxes and Duality in M(atrix)-Theory, Ori J. Ganor, Sanjaye Ramgoolam, Washington Taylor IV, Nucl. Phys. **B492** (1997) 191-204
26. hep-th/9703102, The Incredible Shrinking Torus, W. Fischler, E. Halyo, A. Rajaraman, L. Susskind, Nucl. Phys. **B501** (1997) 409-426
27. hep-th/9702101, Rotational Invariance in the M(atrix) Formulation of Type IIB Theory, Savdeep Sethi, Leonard Susskind, Phys. Lett. **B400 (1997)** 265-268
28. hep-th/9705190, Instantons, Scale Invariance and Lorentz Invariance in Matrix Theory, T. Banks, W. Fischler, N. Seiberg, L. Susskind, Phys. Lett. **B408 (1997)** 111-116
29. hep-th/9705117, Notes on Theories with 16 Supercharges, Nathan Seiberg
30. hep-th/9709091, Schwarzschild Black Holes from Matrix Theory, T. Banks, W. Fischler, I. R. Klebanov, L. Susskind
hep-th/9711005, Schwarzchild Black Holes in Matrix Theory II, T. Banks (Rutgers U.), W. Fischler (U. Texas, Austin), I. R. Klebanov (Princeton U.), L. Susskind (Stanford U.)
hep-th/9709108, Schwarzschild Black Holes in Various Dimensions from Matrix Theory, Igor R. Klebanov, Leonard Susskind (to appear in Physics Letters B)
31. hep-th/9710217, Comments on Black Holes in Matrix Theory, Gary T. Horowitz, Emil J. Martinec
32. hep-th/9705190,Instantons, Scale Invariance and Lorentz Invariance in Matrix Theory, T. Banks, W. Fischler, N. Seiberg, L. Susskind, Phys. Lett. **B408** (1997) 111-116
33. hep-th/9709114, Black Holes in Matrix Theory, M. Li, E. Martinec
34. hep-th/9604042, D-branes and Fat Black Holes, Juan M. Maldacena, Leonard Susskind, Nucl. Phys. **B475** (1996) 679-690

NOTES ON MATRIX AND MICRO STRINGS

R. DIJKGRAAF
Department of Mathematics
University of Amsterdam, 1018 TV Amsterdam

E. VERLINDE
TH-Division, CERN, CH-1211 Geneva 23,
Institute for Theoretical Physics
University of Utrecht, 3508 TA Utrecht

AND

H. VERLINDE
Institute for Theoretical Physics
University of Amsterdam, 1018 XE Amsterdam

1. Introduction

In spite of considerable recent progress, M-theory is as yet a theory without a fundamental formulation. It is usually specified by means of a number of fundamental properties that we know it should posses. Among the most important of these characteristics is that the degrees of freedom and interactions of M-theory should provide a (mathematically consistent) representation of the maximally extended supersymmetry algebra of eleven dimensional supergravity [1]

$$\begin{aligned}\{Q_\alpha, Q_\beta\} &= (\Gamma^m)_{\alpha\beta}P_m + (\Gamma^{mn})_{\alpha\beta}Z^{(2)}_{mn} \\ &\quad + (\Gamma^{m_1\ldots m_5})_{\alpha\beta}Z^{(5)}_{m_1\ldots m_5}\,. \end{aligned} \tag{1.1}$$

Here P_m denotes the eleven dimensional momentum, and $Z^{(2)}$ and $Z^{(5)}$ represent the two- and five-index central charges corresponding to the two types of extended objects present in M-theory, respectively the membrane and the fivebrane. Upon compactification, these extended objects give rise to a rich spectrum of particles, with many quantized charges corresponding to the various possible wrapping numbers and internal Kaluza-Klein momenta.

An alternative but presumably equivalent characterization of M-theory is that via compactification on a circle S^1 it becomes equivalent to ten dimensional type IIA string theory [2]. The particles with non-zero KK-momentum along the S^1 are in this correspondence identified with the D-particles of the IIA model, while the

L. Baulieu et al. (eds.), Strings, Branes and Dualities, 319–356.

string coupling constant g_s emerges as the radius R of the S^1 via

$$\ell_s g_s = R, \qquad \ell_s^2 = \ell_p^3/R. \tag{1.2}$$

Here ℓ_s denotes the string scale and ℓ_p is the 11-dimensional Planck scale. Also upon further compactification, all M-theoretic charges now have well understood interpretations in terms of perturbative and non-perturbative string theoretic configurations, such as wrapped strings, D-branes [3] or solitonic five-branes [4]. The precise and well-defined perturbative rules for determining the interactions among strings and D-branes [3] [5] [6] [7] so far provides the best basis for concrete quantitative studies of M-theory dynamics. String theory still forms the solid foundation on which the new structure of M-theory needs to be built.

From both the above starting points, a convincing amount of evidence has been collected to support the conjecture that toroidal M-theory compactifications exhibit a large group of discrete duality symmetries, that interchange the various internal charges [8]. While some of these dualities are most manifest from the eleven dimensional perspective, others become visible only in the string formulation. In combination, these so-called U-duality symmetries put very strong restrictions on the possible non-perturbative properties of M-theory, and in this way provide a wealth of information about its dynamics.

A very fruitful approach towards unraveling the mysteries of M-theory is provided by the matrix formulation initiated in [9], known as 'Matrix theory.' Its basic postulate is that the full eleven-dimensional dynamics of M-theory can be captured by means of the many-body quantum mechanics of N D-particles of IIA string theory, in the limit where N is taken to infinity. From the eleven dimensional perspective, this limit can be interpreted as applying a Lorentz boost, so that all degrees of freedom end up with a very large momentum along the extra eleventh direction.

One can formally combine this limiting procedure with a decompactification limit of the eleventh direction, so that the IIA string and M-theory compactification manifolds become identified. In this way, a fundamentally new perspective on non-perturbative string theory arises, in which its complete particle spectrum in a given dimension becomes identified with a relatively convenient subset of states compactified to one dimension less. Though in essence a light-cone gauge description, Matrix theory thus manages to provide a rather detailed theoretical framework for exploring the microscopic dynamics of M-theory, while preserving many of the geometrical foundations of string theory. There is now indeed a rather well-developed understanding of how Matrix theory incorporates all perturbative degrees of freedom and interactions of type II strings [10] [11] [12].

In these notes we will review some of the recent developments in this approach to M-theory. In particular we will discuss in detail the recently proposed formulation of M-theory compactified on T^5, based on the 5+1-dimensional world-volume theory of N NS 5-branes [14]. We will pay special attention to the appearance of the M5-brane in this proposal, and with the aim of illuminating the direct correspondence between the framework of [14] and our earlier studies [4]. In addition we will highlight the main ingredients of Matrix string theory [12], and discuss its application to the study of the interacting string-like theory on the 5-brane worldvolume.

2. U-duality and the M5-brane

The degrees of freedom and interactions of M-theory compactified on a five-torus T^5 provide a representation of the maximal central extension of the 5+1 dimensional $\mathcal{N}=4$ supersymmetry algebra, which in a light-cone frame takes the form[1]

$$\begin{aligned}\left\{\widetilde{Q}^a_\alpha,\widetilde{Q}^b_\beta\right\} &= p_+\delta_{\alpha\beta}\mathbf{1}^{ab}\\ \left\{Q^a_\alpha,\widetilde{Q}^b_\beta\right\} &= p_I\gamma^I_{\alpha\beta}\mathbf{1}^{ab}+\delta_{\alpha\beta}Z^{ab}\\ \left\{Q^a_\alpha,Q^b_\beta\right\} &= p_-\delta_{\alpha\beta}\mathbf{1}^{ab}\end{aligned} \tag{2.1}$$

Here the internal labels $a,b=1,\ldots,4$ are $SO(5)$ spinor indices, and $I=1,\ldots,4$ label the four transverse uncompactified space dimensions. The central charge matrix Z^{ab} comprises a total of 16 internal charges, which split up in five quantized Kaluza-Klein momenta p_i with $i=1,\ldots,5$, ten charges $m_{ij}=-m_{ji}$ that specify the winding numbers of the M membranes around the five-torus, and one single charge q_5 that represents the wrapping number of the M5 brane. To make this interpretation more manifest, it is often convenient to expand Z_{ab} in terms of $SO(5)$ gamma matrices as

$$Z_{ab}=q_5\mathbf{1}_{ab}+p_i\gamma^i_{ab}+m_{ij}\gamma^{ij}_{ab}. \tag{2.2}$$

More generally, the decomposition of Z_{ab} into integral charges depends on the moduli of the T^5 compactification, which parametrize the coset manifold

$$\mathcal{M}=SO(5,5)/SO(5)\times SO(5). \tag{2.3}$$

The U-duality group is the subgroup of $SO(5,5)$ rotations that map the lattice of integral charges on to itself, and is thus identified with

$$\mathcal{U}=SO(5,5,\mathbf{Z}). \tag{2.4}$$

From the eleven dimensional viewpoint, the moduli space (2.3) is parametrized by the metric G_{mn} and three-form potential C_{mnk} defined on the internal manifold T^5. After replacing the 3-form C on T^5 by its Hodge-dual 2-form $B=*C$, this parametrization of $\mathcal{M}$ reduces to the familiar one from the toroidal compactification of type II string theory on T^5. This correspondence, which is strengthened by the fact that the U-duality group (2.4) takes the form of the T^5 T-duality group, has become a central theme in recent attempts to formulate the T^5 compactification of M-theory [4, 14].

To study the spectrum of M-theory on T^5, it is natural to start from the maximal dimensional object that can wrap around the internal manifold, which is the

[1]Hence α and β are spinor indices of the transversal $SO(4)$ rotation group.

M5 brane. From its description as a soliton it is known that its low energy collective modes form a tensor multiplet of the $\mathcal{N} = (2,0)$ world-brane supersymmetry, consisting of five scalars, an anti-symmetric tensor with self-dual field strength $T = dU$ and 4 chiral fermions. As shown in [4], this worldvolume field theory can be extended to include all the above 16 charges of M-theory on T^5 via an appropriate identification of flux configurations.

To make this concrete, consider a M5-brane with topology of $T^5 \times \mathbf{R}$ where $\mathbf{R}$ represents the world-brane time τ. To begin with, we can include configurations carrying a non-zero flux

$$m_{ij} = \int_{T^3_{ij}} dU \tag{2.5}$$

through the 10 independent three-cycles T^3_{ij} in T^5. An M5-brane for which these charges are non-zero represents a bound state with a corresponding number of membranes wrapped around the dual 2-cycle. The charges m_{ij} are part of a 16-dimensional spinor representation of the U-duality group $SO(5,5,\mathbf{Z})$, which is isomorphic to the odd (co)homology of the five-torus T^5. This observation motivates us to try to write the other charges q_5 and p_i as fluxes of 5- and 1-form field strengths through the 5-cycle and 1-cycles on T^5.

Before presenting this construction, let us mention that in course we will in fact arrive at a rather new notion of the M5-brane worldvolume theory, that is rather different from the usual one in terms of the physical location of the five-brane soliton in space-time. In particular, the M5-brane worldvolume theory we will arrive at lives on an auxiliary 5+1-dimensional parameter space, denoted by $\widetilde{T}^5$. This auxiliary space is only indirectly related to the embedded M5-brane soliton in physical space-time, via an appropriate geometrical interpretation of the worldvolume fields.

In this new set-up, we can introduce a winding number of the M5-brane via

$$q_5 = \int_{\widetilde{T}^5} dV, \tag{2.6}$$

where the 5-form dV is interpreted as the pull-back to the auxiliary torus $\widetilde{T}^5$ of the constant volume element on the target T^5. We would like to turn the 4-form potential V into an independent field, that is part of the world-brane theory.

To this end we imagine that the (2,0) world-brane tensor multiplet in fact provides an effective *light-cone description* of the M5-brane. Concretely this means that of the five scalars, we will interpret only four as describing transversal coordinates in space-time. This leaves us with one additional scalar field Y. Since the world-brane is 5+1-dimensional, this scalar can be dualized to a 4-form, which we will identify with the 4-form V used in (2.6). Hence we will assume the 5-form field strength $W = dV$ associated with V is normalized such that the associated flux is a non-negative integer.

Quite generally, in a light-cone formalism for extended objects one typically obtains a residual gauge symmetry under volume preserving diffeomorphisms [15]. For the case of the 5-brane these can be used to eliminate all dependence on the five compact embedding coordinates X^i (that map the auxiliary $\widetilde{T}^5$ into the target T^5), except for the volume-form

$$dV = dX^i \wedge dX^j \wedge dX^k \wedge dX^l \wedge dX^m \epsilon_{ijklm}. \tag{2.7}$$

The five spatial components of $*V^i$ can roughly be identified with the embedding coordinates X^i. In this correspondence the linearized gauge-transformation $V \to V + d\Lambda$ of the 4-form potential describes a volume preserving diffeomorphism, since it leaves dV invariant.

We can formalize the duality transformation between V and Y by taking W to be an independent 5-form and introducing Y as the Lagrange multiplier that imposes the Bianchi identity $dW = 0$. The fact that W has integral fluxes implies that Y must be a periodic field, *i.e.* $Y \equiv Y + 2\pi r$ with r integer. The 5 remaining charges p_i can thus be identified with the integer winding numbers of Y around the 5 one-cycles

$$p_i = \oint_{T^1_i} dY. \tag{2.8}$$

Since on-shell $\Pi_V \equiv \delta S/\delta \dot{V} = dY$, we deduce that dY is the canonical momentum for V. From this it is straightforward to show that these operators are indeed the generators of translations along the internal directions on the target space 5-torus. The formula for the M5-brane wrapping number now becomes

$$q_5 = \int_{\widetilde{T}^5} \Pi_Y, \tag{2.9}$$

with Π_Y the conjugate momentum to Y.

In this way all the 16 charges (q, p_i, m_{ij}) have been written as fluxes through the odd homology cycles on the M5-brane. These 16 bosonic zero modes form the world-volume superpartners of the 16 fermionic zero modes, that represent the space-time supersymmetries that are broken by the M5-brane soliton. Based on this result, it was shown in [4] that the (2,0) worldvolume theory of the M5-brane could be used to represent the full maximally extended space-time supersymmetry algebra (2.1).

The distinction between the above and the standard notion of the 5-brane world-volume is perhaps best explained in analogy with the worldsheet description of toroidally compactified string theory. According to the strict definition, the worldsheet of a string refers to its particular location in space-time, and thus also specifies a given winding sector. For strings, however, it is well known to be advantageous to combine into the worldsheet CFT all superselection sectors corresponding to all possible momenta and winding numbers. T-duality for example appears as a true symmetry of the worldsheet CFT only if it includes all sectors. The notion of the M5 worldbrane theory used here is indeed the direct analogue of this: just as for strings, the various wrapping numbers and momenta appear as quantized zero modes of the worldvolume fields. A second important analogy is that, just like the CFT Hilbert space includes unwrapped string states, we can in the above set-up also allow for unwrapped M5 brane states with $q_5 = 0$.

This naturally leads to the further suggestion that it should be possible to extend the M5 worldbrane theory in such a way that U-duality becomes a true symmetry (just like T-duality in toroidal CFT). In [4] it was shown that this can indeed be achieved by assuming that the worldvolume theory on $\widetilde{T}^5$ is described by a 5+1-dimensional second quantized string theory, whose low energy effective modes describe the (2,0) tensor multiplet. Via the $SO(5,5,\mathbf{Z})$ T-duality of this "micro-string theory", this produces a manifestly U-duality invariant description of the BPS states of all M-theory compactification down to 6 or more dimensions [4].

Until quite recently, only limited theoretical tools were available to analyze the detailed properties of this conjectured micro-string theory, or even to confirm its existence. The traditional methods usually start from the physical world volume of a macroscopic M5-brane soliton, and the micro-string theory is in fact not easily visible from this perspective. As seen e.g. from the identifications (2.8) and (2.9), the local physics on the target space torus T^5 is somewhat indirectly related to that on the parameter space $\widetilde{T}^5$ where the micro-string lives. The specification of a particular winding sector q_5, for instance, breaks the U-duality group to geometric symmetry group $SL(5,\mathbf{Z})$, and thus hides the stringy part of the micro-string T-duality symmetry. This illustrates that it may be misleading to think about the micro-strings directly as physical strings moving on the space-time location of the M5-brane soliton.

3. BPS Spectrum

Further useful information about the relation between the auxiliary and target space can be obtained by considering the masses of the various BPS states.

The world-brane theory carries a chiral $\mathcal{N}=2$ supersymmetry algebra. To write this algebra, it is convenient to introduce chiral $SO(4)$ spinor indices α and $\dot\alpha$ for the transverse rotation group (which represents an R-symmetry of the world-volume supersymmetry). We then have

$$\begin{aligned} \left\{Q^\alpha_a, Q^\beta_b\right\} &= \epsilon^{\alpha\beta}(\gamma^0_{ab}H + \gamma^i_{ab}(P_i+W_i)), \\ \left\{Q^{\dot\alpha}_a, Q^{\dot\beta}_b\right\} &= \epsilon^{\dot\alpha\dot\beta}(\gamma^0_{ab}H + \gamma^i_{ab}(P_i-W_i)). \end{aligned} \tag{3.1}$$

Here H is the world-volume Hamiltonian, and P_i is the integrated momentum flux through the five-torus

$$H = \int_{\widetilde{T}^5} T_{00}, \qquad P_i = \int_{\widetilde{T}^5} T_{0i}, \tag{3.2}$$

where T_{mn} denotes the world-volume energy-momentum tensor. Hence H and P_i represent the generators of time and space translation on the 5+1-dimensional parameter space-time. The operators W_i that appear in the algebra (3.1) represent a vector-like central charge, and is a reflection of the (possible) presence of string-like objects in the world-volume theory. This vector charge W_i indeed naturally combines with the momenta P_i into a 10 component vector representation of the $SO(5,5,\mathbf{Z})$ world-volume T-duality group.

The world-brane supercharges represent the unbroken part of space-time supersymmetry algebra. The other generators must be identified with fermionic zero-modes, that are the world-brane superpartners of the bosonic fluxes described above. In order to exactly reproduce the 5+1-dimensional space-time supersymmetry (2.1), however, we must project onto states for which the total momentum fluxes P_i and vector central charges W_i all vanish

$$P_i = 0, \qquad W_i = 0. \tag{3.3}$$

These are the analogues of the usual level matching relations of light-cone string theory. In addition, we must impose

$$H = p_- \tag{3.4}$$

which reflects the identification of the world-volume time coordinate τ with the space-time light-cone coordinate x^+. The mass-shell condition $p_+p_- = m^2$ (we assume for simplicity that all transverse space-time momenta p_I are zero) relates the space-time mass of M5-brane configuration to the world-volume energy H via

$$m^2 = p_+ H \tag{3.5}$$

This relation, together with identification of the central charge Z_{ab} as world-brane fluxes, uniquely specifies the geometric correspondence between the auxiliary torus $\widetilde{T}^5$ and the target space torus T^5. To make this translation explicit, let us denote the lengths of the five sides of each torus by $\widetilde{L}_i$ and L_i respectively, and let us fix the normalization of the fields dU and dY in accordance with the flux quantization formulas (2.5), (2.8), and (2.9). Since the left-hand sides are all integers, this in particular fixes the dimensionalities of the two-form U and scalar Y.

To simplify the following formulas, we will further use *natural units* both in the M-theory target space, as well as on the auxiliary space. In the target space, these natural units are set by the eleven dimensional Planck scale ℓ_p, while on the world-volume, the natural units are provided by the micro-string length scale $\widetilde{\ell}_s$. Apart from the size of the five torus $\widetilde{T}^5$, this string length $\widetilde{\ell}_s$ is indeed the only other length scale present in the 5+1-dimensional world-volume theory.

Now let us consider the physical masses of the various 1/2 BPS states of M-theory. These states can carry either pure KK momentum charge, or pure (non-intersecting) membrane and fivebrane wrapping number. From the target-space perspective, we deduce that these states have the following masses (in 11-dimensional Planck units $\ell_p = 1$)

$$\begin{aligned} &\textit{KK momenta:} && m^2 = (p_i/L_i)^2 \\ &\textit{membranes:} && m^2 = (m_{ij}L_iL_j)^2 \\ &\textit{fivebranes:} && m^2 = (q_5\, V_5)^2 \end{aligned} \tag{3.6}$$

where $V_5 = L_1 \ldots L_5$ denotes the volume of the M-theory five torus T^5 in Planckian units.

In the world-volume language these 1/2 BPS states all correspond to the supersymmetric ground states that carry a single flux quantum number. Their mass is therefore determined via the mass-shell relation (3.5) by the world-volume energy carried by the corresponding flux configuration. To deduce this energy, it is sufficient to note that at very long distances, the world-volume theory reduces to a non-interacting free field theory of the (2,0) tensor multiplet. Hence the flux contribution to the M5-brane Hamiltonian can be extracted from the free field expression[2]

$$H = \frac{1}{\beta} \int_{\widetilde{T}^5} d^5x \left[|dY|^2 + |dU|^2 + |\Pi_Y|^2 \right] \tag{3.7}$$

where β is an arbitrary parameter (to be related later on to the light-cone momentum p_+ in space-time). Now using (2.5), (2.8), and (2.9), we deduce that the 1/2 BPS states have the following worldvolume energies (in micro-string units $\widetilde{\ell}_s = 1$)

$$\begin{array}{lrcl} \textit{KK momenta:} & & & \dfrac{\widetilde{V}_5}{\beta}\left(\dfrac{p_i}{\widetilde{L}_i}\right)^2 \\[2ex] \textit{membranes:} & H & = & \dfrac{(m_{ij}\widetilde{L}_i\widetilde{L}_j)^2}{\beta\widetilde{V}_5} \\[2ex] \textit{fivebranes:} & H & = & \dfrac{q_5^2}{\beta\widetilde{V}_5} \end{array} \tag{3.8}$$

Here $\widetilde{V}_5 = \widetilde{L}_1 \ldots \widetilde{L}_5$ denotes the volume of the auxiliary space $\widetilde{T}^5$ in string units.

From the condition that all these three expressions reproduce the correct mass formulas (3.6) via (3.5), we can now read off the M-theory parameters in terms of the world-volume parameters. First of all we see that the two five tori T^5 and $\widetilde{T}^5$ in fact have the same *shape*, *i.e.* they are related by means of an overall rescaling

$$L_i = v\widetilde{L}_i \tag{3.9}$$

with v some dimensionless number. The remaining M-theory parameters we need to determine are V_5 and p_+, and we need to express them in terms of $\widetilde{V}_5$ and β in (3.7). A little algebra show that

$$V_5 = \frac{1}{\widetilde{V}_5^{2/3}}, \qquad p_+ = \frac{\beta}{\widetilde{V}_5^{1/3}}. \tag{3.10}$$

In particular we find that the wordvolume radii are related to the M-theory radii by

$$\widetilde{L}_i = \frac{L_i}{V_5^{1/2}}. \tag{3.11}$$

[2]Here we have used that the self-duality relation $\Pi_U = dU$. We further left out the transverse scalars X_I since their fluxes are set to zero.

Hence we see in particular that the decompactification limit of M-theory corresponds to the zero volume limit of the auxiliary torus.

Of course, here we have to take into account that T-duality of the micro-string theory can relate different limits of the $\widetilde{T}^5$ torus geometry. This fact plays an important role when we consider partial decompactifications. For example, if we send one of the radii of the M-theory five-torus, say L_5, to infinity, then (3.11) tells us that, after a T-duality on the remaining four compact directions, the worldvolume of $\widetilde{T}^5$ tends to infinity. In this way we deduce that the BPS states of the T^4 compactification of M-theory is described by the zero slope limit of the micro-string theory on a T^5 world-volume. This is of importance in understanding the $SL(5,\mathbf{Z})$ U-duality group of M-theory compactifications on T^4 [16, 17]. Indeed, all U-duality symmetries for T^d compactifications with $d \leq 5$ can be understood in this way as geometric symmetries of the micro-string model.

Finally, let us mention that as soon as one considers states with more general fluxes, they are at most 1/4 BPS and the mass formula becomes more involved. One finds

$$m^2_{\rm BPS} = m_0^2 + 2V_5\sqrt{(\mathcal{K}_i/L_i)^2 + \frac{1}{V_5^2}(\mathcal{W}_i L_i)^2}, \tag{3.12}$$

with

$$\begin{aligned} m_0^2 &= (q_5 V_5)^2 + (p_i/L_i)^2 + (m_{ij}L_iL_j)^2, \\ \mathcal{K}_i &= q_5\, p_i + \frac{1}{2}\epsilon_{ijklm} m_{jk} m_{lm}, \\ \mathcal{W}_i &= m_{ij} p_j. \end{aligned} \tag{3.13}$$

These formulas are completely duality invariant, and were reproduced via a world-volume analysis in [4]. A crucial new ingredient in the description of BPS states with a non-zero value for $\mathcal{K}_i$ is that they necessarily contain worldbrane excitations with non-zero momentum on $\widetilde{T}^5$,[3] while $\mathcal{W}_i \neq 0$ necessarily describes states with non-zero winding number for the micro-strings. Hence, without the micro-string degrees of freedom, the M5-brane would have had no BPS-states with non-zero value for $\mathcal{W}_i$. The geometrical interpretation of the bilinear expressions for $\mathcal{K}_i$ and $\mathcal{W}_i$ will become more clear in the following sections.

4. M-theory from NS 5-branes

The above attempt to unify the complete particle spectrum of a given M-theory compactification in terms of one single world-volume theory has now found a natural place and a far reaching generalization in the form of the matrix formulation of M-theory initiated in [9]. While the previous sections were in part based on

[3]This is implied by the level-matching condition $P_i = 0$, since $\mathcal{K}_i$ in fact represents the zero-mode contribution to P_i. Notice further that for $q_5 \neq 0$, $\mathcal{K}_i$ contains a contribution proportional to q_5 times the space-time KK momentum p_i, which indeed reflects the geometric relation between world-volume and target space translations.

an inspired guess, motivated by U-duality and the form of the space-time supersymmetry algebra, in the following sections we will describe how the exact same structure can be derived from the recently proposed Matrix description [14] of M-theory on T^5. So we will again start from the beginning.

According to the Matrix theory conjecture, one can represent all degrees of freedom of M-theory via an appropriate large N limit of the matrix quantum mechanics of the D0-branes [9]. The true extent this proposal becomes most visible when one combines this large N limit with a decompactification limit $R \to \infty$ of the 11-th M-theory direction. The ratio

$$p_+ = N/R \tag{4.1}$$

represents the light-cone momentum in the new decompactified direction.

U-duality invariance implies that any charge of compactified IIA string theory can in principle be taken to correspond to the eleventh KK momentum N. One can thus reformulate the original matrix conjecture of [9] by applying some duality symmetry of the type II string to map the D0-branes to some other type of branes. Compactification of matrix theory indeed generically proceeds by using the maximal T-duality to interchange the D0 brane charge by that of the maximal dimensional D p-brane that can wrap around the internal manifold. In our specific example of T^5 this would be the D5-brane. Alternatively, as was first suggested in [14] (following up earlier work [16] [17] [4]), one can choose to identify this charge N with the NS five-brane wrapping number around the five-torus T^5.

In the following table we have indicated the chain of duality mappings that provides the dictionary between the original matrix set-up of [9] and the dual language of [14]:

M on T^5	$=$ IIA on T^5	$\xrightarrow{T_5}$ IIB on $\hat{T}^5$	$\xrightarrow{S}$ IIB on $\hat{T}^5$	$\xrightarrow{T_5}$ IIA on $\tilde{T}^5$
N	$D0$	$D5$	$NS5$	$NS5$
p_i	KK	$NS1$	$D1$	$D4$
m_{ij}	$D2$	$D3$	$D3$	$D2$
q_5	$NS5$	$NS5$	$D5$	$D0$

We notice that all 16 quantized charges, that make up the central charge matrix Z_{ab}, have become D-brane wrapping numbers in the last two columns, and furthermore that the $SO(5,5,\mathbf{Z})$ U-duality symmetry is mapped onto the (manifest) T-duality symmetry of the type II theory. Note further that the D-brane charge vector indeed transforms in a spinor representation of the T-duality group.

The string coupling, string scale and dimensions of the five torus are related through this duality chain as follows.[4] The chain consists of a T_5 duality on all five direction of the T^5, an S-duality of the IIB theory, and finally again a maximal T-duality on T^5. The end result is a mapping from large N D-particle quantum dynamics to that of N NS five-branes in IIA theory.

On the left, we have started with the conventional definition of M-theory compactified on a five-torus T^5, with dimensions L_i, $i = 1, \ldots, 5$, as the strong coupling limit of type IIA string theory on the same torus T^5. Under the first T_5 duality, a D0-brane becomes the D5-brane wrapped around the dual torus $\widehat{T}_5$ with dimensions

$$\Sigma_i = \frac{\ell_s^2}{L_i} = \frac{1}{RL_i}. \tag{4.2}$$

Here we used (1.2), and again work in eleven dimensional Planck units $\ell_p = 1$. Following the usual rule of T-duality, the string coupling g_s gets a factor of the volume in string units, so the dual coupling $\widehat{g}_s$ becomes (see again (1.2))

$$\widehat{g}_s = \frac{g_s \ell_s^5}{V_5} = \frac{1}{RV_5} \tag{4.3}$$

where V_5 is again the five-volume of T^5 (in Planck units). The subsequent S-duality transformation maps the D5-brane onto the IIB NS5-brane, inverts the string coupling, and changes the string scale by a factor of $\sqrt{\widehat{g}_s}$. So we have

$$\tilde{g}_s' = V_5 R, \qquad \tilde{\ell}_s^2 = \frac{1}{R^2 V_5}. \tag{4.4}$$

The size of the five torus of course remains unchanged under the S-duality.

Finally we perform the T_5 duality again, so that N counts IIA fivebrane charge. The dimensions of the resulting five-torus $\widetilde{T}^5$ are

$$\tilde{L}_i = \frac{L_i}{RV_5}. \tag{4.5}$$

while the string coupling $\widetilde{g}_s$ of the IIA string after the $T_5 S T_5$ transformation becomes

$$\tilde{g}_s^2 = \frac{R^2}{V_5} \tag{4.6}$$

The string scale of the IIA theory is the same as in (4.4).

[4]For this section we benefited from valuable discussions with G. Moore. A similar presentation can indeed be found in [27].

In [14] convincing arguments were given for the existence of a well-defined limit of type II string theory, that isolates the world volume dynamics of N NS five-branes from the 9+1-dimensional bulk degrees of freedom. The key observation made in [14] is that, even in the decoupling limit $\tilde{g}_s \to 0$ for the bulk string interactions, one is still left with a non-trivial interacting 5+1 dimensional theory that survives on the world volume of the 5-branes. This theory depends on only two dimensionless parameters, namely N and the size of the torus $\tilde{T}^5$ in units of the string length $\tilde{\ell}_s$. If one keeps the last ratio finite, the system will contain string like degrees of freedom. According to the proposal put forward in [14], this world volume theory of N NS 5-branes in the limit of large N may provide a complete matrix theoretic description of M-theory compactified on T^5.

This proposal in fact needs some explanation, since as seen from (4.6) this decoupling limit $\tilde{g}_s \to 0$ in fact contradicts the decompactification limit $R \to \infty$ that is needed for obtaining M-theory on T^5. The resolution of this apparent contradiction should come from the particular properties of the world-volume theory of N 5-branes for large N. As we will see later, via a mechanism familiar e.g. from studies of black holes in string theory [24] or matrix string theory [10][11][12], the system of N wrapped NS 5-branes contains a low energy sector with energies of order $1/N$ times smaller than the inverse size of the five-torus $\tilde{T}^5$. These low energy modes are the ones that will become the M-theory degrees of freedom, and the large N limit must be taken in such a way that only these states survive. This implies that the typical length scale of the relevant configurations is larger by a factor of N than the size of $\tilde{T}^5$.

When one analyzes this low lying energy spectrum for very large N, one discovers that it becomes essentially independent of the string coupling constant $\tilde{g}_s$, provided $\tilde{g}_s$ does not grow too fast with N. It thus seems reasonable to assume that in an appropriate large N limit, the world-volume theory of the NS 5-branes itself becomes independent of $\tilde{g}_s$. This opens up the possibility that the weak coupling limit that was used in [14] for proving the existence of the 5+1-dimensional world-brane theory, in fact coincides with the strong coupling regime that corresponds with the decompactification limit of M-theory.

Before we continue, let us remark that this Matrix theoretic set-up can already be seen to reproduce many qualitative and quantitative features of the framework described in the previous sections. Evidently, the world-volume theory of the N NS 5-branes plays a very analogous role to the proposed M5-brane world-volume theory of sections 2 and 3. Most importantly, in both set-ups the U-duality symmetry of M-theory becomes identified with a T-duality of a 5+1-dimensional worldvolume string theory. In addition, as we will see a little later, all 16 charges of M-theory arise in the present Matrix theory set-up [14] via the exact same identification of fluxes as described in section 2.

Finally, by comparing the formulas (4.4) and (4.5) with those of the previous section, we see that the size and shapes of the respective auxiliary five-tori $\widetilde{T}^5$ and $\tilde{T}^5$ (notice however the subtle difference in notation) are expressed via an identical set of relations in terms of those the M-theory torus T^5. The micro-string torus $\widetilde{T}^5$ of sections 2 and 3 is indeed related the Matrix theory torus $\tilde{T}^5$ used above via a rescaling with a factor N. Since (as will be elaborated later on) the micro-string scale $\widetilde{\ell}_s$ is also N times larger than the fundamental string scale, the two auxiliary five tori have the same size relative to the respective string scales. This fact ensures that the standard T-duality symmetry of the small $\tilde{T}^5$ indeed translates into an isomorphic T-duality group that acts on the large $\widetilde{T}^5$ on which the micro-string lives.

Via this formulation [14], Matrix theory thus naturally incorporates our earlier ideas on the M5-brane. More importantly, however, it also provides a concrete prescription for including interactions, both among the M5-branes as well as for the micro-string theory inside its world-volume.

5. SYM theory on T^5

One possible starting point for concretizing this matrix description of M-theory on T^5 is provided by the large N limit of 5+1-dimensional super Yang-Mills theory

$$S = \frac{1}{g^2}\int d^6x\, \mathrm{tr}\left(\frac{1}{4}F_{\mu\nu}^2 + \frac{1}{2}(D_\mu X^I)^2 \right.$$
$$\left. + \psi\Gamma^i D_i\psi + \psi\Gamma^I[X^I,\psi] + \frac{1}{4}[X^I,X^J]^2\right)$$

Here $I = 1,\ldots,4$ labels the 4 Higgs scalar fields. This theory can be thought of as describing the low energy collective modes of N D5 branes of IIB string theory (the second column in the above table), wrapped around the dual five-torus $\hat{T}^5$. In terms of the parameters above, the Yang-Mills coupling of the D5-brane world-volume theory is then

$$\frac{1}{g_{YM}^2} = \frac{1}{\widehat{g}_s\ell_s^2} = V_5 R^2 \tag{5.1}$$

Alternatively, we can apply an S-duality and view the large N SYM model as the low energy world volume theory of N IIB NS 5-branes. Note that the above YM coupling then becomes equal to the fundamental NS string tension $g_{YM} = \widetilde{\ell}_s$.

According to the prescription of Matrix theory, finite energy states of M-theory on T^5 are obtained by restricting to states of the SYM model with energies of order $1/N$. Since these states are all supported at scales much larger than the string scale, it seems an allowed approximation to discard the stringy corrections to the $U(N)$ SYM description of the NS 5-brane dynamics. An obvious subtlety with making this low-energy truncation is that the 5+1 dimensional SYM model in itself does not represent a well-defined renormalizable field theory. It is conceivable, however, that via the combined large N and IR limit that is needed for Matrix theory, one can still extract well-defined dynamical rules for the subset of the SYM degrees of freedom that are relevant for the description of M-theory on T^5.

A second remark is that the Yang-Mills description in fact seems incomplete. Namely, to describe all possible quantum numbers, one should be able to include all possible bound states of the NS 5-branes with the various D-branes. Generally these are described by turning on appropriate fluxes of the SYM theory. For the

D1-branes and D3-branes, the corresponding winding number are represented via the electric and magnetic flux quantum numbers

$$p_i = \int_{T^4_i} \operatorname{tr} E, \qquad m_{ij} = \int_{T^2_{ij}} \operatorname{tr} F, \tag{5.2}$$

through the dual 4 and 2-cycles of the $\widehat{T}^5$, respectively. The D5 brane wrapping number, however, does not seem to appear naturally as a flux of the SYM theory. This fact has so far been the main obstacle for incorporating the transverse M5 brane into Matrix theory.

Central in the correspondence between the large N SYM theory and M-theory on T^5 is the fact that (apart from this problem of the missing q_5-charge) the maximally extended supersymmetry algebra (2.1) can be represented in terms of the SYM degrees of freedom [20]. The construction is quite standard from the general context of string solitons, and relies on the identification of the world-volume supersymmetry charges of the SYM model

$$\begin{aligned} Q \; = \; & \int \operatorname{tr}\Big[\bar{\theta}\Big(\gamma^i E_i + \gamma^I \Pi_I \\ & -\frac{1}{2}(\gamma^{ij}F_{ij} + \gamma^{iJ}D_i X_J + \gamma^{IJ}[X_I, X_J])\Big)\Big] \end{aligned} \tag{5.3}$$

with the generators the unbroken space-time supersymmetries, while the generators of the broken supertranslations are represented by the fermion zero modes

$$\tilde{Q} = \int \operatorname{tr} \theta. \tag{5.4}$$

The dynamical supersymmetry charges (5.3) generate an algebra of the exact form (3.1), where

$$P_i = \int \operatorname{tr}(E_i F_{ij} + \Pi^I D_i X + \theta^T D_i \theta) \tag{5.5}$$

is the momentum flux and where

$$W_i = \int \operatorname{tr}(F \wedge F)_i \tag{5.6}$$

is the instanton winding number. Hence the SYM field theory explicitly reveals the presence of the string like degrees of freedom in the form of the non-abelian instanton configurations. As we will describe in a later section, the long distance dynamics of these instanton strings may be used as a quantitative description of the interacting micro-string theory.

If k_i denotes the $SU(N)$ Yang-Mills instanton number, we evaluate

$$NW_i = \frac{1}{2}(m \wedge m)_i + Nk_i. \tag{5.7}$$

As we will make more explicit in a moment, the right-hand side can indeed be seen to correspond to a total string winding number. [5] To see this (and to get more insight into how one might recover the M5-brane from the present set-up) it will be useful to consider the world-volume properties of the IIA and IIB NS 5-branes from a slightly different perspective.

6. Winding Microstrings

After the ground breaking work [22] on the entropy of 5+1-dimensional black holes, it became clear that the internal dynamics of the string theory 5-branes are most effectively described in terms of an effective world-volume string theory [4, 24]. These string-like degrees of freedom, which in the following we will also call micro-strings, arise in part due to one-dimensional intersections [22] that occur between D-branes, but also due to fundamental NS strings trapped inside NS 5-branes [21][14]. As indicated in figure 1 , the relation between the world-volume string

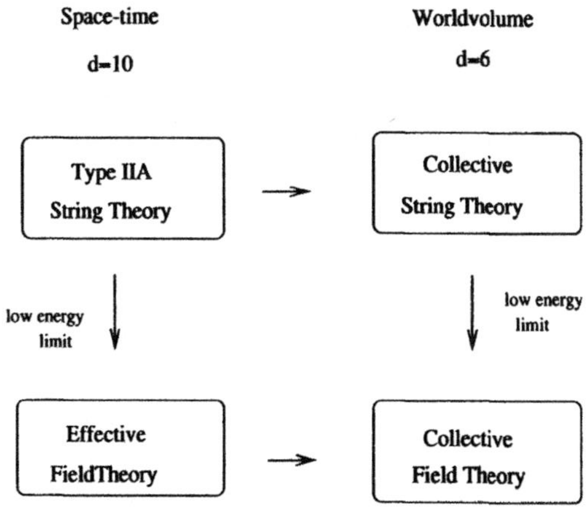

Figure 1. The effective world-volume dynamics of the solitonic NS 5-brane defines a $d = 6$ "collective string theory", whose low-energy degrees of freedom coincide with the usual collective fields.

theory and the low-energy collective modes of the 5-branes indeed parallels that between the fundamental type II strings and the low energy effective field theory in space-time. This correspondence predicts in particular the precise form of the micro-string ground states.

In the past year, this effective string theory has been developed into a remarkably successful framework for describing the statistical and dynamical properties (such as absorption and emission processes) of the 5-brane black holes. Until recently, however, it was unclear to what extent the corresponding 5+1-dimensional

[5] This is true provided the $U(N)$ symmetry is essentially unbroken by the Higgs scalars. More generally, if some of the Higgs scalars have expectation values that are substantially different from zero, W_i will naturally split up into separate contributions corresponding to instantons of the smaller unbroken subgroups.

string could really be isolated as a separate theory, that exists independently from the 10-dimensional type II string. The elegant argumentation of [14] provides convincing evidence that this is indeed the case. We will now collect some useful topological properties of the micro-string theory.

A D-brane in type IIB string theory has odd dimension, and thus a general configuration of D-branes wound around the internal manifold T^5 will be characterized by a 16 component vector $\mathbf{q}^B \in H_{odd}(T^5, \mathbf{Z})$ in the odd integral homology of T^5. Hence the corresponding D-brane intersection strings carry a total winding number given by the intersection form on $H_{odd}(T^5, \mathbf{Z})$, which in terms of our labeling of charges (see Table 1 in section 3) takes the form

$$\frac{1}{2}(\mathbf{q}^B \wedge \mathbf{q}^B)_i = q_5 p_i + \frac{1}{2}(m \wedge m)_i \tag{6.1}$$

Fundamental NS strings can also manifest themselves as string-like objects within the 5-brane world volume. Inside a bound state with N NS 5-branes, each NS string can in fact break up into N separate strings with reduced string tension $1/N\alpha'$. These fractional strings are also intersection strings, and string duality indeed implies that they are indistinguishable (in the sense of statistics) from the above D-brane intersection strings. Denoting the NS string winding charge by k_i, the total number of intersection strings is given by[6]

$$NW_i^B = \frac{1}{2}(m \wedge m)_i + q_5 p_i + N k_i \tag{6.2}$$

This formula generalizes the formula (5.7) for the instanton winding charge to the case where $q_5 \neq 0$. Notice further the correspondence with equation (3.13).

Let us now switch to the IIA perspective. In this case D-branes have even dimensions and thus D-brane configurations are characterized by $\mathbf{q}^A \in H_{even}(T^5, \mathbf{Z})$, and string-like intersections occur between 2- and 4-branes. Thus their total winding number is determined via

$$\frac{1}{2}(\mathbf{q}^A \wedge \mathbf{q}^A)_i = m_{ij} p_j \tag{6.3}$$

which is the intersection form on $H_{even}(T^5, \mathbf{Z})$. Hence, via the same mechanism as described for the IIB case, the total micro-string winding number (including the intersection strings between fundamental strings the N fivebranes) now takes the form

$$NW_i^A = m_{ij} p_j + N k_i \tag{6.4}$$

where k_i denotes the IIA string winding number.

As the IIA theory and IIB model are related via T-duality, the two micro-string theories are actually the same, except that the total winding numbers W_i get interchanged with the total momentum fluxes P_i. From the above formulas it

[6]Here we assume that all strings and branes are at the same location in the non-compact space-time. See also previous footnote.

is easy to see that the flux contribution to e.g. the IIA string winding number W_i^A is indeed identical to the flux contribution to the IIB momentum flux P_i^B given in (5.5). A similar check can also be done for the IIB winding charge W_i^B and the dual IIA momentum flux P_i^A.

Below we have indicated the Matrix-theory interpretation the total momentum flux P_i and micro-string winding number W_i on the auxiliary T^5, as obtained via the $T_5\,S\,T_5$ duality chain:

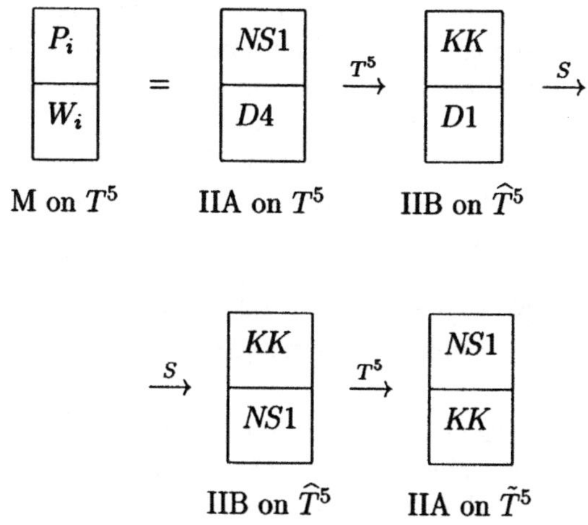

Hence non-zero values of the charges P_i and W_i in Matrix theory indicate the presence of longitudinally wrapped M2 and M5 branes. Finite energy M-theory configurations are therefore selected by imposing level matching constraints $P_i = 0$ and $W_i = 0$, which in the large N NS 5-brane system (the last two columns) project onto the sector with vanishing KK momenta and NS winding number. The same level matching were introduced already in section 3 from the condition that the space-time supersymmetry algebra (2.1) be reproduced.

7. The Matrix M5-brane

Let us now put the description of sections 2 and 3 of the M5-brane in a proper Matrix theoretic perspective. The correspondence will be most clear in the formulation of Matrix theory as the large N limit of N IIA 5-branes.

The fields parametrizing the low energy collective modes of a single IIA 5-brane form an $\mathcal{N}=(2,0)$ tensor multiplet identical to that used in sections 2 and 3. For the compactification on $\tilde{T}^5$, four of the five scalars represent the transverse oscillations of the 5-brane in non-compact directions, while one scalar field Y is extracted from one of the RR fields and is indeed periodic. For N IIA 5-branes, as long as they are far apart, we can separate the worldvolume theory into N different (2,0) tensor multiplets. When they come close together, however, new types of interactions occur. A precise low energy description of this interacting theory is unfortunately not yet available at this point, although some new insights into its structure have been obtained.[7]

Particularly useful information can be extracted by considering the IIA NS 5-branes from the perspective of M-theory[8]. As first pointed out in [26], M2-branes can end on M5-branes via open boundaries in the form of closed strings attached to the M5-brane worldvolume. These membrane boundaries are charged with respect to the self-dual 3-form field strength $T = dU$ on the worldbrane. If we consider an array of N M5-branes, we thus deduce that there are (at least) two basic types of string-like objects on the 5+1-dimensional world volume: strings with variable tension that come from membranes stretched between two different 5-branes, and secondly, strings with fixed tension that correspond to membranes stretched around the 11th direction and returning to the same 5-brane. The first type of strings are charged with respect to (the difference of) the 3-forms of the two 5-branes that are connected by the 2-brane. These charged strings are often referred to as tension-less or self-dual strings. The second type of strings, on the other hand, are neutral with respect to the self-dual 3-form. They represent the micro-strings, *i.e.* the fractionated fundamental IIA strings trapped inside the worldvolume of the N NS fivebranes. The collection of all degrees of freedom together look like an (as yet unknown) non-abelian generalization of the abelian (2,0) theory.

Charge conservation tells us that the charged strings with non-trivial winding numbers around the T^5 are stable against breaking and unwinding. For the neutral strings, on the other hand, there appears to be no similar world-brane argument available that would demonstrate their dynamical stability. From the space-time perspective, however, the micro-string winding number is identified with the total NS string winding number (times N), and therefore represents a genuine conserved charge. This conserved quantity is the vector central charge W_i discussed above.

Among the low energy degrees of freedom of the N IIA 5-brane bound state, we can isolate an overall (2,0) tensor multiplet, that describes the collective center of mass motion of the 5-branes. With the help of this decoupled tensor multiplet, we can include the configurations with non-vanishing D-brane charges by turning on the three types of fluxes m_{ij}, p_i and q_5 as described in section 2. The respective IIA and M-theory interpretation of these fluxes are as indicated in table 1 in section 4. In this way we see that Matrix theory indeed exactly reproduces the description in section 2. In addition, we find that the micro-string degrees of freedom, that were first postulated on the basis of U-duality symmetry, have now found a concrete origin.

Based on this new Matrix perspective, we can now obtain a much more detailed description of the M5-brane theory, based on the following proposed Matrix definition of its world-volume theory:

[7]An (incomplete) list of references include [26][25] [13] [14] [43][44].

[8]Here in fact we do *not* mean the M(atrix) theory perspective related via the $T_5\ S\ T_5$ duality chain of section 3, but rather the M-theory whose S_1 compactification directly gives the IIA theory of the last column of Table 1 in section 4.

The world-volume theory of the M5-brane at light-cone momentum $p_+ = N/R$ is identical to that of a bound state of N IIA NS 5-branes, subject to ten level matching constraints (3.3). The KK momenta, M2 and M5-brane wrapping numbers are in this correspondence identified with respectively the D4, D2 and D0-brane charges.

In essence this identification amounts to a definition of the M5-brane as the collection of all pointlike bound states of Matrix M-theory on $\widetilde{T}^5$, including those with zero fivebrane wrapping number. Via this proposed definition, all the properties postulated in [4] have become manifest. The physical motivation and interpretation of the above proposal has been discussed already in section 2. It would clearly be worthwhile to investigate the correspondence between this characterization of the M5-brane world volume theory and the more conventional low-energy formulation obtained from 11-dimensional supergravity [30].

In the following sections we will to obtain some more detailed dynamical information about the 5+1-dimensional micro-string theory.

8. S_N Orbifolds and 2nd Quantized Strings

Before continuing our journey through the matrix world of M-theory, it will be useful to present a mathematical identity that will be instrumental in establishing a precise link between the present formalism for M-theory and the standard geometric framework of perturbative string theory.

Consider the supersymmetric sigma model on the orbifold target space, defined by the symmetric product space

$$S^N M = M^N / S_N. \tag{8.1}$$

Here M denotes some target manifold, and S_N denotes the symmetric group, the permutation group of N elements. The two-dimensional world-sheet is taken to be a cylinder parametrized by coordinates (σ, τ) with σ between 0 and 2π. The correspondence we will show is that this sigma model, in the limit $N \to \infty$, provides an *exact* light-cone gauge description of the Hilbert space of *second quantized* string theory on the target space-time $M \times \mathbf{R}^{1,1}$.

The S_N symmetry is a gauge symmetry of the orbifold model, that permutes all N coordinate fields $X_I \in M$, where $I = 1, \ldots, N$ labeles the various copies of M in the product M^N. Therefore we can define twisted sectors in which, if we go around the space-like S^1 of the world-sheet, the fields $X_I(\sigma)$ are multivalued, via

$$X_I(\sigma + 2\pi) = X_{g(I)}(\sigma) \tag{8.2}$$

where g denotes an element of the S_N permutation group. A schematic depiction of these twisted sectors is given in figure 2.

The Hilbert space of this S_N orbifold field theory is thus decomposed into twisted sectors labeled by the conjugacy classes of the orbifold group S_N [31],

$$\mathcal{H}(S^N M) = \bigoplus_{\text{partitions } \{N_n\}} \mathcal{H}_{\{N_n\}}. \tag{8.3}$$

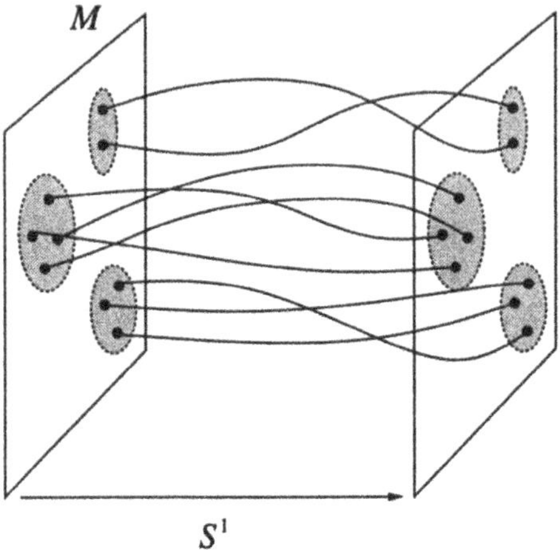

Figure 2. A twisted sector corresponds to a configuration with strings of various lengths.

Here we used that for the symmetric group, the conjugacy classes $[g]$ are characterized by partitions $\{N_n\}$ of N

$$\sum_n nN_n = N, \tag{8.4}$$

where N_n denotes the multiplicity of the cyclic permutation (n) of n elements in the decomposition of g

$$[g] = (1)^{N_1}(2)^{N_2}\ldots(s)^{N_s}. \tag{8.5}$$

In each twisted sector, one must further keep only the states invariant under the centralizer subgroup C_g of g, which takes the form

$$C_g = \prod_{n=1}^{s} S_{N_n} \times \mathbf{Z}_n^{N_n}. \tag{8.6}$$

Here each factor S_{N_n} permutes the N_n cycles (n), while each $\mathbf{Z}_n$ acts within one particular cycle (n).

Corresponding to this factorisation of $[g]$, we can decompose each twisted sector into the product over the subfactors (n) of N_n-fold symmetric tensor products of appropriate smaller Hilbert spaces $\mathcal{H}_{(n)}$

$$\mathcal{H}_{\{N_n\}} = \bigotimes_{n>0} S^{N_n}\mathcal{H}_{(n)} \tag{8.7}$$

where[9]

$$S^N\mathcal{H} = \Big(\underbrace{\mathcal{H}\otimes\ldots\otimes\mathcal{H}}_{N\ times}\Big)^{S_N}. \tag{8.8}$$

[9]Here the symmetrization is assumed to be compatible with the grading of $\mathcal{H}$. In particular for pure odd states S^N corresponds to the exterior product $\wedge^N$.

The spaces $\mathcal{H}_{(n)}$ in the above decomposition denote the $\mathbf{Z}_n$ invariant subsector of the space of states of a single string on $\mathbf{R}^8 \times S^1$ with winding number n. We can represent this space via a sigma model of n coordinate fields $X_I(\sigma) \in M$ with the cyclic boundary condition

$$X_I(\sigma + 2\pi) = X_{I+1}(\sigma), \tag{8.9}$$

for $I \in (1, \ldots, n)$. We can glue the n coordinate fields $X(\sigma)$ together into one single field $X(\sigma)$ defined on the interval $0 \leq \sigma \leq 2\pi n$. Hence, relative to the string with winding number one, the oscillators of the long string that generate $\mathcal{H}_{(n)}$ have a fractional $\frac{1}{n}$ moding. The group $\mathbf{Z}_n$ is generated by the cyclic permutation

$$\omega: \ X_I \rightarrow X_{I+1} \tag{8.10}$$

which via (8.9) corresponds to a translation $\sigma \rightarrow \sigma + 2\pi$. Thus the $\mathbf{Z}_n$-invariant subspace consists of those states for which the fractional left-moving minus right-moving oscillator numbers combined add up to an integer.

It is instructive to describe the implications of this structure for the Virasoro generators $L_0^{(i)}$ of the individual strings. The total L_0 operator of the S_N orbifold CFT in a twisted sector given by cyclic permutations of length n_i decomposes as

$$L_0^{tot} = \sum_i \frac{L_0^{(i)}}{n_i}. \tag{8.11}$$

Here $L_0^{(i)}$ is the usual canonically normalized operator in terms of the single string coordinates $X(\sigma), \theta(\sigma)$ defined above. The meaning of the above described $\mathbf{Z}_n$ projections is that it requires that the contribution from a single string sector to the total world-sheet translation generator $L_0^{tot} - \overline{L}_0^{tot}$ is integer-valued. From the individual string perspective, this means that $L_0^{(i)} - \overline{L}_0^{(i)}$ is a multiple of n_i.

To recover the Fock space of the second-quantized type IIA string we now consider the following large N limit. We send $N \rightarrow \infty$ and consider twisted sectors that typically consist of a finite number of cycles, with individual lengths n_i that also tend to infinity in proportion with N. The finite ratio n_i/N then represents the fraction of the total p_+ momentum (which we will normalize to $p_+^{tot} = 1$) carried by the corresponding string[10]

$$p_+^{(i)} = \frac{n_i}{N}. \tag{8.12}$$

So, in the above terminology, only long strings survive. The usual oscillation states of these strings are generated in the orbifold CFT by creation modes $\alpha_{-k/N}$ with k finite. Therefore, in the large N limit only the very low-energy IR excitations of the S_N-orbifold CFT correspond to string states at finite mass levels.

The $\mathbf{Z}_n$ projection discussed above in this limit effectively amounts to the usual uncompactified level-matching conditions $L_0^{(i)} - \overline{L}_0^{(i)} = 0$ for the *individual* strings,

[10]The fact that the length n_i of the individual strings specifies its light-cone momentum is familiar from the usual light-cone formulation of string theory.

since all single string states for which $L_0 - \overline{L}_0 \neq 0$ become infinitely massive at large N. The total mass-shell condition reads (here we put the string tension equal to $\alpha' = 1$)

$$p_-^{tot} = NL_0^{tot} \tag{8.13}$$

The mass-shell conditions of the individual strings is recovered by defining the individual light-cone momenta via the usual relation

$$p_+^{(i)}p_-^{(i)} = L_0^{(i)}. \tag{8.14}$$

All strings therefore indeed have the same string tension.

In the case that $M = \mathbf{R}^8$, we can choose the fields X^i, θ^a, $\theta^{\dot{a}}$ to transform respectively in the $\mathbf{8}_v$ vector, and $\mathbf{8}_s$ and $\mathbf{8}_c$ spinor representations of the $SO(8)$ R-symmetry group of transversal rotations. The orbifold model then becomes exactly equivalent to the free string limit of the uncompactified type IIA string theory in the Green-Schwarz light-cone formulation. The spacetime supercharges are given by (in the normalization $p_+^{tot} = 1$)

$$\begin{aligned} Q^a &= \frac{1}{\sqrt{N}} \oint d\sigma \sum_{I=1}^{N} \theta_I^a, \\ Q^{\dot{a}} &= \sqrt{N} \oint d\sigma \sum_{I=1}^{N} G_I^{\dot{a}}, \end{aligned}$$

with

$$G_I^{\dot{a}}(z) = \gamma_{a\dot{a}}^i \theta_I^a \partial x_I^i \tag{8.15}$$

In the S_N orbifold theory, the twisted sectors of the orbifold corresponding to the possible multi-string states are all superselection sectors. We can introduce string interactions by means of a local perturbation of the orbifold CFT, that generates the elementary joining and splitting of strings.

9. String Interactions via Twist Fields

Consider the operation that connects two different sectors labeled by S_N group elements that are related by multiplication by a simple transposition. It is relatively easy to see that this simple transposition connects a state with, say, one string represented by a cycle (n) decays into a state with two strings represented by a permutation that is a product of two cycles $(n_1)(n_2)$ with $n_1 + n_2 = n$, or vice versa. So the numbers of incoming and outgoing strings differ by one. Pictorially, what takes place is that the two coinciding coordinates, say X_1 and $X_2 \in M$, connect or disconnect at the intersection point, and as illustrated in figure 3 this indeed represents an elementary string interaction.

In the CFT this interaction is represented by a local operator, and thus according to this physical picture one may view the interacting string theory as obtained

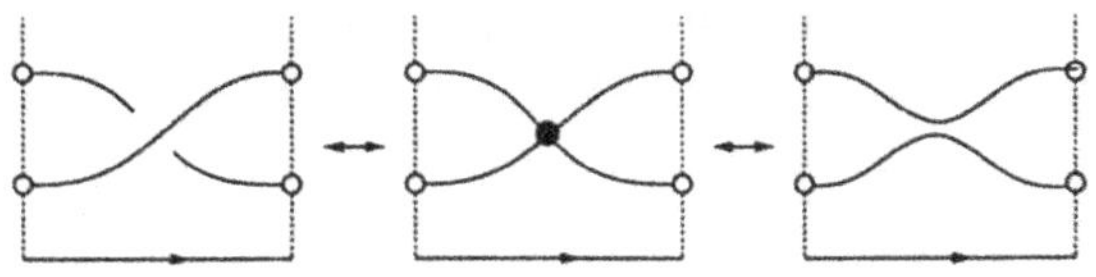

Figure 3. The splitting and joining of strings via a simple permutation.

via a perturbation of the S_N-orbifold conformal field theory. In first order, this perturbation is described via a modification of the CFT action

$$S = S_{CFT} + \lambda \int d^2z\, V_{int} \tag{9.1}$$

where V_{int} is an appropriate twist operator, that generates the just described simple transposition of string coordinates.

Let us consider this operator in more detail for $M = \mathbf{R}^8$. It is clear from the above discussion that the interaction vertex V_{int} will be a twist field that interchanges two eigenvalues, say X_1 and X_2. It acts therefore as a $\mathbf{Z}_2$ reflection on the relative coordinate $X_- = X_1 - X_2$, which has 8 components. So we are essentially dealing with a standard $\mathbf{R}^8/\mathbf{Z}_2$ supersymmetric orbifold. This orbifold has well-known twist fields, which will receive contributions of the bosons and fermions. The bosonic twist operator σ is defined via the operator product

$$\partial X^i_-(z)\, \sigma(0) \sim z^{-\frac{1}{2}} \tau^i(0) \tag{9.2}$$

Since the fermions transform in spinor representation $\mathbf{8}_s$, their spin fields $\Sigma^i, \Sigma^{\dot{a}}$ will transform in the vector representation $\mathbf{8}_v$ and the conjugated spinor representation $\mathbf{8}_c$. They are related via the operator products

$$\theta^a_-(z)\, \Sigma^i(0) \sim z^{-\frac{1}{2}} \gamma^i_{a\dot{a}} \Sigma^{\dot{a}}(0) \tag{9.3}$$

The bosonic twist field σ and the spin fields Σ^i or $\Sigma^{\dot{a}}$ have all conformal dimension $h = \frac{1}{2}$ (i.e. $\frac{1}{16}$ for each coordinate), and the conjugated field τ^i therefore has dimension $h = 1$. For the interaction vertex we propose the following $SO(8)$ invariant operator

$$\tau^i \Sigma^i \tag{9.4}$$

This twist field has weight 3/2 and acts within the Ramond sector. It can be written as the $G_{-\frac{1}{2}}$ descendent of the chiral primary field $\sigma\Sigma^{\dot{a}}$, and furthermore satisfies

$$[G^{\dot{a}}_{-\frac{1}{2}}, \tau^i \Sigma^i] = \partial_z(\sigma \Sigma^{\dot{a}}). \tag{9.5}$$

To obtain the complete form of the effective world-sheet interaction term we have to tensor the left-moving and right-moving twist fields and to sum over the pairs of I, J labeling the two possible eigenvalues that can be permuted by the $\mathbf{Z}_2$ twist

$$V_{int} = \sum_{I<J} \left(\tau^i \Sigma^i \otimes \bar{\tau}^j \bar{\Sigma}^j \right)_{IJ}. \tag{9.6}$$

This is a weight $(\frac{3}{2}, \frac{3}{2})$ conformal field. The corresponding coupling constant λ has therefore total dimension -1 and the interaction will scale linear in g_s just as needed. The interaction preserves the world-sheet $\mathcal{N} = (8,8)$ supersymmetry, as well as the broken space-time supersymmetries Q^a.

It is interesting to compare the above twist field interaction with the conventional formalism of light-cone string theory. In comparison with the Riemann surfaces picture of interacting strings [32], one should note that around the interaction point the usual string world-sheet is actually a double cover of the local coordinate z, while in the NSR language the above operator insertion also represents a picture changing operation.[11]

We expect that for more general transverse target spaces M the interaction vertex can be constructed quite analogously. Its conformal scale dimension depends on the dimension d of M, or equivalently, the central charge of the corresponding $\mathcal{N}=1$ superconformal field theory, via

$$h = \frac{(d+4)}{8}. \tag{9.7}$$

In particular we see that the interaction vertex is irrelevant only as long as $d > 4$, and becomes marginal at $d = 4$.

10. Matrix String Theory

Let us now return to the maximally supersymmetric $U(N)$ Yang-Mills theory (5.1). For generality, let us consider this model in $d+2$-dimensions on a space-time manifold of the form

$$M_{YM} = T^d \times S^1 \times \mathbf{R} \tag{10.1}$$

where $\mathbf{R}$ represents time. Strictly speaking, this SYM theory is well-defined only in dimensions $d+2 \leq 4$. In the following, however, we would like to consider the model in a particular IR limit in the S^1 direction, in which the S^1 is taken much larger than the transverse torus T^d. In addition, we wish to combine this limit with the large N limit, similar to the one discussed in the previous sections. Indeed we would like to claim that the dynamics of the maximally supersymmetric YM theory in this limit exactly reduces to that of a (weakly) interacting type IIA string theory on the target space time

$$M_S = T^d \times \mathbf{R}^{10-d}. \tag{10.2}$$

[11] It should also be noted that a similar formulation of light-cone gauge bosonic string theory can be given in terms of an $S^N R^{24}$ orbifold. In this case the twist field also carries conformal weight (3/2,3/2) [33]. In the recent paper [34] it was shown that this formulation indeed exactly reproduces the Virasoro four-point amplitude. A very similar formulation of string perturbation theory was proposed among others by V.Knizhnik [37].

This particular IR limit effectively amounts to a dimensional reduction, in which the dynamics of the $d+2$ dimensional SYM theory becomes essentially two-dimensional. It is natural and convenient, therefore, to cast the $d+2$ SYM model in the form of a two-dimensional SYM theory. This can be achieved by replacing all fields by infinite dimensional matrices that include the fourier modes in the transversal directions. In particular, the d transverse components A^i of the gauge potential in the original SYM model thus become d additional two-dimensional scalar fields X^i with $i = 1, \ldots, d$ that represent the infinite set of matrix elements in a fourier basis of the corresponding transverse covariant YM-derivatives. The infinite dimensional matrices X^i defined via this procedure satisfy the recursion relations [38]

$$\begin{aligned} X^i_{mn} &= X^i_{(m-1)(n-1)}, \qquad i > d \\ X^i_{mn} &= X^i_{(m-1)(n-1)}, \qquad i \leq d,\ m \neq n \\ X^i_{nn} &= 2\pi L + X^i_{(n-1)(n-1)} \qquad i \leq d \end{aligned}$$

With this definition, the $d+2$ dimensional SYM action can be reduced to the following (quasi) two-dimensional form

$$\begin{aligned} S_{2D} = \int \mathrm{tr} \Big(&(D_\alpha X^I)^2 + \psi \Gamma^\alpha D_\alpha \psi + \lambda^2 F^2_{\alpha\beta} \\ &+ \frac{1}{\lambda} \psi \Gamma^I [X^I, \psi] + \frac{1}{2\lambda^2} [X^I, X^J]^2 \Big) \end{aligned}$$

The interpretation of this $\mathcal{N} = 8$ SYM model as a matrix string theory is based on the identification of the set of eigenvalues of the matrix coordinates X^i with the coordinates of the fundamental type IIA strings. In addition, the coupling λ is given in terms of the string coupling g_s as

$$\lambda^2 = \alpha' g_s^2. \tag{10.3}$$

The dependence on the string coupling constant λ can be absorbed in the area dependence of the two-dimensional SYM model. In this way λ scales inversely with world-sheet length. The free string at $\lambda = 0$ is recovered in the IR limit. In this IR limit, the two-dimensional gauge theory model is strongly coupled and we expect a nontrivial conformal field theory to describe the IR fixed point.

It turns out that we can find this CFT description via the following rather naive reasoning. We first notice that in the $g_s = 0$ limit, the last two potential terms in the 2-dimensional action (10.3) effectively turn into constraints, requiring the matrix fields X and θ to commute. This means that we can write the matrix coordinates X^i in a simultaneously diaganolized form

$$X^i = U \mathrm{diag}(x^i_1, \ldots, x^i_N) U^{-1} \tag{10.4}$$

with $U \in U(N)$. Here U and all eigenvalues x^i_I can of course still depend on the world-sheet coordinates.

Once we restrict ourselves to Higgs field configurations of the form (10.4), the two-dimensional gauge field A_α decouples. To make this explicit, we can use the YM gauge invariance to put $U = 1$ everywhere, which in particular implies that the YM-current $j_{YM} = [X^i, \partial_\alpha X_i]$ vanishes. The only non-trivial effect that remains is that via the Higgs mechanism, the charged components of A_α acquire a mass for as long as the eigenvalues x_i are distinct and thus decouple in the infra-red limit. This leads to a description of the infra-red SYM model in terms of N Green-Schwarz light-cone coordinates formed by the eigenvalues $x_I^i, \theta_I^a, \theta_I^{\dot{a}}$ with $I = 1, \ldots, N$.

The $U(N)$ local gauge invariance implies that there is still an infinite set of discrete gauge transformations that act on the diagonalized configurations (10.4). Part of these discrete gauge transformation act via translations on the eigenvalues x_I^i that represent the transversal components of the gauge potential, turning them into periodic variables. In addition, we must divide out by the action of the Weyl group elements of U(N), that act on the eigenvalues x_I^i via S_N permutations. Thus, via this semi-classical reasoning, we deduce that the IR conformal field theory is the supersymmetric sigma model on the orbifold target space

$$(T^d \times \mathbf{R}^{8-d})^N / S_N. \tag{10.5}$$

When we combine this with the results described in the previous two sections, we deduce that the super-Yang-Mills model in the strict IR limit gives the same Hilbert space as the free light-cone quantization of second quantized type IIA string theory.

The conservation of string number and individual string momenta will be violated if we turn on the interactions of the 1+1-dimensional SYM theory. When we relax the strict IR limit, one gradually needs to include configurations in which the non-abelian symmetry gets restored in some small space-time region: if at some point in the (σ, τ) plane two eigenvalues x_I and x_J coincide, we enter a phase where an unbroken $U(2)$ symmetry is restored. We should thus expect that for non-zero λ, there will be a non-zero transition amplitude between states that are related by a simple transposition of these two eigenvalues.

We now claim that this joining and splitting process is indeed first order in the coupling constant λ as defined in the SYM Lagrangian. Instead of deriving this directly in the strongly coupled SYM theory, we can analyze the effective operator that produces such an interaction in the IR conformal field theory. The identification $\lambda \sim g_s\sqrt{\alpha'}$ requires that this local interaction vertex V_{int} must have scale dimension

$$(h, \overline{h}) = (\frac{3}{2}, \frac{3}{2}) \tag{10.6}$$

under left- and right-scale transformations on the two-dimensional world-sheet. The analysis of the previous section confirms that this is indeed the case.

As a further confirmation of this physical picture, it is relevant to point out that there is no classical barrier against the occurrence of string interactions in the IR limit. For every string world-sheet topology, there exists a solution to the SYM equation of motion of the form (10.4) such that the eigenvalues x_I^i form this string worldsheet via an appropriate branched covering of the cylinder $S^1 \times \mathbf{R}$. (An explicit description of these solutions is given in [35]). The suppression of string interactions in the IR field theory therefore entirely arises due to the quantum fluctuations around these classical field configurations. This suggests that there should exist a close relation between the moduli space of classical configurations of the two-dimensional SYM model in the IR regime and the universal moduli space of all two-dimensional Riemann surfaces.

From this Yang-Mills set-up, it is straightforward to derive the correct scaling dimension of the twist-fields (that create the branch cuts) by computing the one-loop fluctuation determinant around such a non-trivial classical field configuration. To start with, this computation only involves the eigenvalue fields x^i_I, and thereby reduces to the standard analysis of twist field operators in orbifold CFT used above. In the present context, however, the orbifold CFT is automatically provided with an ultra-violet cut-off of the order λ (as defined in (10.3)), since this is the place where the non-abelian SYM dynamics takes over. This explains why the coupling λ in (10.3) also determines the strength of the twist field perturbation away from the orbifold CFT.

It would clearly be interesting to make via this set-up a direct quantitative comparison with the standard perturbative SYM calculations. In [7][9] [36] it was shown that the SYM amplitudes at weak coupling reproduce the tree level graviton exchange amplitude of eleven dimensional supergravity. Non-renormalization theorems of the SYM-model should ensure that these specific amplitudes receive no higher order or non-perturbative corrections, so that we can extrapolate these results the IR limit. Via the comparison with our IR description of the same amplitudes, one can thus uniquely fix the ratio between the orbifold perturbation parameter (*i.e.* the string coupling) and the coupling constant λ in the two-dimensional SYM lagrangian 10.3.

The result (10.6) for the conformal dimension of the interaction vertex V_{int} of course crucially depends on the number of Higgs scalar fields being equal to 8. The fact that for this case V_{int} is indeed irrelevant in the IR limit provides the justification afterwards of the above naive derivation of the orbifold sigma model as describing the IR phase of the $U(N)$ SYM model. Another way of stating the same result is that the $\mathbf{R}^8/\mathbf{Z}_2$ cannot be resolved to give a smooth Calabi-Yau space.

It is indeed interesting to try to repeat the same study for SYM models with less supercharges. Concretely, we could consider the dimensional reduction of the $\mathcal{N}=1$ SYM model in 5+1-dimensions, which will give us a matrix string theory with $d = 4$ transverse Higgs fields. Via the above reasoning we then deduce that in the strict IR limit in the S^1 direction, the string interactions are not turned off, since the interaction vertex V_{int} is now exactly marginal. We will return to this case in the next section.

Finally we remark that, for SYM theories with even less supersymmetries, the naive reasoning applied in this section in fact breaks down, since in this case the interaction vertex V_{int} becomes a relevant perturbation. Hence in this case the correct IR dynamics is no longer accurately represented by a (small) deformation of the naive orbifold IR field theory. Instead one expects that the twist field V_{int}

obtains a non-zero expectation value, and the IR theory presumably takes the form of some Ising type QFT.

11. Microstring Interactions

A relatively concrete model of the micro-string theory can be obtained by applying the reasoning of [14] to the 5+1 SYM formulation of Matrix theory. Namely, in this formulation we can describe longitudinal NS 5-branes (that extend in the light-cone direction) by means of YM instanton configurations [39]. Here the 5-brane number simply coincides with the instanton number k.[12] Similar as in [14], we can then consider the world-volume theory of these k longitudinal NS 5-branes in the limit of zero string coupling. As also explained above, this corresponds to taking the limit where one of the compactification circles becomes very large. The theory thus again reduces to the IR limit a two-dimensional matrix string theory of the form (10.3), but now with matrices X^I that carry a non-zero topological charge [39][20]

$$k = \mathrm{tr} X_{[i} X_j X_k X_{l]} \tag{11.1}$$

representing the instanton number on the transverse T^4. It can be seen that in this case, the IR limit of the matrix string theory (10.3) no longer becomes a free field theory but instead reduces to a sigma model on the moduli space $\mathcal{M}_{N,k}(\hat{T}^4)$ of k $U(N)$ instantons on the four-torus $\hat{T}^4$. Since this instanton moduli space is hyperkähler, the corresponding supersymmetric sigma model is indeed conformally invariant.

In [13] it was proposed that this sigma-model could be used as a concrete way of representing the interacting micro-string theory on the world-volume of the k 5-branes. Crucial for this interpretation is the fact that the moduli space of instanton configurations on T^4 with rank N, instanton number k, and fluxes m can be thought of as a (hyperkähler) deformation of the symmetric product space S^nT^4 with $n = Nk + \frac{1}{2} m \wedge m$. Hence via the results of the previous sections, the sigma model on $\mathcal{M}_{N,k}(\hat{T}^4)$ is indeed expected to behave in many respects as a light-cone description of a 5+1-dimensional interacting string theory.

The interpretation of this sigma model as describing the micro-string dynamics on k five-branes becomes a bit more manifest via the so-called Mukai-Nahm duality transformation. This transformation provides an isometry

$$\mathcal{M}_{N,k}(\hat{T}^4) = \mathcal{M}_{k,N}(T^4) \tag{11.2}$$

between the moduli space of k $U(N)$ instantons on $\hat{T}^4$ to the moduli space of N instantons on a $U(k)$ Yang-Mills theory on the dual four-torus T^4. This dual $U(k)$ SYM theory can be thought of as the world-volume theory of k fivebranes wrapped around T^4, and the N instantons describe the micro-string degrees of

[12]This representation follows from the table in section 7, where the longitudinal M5 branes (via their IIA identification with D4-branes) get mapped by the T^5 duality to D1 strings, which manifests themselves as instantons in the large N D5 SYM theory.

freedom on this world-volume. This dual language is most suitable for discussing the decompactification limit of the M-theory torus, which leads to the description of k uncompactified fivebranes as a non-linear sigma model on $\mathcal{M}_{k,N}(\mathbf{R}^4)$.

The fact that there exists a superconformal deformation of the orbifold sigma model on S^nT^4 is in direct accordance with the results of the previous section, since it is reasonable to suspect that the marginal operator that represents this deformation can be identified with an appropriate $\mathbf{Z}_2$ twist operator that generates the splitting and joining interactions among the microstrings. As the string in this case has 4 transverse dimensions, the twist-field indeed has total dimension 2. Below we would like to make this description more explicit.

To this end, it will be useful to consider the situation in Matrix theory, which describes the k 5-branes in combination with a single IIA matrix string. In fact, to simplify the following discussion, we will give this string the minimal p_+ momentum, that is we will represent it simply by adding one row to the matrix coordinates X^i, so that we are dealing with a $U(N+1)$ matrix theory. Next we write

$$X_i \to \begin{pmatrix} Z_i & Y_i \\ \overline{Y}_i & x_i \end{pmatrix} \tag{11.3}$$

Here the matrix entry x_i represents the coordinate of a propagating type IIA string (with infinitesimal light-cone momentum p_+), and the $U(N)$ part Z_i describes the k 5-branes. Similarly, we can write

$$\theta_a \to \begin{pmatrix} \zeta_a & \eta_a \\ \overline{\eta}_a & \theta_a \end{pmatrix} \tag{11.4}$$

The idea is now to fix Z_i to be some given instanton configuration on T^4 plus a diagonal part $z_I \mathbf{1}_{N\times N}$ that describes the location of the 5-brane in the uncompactified space, and to consider the propagation of the IIA string in its background.

This set-up will be useful in fact for studying two separate regimes, each of which are quite interesting [44] [45]. First there is the micro-string regime, which corresponds to the case where the extra string x is captured inside the k NS 5-branes. In this case, the Y variable need to be thought of as the extra degrees of freedom that extend the moduli space of k $U(N)$ instantons to that of k $U(N+1)$ instantons. Secondly, we can also consider the regime where the IIA string described by the x-coordinate is far enough removed from the NS 5-branes so that it propagates just freely on the background geometry of the 5-brane solitons. This outside regime corresponds to the Coulomb branch of the matrix string gauge-theory, while the micro-string regime represents the Higgs branch.

For both regimes, it will in fact be useful to isolate from the Y degrees of freedom in (11.3) the part that describes the constant zero-modes in the instanton background, satisfying

$$Z_{[i}Y_{j]} = 0, \qquad \gamma^i_{\alpha\beta} Z_i \eta^\beta = 0. \tag{11.5}$$

Since Z_i corresponds to a k-instanton configuration we can use the index theorem to determine that there are k such zero-modes, both for the Y and η field. In the

Coulomb phase they describe the low energy degrees of freedom that mediate the interaction between the string and the 5-brane. In this phase, the Y-modes are massive and thus are naturally integrated out in taking the IR limit on the S^1. In the Higgs phase, on the other hand, they become massless two-dimensional fields, and represent the additional micro-string degrees of freedom that arise from the IIA string trapped inside the k 5-branes. Notice that in this transition from the Coulomb phase to the Higgs phase the single IIA string indeed "fractionates" into k individual micro-strings.

The quadratic part of the action for the Y degrees of freedom takes the form

$$\begin{aligned} S = & \int d^2\sigma \left((DY_i)^2 + |x_I - z_I|^2 |Y_i|^2 \right) \\ & + \int d^2\sigma \left(\overline{\eta}^{\alpha}_a D_+ \eta^{\alpha}_a + \overline{\eta}^{\dot\alpha}_a D_- \eta^{\dot\alpha}_a \right. \\ & \left. + \overline{\eta}^{\alpha}_a \gamma^I_{\alpha\dot\beta} (x-z)_I \, \eta^{\dot\beta}_a \right) \end{aligned} \tag{11.6}$$

Here we suppressed the summation over the index labeling the k zero-modes. The discrete symmetries of the instanton moduli space imply that the model (11.6) possesses a S_k gauge symmetry that acts by permuting these k coordinates. In [42] the presence of the fivebrane background was incorporated in matrix theory by adding an appropriate hypermultiplet to the Matrix action. In our set-up, the role of these additional hypermultiplets is taken over by the off-diagonal fields Y, η.

Let us explain the fermion chiralities as they appear in the Lagrangian (11.6). In the original matrix string action the $SO(8)$ space-time chiralities are directly correlated with the world-sheet chirality, reflecting the type IIA string setup, giving left-moving fermions θ^a and right-moving fermions $\theta^{\dot a}$. The T^4 compactification breaks the $SO(8)$ into an internal $SO(4)$ that acts inside the four-torus times the transversal $SO(4)$ rotations in space-time. The spinors decompose correspondingly as $\mathbf{8}_s \to (\mathbf{2}_+, \mathbf{2}_+) \oplus (\mathbf{2}_-, \mathbf{2}_-)$ and $\mathbf{8}_c \to (\mathbf{2}_+, \mathbf{2}_-) \oplus (\mathbf{2}_-, \mathbf{2}_+)$. The instanton on T^4 with positive charge $k > 0$ allows only zero modes of positive chirality, selecting the spinors that transform as $\mathbf{2}_+$ of the first $SO(4)$ factor. So, through this process the space-time chirality of the fermion zero modes gets correlated with the world-sheet handedness, giving k left-moving fields η^{α}_a and k right-moving fields $\eta^{\dot\alpha}_a$, where $\alpha, \dot\alpha = 1, 2$ denote chiral space-time spinor indices and $a = 1, 2$ label chiral world-volume spinors.

The Higgs phase of the matrix string gauge theory corresponds to the regime where $x_I - z_I \simeq 0$. The above lagrangian of the $4k$ string-coordinates (Y, η) then represents a linearized description of the micro-string motion. inside the fivebrane worldvolume. The linearized world-sheet theory has 4 left-moving and 4 right-moving supercharges

$$\mathcal{F}^{\dot a \alpha} = \oint \partial Y^{\dot a b} \eta^{\alpha}_b, \qquad \overline{\mathcal{F}}^{\dot a \dot\alpha} = \oint \overline{\partial} Y^{\dot a b} \overline{\eta}^{\dot\alpha}_b \tag{11.7}$$

which correspond to the unbroken part of the world-brane supersymmetry and satisfy

$$\{\mathcal{F}^{\dot{a}\alpha},\mathcal{F}^{\dot{b}\beta}\}=2\epsilon^{\dot{a}\dot{b}}\epsilon^{\alpha\beta}\,\mathcal{L}_0, \tag{11.8}$$

Here $\mathcal{L}_0$ is the left-moving world-sheet Hamiltonian.

The ground states of the micro-strings contain the low-energy collective modes of the five-brane. To make this explicit, we note that the left-moving ground state must form a multiplet of the left-moving fermionic zero-mode algebra

$$\{\eta^\alpha_a,\eta^\beta_b\}=\epsilon_{ab}\epsilon^{\alpha\beta} \tag{11.9}$$

which gives 2 left-moving bosonic ground states $|\alpha\rangle$ and 2 fermionic states $|a\rangle$. By taking the tensor product with the right-moving vacua one obtains in total 16 ground states, which represent the massless tensor multiplet on the five-brane. Specifically, we have states $|\alpha\rangle\otimes|\dot{\beta}\rangle$ that describe four scalars $X^{\alpha\dot{\beta}}=\sigma_i^{\alpha\dot{\beta}}X^i$ transforming a vector of the $SO(4)$ R-symmetry, while $|\alpha\rangle\otimes|a\rangle$ and $|a\rangle\otimes|\dot{\alpha}\rangle$ describe the fermions ψ^α_a and $\psi^{\dot{\alpha}}_a$. Finally, the RR-like states $|ab\rangle$ decompose into the fifth scalar Y, and the 3 helicity states of the self-dual tensor field. These are indeed the fields that parametrize the collective excitations of the NS-brane soliton[13].

This 5+1-dimensional string thus forms a rather direct generalization of the 9+1-dimensional Green-Schwarz string. Clearly, however, the linearized theory can not be the whole story, because it would not result in a Lorentz covariant description of the M5 world-brane dynamics. Micro string theory is indeed essentially an interacting theory, without a weak coupling limit. Its interacting dynamics is obtained by including the non-linear corrections to (11.6), obtained by properly taking into account the non-linear geometry of the moduli space $\mathcal{M}_{N,k}(\widehat{T}^4)$ of $U(N)$ k-instantons on the four-torus $\hat{T}^4$.

As argued above, the sigma model on $\mathcal{M}_{N,k}(\widehat{T}^4)$ should be representable in terms of a finite marginal deformation of the above model by means of a twist field. Concretely, the linearized orbifold model has four different twist fields

$$V_{ab}=V_a\otimes\overline{V}_b \tag{11.10}$$

where the left-moving factor V_a is

$$V_a=\gamma^i_{ab}\tau_i\Sigma^b \tag{11.11}$$

and a similar expression holds for the right-moving $\overline{V}_a$. Here Σ^b denotes the fermionic spin field, defined via the operator product relation

$$\eta^\alpha_a(z)\,\Sigma_b(0)\sim z^{-\frac{1}{2}}\epsilon_{ab}\tilde{\Sigma}^\alpha(0) \tag{11.12}$$

with $\tilde{\Sigma}^\alpha(0)$ the dual spin field. Both spin fields have conformal dimension 1/4, while τ_i has dimension 3/4.

[13]A similar CFT description of the fivebrane effective field theory has been proposed in [40].

So, all the above operators describe possible marginal deformations, that at least in first order conformal perturbation theory, preserve the conformal invariance of the sigma model QFT. Theses 4 physical operators V_{ab} are in one-to-one correspondence with the RR-like micro-string ground states $|a\rangle \otimes |b\rangle$, that represent the excitations of $T = dU$ and the scalar field Y. Rotation symmetry on the world-volume, however, selects out the scalar combination

$$V_Y = \epsilon^{ab} V_a \otimes \overline{V}_b \qquad (11.13)$$

that describes the emission of a Y mode, as the only good candidate for representing the string interactions.

Geometrically, the twist fields V_{ab} represent the four independent blow-up modes of a $\mathbf{R}^4/\mathbf{Z}_2$ orbifold singularity. The three symmetric combinations deform the orbifold metric to a smooth ALE space and the three parameters correspond to the different complex structures that one can pick to specify the blow-up. These deformations lead to regular supersymmetric sigma models, but are not rotational invariant. The antisymmetric combination V_Y can be interpreted as turning on a B-field θ-angle dual to the vanishing two-cycle around the $\mathbf{Z}_2$ orbifold point. h The value of this θ-angle away from the pure orbifold CFT, which determines the strength of the micro-string interactions, is in principle fixed by the description of the matrix micro-string as the sigma-model on $\mathcal{M}_{N,k}(\hat{T}^4)$.

The value of the θ-angle modulus can be determined by considering the instanton moduli space around the locus of point-like instantons. This problem was considered in [44], where it is proposed that this naturally fixes its value at $\theta = \pi$.[14] A puzzling aspect of this proposal is that this special value of θ is actually known to lead to a singular CFT [46]. This is problematic, since it would make the sigma model description of the micro-string dynamics inconsistent, or at least incomplete. There could however be a good physical reason for the occurrence of this singular CFT. In the linear sigma model description of the Higgs branch, the singularity seems closely related to the presence of a Coulomb branch on the moduli space that describes the emission of strings from the fivebrane [44]. An alternative physical interpretation, however, is that the singularity of the CFT reflects the possible presence of the tensionless charged strings.

The latter interpretation in fact suggests that the singularity may possibly be resolved by introducing a finite separation between the k fivebranes. In the M-theory decompactification limit in particular, we know that the relevant moduli space $\mathcal{M}_{k,N}(\mathbf{R}^4)$ allows for a number of deformation parameters in the form of the Higgs expectation values of the $U(k)$ SYM model. These Higgs parameters modify the geometry of the instanton moduli space, and indeed have the effect of softening most of its singularities. This is in direct accordance with the micro-string perspective, since for finite Higgs expectation values the charged self-dual strings always have a finite tension.

[14]Here we defined $\theta = 0$ as the value at the free orbifold CFT.

12. Future Directions

To end with, we would like to mention some future directions, where useful new insights may be obtained.

Space-time Geometry.

We have seen that the matrix setup is quite well adapted to extract all the effective degrees of freedom that describe the type IIA fivebrane, and one might hope that the present context of matrix string theory opens up new ways of examining the interaction with their environment. In particular, at weak coupling one would like to be able to recover the classical fivebrane geometry including its anti-symmetric tensor field. This geometry should arise in the CFT description of the Coulomb phase of the matrix string theory, where the string $x(\sigma)$ propagates at a finite distance $|x_I - z_I| > 0$ from the fivebrane soliton.

This description of the string and fivebrane is closely related to the recently studied case of a D-string probe in the background of a D-fivebrane [41]. One can derive an effective action for the x-part of the $U(N)$ matrix by integrating out the off-diagonal components Y and η. In [41] this was shown that the one-loop result for the transversal background metric

$$ds^2 = \left(1 + \frac{k}{|x-z|^2}\right)(dx)^2. \tag{12.1}$$

is indeed the well-known geometry of the NS fivebrane [47].

Most characteristic of the solitonic fivebrane is its magnetic charge with respect to the NS anti-symmetric tensor field B

$$\int_{S_3} H = k \tag{12.2}$$

where $H = dB$ for a three sphere enclosing the brane. This result can be recovered analogously, via a direct adaptation of the analysis of [42]. The transversal rotation group $SO(4) \simeq SU(2)_L \times SU(2)_R$ acts on the relative coordinate field $(x-z)^I$ via left and right multiplication. This group action can be used to decompose the 2×2 matrix $\gamma_I^{\alpha\dot\beta}(x-z)^I$ in terms of a radial scalar field $r = e^{\varphi}$ and group variables g_L, g_R as

$$\gamma_I(x-z)^I = e^{\varphi} g_L g_R^{-1}. \tag{12.3}$$

The combination $g = g_L g_R^{-1}$ labels the $SO(3)$ angular coordinate in the transversal 4-plane. Inserting the above expression for x^I into (11.6), the angles g_L and g_R can be absorbed into η via a chiral rotation, producing a model of k $SU(2)$ fermions coupled to a gauge field $A_+ = g_L^{-1}\partial_+ g_L$, $A_- = g_R^{-1}\partial_- g_R$. Via the standard chiral anomaly argument, one derives that the one-loop effective action includes an $SU(2)$ WZW-model for the field g with level given by the five-brane number k. The background three-form field H in (12.2) is reproduced as the Wess-Zumino term of this action.

We thus recover the CFT description of the type IIA string moving near the fivebrane [47]. This action consists of a level k $SU(2)$ WZW model combined with a Liouville scalar field φ

$$S(x^I) = S_{WZW}(g) + S_L(\varphi). \tag{12.4}$$

with

$$S_L(\varphi) = \int d^2\sigma\Big(|\partial\varphi|^2 + \gamma R^{(2)}\varphi\Big) \tag{12.5}$$

The background charge γ of the scalar φ represents a space-time dilaton field that grows linearly for $\phi \to -\infty$, and is normalized such that the total central charge is $c = 6$.

It is quite meaningful that the radial coordinate $r = e^\varphi$ is described in terms of a Liouville type field theory (12.5). The role of r is indeed very similar to that of the world-sheet metric in two-dimensional quantum gravity: since r represents the mass of the lightest fields that have been integrated out, it also parametrizes the short-distance world-sheet-length scale at which the CFT description is expected to break down. Besides the fact that it acquires a corresponding conformal scale dimension, this also means that it provides the natural cut-off scale at which the short-distance singularities need to be regulated.

In this light, it would be very interesting to study in more detail the transition region between the Coulomb and Higgs phase of this two-dimensional gauge theory, describing the capturing process of the micro-string inside the fivebrane. The relevant space-time geometry in this case should contain all the essential features of a black hole horizon, which initially is located at $r = e^\varphi = 0$. We expect that via the above CFT correspondence, and in particular by studying the limits on its validity, useful new insights into the reliability of semi-classical studies of black hole physics can be obtained [48, 13].

Micro-strings on $K3 \times T^2$.

Interesting new lessons can be learned from the beautiful geometric realization of heterotic string theory via M-theory compactified on $K3$. In this correspondence the perturbative heterotic string becomes a natural part of the M5-brane spectrum wrapped around the $K3$ [49], which thereby provides a far reaching generalisation of its world-sheet dynamics in terms of the above 5+1-dimensional micro-string theory. A matrix formulation of the interactions between different wrapped M5-branes may ultimately provide a description of heterotic string dynamics, in which non-perturbative duality is realized as a geometric symmetry.

As a concrete illustration, we consider the generalization of the one-loop heterotic string partition function. In perturbative string theory, this one-loop calculation proceeds by considering the CFT free energy on a world-sheet with the topology of T^2, *i.e.* the closed world-trajectory of a closed string. In the M-theory realization, the string world-sheet expands to become the world-volume of the M5-brane, while the CFT get replaced by the micro-string theory. The one-loop diagram thus takes the form of a M5-brane world-trajectory with the topology of $K3 \times T^2$. The direct analogue of the perturbative string calculation thus involves the partition function second quantized micro-string theory on this 6-dimensional manifold.

For the restricted case where one considers only the BPS contributions, this calculation can be explicitly done. The BPS restriction essentially amounts to a projection on only the left-moving micro-string modes, and thereby eliminates the effect of its interactions. The M5-brane BPS partition sum can thus be computed by taking the exponent of the relevant one-loop free energy of the micro-string

$$\mathcal{Z}_{M5-brane} = \exp \mathcal{F}_{1-loop}^{micro-string} \tag{12.6}$$

The above one-loop string partition function has in fact been computed already in [51]. Let us briefly describe the result, because it in fact desplays some rather suggestive symmetries.

This partition sum $\mathcal{Z}$ will depend on the moduli that parametrize the shape and size of world-volume manifold $K3 \times T^2$. It turns out that there are three relevant moduli T, U and V, that parametrize the shape and size of the T^2 and a Wilson line expectation value. The expression for the partition function (12.6) now takes the following form[15]

$$\mathcal{Z}(T,U,V) = e^{-2\pi i(T+U+V)} \times \tag{12.7}$$

$$\prod_{(k,l,m)>0} \left(1 - e^{2\pi i(kT+lU+mV)}\right)^{-c(4kl-m^2)}$$

where the coefficients $c(n)$ are defined by the expansion of the $K3$ elliptic genus as

$$\chi_{\tau,z}(K3) = \sum_{h\geq 0,\, m\in \mathbf{Z}} c(4h-m^2)e^{2\pi i(h\tau+mz)} \tag{12.8}$$

and are non-zero for $n \geq -1$. The expression (12.7) can indeed be recognized as a second quantized string partition function [31]. If we expand $\mathcal{Z}$ as a formal Fourier series expansion

$$\mathcal{Z}(T,U,V) = \sum_{k,l,m} D(k,l,m)e^{-2\pi i(kT+lU+mV)} \tag{12.9}$$

the integer coefficients $D(k,l,m)$ represent the total (bosonic minus fermionic) number of states with given total momentum, winding and $U(1)$ charge, as specified by the three integers k, l, m.

The above partition function can indeed be thought as the direct generalization of the genus one partition function of perturbative heterotic string BPS-states. The fact that $\mathcal{Z}$ now depends on three moduli instead of one reflects the fact that the worldsheet CFT is now replace by a world-volume micro-string theory, which in particular also has winding sectors that are sensitive to the size of the T^2. As a consequence, the partition function $\mathcal{Z}$ has the intriguing property that it is symmetric under duality transformations, that in particular interchange the shape and size parameters T and V.

[15]Here $(k,l,m) > 0$ means that $k,l \geq 0$ and $m \in \mathbf{Z}$, $m < 0$ for $k = l = 0$.

The general group of dualities is best exhibited by writing the three moduli into a 2 by 2 matrix

$$\Omega = \begin{pmatrix} T & V \\ V & U \end{pmatrix} \tag{12.10}$$

which takes the suggestive form of a period matrix of a genus 2 curve. The above partition function $\mathcal{Z}$ indeed turns out to be a modular form under the corresponding $SO(3,2;\mathbf{Z})$ modular group, and is in fact *invariant* under a $SL(2,\mathbf{Z})$ subgroup. It is quite tempting to identify this $SL(2,\mathbf{Z})$ symmetry with the electro-magnetic duality symmetry that arises upon further compactification of this theory to four dimensions. A concrete suggestion that was made in [50] is that the coefficients in the expansion (12.9) represent the degeneracy of (bosonic minus fermionic) dyonic BPS states with a given electric charge $q_e \in \Gamma^{22,4}$ and magnetic charge $q_m \in \Gamma^{22,4}$, via the identification

$$d(q_m, q_e) = D(\frac{1}{2}q_m^2, \frac{1}{2}q_e^2, q_e \cdot q_m) \tag{12.11}$$

In [50] it was shown that this formula survives a number of non-trivial tests, such as correspondence with the perturbative heterotic string as well as with the Bekenstein-Hawking black hole entropy formula. If true, this result suggests that the presented formulation of M-theory via the microstring theory and the M5-brane may in fact be extended to compactifications all the way down to four dimensions. Clearly, further study of matrix theory compactifications is needed to confirm or disprove this conjecture.

Acknowledgements

In the course of this work we benefited from useful discussions with L. Alvarez-Gaumé, V. Balasubramanian, T. Banks, M. Becker, C. Callan, M. Douglas, E. Kiritsis, I. Klebanov, D. Kutasov, F. Larsen, J. Maldacena, L. Motl, E. Martinec, G. Moore, N. Seiberg, S. Shenker, E. Silverstein, A. Strominger, L. Susskind, W. Taylor, P. Townsend, E. Witten, and others. This research is partly supported by a Pionier Fellowship of NWO, a Fellowship of the Royal Dutch Academy of Sciences (K.N.A.W.), the Packard Foundation and the A.P. Sloan Foundation.

References

1. P. Townsend, "P-brane democracy," in *Particles, Strings and Cosmology*, eds. J. Bagger e.a. (World Scientific 1996), `hep-th/9507048`.
2. E. Witten, "String Theory Dynamics in Various Dimensions," Nucl. Phys. **B443** (1995) 85–126, `hep-th/9503124`.
3. J. Polchinski, "Dirichlet Branes and Ramond Ramond Charges," `hep-th/951001`.
4. R. Dijkgraaf, E. Verlinde and H. Verlinde, "BPS Spectrum of the Five-Brane and Black Hole Entropy," *Nucl. Phys.* **B486** (1997) 77–88, `hep-th/9603126`; "BPS Quantization of the Five-Brane," *Nucl. Phys.* **B486** (1997) 89–113, `hep-th/9604055`.
5. E. Witten, "Bound States of Strings and *p*-Branes," *Nucl. Phys.* **B460** (1996) 335–350, `hep-th/9510135`.

6. U.H. Danielsson, G. Ferretti, and B. Sundborg, "D-particle Dynamics and Bound States," Int. J. Mod. Phys. **A11** (1996) 5463–5478, hep-th/9603081; D. Kabat and P. Pouliot, "A Comment on Zerobrane Quantum Mechanics," Phys. Rev. Lett. **77** (1996) 1004–1007, hep-th/9603127.
7. M. R. Douglas, D. Kabat, P. Pouliot, and S. H. Shenker, "D-branes and Short Distances in String Theory," *Nucl. Phys.* **B485** (1997) 85–127, hep-th/9608024.
8. C. Hull and P. Townsend, "Unity of Superstring Dualities," *Nucl. Phys.* **B 438** (1995) 109.
9. T. Banks, W. Fischler, S. H. Shenker, and L. Susskind, "M Theory as a Matrix Model: A Conjecture," hep-th/9610043.
10. L. Motl, "Proposals on Non-perturbative Superstring Interactions," hep-th/9701025.
11. T. Banks and N. Seiberg, "Strings from Matrices," hep-th/9702187.
12. R. Dijkgraaf, E. Verlinde, and H. Verlinde, "Matrix String Theory," hep-th/9703030.
13. R. Dijkgraaf, E. Verlinde, and H. Verlinde, "5D Black Holes and Matrix Strings," hep-th/9704018.
14. N. Seiberg, "Matrix Description of M-theory on T^5 and T^5/Z_2," hep-th/9705221
15. J. Hoppe, "Quantum Theory of a Relativistic Surface," Ph.D. Thesis, MIT (1982). B. de Wit, J. Hoppe and H. Nicolai, "On the Quantum Mechanics of Supermembranes," Nucl. Phys. **B305** (1988) 545–581.
16. M. Rozali, "Matrix Theory and U-Duality in Seven Dimensions," hep-th/9702136.
17. M. Berkooz, M. Rozali, and N. Seiberg, "Matrix Description of M theory on T^4 and T^5," hep-th/9704089.
18. W. Fishler, E. Halyo, A. Rajaraman, and L. Susskind, "The Incredible Shrinking Torus," hep-th/9703102.
19. J.M. Maldacena and L. Susskind, "D-branes and Fat Black Holes," *Nucl. Phys.* **B475** (1996) 679, hep-th/9604042.
20. T. Banks, N. Seiberg, and S. Shenker, "Branes from Matrices," hep-th/9612157.
21. H. Verlinde, "The Stringy Fivebrane," talk at Strings '96, Santa Barbara, June 1996.
22. A. Strominger, C. Vafa, "Microscopic Origin of the Bekenstein-Hawking Entropy,"hep-th/9601029.
23. J. Maldacena, "Statistical Entropy of Near Extremal Five-branes," hep-th/9605016.
24. For a recent account of these developments see J. Maldacena and A. Strominger, "Universal Low-Energy Dynamics for Rotating Black Holes" hep-th/9702015, and references therein.
25. E. Witten, "Some Comments on String Dynamics," hep-th/9510135;
26. A. Strominger, "Open P-Branes," hep-th/9512059; P. Townsend, "D-branes from M-branes," hep-th/9512062.
27. A. Losev, G. Moore, S. L. Shatashvili, "M & m's," hep-th/9707250; N. Seiberg and S. Sethi "Comments on Neveu-Schwarz Five-Branes," hep-th/9708085
28. P. Pasti, D. Sorokin, and M. Tonin, "Covariant Action for a D=11 Five-Brane with the Chiral Field," Phys.Lett. **B398** (1997) 41-46, hep-th/9701037; M. Aganagic, J. Park, C. Popescu, J. Schwarz "World-Volume Action of the M Theory Five-Brane," Nucl.Phys. **B496** (1997) 191-214, hep-th/9701166; J. Schwarz, "Remarks on the M5-Brane," talk at Strings '97, hep-th/9707119.
29. D. Sorokin, P. Townsend, " M-theory superalgebra from the M-5-brane," hep-th/9708003
30. E. Witten, "Five-Brane Effective Action In M-Theory" hep-th/9610234
31. R. Dijkgraaf, G, Moore, E. Verlinde, and H. Verlinde, "Elliptic Genera of Symmetric Products and Second Quantized Strings," Commun. Math. Phys. hep-th/9608096.
32. S. Mandelstam, "Lorentz Properties of the Three-String Vertex," *Nucl. Phys.* **B83** (1974) 413–439; "Interacting-String Picture of the Fermionic String," Prog. Theor. Phys. Suppl. **86** (1986) 163–170.
33. S-J. Rey, "Heterotic M(atrix) Strings and Their Interactions," hep-th/9704158.

34. G. Arutyunov and S. Frolov, "Virasoro amplitude from the $S^N\mathbf{R}^{24}$ orbifold sigma model," hep-th/9708129.
35. T. Wynter, "Gauge fields and interactions in matrix string theory," hep-th/9709029
36. K. Becker and M. Becker, "A Two Loop Test of M(atrix) Theory, hep-th/9705091. K. Becker, M. Becker, J. Polchinski, and A. Tseytlin, "Higher Order Scattering in M(atrix) Theory," hep-th/9706072.
37. V. Knizhnik, Sov. Phys. Usp. 32 (1989) 945.
38. W. Taylor, "D-brane Field Theory on Compact Spaces," hep-th/9611042.
39. O.J. Ganor, S. Ramgoolam, and W. Taylor, "Branes, Fluxes and Duality in M(atrix)-Theory," hep-th/9611202.
40. D. Kutasov, "Orbifold and Solitons," hep-th/9512145.
41. M. Douglas, J. Polchinski, and A. Strominger, "Probing Five-Dimensional Black Holes with D-Branes," hep-th/9703031.
42. M. Berkooz and M. Douglas, "Fivebranes in M(atrix) Theory," hep-th/9610236.
43. O. Aharony, M. Berkooz, S. Kachru, N. Seiberg, and E. Silverstein, "Matrix Description of Interacting Theories in Six Dimensions," hep-th/9707079
44. E. Witten, "On the Conformal Field Theory of the Higgs Branch," hep-th/9707093
45. D. Diaconescu and N. Seiberg, "The Coulomb Branch of (4,4) Supersymmetric Field Theories in Two Dimensions," hep-th/9707158
46. P. Aspinwal, "Enhanced Gauge Symmetries and K3 Surfaces," Phys. Lett. **B357** (1995) 329, hep-th/9507012.
47. C. Callan, J. Harvey and A. Strominger, "World-Sheet Approach to Heterotic Solitons and Instantons," *Nucl. Phys.* **359** (1991) 611; I. Antoniadis, C. Bachas, J. Ellis and D. Nanopoulos, "Cosmological String Theories and Discrete Inflation," *Phys. Lett.* **B211** (1988) 393; *Nucl. Phys.* **B328** (1989) 117; S.-J. Rey, "Confining Phase of Superstrings and Axionic Strings," *Phys. Rev.* **D43** (1991) 439.
48. M. Li and E. Martinec, "Matrix Black Holes," hep-th/9703211; talk at *Strings '97*, hep-th/9709114.
49. J. A. Harvey and A. Strominger, "The Heterotic String is a Soliton," *Nucl. Phys.* **B 449** (1995) 535-552; *Nucl. Phys.* B458 (1996) 456-473, hep-th/9504047. S. Cherkis and J.H. Schwarz, "Wrapping the M Theory Five-Brane on K3," hep-th/9703062.
50. R. Dijkgraaf, E. Verlinde, and H. Verlinde, "Counting Dyons in N=4 String Theory," *Nucl. Phys.* **B484** (1997) 543, hep-th/9607026
51. T. Kawai, "$N = 2$ Heterotic String Threshold Correction, $K3$ Surface and Generalized Kac-Moody Superalgebra," hep-th/9512046.

Part II

Seminars

TESTING SEIBERG-WITTEN SOLUTION

A. LOSEV
Institute of Theoretical and Experimental Physics,
117259, Moscow, Russia

N. NEKRASSOV
Institute of Theoretical and Experimental Physics,
117259, Moscow, Russia
Lyman Laboratory of Physics, Harvard University,
Cambridge MA 02138

AND

S. SHATASHVILI
Department of Physics, Yale University,
New Haven CT 06520, Box 208120
on leave of absence from
St. Petersburg Steklov Mathematical Institute

1. Introduction and summary

One of the sources of the recent duality revolution in string theory is the solution of N. Seiberg and E. Witten [1] of $\mathcal{N} = 2$ supersymmetric Yang-Mills theory which has been tested in many indirect ways but never directly. The solution, among other things, predicts the formula for the effective coupling constant of the low-energy theory as a function of the order parameter u (in $SU(2)$ theory):

$$\tau(u) = \frac{2i}{\pi}\log\left(\frac{u}{\Lambda^2}\right) + \sum_{n=1}^{\infty} \tau_n \left(\frac{\Lambda^2}{u}\right)^{2n} \tag{1.1}$$

The coefficients τ_n are claimed to be instanton corrections. The direct test of (1.1) would involve integration of a certain form over the moduli space of instantons on $\mathbb{R}^4$. This integration can be shown to localize onto the space of *point-like instantons* and has a potential divergence in it related to the fact that the space of point-like instantons is non-compact (it is a resolution of singularities of $S^k\mathbb{R}^4$). Although this difficulty is avoidable by appropriate regularization so far no substantial success has been achieved on this route (nevertheless, see the related discussion in [5]). The infrared regularization of the instanton integrals turns out to be tricky

L. Baulieu et al. (eds.), Strings, Branes and Dualities, 359–372.

problem. At first sight it is possible to simply put the periodic boundary conditions and study the theory on a four-torus. Unfortunately the puriest test of (1.1) would involve the computation of the correlation functions of observables which preserve some supersymmetry which are called topological correlation functions. TCF's are almost insensitive to the geometry of the moduli space of vacua of the gauge theory in the case, where the four-manifold X has $b_2^+ > 1$. To make the TCF probe the geometry of the moduli space of vacua one must find such a manifold X which allows for the instanton calculus to be perfomed and has $b_2^+ = 1$. The latter constraint will be explained below. Examples of such manifold X are provided by ruled surfaces, $X = S^2 \times S^2$ among them . The manifold X does not have any covariantly constant spinors and the original supersymmetric theory has no conserved charges. One may study a twisted theory, introduced by E. Witten in [13]. The latter has more freedom in the choice of action then the original physical theory. In particular, one may take the limit of zero coupling in the non-abelian theory which reduces the computations of TFC to the counting intersection numbers of homology classes of compactified moduli space of instantons ion X. Such a counting problem is equivalent to the way S. Donaldson formulated his invariants in the language of problems of "generic position" (for the gauge groups $SU(2)$ and $SO(3)$). The insight coming from the equivalence of Donaldson theory and Witten's twisted version of $\mathcal{N} = 2$ super-Yang-Mills theory is that the same problem may be addressed in the infrared limit. It seems that the well-defined ultraviolet problem in the field-theoretic formulation involves the definition of the compactified moduli space of *distinct* points on the space-time manifold X. The infrared theory contains the information about this compactification in the contact terms between the observables . In the context of gauge theory these contact terms can be studied using severe constraints of modular invariance and ghost number anomalies .

2. The micro/macroscopic theories

The $\mathcal{N} = 2$ super-Yang-Mills theory with the gauge group G has as the basic field the vector multipler of four dimensional $\mathcal{N} = 2$ supersymmetry, which involves a gauge field A^a, a complex scalar ϕ^a and a pair of Weyl fermions $\psi^a_\alpha, \lambda^{\dot{\alpha},a}$. All fields are in the adjoint representation of the gauge group. For the sake of convenience we study the twisted version of the theory where fermions have different Lorentz spins. One should understand once and for all that as long as "physical" questions are asked about the theory formulated on $\mathbb{R}^4$ (euclidean signature) there is no way to distinguish the two models. The distinction appears when the curved gravitational background appears. The twisted theory always has at least one conserved supercharge Q. The square of Q is an infinitesimal gauge transformation with parameter ϕ. The operator Q acts on the Hilbert space of the theory. Its square vanishes on the physical Hilbert space formed by the gauge invariant states. The Hilbert space $\mathcal{H}_S$ is obtaned by quantizing the space of fields on a three-fold S. The Hamiltonian of the theory can be represented as a Q-commutator:

$$H = \{Q, G_r\} \tag{2.1}$$

where G_r is a certain operator which is in fact a twisted version of one of the eight supercharges of the original $\mathcal{N}=2$ susy. There is an operator of ghost number U which can be represented as the integral of the corresponding current J: $U=\int_S J$. The Hamiltonian has ghost charge 0, the operator Q has $U=1$, G_r has $U=-1$, the fields have the following charges:

TABLE 1. Ghost charges, Lorentz spins, Q action

Field	$SU(2)_L \times SU(2)_R$	U charge	Statistics	Q action
A	$(\frac{1}{2},\frac{1}{2})$	0	B	ψ
ψ	$(\frac{1}{2},\frac{1}{2})$	1	F	$D_A\phi$
ϕ	$(0,0)$	2	B	0
χ	$(0,1)$	-1	F	H
H	$(0,1)$	0	B	$[\phi,\chi]$
η	$(0,0)$	-1	F	$[\phi,\bar{\phi}]$
$\bar{\phi}$	$(0,0)$	-2	B	η

Since on the gauge invariant states $Q^2=0$ one may study the cohomology space $H_S=\mathrm{Ker}Q/\mathrm{Im}Q$. The importance of this space is that it is preserved under the evolution and changes of the metric. Indeed, if $Q|\psi_0\rangle=0$ and we identify the states in the evolved space with the help of Hamiltonian H then

$$|\psi_t\rangle = e^{-tH}|\psi_0\rangle = |\psi_0\rangle + Q|\psi_t'\rangle \tag{2.2}$$

that is the cohomology class of a state does not change with time. Now, having got the space H_S one may wonder how many states are there and what are the natural operators which act on H_S. To answer the second question we must figure out what are the operators $\mathcal{O}$ the theory which obey $\{Q,\mathcal{O}\}=0$. The Hamiltonian is one of them but it acts trivially on H_S. What are non-trivial operators? Clearly, these are the ones which cannot be represented as $\{Q,\mathrm{smth}\}$, that is we are interested in Q-cohomology in operators.

The twisted gauge theory has various non-local observables. They are labelled by the following data: a gauge invariant function $\mathcal{P}$ on the Lie algebra $\underline{\mathrm{g}}$ and the number $p=0,\ldots,4$. The p-observable $\mathcal{O}_{\mathcal{P}}^{(p)}$ is a p-form and is supposed to be integrated over a p-cycle in the manifold X. The easiest way to write all p-observables at once is to expand the function $\mathcal{P}(\phi+\psi+F)$ is the forms of varios degree. The observables $\mathcal{O}^{(p)}$ then automatically obey the descend equations

$$d\mathcal{O}^{(p)} = \{Q, \mathcal{O}^{(p+1)}\} \tag{2.3}$$

Of course, the representative of Q-cohomology is not unique. The one we get by expanding $\mathcal{P}(\phi+\psi+F)$ is called *holomorphic*. For some reasons one may prefer a harmonic representative, which is obtained from $\mathcal{P}(\phi)$ by acting with

G_μ on it. Here G_μ are the twisted supercharges, which obey $\{Q,G\} = d$ (we saw one of them in (2.1)). The holomorphic observables can be written quite explicitly: $\mathcal{O}^{(0)}_{\mathcal{P}} = \mathcal{P}(\phi)$, $\mathcal{O}^{(1)}_{\mathcal{P}} = \frac{\partial \mathcal{P}}{\partial \phi^a}\psi^a$, $\mathcal{O}^{(2)}_{\mathcal{P}} = \frac{\partial \mathcal{P}}{\partial \phi^a}F^a + \frac{1}{2}\frac{\partial^2 \mathcal{P}}{\partial \phi^a \partial \phi^b}\psi^a\psi^b$, $\mathcal{O}^{(3)}_{\mathcal{P}} = \frac{1}{2}\frac{\partial^2 \mathcal{P}}{\partial \phi^a \partial \phi^b}\psi^a F^b + \frac{1}{6}\frac{\partial^3 \mathcal{P}}{\partial \phi^a \partial \phi^b \partial \phi^c}\psi^a\psi^b\psi^c$. The top degree observable equals:

$$\mathcal{O}^{(4)}_{\mathcal{P}} = \frac{1}{2}\frac{\partial^2 \mathcal{P}}{\partial \phi^a \partial \phi^b}F^a F^b + \frac{1}{3!}\frac{\partial^3 \mathcal{P}}{\partial \phi^a \partial \phi^b \partial \phi^c}F^a\psi^b\psi^c + \frac{1}{4!}\frac{\partial^4 \mathcal{P}}{\partial \phi^a \partial \phi^b \partial \phi^c \partial \phi^d}\psi^a\psi^b\psi^c\psi^d \tag{2.4}$$

It enters the Seiberg-Witten low-energy effective action, where all the fields are specialized to be abelian. In general, the whole action S equals the sum of the 4-observable, constructed out of the prepotential $\mathcal{F}$ and the Q-exact term:

$$S = \int_X \mathcal{O}^{(4)}_{\mathcal{F}} + \{Q, R\}. \tag{2.5}$$

We can view all observables $\mathcal{O}^{(p)}_{\mathcal{P}}$ for all $\mathcal{P}$ and p as deformations of the theory by adding them to the action (2.5) as follows:

$$S = \sum_{p=0}^{4} T^{p,\alpha} \int_X \left(\mathcal{O}^{(p)}_{\mathcal{F}_p} + \{Q, R^{(p)}\}\right) \wedge e_\alpha \tag{2.6}$$

where e_α runs through a basis in the cohomology group $H^*(X;\mathbb{R})$. In turn, one may expand $T^{p,\alpha}\mathcal{F}_p = \sum_k T^{k,\alpha}\mathcal{P}_k$ with $\mathcal{P}_k$ running over a space of invariant polynomials.

The infrared theory is also a twisted supersymmetric gauge theory with abelian gauge group. The observable $\mathcal{O}^{(p)}_{\mathcal{P},\mathrm{UV}}$ of the ultraviolet theory flows to the observable $\mathcal{O}^{(p)}_{\mathcal{P},\mathrm{IR}}$ of the infrared theory. Since the gauge group of the infrared theory is the maximal torus (extended by the Weyl group) of the ultraviolet theory there is one-to-one correspondence between the gauge invariant functions $\mathcal{P}_{\mathrm{UV}}$ and $\mathcal{P}_{\mathrm{IR}}$. Now suppose that two cycles $C^p \in H_p(X)$ and $C^q \in H_q(X)$ intersect: $C_p \cap C_q = C_{p+q-4} \in H_{p+q-4}(X)$. No matter how large the scale is to get at the intersection point(s) we must look at very short distances. The contribution of the short distance physics is the contact term:

$$\int_{C^p} \mathcal{O}^{(p)}_{\mathcal{P},\mathrm{UV}} \int_{C^q} \mathcal{O}^{(q)}_{\mathcal{Q},\mathrm{UV}} \longrightarrow \int_{C^p} \mathcal{O}^{(p)}_{\mathcal{P},\mathrm{IR}} \int_{C^q} \mathcal{O}^{(q)}_{\mathcal{Q},\mathrm{IR}} + \int_{C^{p+q-4}} \mathcal{O}^{(p+q-4)}_{C(\mathcal{P},\mathcal{Q}),\mathrm{IR}} \tag{2.7}$$

The main problem of the effective usage of low-energy theory is the determination of the contact terms $C(\mathcal{P},\mathcal{Q})$. The important constraint on the function $C(\mathcal{P},\mathcal{Q})$ is that it is independent of p and q. This property was called in [17] *the universality of the contact terms.* It follows from the simple fact that the *relative* geometry of a pair of points on a four-fold X is independent of the dimensions of the cycles the points are confined to. This universality property implies some non-trivial idenities between the modular functions which we describe in the next section. These identities seem to distinguish Seiberg-Witten-like families of abelian varieties and serve as a test of SW technology.

The next contact term is the triple term : suppose three cycles $C^{p_1}, C^{p_2}, C^{p_3}$ are given, then the analogue of (2.7) is:

$$\prod_{i=1}^{3} \int_{C^{p_i}} \mathcal{O}^{(p_i)}_{\mathcal{P}_i,\mathrm{UV}} \longrightarrow \prod_{i=1}^{3} \int_{C^{p_i}} \mathcal{O}^{(p_i)}_{\mathcal{P}_i,\mathrm{IR}} + \sum_{(i,j,k)=(1,2,3)} \int_{C^{p_i}\cap C^{p_j}} \mathcal{O}^{(p_i+p_j-4)}_{\mathcal{C}(\mathcal{P}_i,\mathcal{P}_j),\mathrm{IR}} \int_{C^{p_k}} \mathcal{O}^{p_k}_{\mathcal{P}_k,\mathrm{IR}} + \int_{C^{p_1}\cap C^{p_2}\cap C^{p_3}} \mathcal{O}^{p_1+p_2+p_3-4}_{\mathcal{C}(\mathcal{P}_1,\mathcal{P}_2,\mathcal{P}_3),\mathrm{IR}} \tag{2.8}$$

If the four-observables are involved then we get an infinite sequence of contact functions $\mathcal{C}(\mathcal{P}_1,\ldots,\mathcal{P}_k)$ which are nicely organized in the generating function $\mathcal{F}$:

$$\mathcal{F}(T) = \mathcal{F}_0 + \sum_{k=1}^{\infty} T^1 \ldots T^k \mathcal{C}(\mathcal{P}_1,\ldots,\mathcal{P}_k) \tag{2.9}$$

where T^k's are the deformation parameters, which wil be referred to as times. The first order term in the function $\mathcal{F}$ is $T^k\mathcal{C}(\mathcal{P}_k)$. The "one-contact function" $\mathcal{C}(\mathcal{P}_k)$ is the function $\mathcal{P}_k$ itself.

The claim is finally:

The TCF (= partition function) in the theory with the action (2.6) is equal to that in the low-energy abelian theory with the action

$$S = \int_X \mathcal{F}(a + \psi + F, T^{k,\alpha} e_\alpha) \tag{2.10}$$

In [17] it is shown that the function $\mathcal{F}(T)$ has a very simple geometrical meaning. It describes the deformations of Γ-invaraint Lagrangian submanifolds in a complex vector symplectic space $\mathbb{C}^{2r}$, where Γ is a certain discrete subgroup of linear symplectic group.

Similarly to the studies of two dimensional type **B** topological sigma models a lot of information can be rather easily obtained by working with holomorphic representatives. This is the point of view taken in this lecture. The reader interested in harmonic theory should consult [16], [17], [27].

3. Coulomb branch measure

The general strategy in using the statement (2.10) is to compute the integral over the zero modes of the fields entering abelian multiplets. The final integral is the one over the scalar field a and it involves various modular forms integrated over the modular domain. Without going into tremendous complications of the actual comutations in any realistic setting one may say a lot of non-trivial things about the measure itself. To compare the last sentence with the development of perturbative string theory one may say that the measure of stringy loop computations has been stduied extensively although the actual integral is known only in a limited number of instances.

First of all we must explain what distinguishes the manifolds with $b_2^+ = 1$. This issue is not possible to see working with holomorphic fields only. It turns out that the harmonic representative of the four-observable contains the only one term ($\int_X \eta\chi \wedge F$) which couples to a zero mode of η and survives in the limit of very large X. Thus one needs presicely one zero mode of the field χ otherwise the effective measure vanishes as an inverse power of the size of manifold X.

It turns out that the TCF's are not exactly topological invariants. The reason is that sometimes the enumerative problem of Donaldson's is not well-behaved under the variations of parameters, such as the metric. This occurs precisely when $b_2^+ \leq 1$ due to the abelian instantons[1]. The jumps of the correlations functions/Donaldson invariants for $b_2^+ = 1$ are under control [21], [22],[23],[16].

Below we describe a few techinques of computing the pair contact terms in various theories with simply-laced gauge groups[2]. We use the low-energy effective theory, whose action on $\mathbb{R}^4$ has been computed in [1], and certain aspects of it for the general four-manifold X have been worked out in [12] and also recently in [16], [17], [27].

3.1. LOW-ENERGY THEORY

The low-energy theory contains r $\mathcal{N} = 2$ vector multiplets, which are defined up to Γ- transformation, where Γ is a subgroup of $Sp_{2r}(\mathbb{Z})$, e.g $\Gamma(2)$ or $\Gamma_0(4)$ for $r = 1$. Let us denote the scalar components of the multiplet, which are monodromy invariant (up to a sign) at $u^k = \infty$ by a^i. Then the S-dual ones will be denoted as $a_{i,D}$. The low-energy effective couplings are denoted as $\tau_{ij}(a) = \left(\frac{4\pi i}{g^2_{eff}} + \frac{\theta_{eff}}{2\pi}\right)_{ij} = \frac{\partial a_{i,D}}{\partial a^j} = \tau_{ij,1} + i\tau_{ij,2} \equiv \mathrm{Re}e\tau_{ij} + i\mathrm{Imm}\tau_{ij}$

The Q- transformations of the gauge fields must be consistent with the electric-magnetic duality. It is not clear a priori that such a Q-action exists, since the duality is a non-local operation on the fields, while the Q is a local one. One may try to imitate the twisted version of the supersymmetric duality transformation presented in [1].

We take the approach where only holomorphuic representatives of the fields are used. It allows for a quick check of the modular invariance of the measure in a relatively simple setting. One has to work with (formal) contour integrals. Most of the constructions can be done working with the "holomorphic" fields a, ψ, A only. The only trouble with such a prescription is the absence of Laplacians and non-definiteness of the topological terms like $F \wedge F$. The first problem is avoided in certain cases by working with cohomology (with harmonic forms) while the second may be treated via analytic continuation. In any case such an approach is useful in getting the right structures. Once it is done one may introduce the quartet of the fields $\bar{a}, \eta, \chi, H$ and justify the constructions by working with the standard positive-definite actions.

[1]They are honest solutions of instanton equations with abelian gauge group

[2]The contact term between the two-observables constructed out of quadratic casimirs and certain steps in extending the integrals of [16] to the case of $SU(N)$ gauge group are independently obtained in [27]

Consider the *short superfield:*

$$\mathcal{A}_{i,D} = a_{i,D} + \psi_{i,D} + F_{i,D} \tag{3.1}$$

where $dF_{i,D} = 0$. The operator Q acts as follows: $Qa_{i,D} = 0, Q\psi_{i,D} = da_{i,D}$, $QF_D = d\psi_{i,D}$. We impose (by hands) the condition that $F_{i,D}$ represents the integral cohomology class of the space-time manifold Σ. Thus, $\mathcal{A}_{i,D} \in \Omega^0(\Sigma)_B \oplus \Omega^1(\Sigma)_F \oplus \Omega^2_{\mathbb{Z}}(\Sigma)_B$ Here $\Omega^2(\Sigma)_{\mathbb{Z}}$ is the space of closed two-forms with periods in $2\pi i\mathbb{Z}$. The indices B, F denote the bosonic and fermionic fields respectively. The superfield $\mathcal{A}_{D,i}$ obeys the condition $(Q-d)\mathcal{A}_{D,i} = 0$. One may also fulfill the condition of $Q-d$-closedeness by introducing a complete set of p-forms which we call *the long superfield:* $\mathcal{A}^i = \sum_{p=0}^4 \mathcal{A}^{i,p} \in V = \oplus_{p=0}^4 \Omega^p(X)$ and Q acts as

TABLE 2. Long superfield

Degree of a form	0	1	2	3	4
Field $\mathcal{A}^i =$	$a^i +$	$\psi^i +$	$F^i +$	$\rho^i +$	D^i

$Q\mathcal{A}^{i,p} = d\mathcal{A}^{i,p-1}$. Let $\mathcal{F}_D$ be a holomorphic function on $\mathbb{C}^r$. The "action"

$$S = \int_\Sigma \mathcal{F}_D(\mathcal{A}_D) \tag{3.2}$$

is clearly Q-invariant. The long superfield $\mathcal{A}$ allows reparameterizations:

$$\mathcal{A}^i \mapsto \widetilde{\mathcal{A}^i}(\mathcal{A}) \tag{3.3}$$

induced by the holomorphic maps $a^i \mapsto \tilde{a}^i(a^k)$. Let $\mathcal{L} \subset \mathbb{C}^{2r}$ be a Γ-invariant Lagrangian subvariety. Let $u^k, k = 1, \ldots r$ be the generators of the ring $\mathcal{W}_\mathcal{L}$ of globally defined Γ-invariant holomorphic functions on $\mathcal{L}$. Extend them to the long superfields $\mathcal{U}^k, k = 1, \ldots, r$. Define the measure

$$[\mathcal{DU}] = \prod_{k=1}^r du^k d\psi_u^k dF_u^k d\rho_u^k dD_u^k \tag{3.4}$$

where $(Q-d)\left(u^k + \psi_u^k + F_u^k + \rho_u^k + D_u^k\right) = 0$. The duality transformation proceeds as follows: introduce both $\mathcal{A}_{i,D}$ and $\mathcal{A}^i$ and consider the action $S' = \int_\Sigma \mathcal{A}^i \mathcal{A}_{i,D} - \mathcal{F}(\mathcal{A}^i)$. Let us consider the following (formal) path integral:

$$\int \mathcal{DU}^i \mathcal{D}\mathcal{A}_{i,D} e^{-S}. \tag{3.5}$$

The measure $\mathcal{D}\mathcal{A}_{i,D}$ is defined canonically. The dependence of the measure on $\mathcal{A}^i$ on the choice of the measure $du^1 \wedge \ldots du^r$ is completely parallel to anomaly in Type **B** sigma models in two dimensions [18],[11]. The integral over $\mathcal{A}_{i,D}$ (together with summation over the fluxes of $F_{i,D}$) forces D^i, ρ^i to vanish, while F^i becomes a curvature of a connection A^i. As a result one gets a measure

$$\mathrm{Det}_{ij}\left(\frac{\partial u^i}{\partial a^j}\right)^{\frac{\chi}{2}} \prod_{k=1}^{r} da^k d\psi^k dF^k$$

On the other hand, performing the integral over $\mathcal{U}$ gives us: $a_{i,D} = \frac{\partial \mathcal{F}}{\partial a^i}$, $\psi^i = \left(\tau^{-1}\right)^{ij} \psi_{j,D}$, $F^i = \left(\tau^{-1}\right)^{ij}\left(F_{j,D} - \frac{1}{2}(\tau^{-1})^{lm}(\tau^{-1})^{kp}(\partial^3_{lkj}\mathcal{F})\psi_{m,D}\psi_{p,D}\right)$ with $\tau_{ij} = \frac{\partial^2 \mathcal{F}}{\partial a^i \partial a^j}$. The determinants in this case are slightly involved:

$$\begin{aligned} \mathrm{Det}_{ij}\left(\tfrac{\partial u^i}{\partial a^j}\right)^{\frac{\chi}{2}} (\mathrm{Det}\tau)^{-\mathrm{dim}\Omega^0 + \mathrm{dim}\Omega^1 - \frac{1}{2}\mathrm{dim}\Omega^2} &= \\ (\mathrm{Det}\tau)^{-\frac{\chi}{2}} \mathrm{Det}_{ij}\left(\tfrac{\partial u^i}{\partial a^j}\right)^{\frac{\chi}{2}} &= \\ \mathrm{Det}_{ij}\left(\frac{\partial u^i}{\partial a_{D,j}}\right)^{\frac{\chi}{2}} & \end{aligned} \tag{3.6}$$

Were there no factor $\mathrm{Det}_{ij}\left(\frac{\partial u^i}{\partial a^j}\right)^{\frac{\chi}{2}}$ the duality transformation would be anomalous. This anomaly was already observed in [12] (for $r = 1$). The "action" S' evaluates to (3.2) with the substitution of $\mathcal{F}$ by $\mathcal{F}_D$, which is the Legendre transform of $\mathcal{F}$.

As it has been explained in [12] the low-energy effective action contains the terms which account for the coupling to the background gravitational field:

$$e^{S_{grav}} \sim e^{b(u)\chi + c(u)\sigma}$$

Here $b(u), c(u)$ are the gravitational renormalization coefficients computed for the low-energy $SU(2)$ theory in [12]. We already know that

$$e^{2b} = \mathrm{Det}\frac{\partial u^k}{\partial a^l} \tag{3.7}$$

It remains to compute e^c. Imagine that the manifold X is replaced by a blowup $\widetilde{X}$ at the point P. Geometrically it means that we glue a copy of $\bar{\mathbb{P}}^2$ to X. The Euler characteristics and the signature of the manifold $\widetilde{X}$ are $\chi+1$ and $\sigma-1$ respectively since b_2^- is increased by one in the process ($\bar{\mathbb{P}}^2$ contains a non-contractible two-sphere with self-intersection -1). Consider an integral of the the divergence of the ghost number current over the glued $\bar{\mathbb{P}}^2$ in the instanton sector with the total instanton number k. The instanton charge splits as a sum $k = k_1 + k_2$ where k_1 is what is left on $X - P$, and k_2 is what has gone to $\bar{\mathbb{P}}^2$. The divergence picks up an anomaly:

$$\int_{\bar{\mathbb{P}}^2} dJ = 2\beta_1 k_2 \tag{3.8}$$

where β_1 is the perturbative beta-function. In the case of pure gauge theory it is equal to $-2h^\vee$, $h^\vee$ being the dual Coxeter number. There also could be a gravitational contribution to (3.8) but it vanishes because $b_2^+(\bar{\mathbb{P}}^2) = 0$. Now, using factorization we may replace the glued $\bar{\mathbb{P}}^2$ by a local operator $\mathcal{B}(u^k)$ which must produce the same anomaly as in (3.8). If the theory is asymptotically free, then the right hand side of (3.8) is negative for $k_2 > 0$ and therefore the operator $\mathcal{B} = 1$ and only $k_2 = 0$ contributes. The conclusion is that the blowup does not change the measure. On the other hand the gravitational renormalization together with extra piece of Maxwell partition function combine to

$$e^{b-c}\Theta(\tau) = 1 \tag{3.9}$$

where $\Theta(\tau)$ is a certain theta-constant which is discussed below. Hence:

$$e^{S_{grav}} \sim \mathrm{Det}_{ij}\left(\frac{\partial u^i}{\partial a^j}\right)^{\frac{\chi+\sigma}{2}} \Theta(\tau)^\sigma \tag{3.10}$$

It remains to notice that

$$\mathrm{Det}_{ij}\left(\frac{\partial u^i}{\partial a^j}\right)^4 \Theta^8 = \Delta(u_i)$$

is the modular invariant function on the moduli space of vacua (the discriminant) and rewrite (3.10) as

$$e^{S_{grav}} \sim \mathrm{Det}_{ij}\left(\frac{\partial u^i}{\partial a^j}\right)^{\frac{\chi}{2}} \Delta^{\frac{\sigma}{8}} \tag{3.11}$$

There is another gravitational correction to the effective action, described in [12], namely, if the manifold X is not spin, then there is a term $e^{\frac{1}{2}(w_2(X),F)}$ in the effective measure. We can get rid of it in the course of the study of $SU(2)$ theory by the shift $\tau \to \tau + 1$ thanks to Wu formula. The sign generalizes to

$$e^{(\langle F,\rho\rangle, w_2(X))} \tag{3.12}$$

in the case of general simply-laced group. Suppose we are to integrate the massive W-bosons out. Let $a \in \underline{\mathrm{t}}$ be the scalar in the Cartan part of the vector multiplet. The one-loop determinants on bosons and fermions cancel on non-zero modes leaving purely holomorphic contribution:

$$\prod_\alpha \langle a, \alpha\rangle^{\mathrm{Ind}_{\mathrm{d}\oplus\mathrm{d}^*}(L_\alpha)} \tag{3.13}$$

where α runs over the set of roots, L_α is the line bundle corresponding to the root α, $\mathrm{Ind}_{\mathrm{d}\oplus\mathrm{d}^*}(\mathrm{L}_\alpha)$ is the index of the operator $d \oplus d^*$ coupled to the line bundle L_α (the W_α-boson multiplet). The index formula gives

$$\mathrm{Ind}_{d\oplus d^*}(L_\alpha) = \frac{1}{2}\left(\int_X c_1(L_\alpha)^2 + c_1(X)c_1(L_\alpha)\right) + \frac{\chi+\sigma}{4}.$$

We see that the terms $c_1(L_\alpha)^2$ collect into the perturbative beta-function, the term proportional to $\frac{\chi+\sigma}{4}$ yields the asymptotics of (3.11) in the $\Lambda \to 0$ limit and the odd part $c_1(\Sigma)c_1(L_\alpha)$ survies in the form $\prod_{\alpha>0}(-1)^{c_1(X)c_1(L_\alpha)} \equiv e^{(\langle F,\rho\rangle, w_2(X))}$ since $c_1(X) \equiv w_2(X)\mathrm{mod}2$. We denoted by $c_1(X)$ the first Chern class of the canonical bundle of the almost complex structure which exists on any four-manifold (cf. [12]).

3.2. SPECIFIC COMPUTATIONS OF THE CONTACT TERMS

3.2.1. *Two-observables*

Suppose we are interested in computing $\langle \exp \int_{C_a} \mathcal{O}^{(2)}_{\mathcal{P}_a} \rangle_X$, where C_a are two-cycles on a four-manifold X which may intersect. If we go ahead and write down the effective low-energy measure we immediately face the problem of modular anomaly. Indeed, the low-energy measure contains a theta-function $\Theta_{H_2(X;\mathbb{Z})}(\tau)^3$ associated to an intersection form of X. It comes from the partition function of the Maxwell theory. In the presence of the two-observables this theta-function has an argument: $\Theta(\vec{z}, \tau)$, $z_i = C_a^\vee \frac{\partial \mathcal{P}_a}{\partial a^i}$ and its modular transformation produces a factor

$$\exp \sum_{a,b} \#(C_a \cap C_b)(\tau^{-1})^{ij} \frac{\partial \mathcal{P}_a}{\partial a^i} \frac{\partial \mathcal{P}_b}{\partial a^j} \tag{3.14}$$

This factor must be cancelled by an anomaly of an additional interaction which is developed due to the presence of intersecting densities, i.e. due to contact terms. In order to derive this interaction we again use the trick with blow up of a manifold X this time at the intersection point P. The homology lattice $H_*(\widetilde{X})$ of $\widetilde{X}$ is that of X plus a factor of $\mathbb{Z}$. The intersection form is simply

$$(,)_{\widetilde{X}} = (,)_X \oplus (-1)$$

as the exceptional divisor e (the two-sphere inside $\bar{\mathbb{P}}^2$) has self-intersection -1. Under the isomorphism $H_*(\widetilde{X}) = H_*(X) \oplus \mathbb{Z}$ the inverse images of the cycles in X belong to the component $H_*(X)$ of $H_*(\widetilde{X})$. We shall denote them by the same letters as the cycles in X. To derive the contact term we compare the correlation functions $\langle \int_{C_1} \mathcal{O}^{(2)}_{\mathcal{P}_1} \int_{C_2} \mathcal{O}^{(2)}_{\mathcal{P}_2} \ldots \rangle_X$ and $\langle \int_{\tilde{C}_1} \mathcal{O}^{(2)}_{\mathcal{P}_1} \int_{\tilde{C}_2} \mathcal{O}^{(2)}_{\mathcal{P}_2} \ldots \rangle_{\widetilde{X}}$ where the cycles $\tilde{C}_1, \tilde{C}_2 \in H_*(\widetilde{X})$ do not intersect each other in the vicinity of P and are given by the formulae:

$$\tilde{C}_k = C_k - e, \quad \#\tilde{C}_1 \cap \tilde{C}_2 = \#C_1 \cap C_2 - 1 \tag{3.15}$$

Consider the same ghost number current integral as in (3.8). Due to the presence of two 2-observables $\int_e \mathcal{O}^{(2)}_{\mathcal{P}_{1,2}}$ the anomaly changes to:

$$\int_{\bar{\mathbb{P}}^2} dJ = 2\beta_1 k_2 + U(\mathcal{P}_1) + U(\mathcal{P}_2) - 4 \tag{3.16}$$

where $U(\ldots)$ denotes the ghost charge which equals twice the degree of $\ldots$ for homogeneous $\mathcal{P}_{1,2}$. For simply-laced group and for $k_2 > 0$ the right hand side

[3] in the harmonic approach it becomes Siegel theta-function associated to a lattice of signature (b_2^+, b_2^-)

of (3.16) does not exceed $2(2h-2)-2(2h-T_m)=2T_m-4$, where T_m is the contribution of matter to the perturbative beta-function. Assuming that $T_m<2$ we see that again the operator $\mathcal{B}$ must be equal to one (see [20], [24] for mathematical proof of this result). The net effect of our manipulations is the replacement of the intersecting cycles on the manifold X by the non-intersecting cycles on the manifold $\tilde{X}$. Physically the crucial fact is that under the blowup of a point P a new two-cycle e appears and it leads to the possibility for the gauge field to have a flux through it. In the low-energy effective theory the insertion of the new two-cycle must be reflected in the new factor in the Maxwell partition function, which is the sum over all line bundles on $\overline{\mathbb{P}}^2$ in the presence of two 2-observables $\int_e \mathcal{O}^{(2)}_{\mathcal{P}_{1,2}}$. This partition function is simply:

$$\frac{\partial \mathcal{P}_1}{\partial a^i}\frac{\partial \mathcal{P}_2}{\partial a^j}\frac{\partial}{\partial \tau_{ij}}\log\Theta(\tau) \tag{3.17}$$

where

$$\Theta=\sum_{\lambda\in\Lambda}\exp\left(2\pi i\langle\lambda,\tau\lambda\rangle+\pi i\langle\lambda,\rho\rangle\right) \tag{3.18}$$

with Λ being the set of weights and $\langle,\rangle$ the restriction of the Killing form on $\underline{\mathrm{t}}$ - the Cartan subalgebra of $\underline{\mathrm{g}}$. The term $(-)^{\langle\lambda,\rho\rangle}$ is (3.12) specified to $\bar{\mathbb{P}}^2$ case. Summarizing, we have shown that in the theories with $T_m<2$ the pair contact term is equal to:

$$\mathcal{C}(\mathcal{P}_1,\mathcal{P}_2)=\frac{\partial \mathcal{P}_1}{\partial a^i}\frac{\partial \mathcal{P}_2}{\partial a^j}\frac{\partial}{\partial \tau_{ij}}\log\Theta(\tau) \tag{3.19}$$

3.3. CONTACT TERM OF 0-OBSERVABLES AND 4-OBSERVABLES

Another way of getting the contact term $\mathcal{C}(\mathcal{P}_1,\mathcal{P}_2)$ is by considering 4- and 0-observables. Let us consider asymptotically free theory with massless matter. Let $\mathcal{P}_1=u_1$ be the generator whose fourth decsendant produces the instanton charge (quadratic casimir). The insertion of $e^{t\int_X u_1^{(4)}}$ is equivalent to multiplication of the correlator in the sector with instanton charge k by e^{tk}. It is possible to show that this manipulation is in turn equivalent to rescaling of the fields $u_k\to e^{\frac{d_k}{\beta_1}t}u_k$ and $a^i\to e^{\frac{2t}{\beta_1}}a^i$, where d_k are the weights of the homogeneous generators u_k of the ring of invariant polynomials. This idea yields the following formula [17]:

$$\mathcal{C}(u_1,u_k)=\frac{2}{\beta_1}\left(d_k u_k-a^i\frac{\partial u_k}{\partial a^i}\right) \tag{3.20}$$

Comparing (3.20) and (3.19) in cases where it is possible we get an interesting identity, which singles out the family of SW curves and their generalizations [1], [2],[3]. For concreteness we consider $SU(N_c)$ theory without matter:

$$\frac{\partial u_1}{\partial a^i}\frac{\partial u_k}{\partial a^j}\frac{\partial}{\partial \tau_{ij}}\log\Theta(\tau)=\frac{1}{2N_c}\left((k+1)u_k-a^i\frac{\partial u_k}{\partial a^i}\right) \tag{3.21}$$

where a^i are the A-periods of the differential $x\frac{dz}{z}$ on the curve:

$$z + \frac{1}{z} = x^{N_c} - \sum_{k=1}^{N_c-1} u_k x^{N_c-k-1}$$

It is possible to promote the formulae for the contact terms (3.19) and (3.20) to the *evolution* equations a la Whitham hierarchy (cf. [8], [9]). Details will be shown in a separate publication [4].

4. Remarks on ADHM integration

Consider the moduli space $\mathcal{M}_{k,N}$ of charge k $U(N)$ instantons on $\mathbb{R}^4$. By ADHM construction it is the hyperkahler quotient of linear space of dimension $4k^2 + 4kN$ by the action of $U(k)$. That is, it is the quotient of the space of solutions of certain equations $\vec{\mu} = 0$ by $U(k)$. This representation allows to get the expressions for the integrals of cohomology classes over $\mathcal{M}_{k,N}$ in terms of the contour integrals over the complexified Lie algebra of $U(k)$. The integrals are localized according to the standard equivariant techniques [10] to the fixed points of $U(k)$ action which are nothing but the point-like instantons! The space of those is non-compact - the instanton can run away to infinity. In order to cure this problem let us take into account the natural actions of the Lorentz group $SO(4)$ and gauge group $U(N)$. Suppose we compute the integrals over $\mathcal{M}_{k,N}$ of the $SO(4) \times U(N)$-equivariant cohomology classes $\omega_1(\vec{\epsilon};\vec{a}), \ldots, \omega_p(\vec{\epsilon};\vec{a})$. Here $\vec{\epsilon} = \binom{\epsilon_1}{\epsilon_2}$ is the generator of the Cartan subalgebra of $SO(4)$ and $\vec{a} = (a_1, \ldots, a_N)$ is the generator of $U(N)$ Cartan subalgebra. $\vec{\epsilon}$ generates rotations in two orthogonal planes in $\mathbb{R}^4$. The fixed points of such rotation are formed by the instantons which sit on top of each other in the center of rotation. Now the integral can be evaluated. The classes ω_l are in one-to-one correspondence with the invariant polynomials P_l on the Lie algebra of $U(k)$, i.e. symmetric functions in k variables. We claim that

$$Z(\vec{\epsilon};\vec{a}) := \int_{\mathcal{M}_{k,N}} \prod_l \omega_l(\vec{\epsilon};\vec{a}) =$$

$$\frac{(\epsilon_1+\epsilon_2)^k}{\epsilon_1^k \epsilon_2^k} \oint \bigwedge_{i=1}^{k} \frac{dz_i}{\prod_\lambda (z_i + a_\lambda)(z_i + a_\lambda + \epsilon_1 + \epsilon_2)} \prod_{i\neq j} \frac{z_{ij}(z_{ij}+\epsilon_1+\epsilon_2)}{(z_{ij}+\epsilon_1)(z_{ij}+\epsilon_2)} \quad (4.1)$$

where $z_{ij} = z_i - z_j$.

The formula (4.1) is the only sensible calculation one may perform using ADHM data. To compare it with the calculations involving SW low-energy effective action one needs to develop a formalism which takes into account the isometries of the space-time manifold. It is equivalent to working with $Q + \epsilon^\mu G_\mu$ - cohomology instead of Q-cohomology (cf. [6], [19]). This theory is currently under study [4].

Acknowledgements

We thank G. Moore for many discussions on Donaldson theory and related matters over the years. S.Sh. is grateful to E. Witten for important discussion reviving the interest in the subject. We are grateful to A. Gerasimov, A.Gorsky, A. Lawrence, A. Marshakov, A. Morozov , A. Mironov, N. Seiberg and C. Vafa for useful discussions.

The research of A. L. is supported partially by DOE under grant DE-FG02-92ER40704, by PYI grant PHY-9058501 and RFFI under grant 96-01-01101. The research of N. N. is supported by Harvard Society of Fellows, partially by NSF under grant PHY-92-18167, and partially by RFFI grant 96-02-18046. The research of S.Sh. was supported by DOE grant DE-FG02-92ER40704, by NSF CAREER award, by OJI award from DOE and by A.P. Sloan Foundation. A. L. and N. N. are also supported by grant 96-15-96455 for scientific schools.

References

1. N. Seiberg, E. Witten, "Electric-Magnetic Duality, Monopole Condensation, And Confinement in $N = 2$ Supersymmetric Yang-Mills Theory " Nucl. Phys. B426 (1994) 19-52 (and erratum - ibid. B430 (1994) 485-486)
"Monopoles, Duality and Chiral Symmetry Breaking in N=2 Supersymmetric QCD", hep-th/9408099, Nucl. Phys. B431 (1994) 484-550.
2. P.C. Argyres, A.E. Farragi, "The Vacuum Structure and Spectrum of $N = 2$ Supersymmetric $SU(n)$ Gauge Theory", hep-th/9411057
A. Klemm, W. Lerche, S. Theisen and S. Yankielowicz, "Simple Singularities and $\mathcal{N} = 2$ Supersymmetric Yang-Mills Theory", hep-th/9411048
3. P.C. Argyres, M.R. Plesser and N. Seiberg, "The Moduli Space of Vacua of $N = 2$ SUSY QCD and Duality in $N = 1$ SUSY QCD", hep-th/9603042
4. A. Losev, N. Nekrasov, S. Shatashvili, in progress
5. N. Dorey, V. Khoze, M. Mattis, "Supersymmetry and Multi-Instanton Measure", hep-th/9708036
6. A. Losev, Theor.Math.Phys. 95 (1993) 595
7. K. Saito, Publ. RIMS, Kyoto Univ. 19 (1983) 1231
B. Blok, A. Varchenko, Int.J.Mod.Phys.**A7** (1992) 1467-1490
R. Dijkgraaf, talk on RIMS conference "Infinite dimensional analysis", July 1993
A. Losev, in "Integrable models and Strings ", 172, Proceedings of 1993 Helsinki Conference, Springer-Verlag, 1995
A. Losev, I. Polyubin, Int.J.Mod.Phys. **A10** (1995) 4161
8. A. Gorsky, I. Krichever, A. Marshakov, A. Morozov, A. Mironov, hep-th/9505035, Phys.Lett.B355: 466-474, 1995
9. I. Krichever, in "30 years of the Landau Institute", 477 , I.M. Khalatnikov, V.P. Mineev, V.P. (eds.)
10. M. Atiyah, R. Bott, Topology **23** (1984) 1-28
11. E. Witten, "Introduction to Cohomological Field Theories", Int. J. Mod. Phys. **A6** (1991) 2775.
12. E. Witten, Selecta Mathematica **1** (1995) 383
13. E. Witten, "Topological Quantum Field Theory", Commun. Math. Phys. **117** (1988) 353.
14. E. Witten, "Monopoles and Four-Manifolds", hep-th/9411102
15. E. Witten, "On Quantum gauge theories in two dimensions," Commun. Math. Phys. **141** (1991) 153
"Two dimensional gauge theories revisited", J. Geom. Phys. 9 (1992) 303-368
16. G. Moore, E. Witten, "Integration over u-plane in Donaldson theory", hep-th/9709193
17. A. Losev, N. Nekrasov, S. Shatashvili, "Issues in topological gauge theory" ,hep-th/9711108

18. S. Cecotti, C. Vafa, "Topological Anti-Topological Fusion", Nucl. Phys. **B367** (1991) 359
19. A. Givental, alg-geom/9603021, alg-geom/9701016
20. R. Friedman, J. Morgan, "Smooth Four-Manifolds and Complex Surfaces", Ergebnisse der. Math. und Grenz. **3**, Springer-Verlag, New York, 1994
21. L. Göttsche, alg-geom/9506018
22. L. Göttsche, D. Zagier, alg-geom/9612020
23. R. Borcherds, alg-geom/9609022
24. R. Fintushel and R.J. Stern, Annals of Math. **143**(1996) 529
25. A. Gerasimov, hepth/9305090
26. M. Blau, G. Thompson, Nucl. Phys. **B**408 (1993) 345-390
27. M.Mariño, G. Moore, "Integrating over the Coulomb branch in $\mathcal{N} = 2$ gauge theory", hep-th/9712062

MIRRORS AND PHASES OF N=4 IN D=3

ZHENG YIN
Department of Physics, University of California
Berkeley, CA 94720-7300, U.S.A.
and
Mail Stop 50A-5101
Lawrence Berkeley National Laboratory
1 Cyclotron Road, Berkeley, CA 94720, U.S.A.

1. Introduction

3d mirror symmetry for was first proposed in [1] as a duality between certain pairs of generally different 3d N=4 theories at the infrared limit. Infinite sequences of new mirror pairs and strong field theoretic evidence for them were found in [2]. It is a nonperturbative duality that in particular equates certain quantities receiving large quantum corrections with some that are determined entirely classically. Naturally one asks whether this can be the consequence of some string dualities. There are several different approaches [1, 2, 3, 4, 5, 6, 7], each of which has it own advantage and is related to the others by some sequences of dualities. In this talk I will review one that is particularly intuitive [4, 5]. The global R-symmetry of N=4 theories appears as geometric rotations; R-symmetry breaking part of the moduli space of vacua and the parameters of the field theory are realized as the moduli space of D-branes configurations; mirror symmetry itself is implemented by the S duality of type IIB string theory [4]. This construction allows us to engineer a large class of theories and find their mirror duals [5]. Since then it has been generalized to 3d N=2 theories [8, 9]. Similar ideas of constructing field theories in 4d, reviewed in Eguchi's lecture at this school [10], have also been very fruitful[1]. In this talk, I will focus on 3d N=4 theories. After reviewing the rules for "model building" via "brane engineering," I will show how the mirror pairs emerge from this prescription. As an unexpected reward we can predict an infinite number of 3d field theories without conventional Lagrangian descriptions. Some of them are dual to ordinary Lagrangian theories via mirror symmetry, but the rest are not; yet they can be smoothly connected in the moduli space of brane configurations.

[1]See, for example, [11, 12, 13, 14, 15].

L. Baulieu et al. (eds.), Strings, Branes and Dualities, 373–386.

2. What is 3d Mirror Symmetry?

The structure of N=4 supersymmetric gauge theories in three dimensions can be easily obtained by dimensionally reducing the minimally supersymmetric 6d Yang-Mills Lagrangian [16]. The global R-symmetry of the 3d theory is $SU(2)_{345} \times SU(2)_{789}$. The reason for choosing these subscripts will become clear in the brane realization described later. The field content consists of vector multiplets and hypermultiplets. For each vector multiplet associated with a $U(1)$ factor of the gauge group there are 3 real Fayet-Iliopoulos parameters , $\vec{\zeta}$, which can be thought as coming from the VEV's of a background hypermultiplet. För each hypermultiplet, there are 3 real mass parameters, $\vec{m}$. They are the VEV's of a background vector multiplet. There are also gauge coupling constants, which also come from background vector multiplets. The transformation properties of the parameters and VEV's under R-symmetry are summarized in table 1.

Note that usually the scalars in a hypermultiplet are written as a doublet under $SU(2)_{789}$. However, it is convenient, by a change of variables, to rearrange them into a singlet b and a triplet $\vec{r}$. On the other hand, an interesting feature peculiar to three dimensions is that a vector potential is dual to a scalar by the usual electro-magnetic duality . Of course this duality transformation can be precisely formulated only for a free $U(1)$ gauge fields, but this is what is available for the low energy effective theory at generic points of the vector multiplet branch of moduli space. Therefore it is meaningful to include the dualized scalar in considering the moduli space of vacua. By supersymmetry, the moduli space must be hyper-Kähler for both the hypermultiplets and the (dualized) vector multiplets . Because of their different patterns of global R-symmetry breaking, however, the VEV's of the vector and hyper multiplets respectively are distinct order parameters of the theory, even after taking into account of quantum fluctuations. Vacua of N=4, D=3 gauge theories always contain a vector multiplet branch in which the gauge group is generically broken down to $U(1)^N$ where N is the total rank of the gauge group. This branch is parameterized by the $4N$ real scalars from the N corresponding vector multiplets. If it has a sufficient number of hypermultiplets, there can also be a hypermultiplet branch and/or mixed branches.

The low energy effective action up to two derivatives and four fermions is controlled by the metric of the moduli space, which depends on the parameters of the theory as well as the position on the moduli space. The dependence is constrained by extended supersymmetry: the Kähler potential is the sum of a term that depends only on the hypermultiplet scalar and one only on vector multiplet scalars [17, 18]. So a mixed branch is the direct product of a vector branch $\mathcal{M}_V$ and a "hyper" branch $\mathcal{M}_H$. By reinterpreting the parameters of the theory as the VEV's of background superfields, one can further deduce the effects of tuning them on the metric. The gauge coupling $\frac{1}{g^2}$ lives in a vector multiplet, so it can continuously deform the metric of $\mathcal{M}_V$ but not that of $\mathcal{M}_H$. Since g^2 also plays the role of $\hbar$, one concludes that the the metric of $\mathcal{M}_H$ is determined purely classically whereas that of $\mathcal{M}_V$ in general receives quantum corrections. Mass parameters also live in a vector multiplet, so they also can continuously deform $\mathcal{M}_V$ but not $\mathcal{M}_H$ - they can only affect $\mathcal{M}_H$'s dimensionality by changing the number of

TABLE 1. R charges of the VEV's and parameters

Multiplets/Parameters	Notation	$SU(2)_{345} \times SU(2)_{789}$
Vector	$\vec{a} : a^3, a^4, a^5$	(3,1)
	$A_\mu \to \sigma$	(1,1)
Hyper	$\vec{r} : r^7, r^8, r^9$	(1,3)
	b	(1,1)
Fayet-Iliopoulos	$\vec{\zeta} : \zeta^7, \zeta^8, \zeta^9$	(1,3)
mass	$\vec{m} : m^3, m^4, m^5$	(3,1)
(electric) coupling	e	(1,1)

massless hypermultiplets. Fayet-Iliopoulos parameters, on the other hand, live in hypermultiplet, and therefore can only deform $\mathcal{M}_H$ and change the dimensions of $\mathcal{M}_V$. This is summarized in table 2.

TABLE 2. Parameters' influence on moduli space

Branch	$\vec{m}$	$\vec{\zeta}$	e	m (?)
$\mathcal{M}_V$	deform	reduce	deform	no effect (?)
$\mathcal{M}_H$	reduce	deform	no effect	deform (?)

Looking at the above two tables, the pattern for a possible duality emerges. Starting with some theory that we call A model, if we switch what what we mean by $SU(2)_{345}$ and $SU(2)_{789}$, and at the same time exchange masses with Fayet-Iliopoulos parameters, will we end up with an apparently different theory, model B, that is nonetheless equivalent to A? $\mathcal{M}_V$ of one theory would have to be mapped to $\mathcal{M}_H$ of the other. Classically this is definitely not true: $\mathcal{M}_H$ is a complicated space obtained via the hyper-Kähler quotient construction [19] while $\mathcal{M}_V$ is just a flat space quotiented by the Weyl group for the gauge symmetry. Furthermore, when some of the mass terms $\vec{m}_A$ of, say, A, vanish, enhanced global (flavor) symmetries emerge and act on $\mathcal{M}_H^A$. For B there would have to be corresponding global symmetries emerging at vanishing Fayet-Iliopoulos parameters $\vec{\zeta}_B$ and acting on $\mathcal{M}_V^B$. Classically there is no such symmetry. Therefore this hypothetical duality, named *mirror symmetry in three dimensions* in [1], must be a quantum equivalence. However, a glance at tables 1 and 2 reveals a missing dual of the gauge coupling constant e, whose property as predicted by mirror symmetry is listed in table 2 with question marks. Brane interpretations of this parameter, known as the *magnetic coupling* m (without an arrow), have been proposed [4, 6], but it has yet to be found in any Lagrangian formulation. Given such, mirror symmetry can only manifest for ordinary field theory when e approach a particular value[2]. Being dimensionful (e^2 has the dimension of mass in 3d), the only natural candidates are 0 and ∞. $e = 0$, the classical limit, is already ruled out. So we are left with the strong coupling limit, which, since it is in 3d, is also the infrared limit. This also leads us to one of the most striking aspects of this proposed mirror symmetry: it maps from one theory the metric for $\mathcal{M}_V$, a quantity that receives very large quantum corrections, to, in the dual theory, the metric for $\mathcal{M}_H$, which is given by purely classical expressions. As many physicists have suggested, this may imply the line between quantum and classical physics is more blurred than previously thought.

Because of its strong coupling nature, proving mirror symmetry within the context of field theory will be difficult. Embedding the field theory in the dynamics of branes [20, 21, 22, 23, 24] renders many aspects of mirror symmetry manifest [4, 5], if one assumes the S duality of type IIB string. This will be reviewed extensively in the rest of this talk. Before that, I want to give some example of mirror pairs and one of the many pieces of field theoretic evidence supporting it [2]. They are logically independent of any conjectures about string theory.

A model has gauge group $U(K)$ with N fundamentals and 1 adjoint hypermultiplets. B models has gauge group $U(K)^N$, which we label as $U(K)_\alpha$, $\alpha = 0, \ldots, N-1$. Its hypermultiplets consist of one fundamental charged under $U(K)_1$ and N bi-fundamentals. The latters are each charged respectively under $U(K)_\alpha \times U(K)_{\alpha+1}$ in representation $(\bar{K}, K)$, with cyclic identification $\alpha \sim \alpha + N$. These field contents are nicely encoded in the "quiver" diagrams[25] of figure 1. Each inner node with a number K represents a $U(K)$ gauge group. Each link connecting a pair of them represents a bi-fundamental charged under the pair as (fundamental, fundamental*). An outer node with number N represents fundamentals with multiplicity N charged under the gauge group associated with the inner node to which it is attached.

Quiver for A and B models.

Using the notation given in table 1, the metric for $\mathcal{M}_V^A$ takes the form

$$ds^2 = g_{ij} d\vec{a}_i d\vec{a}_j + (g^{-1})_{ij}(d\sigma_i + \vec{\omega}_{il} \cdot d\vec{a}_l)(d\sigma_j + \vec{\omega}_{jl} \cdot d\vec{a}_l), \tag{2.1}$$

under the constraint

$$\vec{\nabla}_i g_{jl} = \vec{\nabla}_j g_{il} \tag{2.2}$$

$$\frac{\partial}{\partial a_i^x}\omega_{jl}^y - \frac{\partial}{\partial a_j^y}\omega_{il}^x = \epsilon_{xyz}\frac{\partial}{\partial a_i^z} g_{jl}. \tag{2.3}$$

[2]However, it is possible, and the brane realization discussed later strongly suggests, that m can be a field theory parameter that has no Lagrangian representation.

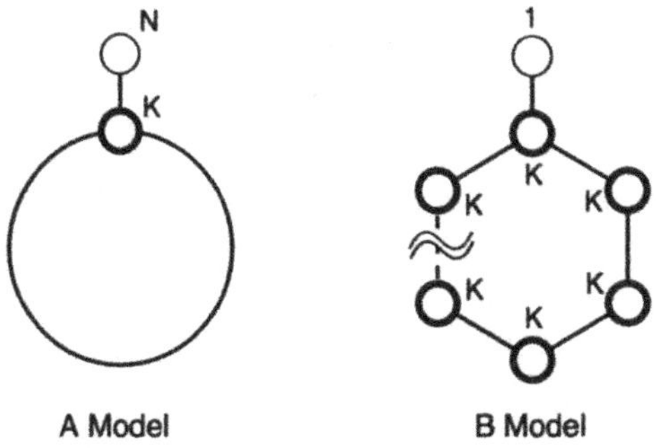

Figure 1.

Here $i, \ldots = 0, \ldots, k-1$ index the Cartan of $U(k)$; $x, \ldots = 3,4,5$. In [2], g_{ij} is computed. Perturbatively it is one-loop exact:

$$
\begin{aligned}
g_{i \neq j} &= \frac{2}{|\vec{a}_i - \vec{a}_j|} - \frac{1}{|\vec{a}_i - \vec{a}_j + \vec{m}_{\text{adj}}|} - \frac{1}{|\vec{a}_i - \vec{a}_j - \vec{m}_{\text{adj}}|}, \\
g_{ii} &= \frac{1}{e^2} + \sum_{\alpha=0}^{N-1} \frac{1}{|\vec{a}_i - \vec{m}_\alpha|} \qquad (2.4) \\
&+ \sum_{j=1\ldots K}^{j \neq i} \left(\frac{-2}{|\vec{a}_i - \vec{a}_j|} + \frac{1}{|\vec{a}_i - \vec{a}_j + \vec{m}_{\text{adj}}|} + \frac{1}{|\vec{a}_i - \vec{a}_j - \vec{m}_{\text{adj}}|} \right). \qquad (2.5)
\end{aligned}
$$

Here $\vec{m}_{\text{adj}}$ is the triplet mass for the adjoint hypermultiplet; $\vec{m}_\alpha$ those of the fundamentals, indexed by $\alpha = 1, \ldots, N$. When $\vec{m}_{\text{adj}} = 0$, there is no instanton correction to the metric and (eq. 2.4) is also nonperturbatively exact. For illustration, consider this simpler case, so that $g_{i \neq j}$ vanishes. (eq. 2.1) and (eq. 2.4) state that $\mathcal{M}_V^A$ is the direct product of K multi-Taub-NUT space with charge N. After quotiented by the Weyl group of $U(K)$, the direct product becomes a symmetric product.

It turns out that setting $\vec{m}_{\text{adj}} = 0$ for A model is mapped by mirror symmetry to a condition on the Fayet-Iliopoulos parameters $\vec{\zeta}_\alpha$ of B model:

$$\sum_\alpha \vec{\zeta}_\alpha = 0.$$

In this case, $\mathcal{M}_H^B$ is given by the symmetric product of K ALE spaces with A_{N-1} singularity [26]. The metric of each ALE space is given by

$$ds^2 = g d\vec{r}^2 + \frac{1}{g}(db + \vec{\omega}(\vec{r}) \cdot d\vec{r})^2$$

with

$$g(\vec{r}) = \sum_{\alpha=0}^{N-1} \frac{1}{\left|\vec{r} - \sum_{\beta=0}^{\alpha} \vec{\zeta}_\beta\right|}.$$

Now the metric for each Taub-NUT factor of $\mathcal{M}_V^A$ can be read from (eq. 2.4) after setting $\vec{m}_{\text{adj}} = 0$.

$$g(\vec{a}) = \frac{1}{e^2} + \sum_{\alpha=0}^{N-1} \frac{1}{|\vec{a} - \vec{m}_\alpha|}.$$

It is clear that $\mathcal{M}_V^A = \mathcal{M}_H^B$ if and only if $e = \infty$ and one makes the following identification[3]

$$\vec{\zeta}_\alpha = \vec{m}_\alpha - \vec{m}_{\alpha-1}.$$

3. Set the Branes to work

Consider in type IIB string theory, a configuration that includes 3 types of branes, whose worldvolume configurations are given in table 3 [4]. Such a configuration

TABLE 3. Configurations of the branes

	0	1	2	3	4	5	6	7	8	9
D3-brane	×	×	×				×			
D5-brane	×	×	×					×	×	×
NS5-brane	×	×	×	×	×	×				

can preserve up to 8 supercharges, corresponding to N=4 in 3d. Of the original $Spin(1,9)$ Lorentz symmetry, only the (1+2)d Lorentz group $SL(2,R)_{012}$ and the R-symmetry group $SU(2)_{345} \times SU(2)_{789}$ remain manifest. I will now review some basic rules for "model building" from branes and their justifications. Many of them first appeared in [4].

- By sending M_{planck} to ∞, the worldvolume theory on the branes decouple from the bulk fields such as graviton.
- D3-branes can break and end on D5 or NS5-branes without violating RR charge conservation [27, 28] by becoming a magnetic "monopole" on the 5-branes, as depicted on the left in figure 2. There the horizontal lines represent D3-branes; solid and dashed vertical lines represent NS and D5-branes respectively. Conversely, two D3-branes ending on the same 5-brane from opposite sides can rejoin. D3-brane ending on 5-branes and forbidden configurations.

[3]To see the counting of parameters matches, note that the "center" of the mass parameters can be absorbed by shifting the origin of $\mathcal{M}_V$ on both sides. Here this is used to set $\vec{m}_{N-1} = 0$ for A model.

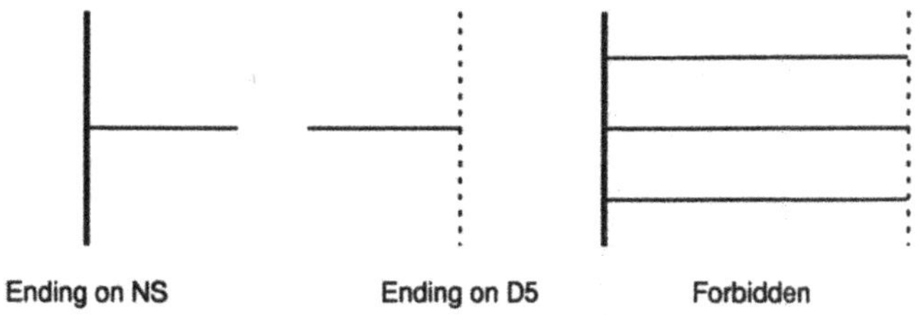

Figure 2.

- To the worldvolume theory on the D3-branes, breaking and ending are tantamount to imposing boundary conditions, reducing N=4, D=4 supermultiplets to N=4, D=3 supermultiplets.
- By taking the appropriate scaling limit, the Kaluza-Klein modes along the 6th direction can be kept massive and integrated out. The effective infrared theory on the D3-brane is therefore a (1+2)d QFT.
- The worldvolume theories on the 5-branes are weakly coupled in the infrared. The fields on them have an infinite volume coefficient in their kinetic terms as compared to D3-brane fields due to their relative sizes. As a result they are frozen as background fields and their VEV's are parameters of the effective 3d theory. Gauge symmetries on the 5-branes become global symmetries.
- When a NS5-brane crosses a D5-brane, a D3-brane is created and stretched in between. This is a consequence of charge conservation [4][4].
- Certain configurations are believed to be forbidden. They involve more than one D3 brane stretched between the same NS-D5 pair, such as the one in the right of figure 2.

Following the above rules one can build brane configurations of arbitrary complexity. To read off the contents of the resulting field theory, one also needs to know the following.

- Dynamical fields of the decoupled (1+2)d theory arise out of the lightest excitations of open strings starting and ending on the D-branes. There are three types: open strings connecting between D3-branes in the same "cubicle" (figure 3a) give vector multiplets; those between D3-branes in adjacent "cubicle" give bi-fundamental hypermultiplets (figure 3b); while open strings connecting between D3 and D5 give fundamental hypermultiplets (figure 3c). These can be read off from perturbative open string quantization and the boundary conditions. Their transformation properties under R-symmetry agree with the assignment given in the last section. Their end points are electric sources on the D3-branes. Field content from open fundamental strings.
- Configuration of the D3-branes selects the gauge symmetries and the vacuum, as illustrated in figure 4. D3-brane configuration selects gauge group and vacuum

[4]For some other treatments of this phenomenon, see [29] and the references therein.

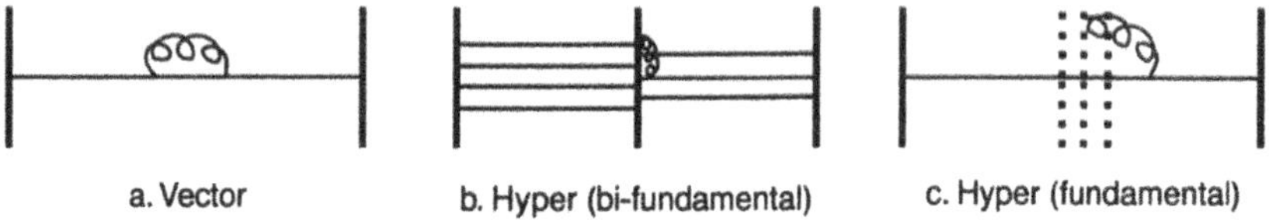

Figure 3.

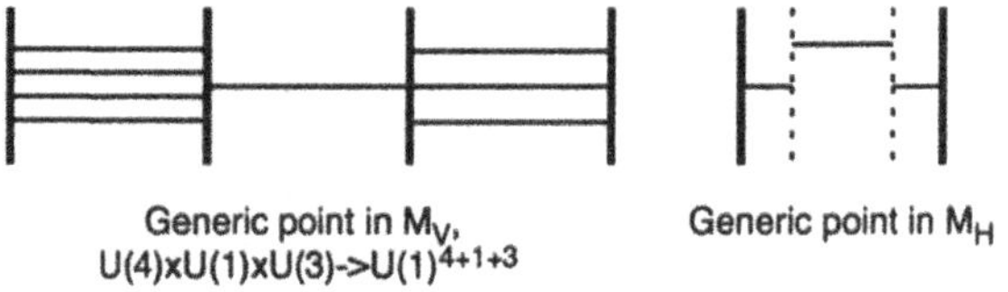

Figure 4.

- Configuration of the 5-branes determines the parameters (figure 5) and global symmetries of the 3d QFT. Note the identification of the magnetic coupling constant as the inverse square root of the distance between adjacent D5-branes along the 6th direction. The frozen gauge dynamics of the D5-brane gives rise to the flavor global symmetry acting on the fundamental hypermultiplets. Since in conventional Lagrangian field theories, this global symmetry is restored by setting the masses to zero, the magnetic coupling is fixed to be infinite. At the same time, the global symmetry resulting from the NS5-branes is broken unless they coincide, which amounts to setting Fayet-Iliopoulos to zero and e to infinity — it can only appear at a nontrivial infrared fixed point. 5-brane configuration fix parameters and global symmetries

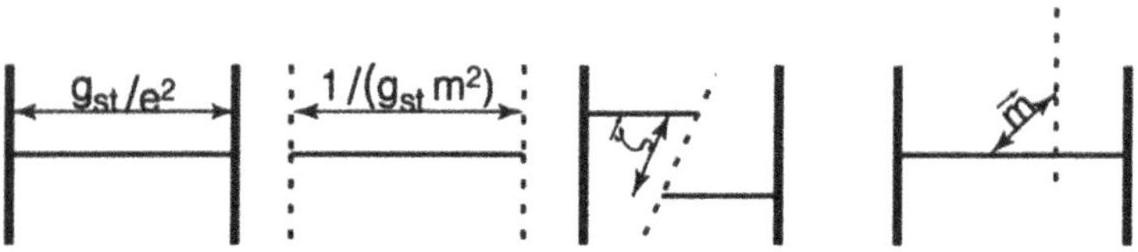

Figure 5.

- Type IIB string theory has a nonperturbative S duality [30] that inverts the string coupling constant and exchange NS5-branes with D5-branes as well as fundamental strings with D-strings. It leaves invariant D3-branes but acts on their worldvolume theories as the S duality for N=4, D=4 SYM [31].
- We therefore should also consider excitations of open D-strings starting and ending on D3-branes and/or NS5-branes. They are simply obtained from the S dual of figure 3. However, on D3-branes, rather than generating additional degree of freedom, they are related to the open string fields nonlocally — end

points of these two types of string on D3-branes are respectively magnetic and electric sources. Each appear as solitonic excitations of the other and are exchanged by the field theoretic S duality.

- Therefore there are two descriptions of the same 3d theory: one uses open fundamental string fields as the canonical variables while the other uses open D-string fields. Their equivalence is the 3d mirror symmetry, and it follows from the S duality type IIB string theory.
- It is convenient to rephrase this duality as an operation combing an S duality transformation with the interchange of 345 and 789 directions [4]. This makes explicit the interchange of the two R-symmetry factors, $\mathcal{M}_V$ with $\mathcal{M}_H$, masses with Fayet-Iliopoulos parameters, and e with m. For reasons just stated, this is reflected on ordinary Lagrangian field theories only in their infrared limit.

4. Mirror Pairs

Now I will present a few examples of mirror pairs constructed in [5]. Their field contents are best described by the type of quiver diagrams introduced earlier. They are obtained using the brane-engineering rules outlined above, but with a compactified 6th direction, i.e. with periodic identification along X^6. As a warm-up, let's look at the A and B models given in figure 1. The corresponding brane configurations are sketched in figure 6 Brane configurations for the quivers in

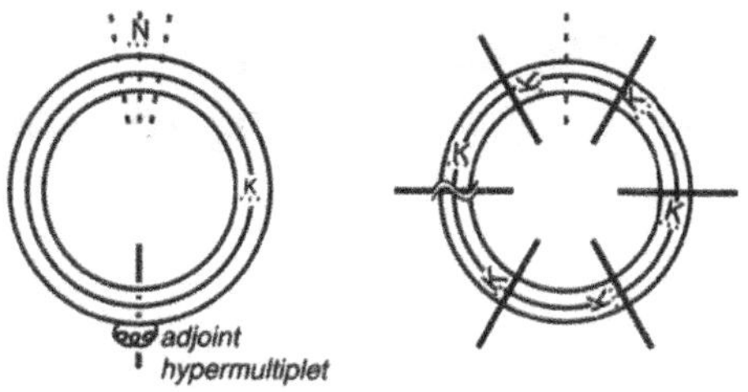

Figure 6.

figure 1.

For A model, the N fundamentals of $U(K)$ originate from open strings stretched between the N D5 and the K D3-branes. A special feature of this configuration is that the bi-fundamental, coming from open string stretching between "nearest-neighbor" D3-branes, become the adjoint of $U(N)$ because there is only one gauge groups. For B model, more generic situation prevails and there are N bi-fundamentals. The correspondence between the moduli spaces of vacua of A and B as well as the identification (eq. 2) is evident. As it is, there is one constraint on the field theory parameters, namely

$$\vec{m}_{\text{adj}} = 0 = \sum_{\alpha} \vec{\zeta}_{\alpha}.$$

A way to relax this condition has been given in [12].

Now let's look at mirror pairs of more complex theories. As depicted by the quiver diagrams in figure 7, model A again has gauge group $U(K)^N$ and N bi-fundamentals, but each $U(K)$ now has fundamentals with an arbitrary number of flavors w_i. Its mirror, model B, has gauge group $U(K)^M$ with

$$M = \sum_i w_i.$$

Besides the M bi-fundamentals, it has N fundamentals arranged as shown in figure 7. If all $w_i > 0$, each $U(K)$ factor has at most one fundamental. If some $w_i = 0$, the corresponding nodes in B model's quiver coalesce and give rise to fundamentals of higher flavor. Such mirror pairs are again constructed via the S duality of type IIB string theory [5]. The mirror map relates the moduli spaces of the two theories, as well as their parameters, in the same manner as the simpler case discussed above. Arbitrary flavor of fundamentals and the mirror

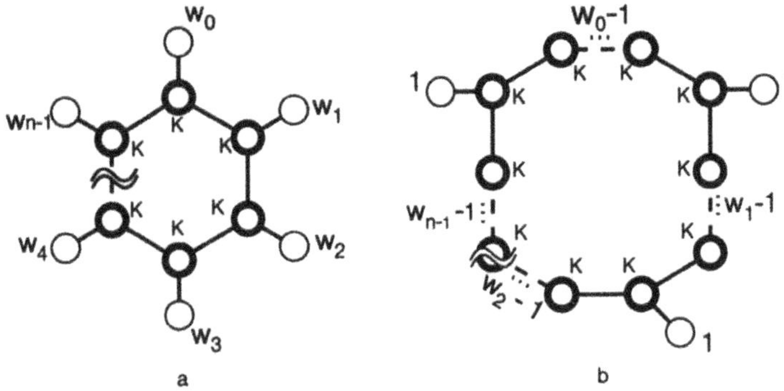

Figure 7.

It is natural to generalize to the cases with A model having gauge group $\prod_{i=0}^{N-1} U(K_i)$, N bi-fundamentals, and arbitrary fundamentals. Its quiver is shown in figure 8, along with its brane realization. Quiver and brane realization of $\prod_i U(K_i)$ However, here one encounters an important subtlety. Although one can always perform a S duality transformation and obtain the mirror configuration, the result does not always correspond to a gauge theory. To see this, recall that mirror symmetry exchanges $\mathcal{M}_V$ with $\mathcal{M}_H$. A universal property of N=4, D=3 super-Yang-Mills theories is the existence of the Coulomb phase, a branch of $\mathcal{M}_V^B$ with $4r$ dimensions, where r is the total rank of the gauge group. This is mapped under mirror symmetry to a branch of $\mathcal{M}_H^A$: the completely Higgsed phase. Therefore a necessary condition for model B to have an ordinary gauge theoretic Lagrangian description is for model A to admit complete Higgsing. This amounts to requiring [32, 33, 2, 5]:

$$2k_i - k_{i-1} - k_{i+1} \leq w_i. \tag{4.1}$$

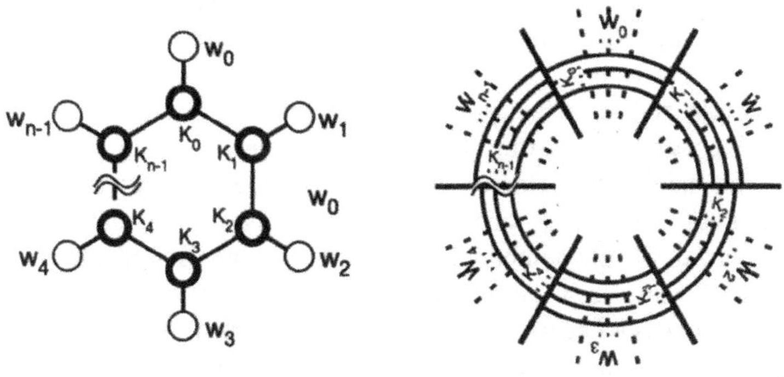

Figure 8.

When this is satisfied, the mirror gauge theory can be constructed along the same vein as before. The details become complicated and can be found in [5].

5. Phases and Transitions

What happens if (eq. 4.1) is not satisfied? S duality still gives a mirror configuration, but one without a gauge theoretic description. An example of this is shown in figure 9a. Brane configurations giving rise to novel field theories

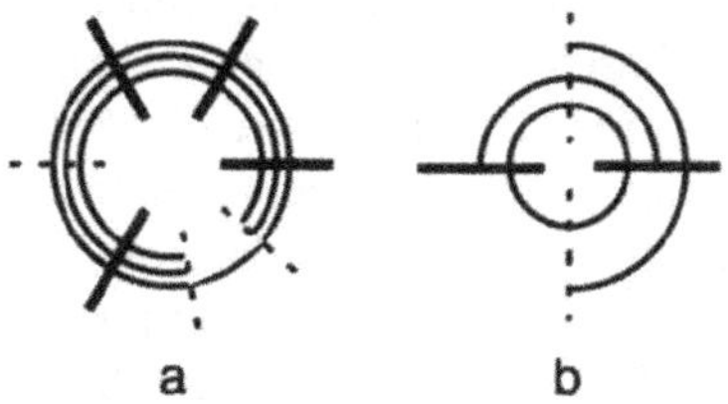

Figure 9.

To understand this phenomenon, note that from the field theory perspective, mirror symmetry corresponds to the Z_2-wise freedom in labeling the two $SU(2)$ R-symmetries. While the vector and hypermultiplets transform distinctly though somewhat symmetrically under them, their interactions enter in the Lagrangian in rather different form. The Lagrangian descriptions of a theory and its dual would in general be quite different, as the examples above show. Actually, there is no reason a priori to expect that both sides of a mirror pair have Lagrangian descriptions at all. It is natural to conjecture this is the what is happening here[5].

[5]In fact, barring an unexpected way to incorporate the mysterious magnetic coupling into a Lagrangian formulation, even for the cases in which both of the dual pair have a Lagrangian description, mirror symmetry is manifest only at the infinite coupling, i.e.

3d N=4 theories with Lagrangian description can therefore be classified into those that have two, related by mirror symmetry in the infrared, and those that have only one[6].

An example of the latter type, in which the non-Lagrangian description is that with E_n tensionless string [34, 35], was conjectured already in [1]. It has recently been explicitly engineered using one of the alternative formulations of mirror symmetry from string theory [6, 7]. Here we have a very simple prescription for engineering an infinite number of such Lagrangian-non-Lagrangian mirror pairs. As noted in [1], these are local quantum field theories, simply because on one side of the mirror there is a Lagrangian description that flows to it. However, experience in 2, 4, 5, and 6 dimensions has shown a Lagrangian description, though convenient in many ways, may not be a necessary condition for a local quantum field theory (see, for example, [36, 37, 38, 39]. Indeed, one can easily engineer using branes a third class of theories that have *no* Lagrangian description on either side of the mirror. An example is shown in figure 9b. Since the decoupling of bulk as well as Kaluza-Klein modes works just as in the more mundane cases, they should still be interacting local quantum field theories, but with no known Lagrangian description flowing to it.

Such interesting phenomena deserve an explanation from string theory. In that context mirror symmetry is the equivalence of two descriptions of the same physics related by S duality. The degrees of freedom on the D3-brane theory can be captured both by open fundamental string and by open D-string fields. Starting with, say, a description using only open fundamental string fields, one can employ standard string perturbation theory to obtain their interactions and write a Lagrangian for the field theory modes in the decoupling limit. The same prescription goes through for a description based solely on open D-string fields. Suppose, however, that there is no way to capture the full degrees of freedom by using only, say, open D-string fields. The description on the B model side may not be in the form of a local action, as open D-strings and open fundamental strings ending on D3-branes are mutually nonlocal. If neither open fundamental string fields nor open D-string fields can account for the full degrees of freedom by themselves respectively, we end up with the third class of theories.

Transition between different classes of field theories via brane motion. Remarkably, with branes one can not only engineer examples of all three types of theories, but also interpolate between them continuously by moving the 5-branes around. Shown in figure 10a, b is the mirror of figure 9a. It is a theory of the second type (Lagrangian-non-Lagrangian mirror pair). By moving one NS5-branes past another, we arrive at the theory depicted in figure 10c, d, which is of the first type (Lagrangian-Lagrangian mirror pair). Similar transition can be engineered between

infrared limit, as mentioned earlier. Here I shall be referring to that limit implicitly unless otherwise stated.

[6]One should note that the lack of a completely higgsed phase is necessary but not sufficient for a non-Lagrangian dual. For example, a free $U(1)$ vector multiplet, which obviously does not have any Higgs phase, is dual to a free hypermultiplet. However, such free theories cannot help explain interacting fixed points, e.g. when the D3-branes in figure 9a or b coincide.

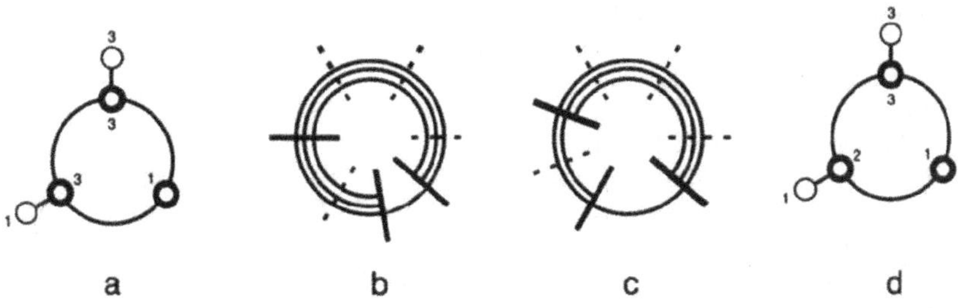

Figure 10.

theories of the third class and the first two as well. To appreciate the meaning of such a process, recall that the electric and magnetic coupling constants are inversely proportional to the square root of the distance between adjacent NS5 and adjacent D5-branes respectively. Moving 5-branes of the same type past each other effectively makes coupling constants of the corresponding type imaginary. This is indicative of a change of the effective degrees of freedom describing the system, namely a phase transition. That smooth movements in the moduli space of brane configurations can connect distinct classes of field theories via some type of phase transitions is one of the most important lessons to be learned from this work.

Acknowledgements

I would like to thank the organizers of this summer institute for invitation, financial support, and the opportunity to present this talk, as well as for making the whole school a rewarding experience. I am also indebted to J. de Boer, K. Hori, H. Ooguri, and Y. Oz for collaboration and many useful discussions. I would like to especially acknowledge H. Ooguri for encouragement. My research is supported in part by DOEAC03-76SF00098 and a grant from the Dean of Graduate Division, U. C. Berkeley.

References

1. K. Intriligator, N. Seiberg, *Phys. Lett.* **387B** (1996) 513, hep-th/9607207.
2. J. de Boer, K. Hori, H. Ooguri, Y. Oz, *Nucl. Phys.* **B493** (1997) 101, hep-th/9611063.
3. M. Porrati, A. Zaffaroni, *Nucl. Phys.* **B490** (1997) 107, hep-th 9611201.
4. A. Hanany, E. Witten, *Nucl. Phys.* **B492** (1997) 152, hep-th/9611230.
5. J. de Boer, K. Hori, H. Ooguri, Y. Oz, Z. Yin, *Nucl. Phys.* **B493** (1997) 148, hep-th/9612131.
6. K. Hori, H. Ooguri, C. Vafa, *Nucl. Phys.* **B504** (1997) 147, hep-th/9705220.
7. S. Katz, P. Mayr, C. Vafa, hep-th/9706110.
8. J. de Boer, K. Hori, Y. Oz, Z. Yin, *Nucl. Phys.* **B502** (1997) 107, hep-th/9702154.
9. O. Aharony, A. Hanany, K. Intriligator, N. Seiberg, M. J. Strassler, *Nucl. Phys.* **B499** (1997) 67, hep-th/9703110.
10. T. Eguchi, lectures given at this school.
11. S. Elitzur, A. Giveon, D. Kutasov, *Phys. Lett.* **400B** (1997) 269, hep-th/9702014.

12. E. Witten, *Nucl. Phys.* **B500** (1997) 3, hep-th/9703166.
13. K. Hori, H. Ooguri, Y. Oz, hep-th/9706082.
14. E. Witten, hep-th/9706109.
15. J. de Boer, K. Hori, H. Ooguri, Y. Oz, hep-th/9711143.
16. N. Seiberg, E. Witten, hep-th/9607163.
17. B. de Wit, P. G. Lauwers, A. van Proeyen, *Nucl. Phys.* **B255** (1985) 569.
18. P. Argyres, M. R. Plesser, N. Seiberg, *Nucl. Phys.* **B471** (1996) 159, hep-th/9603042.
19. N. Hitchin, A. Karlhede, U. Lindström, M. Roček, *Comm. Math. Phys.* **108** (1987) 535.
20. M. R. Douglas, hep-th/9604198.
21. A. Sen, *Nucl. Phys.* **B475** (1996) 562, hep-th/9605150.
22. T. Banks, M. R. Douglas, N. Seiberg, *Phys. Lett.* **387B** (1996) 278, hep-th/9605199.
23. M. R. Douglas, D. Kabat, P. Pouliot, S. H. Shenker, *Nucl. Phys.* **B485** (1997) 85, hep-th/9608024.
24. N. Seiberg, *Phys. Lett.* **384B** (1996) 81, hep-th/9606017.
25. M. R. Douglas, G. Moore, hep-th/9603167.
26. P. B. Kronheimer, H. Nakajima, *Math. Ann.* **288** (1990) 263.
27. A. Strominger, *Phys. Lett.* **383B** (1996) 44, hep-th/9512059.
28. P. K. Townsend, *Phys. Lett.* **373B** (1996) 68, hep-th/9512062.
29. I. Klebanov, hep-th/9709160.
30. J. H. Schwarz, *Phys. Lett.* **360B** (1995) 13, hep-th/9508143.
31. M. B. Green, M. Gupertle, *Phys. Lett.* **377B** (1996) 28.
32. H. Nakajima, *Duke Math. J.* **76** (1994) 365.
33. H. Nakajima, Quiver Varieties and Kac-Moody algebras, preprint.
34. O. J. Ganor, A. Hanany, *Nucl. Phys.* **B474** (1996) 122, hep-th/9602120.
35. N. Seiberg, E. Witten, *Nucl. Phys.* **B471** (1996) 121.
36. P. C. Argyres, M. R. Douglas, *Nucl. Phys.* **B448** (1995) 93, hep-th/9505062.
37. N. Seiberg, *Phys. Lett.* **388B** (1996) 753, hep-th/9608111.
38. N. Seiberg, *Phys. Lett.* **390B** (1997) 169, hep-th/9609161.
39. N. Seiberg, hep-th/9705117.

INTERSECTION RULES AND OPEN BRANES

RICCARDO ARGURIO
Service de Physique Théorique
Université Libre de Bruxelles, Campus Plaine, C.P.225
Boulevard du Triomphe, B-1050 Bruxelles, Belgium

1. Introduction

Classical solutions of various supergravities are very interesting to study in the context of string theories and M-theory, because they are often essential in establishing or corroborating the existence of dualities relating (some compactified versions of) the above-mentioned theories [1, 2, 3]. These p-brane solutions (for some reviews on p-branes, see e.g. [4, 5, 6]) provide us with informations on the long-range, low-energy fields produced by objects which live in a more complete theory, i.e. a theory of superstrings or the 'would-be' M-theory. Solutions involving several (classical) branes are thus useful in determining some characteristics of the interactions between the quantum objects, and in putting forward conjectures about the quantum dynamics of the underlying theory.

The problem of studying the interactions between different branes in string and M-theory will be addressed in this contribution considering supergravity solutions which involve intersecting branes. More precisely, we will be concerned with orthogonal intersections of extremal branes. This means that each constituent brane saturates a BPS bound, its mass being equal to its charge in the relevant units. Considered on its own, such a single brane solution would preserve half of the space-time supersymmetries. The intersecting brane solutions that we will consider are such that the full solution still preserves some (lower) fraction of supersymmetry. This is related to the fact that the binding energy of these configurations vanishes. As we will show hereafter, these solutions are relevant to the study of black hole physics, since they will provide, in the reduced space-time, black holes with non-vanishing Bekenstein-Hawking entropy, but which are nevertheless still supersymmetric and thus much more tractable.

Historically, solutions allowing for a single p-brane were presented most generally in [7]. The most important feature of these solutions, in their extreme limit, is that they are characterized entirely by a single harmonic function, which depends on the coordinates transverse to the brane. In spring '96, Papadopoulos and Townsend [8] reinterpreted some 11 dimensional supergravity solutions found by

L. Baulieu et al. (eds.), Strings, Branes and Dualities, 387–397.

Güven [9] as intersecting branes and then used this interpretation to build new solutions. Soon after, Tseytlin further generalized in [12] these solutions to include an independent harmonic function for each (non-parallel) brane in the solution. The application of dualities to these particular solutions then predicted a lot of new configurations involving all sorts of branes. For D-branes, these new solutions were compatible with the supersymmetric intersections derived in string theory (see [10, 11]). The main common feature, besides the appearance of the harmonic function associated to each brane, was the vanishing of the binding energy. The rule to build such solutions was simply to 'superpose' the single brane solutions. This led to the formulation by Tseytlin of the 'harmonic superposition rule' for orthogonally intersecting branes. Still, the dimension of the intersection had to be determined case by case, from supersymmetry arguments and/or by duality. The nature of the argument strongly depended on the type of brane considered (NS-, D- or M-brane).

The outline of the rest of my contribution is as follows. We will first show how to derive the harmonic superposition rule from the equations of motion of a general theory, which models supergravity. Supersymmetry will not be an ingredient of this derivation, though it will (remotely) motivate some of the ansätze made in order to solve the equations. Almost as a byproduct, some of the equations of motion will reduce to a set of algebraic equations determining the dimension of the pairwise intersections of the branes in the configurations (i.e. the 'intersection rules'). We will then proceed to the tentative deduction of some brane dynamics from these solutions. In the case at hand we will consider the possibility for some branes to open, with boundaries tied to some other brane. For this to work, we have to check that the charge of the open brane is still conserved. The mechanism by which this is done sheds some light on the world-volume effective theory of the brane on which the open brane ends. In the end we speculate on the relevance of closed brane emission by other branes. This talk is based on the two papers [13, 14], written in collaboration with F. Englert, L. Houart and P. Windey.

2. The Harmonic Superposition and the Intersection Rules

In this section we will derive the harmonic superposition rule and the dimension of the intersection of extreme branes simply solving a set of bosonic equations of motion, provided some particular ansätze are made. A more detailed and step-by-step derivation can be found in [13]. As a starting point we take a general action in D dimensions, which can be the bosonic part of a supergravity action:

$$I = \frac{1}{16\pi G_N^{(D)}} \int d^D x \sqrt{-g} \left(R - \frac{1}{2}(\partial\phi)^2 - \sum_I \frac{1}{2n_I!} e^{a_I \phi} F_{n_I}^2 \right), \quad I = 1 \dots \mathcal{M}. \tag{2.1}$$

$\mathcal{M}$ is the number of antisymmetric tensor fields, and we take $n_I \leq D/2$ for all field strengths. The metric is written in the Einstein frame, and the coupling of the forms to the dilaton in this frame is entirely governed by the constants a_I. Note that since there is only one scalar, this theory is most suitable to model 10 or 11 dimensional supergravities. Generalizations to include several scalars can be found in the literature. Also, we did not include for the moment Chern-Simons terms. A posteriori, they can be shown to play no role in the determination of the classical solutions considered here, but their presence will be crucial when we will discuss the opening of the branes.

We begin now to simplify the problem taking the metric to be of a particular, diagonal, form:

$$ds^2 = -B^2dt^2 + C_1^2dy_1^2 + \ldots + C_p^2dy_p^2 + G^2dx_adx_a, \qquad a = 1\ldots D-p-1. \quad (2.2)$$

We have an $SO(D-p-1)$ symmetry left in the 'overall transverse' space of the x's, which are taken to be the non-compact directions. Note that there is no a priori $SO(p,1)$ symmetry, and that we will not necessarily have a p-dimensional brane in the solution. The y's directions will be eventually compactified. Since the metric is diagonal, we exclude for the moment solutions involving KK waves and monopoles. All functions in the problem depend only on the x_a's.

For the n-form field strengths, we have the choice between two different ansätze:

$$\text{Electric} \qquad F_{ty_1\ldots y_{q_A}a} = \partial_a E_A, \quad (2.3)$$

$$\text{Magnetic} \qquad \tilde{F}_{ty_1\ldots y_{q_A}a} = \partial_a E_A, \quad (2.4)$$

where we have defined the dual field strength by:

$$\tilde{F}_{\mu_1\ldots\mu_{D-n}} = \sqrt{-g}e^{a\phi}\epsilon_{\mu_1\ldots\mu_D}F^{\mu_{D-n+1}\ldots\mu_D}.$$

The space-time charges are thus respectively defined by:

$$Q_A^{el} \sim \int *F_{q_A+2}, \qquad Q_A^{mag} \sim \int F_{D-q_A-2}. \quad (2.5)$$

$A = 1\ldots\mathcal{N}$, where $\mathcal{N}$ is the total number of different (non-parallel) branes, electric and magnetic, in the solution. This number can of course exceed the number of different n-forms.

We can now take the key steps which will enable us to solve quite straightforwardly the equations of motion derived from the theory above. These are the following two ansätze:

- Extremality, which (by experience) is enforced on the metric by the condition:

$$BC_1\ldots C_pG^{D-p-3} = 1. \quad (2.6)$$

- No-force condition between the constituent branes (in other words, the requirement that the branes form a BPS marginal bound state). This is translated in our problem in the statement that to each brane is associated one independent harmonic function, and that the solution is completely characterized by these $\mathcal{N}$ harmonic functions.

These two conditions could in principle be found asking the solution to preserve some supersymmetries, i.e. demanding that the equation $\delta_{SUSY}\psi = 0$ has non-trivial solutions. However this cannot be done in this generic set up, i.e. for arbitrary D.

The mathematical implementation of the second ansatz can be motivated as follows. We know that a single brane solution is entirely determined by only one harmonic function. If there are $\mathcal{N}$ branes in the configuration, but there is no binding energy, nothing prevents us from pulling one of the branes apart from the others. Then the fields near that brane should be a good approximation to the fields in the single brane solution. Thus we see that we should expect exactly $\mathcal{N}$ independent functions in the solution.[1]

These $\mathcal{N}$ independent functions are taken to be H_A such that the E_A's in (2.3) and (2.4) satisfy:

$$E_A \sim H_A^{-1}. \tag{2.7}$$

Then the equations of motion impose (see [13] for the details) $\partial_a \partial_a H_A = 0$, which gives:

$$H_A = 1 + \sum_k \frac{c_A Q_{A,k}}{|x^a - x_k^a|^{D-p-3}}. \tag{2.8}$$

Solving for the Einstein and the dilaton equations gives the following metric and dilaton:

$$ds^2 = -\prod_A H_A^{-2\frac{D-q_A-3}{\Delta_A}} dt^2 + \sum_i \prod_A H_A^{-2\frac{\delta_A^{(i)}}{\Delta_A}} dy_i^2 + \prod_A H_A^{2\frac{q_A+1}{\Delta_A}} dx_a dx_a, \tag{2.9}$$

$$e^{\phi} = \prod_A H_A^{\varepsilon_A a_A \frac{D-2}{\Delta_A}}, \tag{2.10}$$

where $\Delta_A = (q_A + 1)(D - q_A - 3) + \frac{1}{2}a_A^2(D-2)$, $\varepsilon_A = +(-)$ if the corresponding brane is electrically (magnetically) charged and $\delta_A^{(i)} = D - q_A - 3$ or $-(q_A + 1)$ depending on whether the direction of y_i is parallel or perpendicular to the q_A-brane. Note that $\Delta_A = 16$ and 18 for all the branes of, respectively, 10 and 11 dimensional supergravities. To recapitulate, in order to build up a metric according to the harmonic superposition rule, we have to include a factor of $H_A^{-2\frac{D-q_A-3}{\Delta_A}}$ in front of each coordinate longitudinal to the q_A-brane (including the time direction), and a factor of $H_A^{2\frac{q_A+1}{\Delta_A}}$ in front of each transverse coordinate, and this has to be done for each brane in the configuration.

In the process of finding (2.9) and (2.10), we did not use the $R^a{}_b$ off-diagonal components of the Einstein equations. These have by now reduced to a set of algebraic conditions, that for consistency impose the following pairwise intersection rule for $\bar{q} = \dim(\cap)$:

$$\bar{q} + 1 = \frac{(q_A + 1)(q_B + 1)}{D-2} - \frac{1}{2}\varepsilon_A a_A \varepsilon_B a_B. \tag{2.11}$$

We now point out some remarks.

[1]See [15] for a derivation of the intersection rules based on the application of the no-force ansatz on the effective brane actions, and see [16] for a detailed discussion of the second ansatz and its extension to non-extreme intersecting branes.

The formulae above (2.9), (2.10) and (2.11) hold for $D - p > 3$, in which case the space is asymptotically flat, as well as for $D - p = 2$ or 3, where the equation (2.8) does not hold any more, i.e. the H_A's do not tend to a finite value at infinity. In that cases the solutions have to be considered rather formally. Also for a Euclidean signature the same formulae hold, without the obligation for the time coordinate to be always longitudinal to all the branes (however the electric fields have to be imaginary).

The total mass of these configuration is, as expected, the sum of the masses of each constituent brane, which are equal to the charges: $M = \sum M_A = \sum Q_A$.

All the solution above have a functional dependence restricted to the overall transverse space. Some configurations which exist in string theory, as the two D5-branes intersecting over a string (the $\nu = 8$ configurations in [11]) and their duals, are thus excluded since they correspond to the solutions discussed in [17], where the functions depend on the 'relative transverse' coordinates.

As already stated above, we did not consider for simplicity non-diagonal metrics. One can nevertheless find the solutions involving KK travelling waves and KK monopoles applying some duality transformation on the solutions above, since all KK charges are related by U-duality to the RR and NSNS charges. A classification of the intersections involving also KK branes can be found in [18].

3. Intersections in String and M-theory and Black Hole Entropy

We can now specialize the formula above (2.11) to the case of $D = 10$ and 11 maximal supergravities. Actually, this is done straightforwardly specifying D and the dilaton couplings a_A. For $D = 11$ we simply have $a = 0$ since there is no dilaton. For $D = 10$ IIA and IIB theories, we have $a = -1$ for the NSNS 3-form field strength and $\varepsilon a = \frac{1}{2}(3 - q)$ for a q-brane carrying RR electric or magnetic charge.

As a first application, we will use the metric (2.9) to derive the number of charges one needs to build up a (supersymmetric) extreme black hole with non-vanishing horizon area in a definite number of non-compact dimensions. Let us define $\bar{D} = D - p$ ($D = 10, 11$). Then in the Einstein frame the Bekenstein-Hawking entropy is proportional to the horizon area defined as follows:

$$S \sim V(\text{compact space}) A(S^{D-p-2})|_{r=0}.$$

Using (2.9) in the case where all the harmonic functions are centered at the same point $r = 0$ (the horizon), we have:

$$\begin{aligned} S &\sim \prod_A H_A^{1/2} r^{\bar{D}-2} \\ &\sim \prod_A Q_A^{1/2} r^{-\frac{1}{2}\mathcal{N}(\bar{D}-3)+\bar{D}-2} \\ &\sim \prod_A Q_A^{1/2}, \end{aligned}$$

the last relation being true only provided the following relation between $\mathcal{N}$ and $\bar{D}$ holds:

$$\mathcal{N} = 2\frac{\bar{D}-2}{\bar{D}-3}. \qquad (3.1)$$

This relation has only two integer solutions, which are:

$$\bar{D} = 5, \quad \mathcal{N} = 3 \qquad \text{and} \qquad \bar{D} = 4, \quad \mathcal{N} = 4. \tag{3.2}$$

This also proves that there are no (stringy) extreme black holes with non-zero entropy in $\bar{D} \geq 6$.

All the solution described in (2.9)–(2.11) can be shown to be supersymmetric, and the fraction of preserved supersymmetry is in general at least $1/2^{\mathcal{N}}$. For instance, the $\mathcal{N} = 3$ and $\mathcal{N} = 4$ solutions discussed just above both preserve 1/8 of supersymmetry, thus providing an example and a counter-example to the '$1/2^{\mathcal{N}}$ rule'.

We can now summarize all the possible pairwise intersections between the branes which appear in string/M-theory. We use the notation $q_A \cap q_B = \bar{q}$. This rules appeared case by case in the literature, following from rather different arguments, in [19, 20, 15].

In $D = 11$, we have for the M-branes:

$$2 \cap 2 = 0, \qquad 2 \cap 5 = 1, \qquad 5 \cap 5 = 3. \tag{3.3}$$

In $D = 10$, for the intersections between D-branes we have generically $q_1 \cap q_2 = \frac{1}{2}(q_1 + q_2 - 4)$, which gives the following three cases:

$$q \cap q = q - 2 \tag{3.4}$$
$$(q-2) \cap q = q - 3 \tag{3.5}$$
$$(q-4) \cap q = q - 4 \tag{3.6}$$

The last case (3.6) can be interpreted as a D$(q-4)$-brane within a Dq-brane as in [21].

The intersections involving NSNS branes are:

$$1_F \cap 5_S = 1 \tag{3.7}$$
$$1_F \cap q_D = 0 \tag{3.8}$$
$$q_D \cap 5_S = q - 1, \qquad 1 \leq q \leq 6 \tag{3.9}$$

where the subscripts F, S and D denote respectively fundamental strings, solitonic 5-branes and D-branes.

It is interesting to see how all these intersections come on an equal footing in this framework, while they have a very different origin in the underlying theories.

4. When the Intersection is Actually a Boundary

The second case in (3.3) and the cases (3.5), (3.8) and (3.9) all have a common feature: the intersection has the same dimension as the (would-be) boundary of one of the two branes. Are we allowed to consider each $p-1$ dimensional intersection as the boundary of an open p-brane tied to the world-volume of the other brane? The case (3.8) is effectively consistent with the picture of fundamental strings ending on D-branes, but for the other cases we do not have any quantum description of the phenomenon.

From the supergravity point of view, all these branes appear to be closed. The intersection is not localized in the compact space, rather all the branes are 'smeared' over the transverse compact directions. In this sense there is little distinction between closed branes and open branes with both ends joined.

If we want to go deeper into the consideration of open branes, we need some additional input. For the opening of the branes to be consistent, we need a conservation law for the charge carried by the open brane. We will now see that such a conservation law exists, provided the boundaries of the open brane are constrained to live on another brane.

In words, the mechanism goes as follows. The charge carried by the brane is conserved when the brane is open if the boundary carries itself a charge in the effective theory on the world-volume of the brane on which it is constrained. In this way, each brane which can act as a D-brane for other branes (including strings) has an effective world-volume theory whose field content is determined by this mechanism.

It has to be noted that branes ending on other branes were used in [22, 23] and all the following literature to study field theory phenomena, such as dualities, from the dynamics of brane configurations.

There are two complementary approaches to the charge conservation. One is due to Townsend [24] and crucially makes use of the Chern-Simons terms in the supergravity equations of motion. The other is based on the gauge invariance of the open brane world-volume action. We work out here both approaches for a definite example, a D2-brane ending on a D4-brane in IIA string theory. The general case is described in [14].

The equation of motion for the 4-form field strength has to be supplemented by the Chern-Simons term present in the full IIA supergravity action, and by the source due to the presence of the D2-brane. The equation thus reads, neglecting the dilaton and all numerical factors:

$$d * F_4 = F_4 \wedge H_3 + Q_2 \delta_7. \tag{4.1}$$

Since there is also a D4-brane in the configuration, the Bianchi identity for F_4 is also modified by a source term:

$$dF_4 = Q_4 \delta_5. \tag{4.2}$$

On the other hand, due to the absence of NS5-branes, H_3 can be globally defined as $H_3 = dB_2$. The equation (4.1) can be rewritten as:

$$d(*F_4 - F_4 \wedge B_2) = Q_2 \delta_7 + Q_4 \delta_5 \wedge B_2. \tag{4.3}$$

We can now integrate both sides of this equation over a 7-sphere S^7 which intersects the D2-brane only once (this is possible only if the D2-brane is open). The result is:

$$0 = Q_2 + Q_4 \int_{S^2} \hat{B}_2, \tag{4.4}$$

where the hat denotes the pull-back to the world-volume of the D4-brane of a space-time field.

We see that the Chern-Simons term indicates the presence on the world volume of the D4-brane of a 2-form field strength, for which the (string-like) boundaries of the D2-branes act as magnetic charges. As we will discuss shortly, the gauge invariant combination is $\mathcal{F}_2 = dV_1 - \hat{B}_2$. The presence of the Chern-Simons term in (4.1) ensures the consistency between charge conservation of the open D2-brane and gauge invariance of the world-sheet action of the open fundamental string. It is essential to note that a CS term exists for each case mentioned above which could lead to the opening of one brane.

Considering now the world-volume action of the D2-brane, we know that there is a minimal coupling to the RR 3-form potential:

$$I_{D2} = Q_2 \int_{W_3} \hat{A}_3 + \ldots$$

When the D2-brane is open, the gauge transformation $\delta A_3 = d\Lambda_2$ becomes anomalous:

$$\delta I_{D2} = Q_2 \int_{(\partial W)_2} \hat{\Lambda}_2.$$

The standard way to cancel this anomaly is by constraining the boundary $(\partial W)_2$ to lie on the D4-brane world-volume where a 2-form gauge potential V_2, transforming as $\delta V_2 = \hat{\Lambda}_2$, couples to it. The boundary of the D2-brane is now an electric source for the 3-form field strength built out from this potential. Again, the gauge invariant combination is given by $\mathcal{G}_3 = dV_2 - \hat{A}_3$.

The analysis of the Goldstone modes of broken supersymmetry and of broken translation invariance, and the requirement that these bosonic and fermionic modes still fit into a representation of the unbroken supersymmetries, forces us to identify the two field strengths $\mathcal{F}_2$ and $\mathcal{G}_3$ by an electric-magnetic duality on the D4-brane world-volume:

$$\mathcal{F}_2 = \star\mathcal{G}_3.$$

Moreover, we could have analyzed instead the (more familiar) configuration of a fundamental string ending on the D4-brane. We would have found that its end point behaves like an electric charge for the 2-form field strength and like a magnetic charge for the 3-form field strength. Thus we conclude that the boundaries of the string and of the membrane are electric-magnetic dual objects on the world-volume of the D4-brane.

Let us now review the outcome of the analysis above when applied to all the cases discussed at the beginning of this section.

- All the Dp-branes have a world-volume effective theory which can be formulated in terms of a 2-form field strength $\mathcal{F}_2$. The electric charges are the end points of the fundamental strings, while the magnetic charges are the boundaries of the D$(p-2)$-branes (as in [19]). Note the interesting case of the D3-brane on the world-volume of which the S-duality between fundamental strings and D-strings becomes electric-magnetic duality between their end points. The presence of the 2-form field strength is in this case supported by the quantum stringy computation which gives super-Yang-Mills as the low energy effective action of the D-branes.

- On the world-volume of the IIA NS5-brane, we can have the boundaries of the D2- and D4-brane. The boundary of the D2-brane is self-dual and thus couples to a self dual 3-form field-strength, while the boundary of the D4-brane couples magnetically to a scalar potential. This scalar potential is nothing else than the 11th direction which remains after reduction of the M5-brane action. There is also a limiting case here: from (3.9) we can see that a D6-brane can end on a NS5-brane. In this case however the NS5-brane *is* the boundary of the D6-brane, much in the same way as the D0-branes are the end points of fundamental strings. The charge conservation in these cases has to be treated in a somewhat different way, e.g. one needs the presence of D8-branes resulting in a non-vanishing cosmological constant, see [25].
- For the IIB NS5-brane, again all the IIB D-branes can have boundaries on it. The D1- and the D3-brane boundaries are respectively the electric and magnetic charge related to a 2-form field strength $\tilde{\mathcal{F}}_2 = d\tilde{V}_1 - \hat{A}_2^{RR}$ which can be considered the S-dual of the $\mathcal{F}_2$ field on the D5-brane. The boundary of the D5-brane couples electrically to a (non-propagating) 6-form field strength $\mathcal{G}_6$. This 6-form should be related to the mass term in the IIB NS5-brane action as discussed in [26], and could play a role in the definition of an $SL(2, Z)$ invariant IIB 5-brane action. Indeed, by S-duality we should also have the possibility of a NS5-brane ending on a D5-brane.

Let us conclude this contribution with some speculations about the emission of closed branes. The idea is to revert to argument which leads, from the intersecting configurations, to the open brane configurations.

Suppose that we have an open q-brane with both boundaries tied to the same closed p-brane. The world-volume of the open q-brane wraps some transverse compact directions in order to have a definite space-time charge. The two boundaries are, from the point of view of the world-volume of the closed p-brane, two opposite charges. This has as a consequence that the configuration is not BPS, and therefore not supersymmetric, since there is a force between these two charges. Now the two opposite charges can meet and annihilate, which from the q-brane point of view means that the two boundaries meet and reconnect. The q-brane is now closed, and moreover the bound state it constitutes with the closed p-brane has vanishing binding energy. Nothing then prevents it to leave the p-brane (there is some energy left due to the attraction of the two opposite charges). This is thus a very rough picture of how a p-brane could emit other closed branes.

This mechanism has been shown for the emission of closed fundamental strings by D-branes [27]. To describe in a more detailed way the emission of higher branes, a quantum theory of the latter is still lacking. Matrix theory [28] could be a suitable framework to treat this problem (see e.g. [29] for a proposal on open membranes in Matrix theory).

The interest to study these processes is certainly very high. Brane emission could be the dominant process when Hawking radiation is considered at strong coupling, $g_s \gg 1$, or for the emission from black NS 5-branes.

Acknowledgements

I would like to thank my collaborators in the work which was the subject of this talk, F. Englert, L. Houart and P. Windey, and the organizers of the Cargèse summer school for giving me the opportunity to present this work. The author is a research assistant (Aspirant) of the Fonds National de la Recherche Scientifique (Belgium).

References

1. Hull, C. M. and Townsend, P. K. (1995) Unity of Superstring Dualities, *Nucl. Phys.*, **B438**, p. 109; hep-th/9410167.
2. Witten, E. (1995) String Theory Dynamics in Various Dimensions, *Nucl. Phys.*, **B443**, p. 85; hep-th/9503124.
3. Townsend, P. K. (1995) P-Brane Democracy, proceedings of the March 95 PASCOS/John Hopkins Conference; hep-th/9507048.
4. Stelle, K. S. (1997) Lectures on Supergravity p-branes, Preprint, hep-th/9701088.
5. Gauntlett, J. P. (1997) Intersecting Branes, Preprint, hep-th/9705011.
6. Youm, D. (1997) Black Holes and Solitons in String Theory, Preprint, hep-th/9710046.
7. Horowitz, G. T. and Strominger, A. (1991) Black Strings and p-Branes, *Nucl. Phys.*, **B360**, p. 197.
8. Papadopoulos, G. and Townsend, P. K. (1996) Intersecting M-branes, *Phys. Lett.*, **B382**, p. 65; hep-th/9603087.
9. Güven, R. (1992) Black p-Brane Solutions of D=11 Supergravity Theory, *Phys. Lett.*, **B276**, p. 49.
10. Polchinski, J. (1996) TASI Lectures on D-Branes, Preprint, hep-th/9611050.
11. Green, M. B. , Lectures in Cargèse '97 summer school.
12. Tseytlin, A. A. (1996) Harmonic superpositions of M-branes, *Nucl. Phys.*, **B475**, p. 149; hep-th/9604035.
13. Argurio, R., Englert, F. and Houart, L. (1997) Intersection Rules for p-Branes, *Phys. Lett.*, **B398**, p. 61; hep-th/9701042.
14. Argurio, R., Englert, F., Houart L. and Windey, P. (1997) On the Opening of Branes, *Phys.Lett.*, **B408**, p. 151; hep-th/9704190.
15. Tseytlin, A. A. (1997) No-force condition and BPS combinations of p-branes in 11 and 10 dimensions, *Nucl.Phys.*, **B487**, p. 141; hep-th/9609212.
16. Ohta, N. (1997) Intersection Rules for Non-Extreme p-Branes, *Phys.Lett.*, **B403**, p. 218; hep-th/9702164.
17. Gauntlett, J. P., Kastor, D. A. and Traschen, J. (1996) Overlapping Branes in M-Theory, *Nucl. Phys.*, **B478**, p. 544; hep-th/9604179.
18. Bergshoeff, E., de Roo, M., Eyras, E., Janssen B. and van der Schaar, J. P. (1997) Intersections Involving Monopoles and Waves in Eleven-Dimensions, *Class.Quant.Grav.*, **14**, p. 2757; hep-th/9704120.
19. Strominger, A. (1996) Open P-Branes, *Phys. Lett.*, **B383**, p. 44; hep-th/9512059.
20. Townsend, P. K. (1996) D-branes from M-branes, *Phys. Lett.*, **B373**, p. 68; hep-th/9512062.
21. Douglas, M. R. (1995) Branes within Branes, Preprint hep-th/9512077.
22. Hanany, A. and Witten, E. (1997) Type IIB Superstings, BPS Monopoles, and Three-Dimensional Gauge Dynamics, *Nucl. Phys.*, **B492**, p. 152; hep-th/9611230.
23. Elitzur, S., Giveon, A. and Kutasov, D. (1997) Branes and N=1 Duality in String

Theory, *Phys.Lett.*, **B400**, p. 269; hep-th/9702014.

24. Townsend, P. K. (1996) Brane Surgery, in Proceedings of the European Research Conference on 'Advanced Quantum Field Theory', La Londe les Maures, France, Sept. 1996, Preprint hep-th/9609217.
25. Polchinski, J. and Strominger, A. (1996) New Vacua for Type II String Theory, *Phys.Lett.*, **B388**, p. 736; hep-th/9510227.
    Hanany, A. and Zaffaroni, A. (1997) Chiral Symmetry from Type IIA Branes, Preprint hep-th/9706047.
26. Townsend, P. K. (1997) Membrane tension and manifest IIB S-duality, Preprint hep-th/9705160.
27. Callan, C. G. and Maldacena, J. M. (1996) D-Brane Approach to Black Hole Quantum Mechanics, *Nucl. Phys.*, **B472**, p. 591; hep-th/9602043.
    Hashimoto, A. and Klebanov, I. R. (1996) Decay of Excited D-Branes, *Phys. Lett.*, **B381**, p. 437; hep-th/9604065.
    Hashimoto, A. (1996) Perturbative Dynamics of Fractional Strings on Multiply Wound D-strings, Preprint hep-th/9610250.
28. Banks, T., Fischler, W., Shenker, S. H. and Susskind, L. (1997) M Theory as a Matrix Model: a Conjecture, *Phys. Rev.*, **D55**, p. 5512; hep-th/9610043.
29. Li, M. (1997) Open Membranes in Matrix Theory, *Phys.Lett.*, **B397**, p. 37; hep-th/9612144.

HOW TO COUNT THE STATES OF EXTREMAL BLACK HOLES IN $N = 8$ SUPERGRAVITY

VIJAY BALASUBRAMANIAN
Harvard University,
Lyman Laboratory of Physics,
Cambridge, MA 02138,
U.S.A.

1. Introduction

The basic technique for counting the states of extremal black holes in string theory is to represent the black hole as a bound state of p-brane solitons. The horizon area of the resulting supergravity solution is related to an entropy by the Bekenstein-Hawking formula $S = A/4G_N$. At weak coupling, p-branes charged under the Ramond-Ramond fields of Type II string theory are easily quantized as D-branes [1]. This fact was exploited in the seminal paper of Strominger and Vafa who showed that for some black holes corresponding at weak coupling to systems of D-branes, the collective coordinate degeneracy of the bound state can be easily calculated [2]. If the bound state in question is a BPS state, this degeneracy can be extrapolated back to strong coupling since the representation theory of supersymmetry protects it from corrections. To summarize, the program for counting the states of black holes amounts to treating the black hole as a bound state of solitons, quantizing the solitons at weak coupling, and extrapolating the resulting collective coordinate degeneracy back to strong coupling via an appeal to supersymmetry.

In the present paper we apply this method of counting states (see Figure 1) to supersymmetric black holes in $N = 8$ supergravity, the low energy effective theory of Type II string theory compactified on a 6-torus. In the bulk of the paper we will work with IIA string theory which has 0-, 2-, 4- and 6-brane solitons. Charged black holes can be constructed by wrapping these branes on the 6-torus. We are interested in supersymmetric black holes because the weak-to-strong coupling transformation used in Fig. 1 generally preserves only the degeneracy of states that are annihilated by some of the supercharges. Such supersymmetric black holes are also extremal because they will satisfy a BPS mass formula relating the mass of the black hole to the charges it carries.

We begin in Sec. 2 by discussing the general prediction for the entropy of

L. Baulieu et al. (eds.), Strings, Branes and Dualities, 399–410.

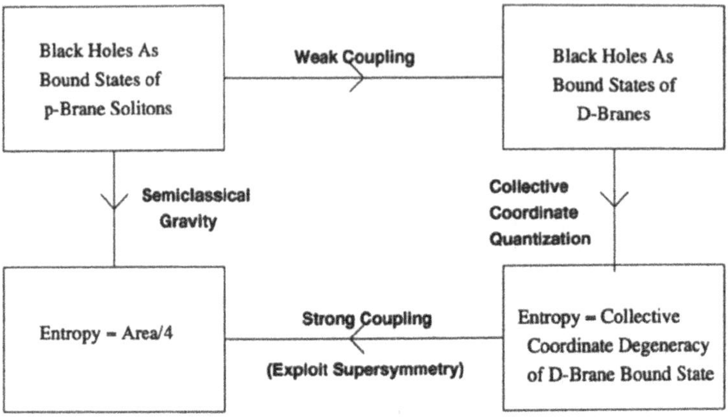

Figure 1. How to count black hole states

extremal black holes in terms of the quartic invariant of the $E(7,7)$ duality group of $N = 8$ supergravity. We want to construct extremal black holes out of bound states of branes. So, in Sec. 3, we show how to create general supersymmetric bound states of 2-branes at angles and 6-branes wrapped on the internal 6-torus. These are T-dual to bound states of 4-branes at angles and 0-branes. Sec. 4 finds the classical black hole solutions to $N = 8$ supergravity corresponding to such branes at angles. Their classical entropy $S = A/4G_N$ precisely matches the prediction given by the E_7 quartic invariant in Sec. 2. Then, in Sec. 5, we count the degeneracy of the microscopic bound state of D-branes and demonstrate detailed agreement with the classical entropy.

In general, extremal black holes in $N = 8$ supergravity can carry 56 different $U(1)$ charges. As discussed in Sec. 2, this space of black holes can be generated by dualities from a 5-parameter generating solution [3]. The known generating solution of Cvetic and Tseytlin [4] contains NS 5-branes and fundamental strings and it is difficult to count its states microscopically. On the other hand, all treatments in the literature of four dimensional black hole entropy using D-branes correspond to 4 parameter generating solutions (for example, see [5, 6]). In the discussion concluding this paper, we show how a 5 parameter generating solution containing only D-branes can be constructed by intersecting 3-branes at angles. This solution can be dualized into a system of 4-branes, 2-branes and 0-branes, and clearly displays in D-brane language the final parameter missing from the discussion of extremal black hole entropy in $N = 8$ supergravity. Counting the states of these black holes introduces some interesting new features.

2. Black Hole Entropy in $N = 8$ Supergravity

The no hair theorem for black holes says that a non-rotating black hole is completely characterized by its mass and charges. The mass of a supersymmetric black hole is further related to its charges by a BPS bound. It follows from this that the area and entropy of an extremal black hole in $N = 8$ supergravity should be completely specified by its charges under the 56 $U(1)$ fields of the theory. $N = 8$ supergravity has a duality group - $E(7,7)$ - that mixes up these charges and dresses them with the moduli associated with the shape of the internal T^6 at infinity. However, the entropy, $S = A/4G_N$, should be invariant under duality since we do not expect the degeneracy of supersymmetric states to change in a dual description [7, 8, 9]. So the entropy of a supersymmetric black hole is expected to be a duality invariant.

The charges of $N = 8$ supergravity rotate in the 56 dimensional antisymmetric tensor representation of the $E(7,7)$ duality symmetry.[1] These charges are associated with various solitons wrapped on the internal T^6 - for example, there are 15 charges coming from 2-branes wrapped on different cycles and 15 from 4-branes. The integral quantized charges are generally "dressed" by moduli scalar fields that represent the shape of the asymptotic internal torus and therefore parametrize inequivalent vacua. The moduli parametrize the coset $E(7,7)/SU(8)$ and therefore $SU(8)$ is the part of the duality group that mixes the charges in a nontrivial way.

For our purposes it is convenient to work in the $SO(8)$ formalism where the 56 charges are assembled in an 8 by 8 antisymmetric tensor whose indices rotate in the vector representation of $SO(8) \in SU(8)$. Then the charge matrix is:

$$z_{ij} = \frac{1}{\sqrt{2}}(x_{ij} + iy_{ij}) \tag{2.1}$$

For solutions that contain only Ramond-Ramond charges associated with D-branes, the x and y variables are directly related to wrapped branes. Taking q_{ij} and p_{ijkl} to be the charges of 2-branes and 4-branes wrapped on the (ij) 2-cycle and the $(ijkl)$ 4-cycle, $x_{ij} = \epsilon^{ijklmn}p_{klmn}/\sqrt{2}$ and $y_{ij} = q_{ij}/\sqrt{2}$ for $i, j \leq 6$. Finally, if Q_0 and Q_6 are the 0-brane and 6-brane charges in the system, $x^{78} = Q_0/\sqrt{2}$ and $y^{78} = -Q_6/\sqrt{2}$.

By the above arguments, the area of a black hole must be given in terms of a duality invariant constructed from the charges. After accounting for dimensions, this was identified in [9] as $A = 4\pi\sqrt{J_4}$ where J_4 is the quartic invariant [10]:

$$-J_4 = x^{ij}y_{jk}x^{kl}y_{li} - \frac{x^{ij}y_{ij}x^{kl}y_{kl}}{4} + \frac{\epsilon_{ijklmnop}}{96}(x^{ij}x^{kl}x^{mn}x^{op} + y^{ij}y^{kl}y^{mn}y^{op}) \tag{2.2}$$

This gives a prediction from duality for the area of four dimensional black holes in string theory. In this paper we will construct rather complicated bound states made from arbitrary numbers of 2-branes and 6-branes and show that the above prediction is verified in detail.

2.1. GENERATING SOLUTIONS

Since a supersymmetric black hole is completely specified by its charges, and duality mixes up these charges, the entire 56 dimensional spectrum of black holes

[1]The useful reference for the remainder of this section is Cremmer and Julia [10].

can be generated by duality from a much smaller space of generating solutions. In fact, the generating solution must have five independent parameters [3]. To see this it is convenient to rewrite the charge matrix in $SU(8)$ formalism [10]:

$$\mathcal{Z}_{AB} = -\frac{1}{4} z_{ij} (\Gamma^{ij})_{AB} \tag{2.3}$$

where Γ^{ij} are generators of $SO(8)$. Now $\mathcal{Z}$ transforms under $SU(8)$ as $\mathcal{Z} \to U\mathcal{Z}U^T$. The $SU(8)$ transformations can be used to skew diagonalize $\mathcal{Z}$ giving four complex eigenvalues:

$$\mathcal{Z} = \begin{pmatrix} \lambda_1 e^{i\theta_1}\tau & 0 & 0 & 0 \\ 0 & \lambda_2 e^{i\theta_2}\tau & 0 & 0 \\ 0 & 0 & \lambda_3 e^{i\theta_3}\tau & 0 \\ 0 & 0 & 0 & \lambda_4 e^{i\theta_4}\tau \end{pmatrix} \quad ; \quad \tau = \begin{pmatrix} 0 & 1 \\ -1 & 0 \end{pmatrix} \tag{2.4}$$

We can perform further $SU(8)$ transformations that eliminate the phase of the eigenvalues in first three blocks, by adding compensating phases in the last block. So, in general, we can diagonalize to get 1 complex and 3 real skew eigenvalues for a total of 5 parameters. Turning this around, in order to obtain the full spectrum of black holes by duality transformations of a generating solution, the latter must have five parameters.

The invariant phase that the $SU(8)$ transformations are unable to get rid of is associated with the presence of electric and magnetic dual objects at the same time.[2] It is difficult to turn on such dual charges supersymmetrically and so the only five parameter generating solution available at present is that of [4]. The bulk of this paper deals with solutions carring pure electric or magnetic charge; under duality these transform into a 55 parameter family of solutions, one short of the desired total. The conclusion discusses a new five parameter generating configuration made purely from D3-branes at angles.

3. Supersymmetry and Branes at Angles

In the previous section we discussed the prediction from duality for the area of extremal black holes in $N = 8$ supergravity: $A = 4\pi\sqrt{J_4}$. To test this prediction we will construct complicated black holes containing many microscopic branes and compute their area. We begin, in this section, on the microscopic side (the upper right corner of Fig. 1) by identifying the most general supersymmetric bound states of branes that can be made from 2-branes and 6-branes. This can be done using the techniques of [11, 12].

Working in lightcone frame, Q and $\tilde{Q}$, the two supercharges of Type II string theory, are 16 component chiral $SO(8)$ spinors. A Dp-brane imposes the projection [13, 11] [3]:

$$Q \pm \Omega_p(\gamma)\tilde{Q} = 0 \tag{3.1}$$

[2]For example, consider a system with 6-branes, and 2-branes on the (12), (34) and (56) cycles. Then if we could add 4-brane charge on the (3456) cycle, z_{12} would be complex. The resulting phase in $\mathcal{Z}$ could not be removed by duality.

[3]We take $\Gamma_{11}\tilde{Q} = +\tilde{Q}$.

Here Ω is the volume form of the brane $\Omega_p(\gamma) = \frac{1}{(p+1)!}\epsilon_{i_0\cdots i_p}\gamma^{i_0}\cdots\gamma^{i_p}$, and is normalized to be a projection. The $\pm$ signs in Eq. 3.1 distinguish between branes and anti-branes or, equivalently, between opposite orientations.[4]

In order to study simultaneous projections imposed by many 2-branes we introduce complex coordinates (z_1, z_2, z_3) on the 6-torus, related to the real coordinates via $z_\mu = (y_{2\mu-1} + iy_{2\mu})/\sqrt{2}$, $\mu = 1, 2, 3$. The corresponding complexified Gamma matrices are $\Gamma^\mu = (\gamma^{2\mu-1} + i\gamma^{2\mu})/2$ and their complex conjugates $\bar{\Gamma}^{\bar{\mu}} = (\gamma^{2\mu-1} - i\gamma^{2\mu})/2$. Take z_4 and $\bar{z}_4$ to be complex coordinates for the remaining two dimensions in lightcone frame and define Γ^4 and $\bar{\Gamma}^4$ accordingly. These complexified matrices obey a Clifford algebra and so define a Fock basis $|n_1, n_2, n_3\rangle \otimes |n_4\rangle$ on which Γ_μ ($\bar{\Gamma}_{\bar{\mu}}$) and Γ_4 ($\bar{\Gamma}_{\bar{4}}$) act as annihilation (creation) operators. Specifically:

$$\bar{\Gamma}_{\bar{\mu}}\Gamma_\mu|n_1, n_2, n_3\rangle \otimes |n_4\rangle = n_\mu|n_1, n_2, n_3\rangle \otimes |n_4\rangle \tag{3.2}$$

where the n_μ and n_4 take values 0 and 1.

3.1. 2-BRANES AT ANGLES AND 6-BRANES

Using the above complex notation, a 2-brane wrapped on the (y_1, y_2) cycle imposes the projection:

$$\gamma^0 Q = \gamma^1\gamma^2\tilde{Q} = -i(2\bar{\Gamma}^{\bar{1}}\Gamma^1 - 1)\tilde{Q} \tag{3.3}$$

Consider a set of 2-branes rotated relative to this reference brane on the 6-torus. The ith brane is rotated by some $R_i \in SO(6)$ and imposes the projection:

$$\gamma^0 Q = (R_i\gamma)^1(R_i\gamma)^2\tilde{Q} \tag{3.4}$$

where R_i is in the fundamental representation of $SO(6)$.[5] We can also write this as:

$$\gamma^0 Q = -iS_{(R_i)}\,(2\bar{\Gamma}^{\bar{1}}\Gamma^1 - 1)\,S^\dagger_{(R_i)}\,\tilde{Q} \tag{3.5}$$

where $S_{(R_i)}$ is the spinor representation of the rotation. The Fock space elements $|n_1n_2n_3\rangle \otimes |n_4\rangle$ which form a basis for the spinors $\tilde{Q}$ are eigenstates of the Γ-matrix projections in Eq. 3.3 and Eq. 3.5: $-i(2\bar{\Gamma}^{\bar{j}}\Gamma^j - 1)|n_1n_2n_3\rangle \otimes |n_4\rangle = i(1 - 2n_j)|n_1n_2n_3\rangle \otimes |n_4\rangle$. So there are simultaneous solutions of Eq. 3.3 and all the Eqs. 3.5 for each i, if there exist some $\tilde{Q}$ which are singlets under *all* the rotations: $S_{(R_i)}\tilde{Q} = S^\dagger_{(R_i)}\tilde{Q} = \tilde{Q}$.

Given such a collection of $R_i \in SO(6)$ that leave some $\tilde{Q}$ invariant, the set of products of R_i and their inverses form a subgroup of $SO(6)$. So the problem of finding supersymmetric relative rotations is reduced to one of finding subgroups of $SO(6)$ that leave some $\tilde{Q}$ invariant. Examining the explicit Fock basis in Eq. 3.2, it is clear that $|n_4\rangle$ is inert under $SO(6)$ rotations so that $\tilde{Q}$ decomposes as $\mathbf{16} \to \mathbf{8}_0 + \mathbf{8}_1$. The $\mathbf{8}$ indicates the 8 dimensional spinor representation

[4] We will take the torus to have unit moduli. See [11] for extensions to a general torus.
[5] Of course, only a discrete subgroup of $SO(6)$ is allowed so that winding numbers remain finite.

of $SO(6)$ and the subscripts indicate the eigenvalue of n_4 in each representation. The largest subgroup of $SO(6)$ under which spinors transform as singlets is $SU(3)$, with the decomposition $\mathbf{8} \to \mathbf{1} + \mathbf{3} + \bar{\mathbf{3}} + \mathbf{1}$. So an arbitrary collection of 2-branes related by $SU(3)$ rotations is supersymmetric. In fact, the branes can be related by $U(3)$ rotations because the $U(1)$ factor in $U(3) = SU(3) \times U(1)$ cancels between $\mathcal{S}_{(R_i)}$ and $\mathcal{S}^\dagger_{(R_i)}$ in Eq. 3.5. Concisely, a collection of 2-branes may be wrapped supersymmetrically on arbitrary $(1,1)$ cycles relative to some complex structure. Further global rotations of the entire configuration by $SO(6)/U(3)$ rotations may be used to turn on 2-brane charges on arbitrary 2-cycles. It is argued in [12] that the general supersymmetric bound state of 2-branes on a 6-torus has this form.

We can determine the amount of supersymmetry surviving the presence of $U(3)$ rotated D2-branes by looking for $U(3)$ invariant spinors $\tilde{Q}$. Given the reference configuration Eq. 3.3 and the Fock basis discussed above, it is readily shown that the $U(3)$-invariant spinors are $\tilde{Q} = \{|000\rangle \otimes |n_4\rangle, |111\rangle \otimes |n_4\rangle\}$ where $n_4 = \{0,1\}$. These four solutions give the equivalent of $N=1, d=4$ supersymmetry. D6-branes can be added without breaking any additional supersymmetry so long as we pick the orientation associated with the minus sign in Eq. 3.1. The 6-brane imposes the condition:

$$\gamma^0 Q = -i(2\Gamma^{\bar{1}}\Gamma^1 - 1)(2\Gamma^{\bar{2}}\Gamma^2 - 1)(2\Gamma^{\bar{3}}\Gamma^3 - 1)\tilde{Q} \tag{3.6}$$

which is solved by the same spinors that survive the presence of the 2-branes. T-duality of the entire 6-torus converts the 6-branes into 0-branes. Furthermore, 2-branes on some $(1,1)$ cycle turn into 4-branes on the dual $(2,2)$ cycle. So we learn that 4-branes wrapped on arbitrary $(2,2)$ cycles are supersymmetric since they are related by relative $U(3)$ rotations, and 0-branes may be bound to them without breaking supersymmetry.

4. Classical Solutions for Branes at Angles

Having identified supersymmetric microscopic configurations of 2-branes and 6-branes, we would like to find the corresponding classical solutions. These will turn out to be four dimensional extremal black holes, giving us an opportunity to test the $E(7,7)$ invariant prediction for black hole entropy.

Choosing complex coordinates $z^j = (x^{2j-1} + ix^{2j})/\sqrt{2}$ on the T^6, the Kähler form and volume of the asymptotic torus are $k = i\sum_{J=1}^{3} dz^J \wedge d\bar{z}^{\bar{J}}$ and $\mathrm{Vol}(T^6) = \int_{T^6} \mathrm{dVol} = \int_{T^6} k \wedge k \wedge k/3!$. The volume is set to 1 by taking the asymptotic moduli to be unity.[6] Following the previous section, we consider a supersymmetric collection of 2-branes wrapped on $(1,1)$ cycles. Thinking of each brane as being $U(3)$ rotated relative to the given complex structure, the jth brane is characterized by the $(1,1)$ volume form $\omega_j = i\,(R_{(j)})^1_J\,(R^*_{(j)})^1_K\, dz^J \wedge d\bar{z}^{\bar{K}}$. Such wrapped branes produce a pressure on the geometry causing it to deform between the position of the branes and the flat space at infinity.

[6]We follow the conventions of [12] for the normalization of forms, wedge products and Hodge dual.

Remarkably, the full solution can be understood in terms of a single (1,1) form ω characterizing the ensemble of branes: $\omega = \sum_j X_j \omega_j$ Here the X_j are harmonic functions in the transverse space $X_j = P_j/r$ and P_j is the charge of the jth 2-brane. [7] Since the classical solutions will only depend on ω, many different microscopic configurations of branes will have the same macroscopic solution - this is a reflection of the no-hair theorem. It is also useful to define the intersection numbers $C_{ij} = (1/\mathrm{Vol}(T^6)) \int_{T^6} k \wedge \omega_i \wedge \omega_j$ and $C_{ijk} = (1/\mathrm{Vol}(T^6)) \int_{T^6} \omega_i \wedge \omega_j \wedge \omega_k$. Then C_{ijk} is proportional to the number of points at which a T-dual collection of 4-branes intersect on the 6-torus. This connection will be used in Sec. 5 in computing the degeneracy of the solutions constructed in this section.

A classical solution corresponding to a collection of 6-branes and 2-branes at angles on a 6-torus is completely described in terms of the metric, the dilaton, the RR 3-form gauge field and the RR 7-form gauge field. The solution in string metric is:

$$\begin{aligned} ds^2 &= (F_2F_6)^{\frac{1}{2}} dx_\perp^2 + (F_2F_6)^{-\frac{1}{2}} \left[-dt^2 + (h_{\mu\bar{\nu}}\, dz^\mu\, d\bar{z}^{\bar{\nu}} + h_{\bar{\mu}\nu}\, d\bar{z}^{\bar{\mu}}\, dz^\nu)\right] \\ A_{(3)} &= \frac{1}{F_2}\, dt \wedge K \qquad ; \qquad A_{(7)} = -\frac{1}{F_6}\, dt \wedge \mathrm{dVol} \qquad (4.1) \\ e^{-2\Phi} &= \sqrt{\frac{F_6^3}{F_2}} \qquad (4.2) \end{aligned}$$

where the 2-form K is:

$$K \equiv *\frac{(k+\omega)\wedge(k+\omega)}{2!} \tag{4.3}$$

and is simply proportional to the internal Kähler metric in the presence of 2-branes:

$$G = i g_{\mu\bar{\nu}}\, dz^\mu \wedge d\bar{z}^{\bar{\nu}} = \frac{i}{\sqrt{F_2F_6}} h_{\mu\bar{\nu}}\, dz^\mu \wedge d\bar{z}^{\bar{\nu}} \equiv \frac{K}{\sqrt{F_2F_6}} \tag{4.4}$$

The functions F_2 and F_6 have simple expressions:

$$\begin{aligned} F_2 &= \frac{\int_{T^6}(k+\omega)^3}{3!\mathrm{Vol}(T^6)} = 1 + \sum_i X_i + \sum_{i<j} X_i X_j C_{ij} + \sum_{i<j<k} X_i X_j X_k C_{ijk} \\ F_6 &= 1 + \frac{Q_6}{r} \qquad (4.5) \end{aligned}$$

Here $dx_\perp^2 = dx_7^2 + dx_8^2 + dx_9^2$ refers to the noncompact part of the space.

In [12] it is explicitly shown that this solution satifies the spacetime equations of motion and is supersymmetric.[8] The four dimensional Einstein metric is related to the string metric by $e^{-2\Phi_4} = e^{-2\Phi}\sqrt{\det g_{int}} = \sqrt{F_6^3/F_2}\sqrt{F_2/F_6^3} = 1$. So the

[7] In general, we could separate the branes in the transverse space by choosing $X_j = P_j/|\vec{r}-\vec{r}_j|$ but will not do so here since we are interested in four-dimensional black holes.

[8] There is an extensive literature on intersecting branes. See [14], for example. Classical branes at angles are discussed, among other articles, in [15]. More references are provided in [12].

Einstein metric is: $ds_4^2 = (F_2F_6)^{-1/2}(-dt^2) + (F_2F_6)^{1/2}(dr^2 + r^2d\Omega^2)$ which describes a black hole with horizon at $r = 0$. The area of the sphere at radius r is $A = 4\pi r^2(F_2F_6)^{1/2}$. In the limit $r \to 0$ this gives the area:

$$A = 4\pi(Q_6 \sum_{i<j<k} P_iP_jP_kC_{ijk})^{1/2}. \tag{4.6}$$

In order to compare this with the prediction from duality it is convenient to project the brane charges P_j onto the different 2-cycles of the torus. Defining a basis for $(1,1)$ forms $\Omega^{a\bar{b}} = i2dz^a \wedge d\bar{z}^{\bar{b}}$, the charge on the $(a\bar{b})$ cycle is given by $q_{a\bar{b}} = \sum_i P_i\alpha_{ia\bar{b}}$ where $\omega_i = \sum_{a\bar{b}} \alpha_{ia\bar{b}}\Omega^{a\bar{b}}$. In terms of this charge matrix, the area in Eq. 4.6 can be rewritten as:

$$A = 4\pi\sqrt{Q_6 \mathrm{det} q} \tag{4.7}$$

Now consider the prediction from duality that $A = 4\pi\sqrt{J_4}$. Comparing with Eq. 2.2 for J_4, all the x_{ij} vanish since there are no 4-branes or 0-branes. Then, y_{ij} for $i, j \leq 6$ represent 2-brane charges on different cycles while y_{78} is the 6-brane charge. Transforming the J_4 symbol into complex coordinates as in this section, it becomes $J_4 = Q_6 \mathrm{det} q$ matching Eq. 4.7 for an arbitrary number of 2-branes at angles in the presence of 6-branes.

5. Counting the States of Our Black Holes

The first step towards a microscopic understanding of the entropy of our black holes is to understand how the physical charges P_i and Q_6 are related to the wrapping numbers of branes on different cycles. Following [12], the quantization condition for 6-branes is $Q_6 = (l_sg/4\pi)N_6$ where N_6 is the number of times the 6-branes wrap the entire torus and $l_s = 2\pi\sqrt{\alpha'}$ is the string length. Similarly, if P_j and ω_j are the physical charge and volume form characterizing the jth 2-brane, we have $P_j\omega_j = (l_sg/4\pi)M_jv_j$ where M_j is the wrapping number and v_j is the element of *integral* cohomology characterizing the cycle on which the brane is wrapped.

We then define $N_{ijk} = \frac{1}{\mathrm{Vol}(T^6)} \int_{T^6} v_i \wedge v_j \wedge v_k$. T-duality of the entire 6-torus turns 2-branes wrapped on v_i into 4-branes wrapped on $*v_i$. Three 4-branes on a 6-torus generically intersect at a point and the N_{ijk} are integers counting the number of intersection points. Using the quantized charges we find that $\sum_{i<j<k} P_iP_jP_kC_{ijk} = (l_sg/4\pi)^3 \sum_{i<j<k}(M_iM_jM_k)N_{ijk}$ Finally, the entropy of the black hole is given by $S/4G_N$ and the Newton coupling is $G_N = g^2\alpha'/8 = g^2l_s^2/32\pi^2$. Putting everything together we find that:

$$S = 2\pi\sqrt{N_6 \sum_{i<j<k} (M_iM_jM_k)N_{ijk}} \tag{5.1}$$

We will now explain the microscopic origin of this entropy.

To count the states of our black holes it is easiest to dualize the entire 6-torus so that the 6-branes become 0-branes, and the 2-branes become 4-branes wrapped on the cycles $*v_j$. Of course, the entropy in Eq. 5.1 remains the same, with N_{ijk}

counting the number of points at which the ith, jth and kth 4-branes intersect. In fact, it is convenient to lift the system into 11 dimensions and view the 4-branes as M-theory 5-branes intersecting along the 11th circle. Imagine making the torus small and the 11th circle big. Then the mutual intersection of any triple of 5-branes is a string that has $(0,4)$ supersymmetry [6, 16, 12, 17]. Putting momentum along this effective string would induce 0-brane charge from the 10 dimensional perspective. However, since we want to preserve supersymmetry we can only introduce momentum in the left-moving direction. From the D-brane point of view, this means that only one of the two signs in Eq. 3.1 is allowed, as observed for the 6-branes at the end of Sec. 3.

So the entropy of the black hole arises from the number of ways in which N_6 units of left-moving momentum can be distributed amongst the effective intersection strings. It was argued heuristically in [6, 16, 18] that the left moving, non-supersymmetric sector of the effective string has a central charge $c = 6$. As shown more carefully in [17] there are indeed 6 bosons in the left moving sector, corresponding to the position of the effective string on the T^6. Actually, the authors of [17] count the states of black holes by moving away from the singular limit of intersecting branes. Instead they consider a single M5-brane wrapped on S^1 times a complicated 4-cycle in a Calabi-Yau 3-fold and count its fluctuations. Their formulae do not directly apply to tori which introduce some new features. So we will not use their techniques here and will remain in the intersecting brane limit following [12].

First consider a situation with $N_6 \gg \sum N_{ijk} M_i M_j M_k$. Each of the $N_{int} = \sum N_{ijk} M_i M_j M_k$ effective strings has 6 bosons in its leftmoving sector. So the problem is simply to distribute N_6 units of momentum amongst $6N_{int}$ bosons. This is identical to the problem of computing the density of states of a string with central charge $c_{\text{eff}} = 6N_{int}$ and for large N_6 this can be read off from [19]: $d(N_6) = \exp 2\pi\sqrt{N_6 c_{\text{eff}}/6}$. Taking the logarithm we find the entropy:

$$S = 2\pi \sqrt{N_6 \sum_{i<j<k} (M_i M_j M_k) N_{ijk}} \tag{5.2}$$

which exactly matches Eq. 5.1. Here we assumed that at a given intersection point of the ith, jth and kth branes there are $M_i M_j M_k$ effective strings singly wound around the 11th circle. We can relax the condition $N_6 \gg \sum N_{ijk} M_i M_j M_k$ by including the multiply wound strings arising from multiply wound 5-branes.

The easiest way to deal with this is to observe that a given triple intersection of 5-branes with wrapping numbers M_i, M_j and M_k will be described by a $(0,4)$ supersymmetric sigma model on the orbifold target:

$$\mathcal{M} = \frac{(T^6)^{M_i M_j M_k}}{S(M_i M_j M_k)} \tag{5.3}$$

We orbifold by the symmetric group $S(M_i M_j M_k)$ to account for symmetry under exchange of the 5-branes.[9] We are interested in putting momentum on the left-moving side where there are 6 bosons characterizing the position on T^6. In our

[9]One might have naively supposed that the appropriate orbifold group would be

case, each triplet of 5-branes intersects in N_{ijk} locations giving rise to N_{ijk} effective strings. So from the M-theory perspective, our solutions are described in the small torus limit by $N_{tot} = \sum_{i<j<k} N_{ijk}$ effective strings, each propagating on an orbifold like Eq. 5.3. The appropriate total effective conformal field theories have central charges that are the sum of contributions from many effective strings, and the resulting degeneracy exactly matches our Eq. 5.1 following [20, 21].

6. Discussion: 5 Parameter Generating Solution

So far we have shown that duality prediction for the entropy of a black hole is verified in detail for general supersymmetric systems of 2-branes and 6-branes or 4-branes and 0-branes. Classical solutions containing arbitrary numbers of branes at angles have the appropriate horizon area and it is possible to count the states microscopically. However, as discussed in Sec. 2.1, the orbit under duality of the configurations discussed so far does not account for the full 56 charge spectrum of four dimensional black holes. One way to see this is to observe that the entropy in Eq. 4.7 is characterized by four invariant parameters - Q_6 and the 3 real eigenvalues of the charge matrix q. Equivalently, in the notation of Sec. 2, we have studied configurations in which the charge matrix z_{ij} is either pure real or pure imaginary.

The additional invariant phase in the charge matrix that is absent in our solutions is associated with the presence of electric-magnetic dual pairs at the same time. It is therefore natural to expect that the desired generating solution can be constructed solely out of self-dual 3-branes.[10] Due to lack of space, the corresponding classical solutions and the counting of states will be discussed elsewhere. Here we show how 3-branes at angles can be a convenient generating configuration and derive the duality prediction for the entropy.

Let $z^i = (x^{2i-1} + ix^{2i})/\sqrt{2}$ be complex coordinates for a 6-torus. Consider any collection of 3-branes wrapped on cycles $\mathrm{Re}(\tilde{z}^1, \tilde{z}^2, \tilde{z}^3)$, with $\tilde{z}^i = R^i_j z^j$ where $R \in SU(3)$. Such a collection of 3-branes is supersymmetric [11]. We consider diagonal $SU(3)$ rotations $R = \mathrm{Diag}(e^{-i\theta_1}, e^{-i\theta_2}, e^{-i\theta_3})$ that rotate the branes separately on the (12), (34) and (56) tori. The generating configuration has the following branes at angles:

Here θ_i is the angle on the (x^{2i-1}, x^{2i}) torus and N_j are the wrapping numbers on the rotated 3-cycles. In order that all wrapping numbers are finite, $\cot\theta = p/q$ is rational. If $N(stu)$ is the induced 3-brane charge on the (stu) cycle we have: $N(135) = N_1 + N_4 p^2$, $N(146) = -N_2$, $N(236) = -N_3$, $N(245) = -N_4 q^2$, $N(145) = -pqN_4$, $N(235) = pqN_4$.

To show that this configuration displays the five parameters required in a generating solution, we T-dualize on the 1, 3 and 5 cycles. The first three sets of branes turn into 0-branes and 4-branes, and the last set turns into 4-branes with

$S(M_i)S(M_j)S(M_k)$. This would account for the exchange symmetry of each kind of 5-brane. However, the analysis of [2, 20] indicates that the appropriate group is $S(M_1 M_2 M_3)$. Indeed, this is the orbifold that is consistent with T-duality.

[10] In writing this section I have benefitted from Finn Larsen's unpublished notes concerning the E_7 quartic invariant and black holes.

TABLE 1. Generating Configuration

Wrapping Numbers	θ_1	θ_2	θ_3
N_1	0	0	0
N_2	0	$\pi/2$	$-\pi/2$
N_3	$-\pi/2$	0	$\pi/2$
N_4	θ	$-\theta$	0

worldvolume fluxes turned on. The worldvolume fluxes induce additional 0-brane and 2-brane charges via the Chern-Simons couplings on the 4-brane [22]. The net charges can be read off from the $N(stu)$ 3-brane charges. We have $Q_0 = N_1 + N_4 p^2$ 0-branes, and $-N_4 q^2$ 4-branes on the (1234) cycle. There are $-N_3$ and $-N_2$ 4-branes on the (1256) and (3456) cycles with $-N_4 pq$ and $N_4 pq$ 2-branes on the dual (34) and (12) cycles. Assembling these charges into the J_4 invariant in Eq. 2.2 we find:

$$J_4 = Q_0 Q_2 Q_3 Q_4 - \frac{Q_4^2 \cot^2\theta}{4}(Q_2 + Q_3)^2 \tag{6.1}$$

where $Q_2 = N_2$, $Q_3 = N_3$. $Q_4 = N_4 q^2$ and $\cot\theta = p/q$. The second term in this equation arises from the mixing of the real and imaginary parts of the charges z_{ij}. This was not probed by the configurations studied in earlier sections.

To verify that this is a five parameter generating configuration we transform to the $SU(8)$ representation of the charge matrix following Eq. 2.3. It is helpful to choose variables $P_L = Q_3 - Q_2$, $P_R = -(Q_3 + Q_2)$, $Q_L = Q_0 - Q_4$, $Q_R = Q_0 + Q_4$ and $\beta = Q_4 \cot\theta$. Then the $SU(8)$ form of the charge matrix is:

$$Z_{12} = \frac{1}{4}(Q_L - P_L - 2i\beta) \quad ; \quad Z_{56} = \frac{1}{4}(Q_R + P_R) \tag{6.2}$$

$$Z_{34} = \frac{1}{4}(Q_L + P_L + 2i\beta) \quad ; \quad Z_{78} = \frac{1}{4}(Q_R - P_R) \tag{6.3}$$

The resulting J_4 invariant is:

$$J_4 = \left[\left(\frac{Q_R^2 - Q_L^2}{4}\right)\left(\frac{P_R^2 - P_L^2}{4}\right) - \left(\frac{\beta^2 P_R^2}{4}\right)\right] \tag{6.4}$$

Written this way, the charge matrix and J_4 invariant are precisely the same as those of the NS-NS, five parameter generating solution discussed in [4, 3]. This shows that the classical solution for the configuration of 3-branes at angles in Table 1 is also a generating solution.

A definitive way of counting the states of this generating configuration is to extend the analysis of [17] to tori. The 4-branes then become M5-branes wrapped on a circle and the 2-brane charges become flux on the 5-brane. We hope to report on this elsewhere. Counting the states of our generating configuration accounts, after duality, for the entropy of the entire 56 parameter space of extremal black holes in $N = 8$ supergravity.

Acknowledgements

I am grateful to R. Gopakumar, F. Larsen and A. Strominger for useful conversations and comments. Much of this work was done in collaboration with F. Larsen and R. Leigh. I thank the organizers of the 1997 Cargese School for the opportunity to present it. My references are not intended to be comprehensive and I apologize for omissions. I am supported by the Harvard Society of Fellows and by the NSF under grant NSF-Phy-91-18167.

References

1. J. Polchinski. *Phys. Rev. Lett.*, 75:4724, 1995. hep-th/9510017.
2. A. Strominger and C. Vafa. *Phys. Lett. B*, 379:99–104, 1996. hep-th/9601029.
3. M. Cvetič and C. Hull, *Nucl. Phys. B*, 480:296–316, 1996, hep-th/9606193. M. Cvetič and D. Youm, *Phys. Rev. D*, 53:584–588, 1996, hep-th/9507090.
4. M. Cvetič and A. Tseytlin. *Phys. Rev. D*, 53:5619–5633, 1996. hep-th/9512031.
5. A. Strominger and J. Maldacena, *Phys. Rev. Lett.*, 77:428–429, 1996, hep-th/9603060. C.V. Johnson, R.R. Khuri, and R.C. Myers, *Phys. Lett. B*, 378:78–86, 1996, hep-th/9603061.
6. V. Balasubramanian and F. Larsen, *Nucl. Phys.B*, 478:199, 1996, hep-th/9604189. I. Klebanov and A. Tseytlin, *Nucl. Phys.B*, 475:179–192, 1996, hep-th/9604166.
7. F. Larsen and F. Wilczek. *Phys. Lett. B*, 375:37–42, 1996. hep-th/9511064.
8. S.Ferrara and R. Kallosh. *Phys. Rev. D*, 54:1525–1534, 1996. hep-th/9603090.
9. R. Kallosh and B. Kol. *Phys. Rev. D*, 53:5344–5348, 1996. hep-th/9602014.
10. Cremmer and Julia. *Nucl. Phys.B*, 159:141, 1979.
11. V. Balasubramanian and R.G. Leigh, *Phys. Rev. D*, 55:6415–6422, 1997, hep-th/9611165. M. Berkooz, M.R. Douglas, and R.G. Leigh, *Nucl. Phys.B*, 480:265–278, 1996, hep-th/9606139.
12. V. Balasubramanian, F. Larsen, and R.G. Leigh. hep-th/9704143, to appear in Phys. Rev. D.
13. J. Polchinski. hep-th/9611050.
14. G. Papadopoulos and P.K. Townsend, *Phys. Lett. B*, 380:273–279, 1996, hep-th/9603087.< A.A. Tseytlin, *Nucl. Phys.B*, 475:149–163, 1996, hep-th/9604035. J.P. Gauntlett, D.A. Kastor, and J. Traschen, *Nucl. Phys.B*, 478:544–560, 1996, hep-th/9604035. M.J. Duff and J. Rahmfeld, Nucl. Phys. B, 481:332-352,1996, hep-th/9605085. R. Argurio, hep-th/9712170.
15. J.C. Breckenridge, G. Michaud, and R.C. Myers, *Phys. Rev. D*, 56:5172–5178, 1997, hep-th/9703041. J. P. Gauntlett, G. W. Gibbons, G. Papadopulos, and P. K. Townsend, *Nucl. Phys. B*, 500:133–162, 1997, hep-th/9702202. M.S. Costa and M. Cvetič, Phys. Rev. D, 56:4834-4843, 1997, hep-th/9703204.
16. J. Maldacena. *Phys. Lett. B*, 403:20, 1997. hep-th/9611163.
17. J. Maldacena, A. Strominger, and E. Witten. hep-th/9711053.
18. K. Behrndt and T. Mohaupt. *Phys. Rev. D*, 56:2206, 1997. hep-th/9611140.
19. M.B. Green, J.H. Schwarz, and E. Witten. Cambridge University Press, 1987.
20. R. Dijkgraaf, G. Moore, E. Verlinde, and H. Verlinde. hep-th/9608096.
21. R. Dijkgraaf, E. Verlinde, and H. Verlinde. *Nucl. Phys. B*, 486:77–88, 1996, hep-th/9603126. *Nucl. Phys. B*, 484:543–561, 1997, hep-th/9607026.
22. M.R. Douglas, hep-th/9512077. M. Green, J.A. Harvey, and G. Moore, *Class. Quant. Grav.*, 14:47–52, 1997. hep-th/9605033.

ASPECTS OF D-INSTANTONS

MICHAEL GUTPERLE
Joseph Henry Laboratories,
Princeton University,
Princeton, New Jersey 08544, USA

1. Introduction

D-instantons are responsible for several important nonperturbative effects in superstring theory.

Firstly, they appear as euclidian Dirichlet p-branes whose $p+1$ dimensional worldvolume wraps a supersymmetric $p+1$ cycle in the compactification manifold. Such configurations were discussed as euclidian membranes and fivebranes in Calabi-Yau compactifications of M-theory [1], instanton corrections to the hypermultiplet geometry near the conifold in IIA on a Calabi-Yau manifold [2], D-string instanton corrections to R^4 and F^4 terms in type I string theory which are important for type I/heterotic duality [3, 4], threshold corrections to R^4 terms and U-duality in toroidally compactified type II theories [5].

Secondly, even in uncompactified ten dimensional IIB superstring theory they make an appearance as $p=-1$ branes [6], i.e D-branes with worldvolumes which are points in spacetime.

Indeed in the context of bosonic string theory such D-instantons were the first 'Dirichlet' objects considered in the past. The introduction of boundaries where the embedding coordinates satisfy Dirichlet boundary conditions in all directions was used to implement point-like partonic behavior in dual model amplitudes which is a desirable feature for a string theory of strong interactions [7]. The fact that momentum is not conserved in the presence of Dirichlet boundaries was used to define off-shell continuations of string scattering amplitudes [8]. It was also observed that the characteristic softness [9] of string amplitudes in the high energy fixed angle scattering regime, does not hold in the presence of D-instantons [10]. Polchinski [11] showed in the context of bosonic strings that the new singularities which arise because of the pointlike nature of the open string boundary conditions cancel in a generalization of the Fischler-Susskind mechanism. In this scheme the combinatorics of string perturbation theory in the presence of D-instantons changes and one has to include disconnected worldsheets of different topology whose boundaries are mapped into the same spacetime point.

L. Baulieu et al. (eds.), Strings, Branes and Dualities, 411–422.

In the context of IIB superstrings D-instantons have several new properties. A D-instanton is a BPS solution which preserves half of the spacetime supersymmetries. In [12] the supergravity solution corresponding to a D-instanton was found. The broken supersymmetries induce fermionic zero modes which have to be integrated over in order to restore spacetime supersymmetry. This induces new effective (t'Hooft) vertices in the low energy effective action. In [13] the instanton induced interactions were derived and the $SL(2,Z)$ duality symmetry of IIB superstring theory was used to conjecture a modular invariant completion of the R^4 term which contains tree level, one loop and instanton corrections. In [14, 15, 16] these expressions were related to similar terms in M-theory and in particular it was argued that they arise from a one loop calculation in eleven dimensions . The material presented in this talk was obtained in collaboration with M.B. Green, H.H. Kwon and P. Vanhove and the interested reader is referred to the original papers for a more detailed discussion.

2. IIB supergravity

The massless fields of IIB superstring theory are given by the bosonic fields : a complex scalar ρ, two second rank AST potentials $B_a^{(2)}$ $a = 1,2$, the zehnbein e_μ^a and a fourth rank AST potential $C^{(4)}$ with selfdual field strength and the fermionic fields: a complex spin 1/2 dilatino λ and a complex spin 3/2 gravitino ψ_μ of opposite chirality. As usual in extended supergravity theories the scalar fields parametrize a coset, which for IIB supergravity is $SL(2,R)/U(1)$. Under $SL(2,R)$ transformations only the second rank AST transform as doublets. The fermions λ and ψ_μ transform under the local $U(1)$ with weight 3/2 and 1/2 respectively. The metric and the fourth rank AST are inert under both transformations. The scalars can be parameterized by a zweibein [17]

$$V = \begin{pmatrix} V_+^1 & V_-^1 \\ V_+^2 & V_-^2 \end{pmatrix} = \frac{1}{\sqrt{2i\rho_2}} \begin{pmatrix} \rho e^{i\phi} & \bar{\rho} e^{-i\phi} \\ e^{i\phi} & e^{-i\phi} \end{pmatrix}. \tag{2.1}$$

The group $SL(2,R)$ acts by matrix multiplication from the left and the local $U(1)$ acts from the right and induces a shift $\phi \to \phi + \alpha$. We will use the $U(1)$ gauge invariance to set $\phi = 0$. A compensating gauge transformation accompanies an $SL(2,R)$ transformation in order to maintain the gauge condition. In this case a $SL(2,R)$ transformation of the complex scalar and the compensating $U(1)$ transformation are given by

$$\rho \to \frac{a\rho + b}{c\rho + d}, \quad e^{i\alpha} = \left(\frac{c\bar{\rho} + d}{c\rho + d}\right)^{\frac{1}{2}}. \tag{2.2}$$

It is useful to define a scalar field strength P_μ and a complex combination of the AST H which have $U(1)$ weight 2 and 1 respectively.

$$P_\mu = -\epsilon_{ab} V_+^a \partial_\mu V_+^b = \frac{i}{2}\frac{\partial_\mu \rho}{\rho_2}, \quad H_{+\mu\nu\rho} = V_{+a} \partial_{[\mu} B^a_{\mu\rho]}. \tag{2.3}$$

3. D-instanton in IIB supergravity

The classical D-instanton is a solution of the euclidian IIB supergravity [12] which preserves half the euclidian supersymmetry. Only the metric and the scalar fields are nontrivial in this background so the nontrivial part of the action is given by

$$S = \int d^{10}x\sqrt{-g}\left\{R - \frac{1}{2\rho_2^2}\partial_\mu\rho\partial^\mu\bar{\rho}\right\}. \tag{3.1}$$

The BPS constraint on the solution requires that

$$\delta\lambda^* = i\gamma^\mu P_\mu\epsilon + .. = 0, \quad \delta\psi_\mu = D_\mu\epsilon + .. = 0. \tag{3.2}$$

The classical D-instanton solution is given, in the Einstein frame, by setting $g^E_{\mu\nu} = \eta_{\mu\nu}$. If the R⊗R scalar is expressed as the sum of a constant term and an euclidian fluctuation, $\rho_1 = \chi + if(x)$, the BPS condition (3.2) reduces to $de^{-\phi} = -df$. The dilaton equation of motion is solved by

$$e^{\phi(x)-\phi_\infty} = 1 + \frac{Ce^{-\phi_\infty}}{|x-y|^8}, \tag{3.3}$$

where y is the position of the D-instanton in space time. The string coupling constant is identified with the expectation value of the dilaton at $|x| \to \infty$, $g = e^{\phi_\infty} = 1/\rho_2$. The constant C is quantized and given by $C = 2N/\pi^{3/2}$ where $N \in Z$. The action of a charge N D-instanton is given by $S = 2\pi i N\rho$. The charge N D-instanton configuration preserves sixteen of the thirty-two supersymmetries. The broken supersymmetries generate fermionic zero modes in the instanton background which have to be integrated over and they induce new vertices which soak up the zero modes

$$\int d^{10}y\, d^{16}\epsilon_0\, \langle\psi^1(x_1)\cdots\psi^n(x_n)\rangle_N = e^{2\pi i N\rho}\int d^{10}y\, d^{16}\epsilon\, \langle\psi\rangle_{s_1}\cdots\langle\psi\rangle_{s_n}, \tag{3.4}$$

where $\langle\psi\rangle_{s_i}$ defines a tadpole of a field of IIB supergravity which soaks up s_i fermionic zero modes and $\sum_i s_i = 16$.

4. Instanton induced interactions

In closed string perturbation theory processes in the presence of D-instantons are formulated by including worldsheets with boundaries, where the boundary is mapped to a point in spacetime corresponding to the position of the instanton. Because the boundary conditions relate leftmoving and rightmoving worldsheet fields, a D-instanton preserves half of the closed string supersymmetry, i.e is a BPS state. The broken spacetime supersymmetry generators have the form of zero momentum open string fermion vertex operators and have a natural interpretation as generating fermionic zero modes in the instanton background. The simplest open-string world-sheet that arises in a D-brane process is the disk diagram (*cf.* Fig.4).

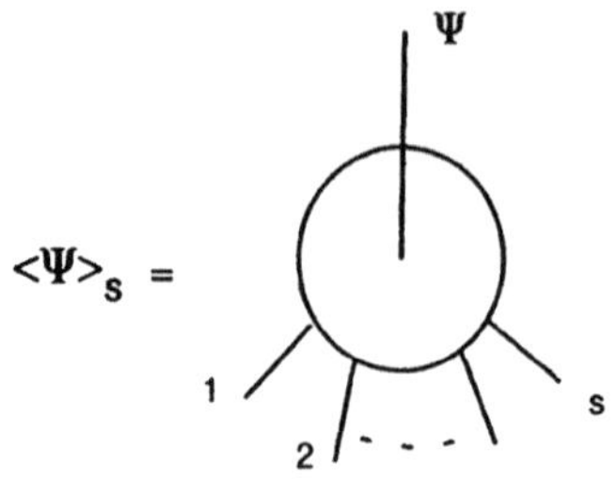

Figure 1. An on-shell closed-string state, Ψ, coupling to s open-string fermions on a disk with Dirichlet boundary conditions.

A tadpole diagram with one fermionic zero mode attached couples to the dilatino and the calculation in A.1 gives

$$\langle\lambda\rangle_1 = \langle cF_{-\frac{1}{2}}(x)c\tilde{c}V_{(\lambda)}(z,\bar{z})\rangle = \bar{\zeta}_\lambda\epsilon_0, \tag{4.1}$$

where $\zeta^a_\lambda = \gamma^\mu_{ab}(\zeta^b_\mu{}^{RN} + i\zeta^b_\mu{}^{NR})$ is the holomorphic combination of the two dilatinos. Similarly one can evaluate the one point functions of the complex combination of the NS $\otimes$ NS and R $\otimes$ R two-forms, $B = B_{NS} + iC^{(2)}$ with two fermionic zero modes attached $\langle B\rangle_2$ and the tadpole $\langle\psi\rangle_3$ holomorphic gravitino with three fermionic zero modes attached. Of particular interest is the disk with four fermionic zero modes attached which couples to the graviton and the calculation in A.4 gives

$$\langle h\rangle_4 = \langle cF_{\frac{1}{2}}(x_1)\int dx_2dx_3dx_4F(x_2)_{\frac{1}{2}}F_{-\frac{1}{2}}(x_3)F_{-\frac{1}{2}}(x_4)c\tilde{c}V^{NN}(z,\bar{z})\rangle$$
$$= \bar{\epsilon}_0\gamma^{\rho\mu\tau}\epsilon_0\;\bar{\epsilon}_0\gamma^{\lambda\nu\tau}\epsilon_0\;\zeta_{\mu\nu}k_\rho k_\lambda.$$

The results of the calculation in the appendix of tadpoles for the massless fields which absorb up to four fermionic zero modes are summarized with their weight under $U(1)$ and the coupling in the following table

Field	U(1)-wt	no. z.m.	tadp
$\langle\lambda\rangle$	3/2	1	$\bar{\lambda}\epsilon$
$\langle H\rangle$	1	2	$\bar{\epsilon}\gamma^{\mu\nu\rho}\epsilon H_{\mu\nu\rho}$
$\langle\psi\rangle$	1/2	3	$\bar{\epsilon}\gamma^{\mu\nu\rho}\epsilon\bar{\psi}_\rho\gamma_\mu\epsilon k_\nu$
$\langle h\rangle$	0	4	$\bar{\epsilon}\gamma^{\rho\mu\tau}\epsilon\bar{\epsilon}\gamma^{\lambda\nu\tau}\epsilon\zeta_{\mu\nu}k_\rho$

There are a whole variety of instanton induced vertices which are in fact related to each other by supersymmetry . The simplest case is given by a sixteen fermion term which is induced by sixteen disks with one holomorphic dilatino λ and one zero mode ϵ attached to each disk.

$$\int d^{10}y\int d^{16}\epsilon_0\prod_{i=1}^{16}\langle\lambda\rangle_i = \epsilon^{a_1\cdots a_{16}}\lambda^{a_1}\ldots\lambda^{a_{16}}. \tag{4.2}$$

Among many other terms we focus on a R^4 term which arises from four disks with a graviton and four fermionic zero modes attached to each disk.

$$\begin{aligned} A_4(\{\zeta_h^{(r)}\}) &= \int d^{10}y d^{16}\epsilon_0 \langle h^{(1)}\rangle_4 \langle h^{(2)}\rangle_4 \langle h^{(3)}\rangle_4 \langle h^{(4)}\rangle_4 \\ &= \int d^{10}y d^{16}\epsilon_0 \prod_{r=1}^{4} \left(\bar\epsilon_0 \gamma^{\mu_r\sigma_r\rho}\epsilon_0 \; \bar\epsilon_0\gamma^{\nu_r\tau_r\rho}\epsilon_0 \zeta^{(\mu_r}\tilde\zeta^{\nu_r)} \; k_r^{\sigma_r} k_r^{\tau_r}\right), \\ &= \int d^{10}y \left(\hat t^{i_1 j_1 \cdots i_4 j_4} \hat t_{m_1 n_1 \cdots m_4 n_4} - \frac{1}{4}\epsilon^{i_1 j_1\cdots j_4 j_4}\epsilon_{m_1 n_1\cdots m_4 n_4}\right) \\ &\quad \times R^{m_1 n_1}_{i_1 j_1} R^{m_2 n_2}_{i_2 j_2} R^{m_3 n_3}_{i_3 j_3} \end{aligned}$$

The close connection between $t^8 t^8 R^4$ and λ^{16} terms is obviously a consequence of their common superspace origin which can be seen already in the rigid limit of the linearized theory. Consider a linear superfield $\Phi(x,\theta)$ (where θ is a complex chiral $SO(9,1)$ Grassmann spinor) that satisfies the holomorphic constraint $D^*\Phi = 0$ and the on-shell condition $D^4\Phi = D^{*4}\Phi^*$ [18] where

$$D_A = \frac{\partial}{\partial\theta^A} + 2i(\Gamma^\mu\theta^*)_A\partial_\mu, \qquad D^*_A = -\frac{\partial}{\partial\theta^{*A}} \tag{4.3}$$

(recall that $\bar\theta = \theta\Gamma^0$ does *not* involve complex conjugation) are the holomorphic and anti-holomorphic covariant derivatives that anticommute with the rigid supersymmetries

$$Q_A = \frac{\partial}{\partial\theta^A}, \qquad Q^*_A = -\frac{\partial}{\partial\theta^{*A}} + 2i(\bar\theta\Gamma^\mu)_A\partial_\mu. \tag{4.4}$$

The field Φ has an expansion in powers of θ (but not θ^*), describing the 256 fields in an on-shell supermultiplet.

$$\begin{aligned} \Phi &= \rho_0 + \Delta \\ &= \rho_0 + a - \frac{2i}{g}\bar\theta\lambda - \frac{1}{24g}\bar\theta\Gamma^{\mu\nu\sigma}\theta G_{\mu\nu\sigma} + \frac{i}{6g}\bar\theta\Gamma^{\mu\nu\sigma}\theta\bar\theta\Gamma_\nu\partial_\sigma\psi_\mu \\ &\quad - \frac{i}{48g}\bar\theta\Gamma^{\mu\nu\eta}\theta\bar\theta\Gamma_\eta{}^{\sigma\tau}\theta R_{\mu\nu\sigma\tau} + \cdots, \end{aligned} \tag{4.5}$$

where Δ is the linearized fluctuation around a flat background with a constant scalar, $\rho_0 = \rho - a = C_0^{(0)} + ig^{-1}$. Note that because of the constraint $D^4\Phi = D^{*4}\Phi^*$ the expansion terminates at θ^8. The coefficients of the component fields are consistent with the conventions used in [17]. The terms indicated by $\cdots$ fill in the remaining members of the ten-dimensional $N = 2$ chiral supermultiplet, comprising (in symbolic notation) $\partial dC^{(4)}$, $\partial\psi^*_{\mu\nu}$, $\partial^2 G^*_{\mu\nu\sigma}$, $\partial^3\lambda^*$ and $\partial^4\rho^*$. The linearized supersymmetric on-shell action has the form [16]

$$S' = \mathrm{Re}\int d^{10}x d^{16}\theta\, g^4 F[\Phi], \tag{4.6}$$

which is manifestly invariant under the rigid supersymmetry transformations, (4.4). The various component interactions contained in (4.6) are obtained from the θ^{16} term in the expansion,

$$F[\Phi] = F(\rho_0) + \Delta \frac{\partial}{\partial \rho_0} F(\rho_o) + \tfrac{1}{2}\Delta^2 \left(\frac{\partial}{\partial \rho_0}\right)^2 F(\rho_o) + \cdots. \tag{4.7}$$

All terms are F-terms, i.e they are realized as integrals over half the superspace.

5. Modular Invariance

The fact that the Lorentz index structure of the $t_8 t_8 R^4$ term in the instanton induced effective action is of the same form as the tree level [19] and one loop [20] perturbative terms together with invariance of the graviton under the $SL(2,Z)$ duality of IIB string theory make it very plausible to conjecture for the exact nonperturbative R^4 term [13].

$$S_{R^4} = \int d^{10}x \; t_8 t_8 \; R^4 \; f(\rho, \bar{\rho}), \tag{5.1}$$

where the nonholomorphic modular function f is given by

$$f(\rho,\bar{\rho}) = \sum_{(p,n)\neq(0,0)} \frac{\rho_2^{3/2}}{|p+n\rho|^3}, \tag{5.2}$$

In the string frame the large ρ_2 expansion gives

$$\begin{aligned} \rho_2^{1/2} f(\rho,\bar{\rho}) &= 2\zeta(3)(\rho_2)^2 + \frac{2\pi^2}{3} + 4\pi^{3/2} \sum_N (N\rho_2)^{1/2} \sum_{N|\hat{m}} \frac{1}{\hat{m}^2} \\ &\times \left(e^{2\pi i N\rho} + e^{-2\pi i N\bar{\rho}}\right)\left(1 + \sum_{k=1}^{\infty} (4\pi N \rho_2)^{-k} \frac{\Gamma(k-1/2)}{\Gamma(-k-1/2)k!}\right) \end{aligned} \tag{5.3}$$

There are several interesting properties of this expansion. There are terms corresponding to tree level, one loop and instanton contributions. The tree and one loop terms exactly agree with results in string perturbation theory. The fact that no higher perturbative corrections appear in (5.3) together with the fact that all such terms have the form of integral over half the superspace, lead to the conjecture that the R^4 and related terms are protected from higher loop perturbative corrections by a nonrenormalization theorem. Recently substantial evidence for the validity of such a theorem has been obtained [21, 22]. It is also interesting to note that it is possible to infer the value of certain integrals which are important for the calculation of the Witten index for bound states of D0-branes [23] from the modular function f [24]. The nonholomorphic function f is invariant under $SL(2,Z)$ modular transformations; this is because the graviton and hence $t_8 t_8 R^4$ is invariant under $SL(2,Z)$. For other terms which are related to R^4 like the λ^{16}

the modular function has to be generalized, since such terms transform with a nontrivial weight under the induced $U(1)$ R-symmetry. For a term in the effective action which transforms with weight $2k$ a natural generalization of (5.2) is [1],

$$f^{(k)} = \sum_{(m,n)\neq(0,0)} \frac{\rho_2^{3/2}(m+n\rho)^{2k}}{|m+n\rho|^{2k+3}}. \tag{5.4}$$

This function transforms under $SL(2,Z)$ with a $U(1)$ weight k

$$f^{(k)}(\frac{a\rho+b}{c\rho+d}) = \left(\frac{c\bar\rho+d}{c\rho+d}\right)^k f^{(k)}(\rho). \tag{5.5}$$

The functions $f^{(k)}$ with different k are related to $f^{(0)}$ by

$$f^{(k)} = \rho_2^k D_{2(k-1)}\cdots D_2 D_0 f^{(0)}, \tag{5.6}$$

where

$$D_l = i(\frac{\partial}{\partial\rho} - \frac{l}{\rho-\bar\rho}), \tag{5.7}$$

is the nonholomorphic covariant derivatives which maps modular forms of weight l into modular forms of weight $l+2$.

Comparing the relations in (4.6) with (5.6) it becomes clear that the linearized superspace incorporated the action of the ordinary derivatives $\partial/\partial\rho$ by expanding the function $F[\Phi]$, but does not provide a covariantization. One indication that the full nonlinear superspace should produce a structure like (5.6) can be obtained by considering the perturbative fluctuations around the instanton configurations which are supressed by powers of ρ_2. For the λ^{16} term the lowest order contribution in the charge one instanton sector comes from sixteen disks with one dilatino and one fermionic zero mode attached. Because the genus of the disk is -1 and the insertion of a closed string vertex operator contributes one power of g such a term contributes to order g^0. A term where two dilatini are on the same disk with two fermionic zero modes attached is of order g and corresponds to a perturbative fluctuation. There should be a contact term which contributes when the vertex operator of the two dilatini come close to each other. Such contact terms are suppressed by $1/\rho_2 = g$ and should be responsible for the covariantization of the leading D-instanton terms. Such terms with more than one closed string vertex operator on a disk together with disks with handles attached should produce the perturbative fluctuations around the instanton configuration 5.3.

6. One loop in eleven dimensions

In this section we briefly review the relation of $t^8t^8R^4$ term (5.1) by evaluating the four-graviton one loop amplitude of eleven-dimensional supergravity compactified on a torus, T^2, in the directions 9, 11. The volume of the torus is $\mathcal{V} = R_9R_{11}$ and its complex structure, $\Omega = \Omega_1 + i\Omega_2$, may be expressed in terms of the components

[1]Similar considerations can be found in [25]

eleven-dimensional metric $G_{\hat{\mu}\hat{\nu}}$ ($\hat{\mu} = 0, \cdots, 9, 11$) as $\Omega_2 = G_{9\,9}/G_{11\,11} = R_9/R_{11}$ and $\Omega_1 = G_{9\,11}/G_{11\,11} = C^{(1)}$, where $C^{(1)}$ is the component of the IIA one-form along the direction x^9. M-theory on T^2 is identified with IIB on S^1 via setting $\Omega = \rho$. The expression for the amplitude, obtained rather efficiently by considering a first-quantized super-particle in the light-cone gauge [15], has the structure,

$$A_{R^4} = \int d^9p \frac{1}{\mathcal{V}} \sum_{m,n} \int \prod_{r=1}^{4} d\tau_r \mathrm{tr}(V_h^{(1)}(k^1) V_h^{(2)}(k^2) V_h^{(3)}(k^3) V_h^{(4)}(k^4)), \qquad (6.1)$$

where $V_h^{(r)}(k^r)$ is the graviton vertex operator for the rth graviton with momentum k^r and is evaluated at a proper time τ_r around the loop (and dimensional quantities are measured in units of the eleven-dimensional Planck scale). The trace is over the fermionic zero modes and the loop momentum in the directions $0, \cdots, 8$ and the integers m and n label the Kaluza-Klein momenta in the directions 9, 11 of T^2. The trace over the fermionic modes gives an overall factor, K, which contains eight powers of the external momenta and is the linearized form of $t^8 t^8 R^4$. The $t^8 t^8 R^4$ term in (6.1) is obtained by setting the momenta equal to zero in the remainder of the loop integral. It is convenient to perform a double Poisson resummation in order to rewrite the sum over the Kaluza–Klein momenta as a sum over windings, $\hat{m}$ and $\hat{n}$, of the loop around the cycles of T^2, giving,

$$\begin{aligned} \mathcal{V} A_{R^4} &= K\mathcal{V} \sum_{\hat{m},\hat{n}} \int d\tau \tau^{\frac{1}{2}} e^{-\tau \mathcal{V} \frac{|\hat{m}+\hat{n}\Omega|^2}{\Omega_2}} \\ &= K\mathcal{V} C + \pi^2 K \mathcal{V}^{-\frac{1}{2}} f(\Omega, \bar{\Omega}), \end{aligned} \qquad (6.2)$$

where

$$f(\Omega, \bar{\Omega}) = \pi^{-2} \Gamma(3/2) \sum_{(m,n) \neq (0,0)} \frac{\Omega_2^{3/2}}{|m + n\Omega|^3} \qquad (6.3)$$

is a modular function. The cubic ultraviolet divergence of A_{R^4} is contained in the zero winding term, $\hat{m} = 0 = \hat{n}$, which has the divergent coefficient, C. It was argued in [14] that in any regularization that is consistent with T-duality between the IIA and IIB string theories in nine dimensions this has to be replaced by a regularized finite value, $C = \pi/3$ [15]. Presumably a microscopic description of M-theory (such as matrix theory) would reproduce this value. In the limit of zero volume, $\mathcal{V} \to 0$, this term disappears and only the term with coefficient $\mathcal{V}^{-1/2}$ survives. Similar results for the λ^{16} and related terms and the corresponding modular functions f_k have been obtained in [16] by evaluating different one loop amplitudes of M compactified on T^2.

Acknowledgements

I would like to thank M.B. Green, H.H. Kwon and P. Vanhove for collaboration and discussions on matters presented in this talk. I gratefully acknowledge the partiall support by a DOE grant DE-FG02-91ER40671, NSF grant PHY-9157482 and a James S. McDonnell grant 91-48.

A. One point functions in the upper half plane

We give two examples of the calculation of one point functions on the half plane which give the tadpoles $\langle\Psi\rangle$ of closed string fields used in section 4.

A.1. DILATINO

Attaching one fermionic zero mode to the boundary leads to a non-vanishing one-point function for a combination of the two dilatinos from the NS $\otimes$ R

$$\begin{aligned}\langle\lambda^{NR}\rangle_1 &= \langle c\bar{c}V_{(-1,-\frac{1}{2})}(v,z)cV_{-\frac{1}{2}}(\epsilon,x_1)\rangle \\ &= v_{\mu a}\epsilon_b\langle c\bar{c}e^{-\phi}\psi^{\mu}e^{-\frac{1}{2}\bar{\phi}}\bar{S}^a e^{ikX}(z,\bar{z})ce^{-\frac{1}{2}\phi}S^b(x_1)\rangle \\ &= iv^{\mu}_{a}\epsilon_b(z-x_1)(\bar{z}-x_1)(z-\bar{z})\frac{\gamma^{ab}_{\mu}}{(z-x_1)(\bar{z}-x_1)(z-\bar{z})} \\ &= i\bar{\epsilon}\lambda \qquad\qquad (A.1)\end{aligned}$$

and similarly for the R $\otimes$ NS dilatino

$$\langle\lambda^{RN}\rangle_1 = \langle c\bar{c}V_{(-\frac{1}{2},-1)}(v,z)cV_{-\frac{1}{2}}(\epsilon,x_1)\rangle = \bar{\epsilon}\lambda. \qquad (A.2)$$

The calculation of such correlation functions on the upper half plane is most easily done by doubling. One uses the boundary conditions for the D-instanton to map the antiholomorphic fields of the vertex operators living in the upper half plane to holomorphic fields in the lower half plane and evaluates the correlation function using holomorphic correlators in the complex plane. Putting (A.1) and (A.2) together the coupling of the dilatinos is given by

$$\langle\lambda\rangle_1 = (\zeta^{RN}_{a\mu} + i\zeta^{NR}_{a\mu})(\gamma^{\mu})^a_b\epsilon^b_0. \qquad (A.3)$$

A.2. NS $\otimes$ NS ANTISYMMETRIC TENSOR

Attaching two fermionic zero modes produces a coupling of the D-instanton to the antisymmetric rank 2 tensor fields from the NS $\otimes$ NS and R $\otimes$ R sector. For the NS $\otimes$ NS antisymmetric tensor the relevant amplitude is

$$\begin{aligned}\langle B^{NN}_{\mu\nu}\rangle_2 &= \langle c\bar{c}V_{(0,-1)}(v,z)cV_{-\frac{1}{2}}(\epsilon,x_1)\int dx_2 V_{-\frac{1}{2}}(\epsilon,x_2)\rangle \\ &= \zeta^{NN}_{\mu\nu}\epsilon^1_a\epsilon^2_b\Big\langle c\bar{c}e^{-\phi}\psi^{\mu}(\bar{\partial}X^{\nu}+ik_{\lambda}\bar{\psi}^{\lambda}\bar{\psi}^{\nu})e^{ikX}(z,\bar{z})ce^{-\frac{1}{2}\phi}S^a(x_1) \\ &\quad\times\int dx_2 e^{-\frac{1}{2}\phi}S^b(x_2)\Big\rangle \qquad\qquad (A.4)\end{aligned}$$

Note that the $\bar{\partial}X^{\mu}$ cannot be contracted into anything, the $\bar{\psi}^{\lambda}\bar{\psi}^{\nu}(\bar{z})$ in $V_{(0,0)}$ acts as a current $\bar{j}^{\lambda\nu}$ insertion which can be evaluated,

$$\langle B_{\mu\nu}\rangle_2 = ik_{\lambda}\zeta^{NN}_{\mu\nu}\epsilon^a_1\epsilon^b_2\int dx_2\frac{(z-\bar{z})(\bar{z}-x_1)}{(z-x_2)(x_1-x_2)}$$

$$
\begin{aligned}
&\times\left(\frac{\gamma_{ab}^{\lambda\nu\mu}}{\bar z - x_1} - \frac{\gamma_{ab}^{\lambda\nu\mu}}{\bar z - x_2} + \frac{\eta^{\lambda\mu}\gamma_{ab}^{\nu} - \eta^{\nu\mu}\gamma_{ab}^{\lambda}}{z-\bar z}\right)\\
&= ik_\rho\zeta_{\mu\nu}^{NN}\bar\epsilon\gamma^{\mu\nu\rho}\epsilon\int dx_2\frac{(z-\bar z)}{(z-x_2)(\bar z - x_2)}\\
&= \pi H_{\mu\nu\rho}^{NN}\bar\epsilon\gamma^{\mu\nu\rho}\epsilon. \qquad (A.5)
\end{aligned}
$$

Where we defined $H_{\mu\nu\rho}^{NN} = ik_\rho\zeta_{\mu\nu}^{NN}$. Only the first two terms proportional to $\gamma^{\mu\nu\rho}$ contribute because the term proportional to just one γ matrix in the first line of (A.5) $\bar\epsilon_1\gamma^\mu\epsilon_2$ vanishes after the spinor wavefunctions ϵ^1 and ϵ^2 are antisymmetrized. In the last line we used the fact that

$$
\int dx\frac{(z-\bar z)}{(z-x)(\bar z - x)} = \pi \qquad (A.6)
$$

which is true for any z with non vanishing imaginary part by the residue-theorem.

A.3. R ⊗ R ANTISYMMETRIC TENSOR

For the R ⊗ R antisymmetric tensor the relevant amplitude is given by

$$
\begin{aligned}
\langle B^{RR}\rangle_2 &= \langle c\bar c V_{(-\frac12,-\frac12)}(F^{(1)},z)cV_{-\frac12}(\epsilon,x_1)\int dx_2 V_{-\frac12}(\epsilon,x_2)\rangle\\
&= \frac{i}{4}k_{\mu_1}\zeta_{\mu_2\mu_3}^{RR}\epsilon_c^1\epsilon_d^2\Big\langle c\bar c e^{-\frac12\phi}e^{-\frac12\bar\phi}S^a\gamma_{ab}^{\mu_1\mu_2\mu_3}\bar S^b e^{ikX}\\
&\qquad \times ce^{-\frac12\phi}S^c(x_1)\int dx_2 e^{-\frac12\phi}S^d(x_2)\Big\rangle\\
&= -\frac14 k_{\mu_1}\zeta_{\mu_2\mu_3}^{RR}\epsilon_c^1\epsilon_d^2\int dx_3\\
&\qquad\times\left((\bar z_1 - x_2)\frac{\mathrm{tr}(\gamma^\rho\gamma^{\mu_1\mu_2\mu_3})\gamma_\rho^{cd}}{(\bar z - x_2)(x_1-x_2)} - (z-\bar z)\frac{(\gamma^\rho\gamma^{\mu_1\mu_2\mu_3}\gamma_\rho)^{cd}}{(z-x_2)(\bar z - x_2)}\right)\\
&= i\pi H_{\mu\nu\rho}^{RR}\bar\epsilon\gamma^{\mu\nu\rho}\epsilon \qquad (A.7)
\end{aligned}
$$

Again $H_{\mu\nu\rho}^{RR} = ik_\rho\zeta_{\mu\nu}^{RR}$. Note that the first term vanishes because of the trace of γ-matrices and Bose-Fermi symmetry of ϵ_1,ϵ_2. To evaluate the second term we used the identity $\gamma^\rho\gamma^{\mu_1\mu_2\mu_3}\gamma_\rho = -4\gamma^{\mu_1\mu_2\mu_3}$ for ten dimensional γ-matrices and the integral (A.6).

Putting together (A.5) and (A.7) we see that the following couples to the D-instanton with two fermionic zero modes attached.

$$
\langle B\rangle_2 = \bar\epsilon_0\gamma^{\mu\nu\rho}\epsilon_0 ik_{[\mu}(\zeta_{\nu\rho]}^{NN} + i\zeta_{\nu\rho]}^{RR}) \qquad (A.8)
$$

A.4. GRAVITON

Attaching four fermionic zero modes gives a coupling of the self-dual R ⊗ R four form and the graviton to the D-instanton. We focus here on the graviton where the relevant amplitude is given by

$$\begin{aligned}\langle h\rangle_4 &= \langle c\bar{c}V_{(0,0)}(\zeta,z)cV_{-\frac{1}{2}}(x_1)\int dx_2dx_3dx_4V_{-\frac{1}{2}}(x_2)V_{-\frac{1}{2}}(x_3)V_{-\frac{1}{2}}(x_4)\rangle \\ &= \zeta_{\mu\nu}\epsilon_1^a\epsilon_2^b\epsilon_3^c\epsilon_4^d\,\langle c\bar{c}(\partial X^\mu + ik_\rho\psi^\rho\psi^\mu)(\bar{\partial}X^\nu + ik_\lambda\bar{\psi}^\lambda\bar{\psi}^\nu)e^{ikX}(z,\bar{z}) \\ &\quad\times ce^{-\frac{\phi}{2}}S^a(x_1)\int dx_2dx_3dx_4e^{-\frac{\phi}{2}}S^b(x_2)e^{-\frac{\phi}{2}}S^c(x_3)e^{-\frac{\phi}{2}}S^d(x_4)\Big\rangle.\end{aligned}$$

Using doubling and the well known correlation function for the superghosts and spin fields, in particular

$$\begin{aligned}&\langle j^{\rho\mu}(z)j^{\lambda\nu}(\bar{z})S^a(x_1)S^b(x_2)S^c(x_3)S^d(x_4)\rangle \\ &= \sum_{i=1}^{4}\frac{M^{\rho\mu}(i)}{z-x_i}\sum_{j=1}^{4}\frac{M^{\lambda\nu}(i)}{\bar{z}-x_i}\frac{x_{14}x_{23}\gamma^\mu_{ab}\gamma_{\mu\,cd} - x_{12}x_{34}\gamma^\mu_{ad}\gamma_{\mu\,bc}}{(x_{12}x_{13}x_{14}x_{23}x_{24}x_{34})^{3/4}}.\end{aligned} \tag{A.9}$$

The operator $M(i)$ acts on the spinor index associated with the i-th spin field as a Lorentz transformation. As above only terms like $\epsilon_i\gamma^{\mu_1\mu_2\mu_3}\epsilon_j$ are nonzero after antisymmetrization over the spinor wavefunctions ϵ_i.

For example the contributions of the form $\bar{\epsilon}_1\gamma^{\rho\mu\tau}\epsilon_2\,\bar{\epsilon}_3\gamma^{\lambda\nu\tau}\epsilon_4$ is given by $M^{\rho\mu}$ acting on γ^τ_{ab} and $M^{\lambda\nu}$ acting on γ^τ_{cd}, which results in

$$\begin{aligned}&\bar{\epsilon}_1\gamma^{\rho\mu\tau}\epsilon_2\,\bar{\epsilon}_3\gamma^{\lambda\nu\tau}\epsilon_4\int dx_2dx_3dx_4\frac{(z-\bar{z})(z-x_1)(\bar{z}-x_1)}{(x_{12}x_{13}x_{24}x_{34})} \\ &\qquad\times\left(\frac{1}{z-x_1}-\frac{1}{z-x_2}\right)\left(\frac{1}{\bar{z}-x_3}-\frac{1}{\bar{z}-x_4}\right) \\ &= \bar{\epsilon}_1\gamma^{\rho\mu\tau}\epsilon_2\,\bar{\epsilon}_3\gamma^{\lambda\nu\tau}\epsilon_4\int dx_2dx_3dx_4\frac{(z-\bar{z})(\bar{z}-x_1)}{x_{12}x_{34}(z-x_2)(\bar{z}-x_3)(\bar{z}-x_4)} \\ &= \bar{\epsilon}_1\gamma^{\rho\mu\tau}\epsilon_2\,\bar{\epsilon}_3\gamma^{\lambda\nu\tau}\epsilon_4\int dx_2\frac{(z-\bar{z})}{(z-x_2)(\bar{z}-x_2)} \\ &= \pi\bar{\epsilon}_1\gamma^{\rho\mu\tau}\epsilon_2\,\bar{\epsilon}_3\gamma^{\lambda\nu\tau}\epsilon_4.\end{aligned} \tag{A.10}$$

There are three other contributions in (A.9) given by

$$\bar{\epsilon}_1\gamma^{\lambda\nu\tau}\epsilon_2\,\bar{\epsilon}_3\gamma^{\rho\mu\tau}\epsilon_4,$$

$$\bar{\epsilon}_1\gamma^{\lambda\nu\tau}\epsilon_4\,\bar{\epsilon}_2\gamma^{\rho\mu\tau}\epsilon_3$$

and

$$\bar{\epsilon}_1\gamma^{\rho\mu\tau}\epsilon_4\,\bar{\epsilon}_2\gamma^{\lambda\nu\tau}\epsilon_3.$$

After antisymmetrization over all $\epsilon_i, i = 1,2,3,4$ these terms add up to give like the result for the graviton four point function

$$\langle h\rangle_4 = \pi\zeta_{\mu\nu}k_\rho k_\lambda\bar{\epsilon}\gamma^{\rho\mu\tau}\epsilon\,\bar{\epsilon}\gamma^{\lambda\nu\tau}\epsilon \tag{A.11}$$

References

1. K. Becker, M. Becker and A. Strominger, Fivebranes, membranes and nonperturbative string theory, hep-th/9507158, *Nucl. Phys.* **B456** (1995) 130.

2. H. Ooguri and C. Vafa, Summing up D instantons, hep-th/9608079.
3. C. Bachas et al., Heterotic / type I duality and D-brane instantons, hep-th/9707126;E. Kiritsis and N. A. Obers, Heterotic type I duality in D $\leq$ 10-dimensions, threshold corrections and D instantons, hep-th/9709058.
4. C. Bachas, *Heterotic versus Type I*, hep-th/9710102;
C. Bachas and P. Vanhove, *The Rules of D-brane Calculus*, CPTH-S587.1297.
5. E. Kiritsis and B. Pioline, On R^4 threshold corrections in IIb string theory and (p,q) string instantons, hep-th/9707018.
6. M.B. Green, Point-like states for type IIB superstrings, hep-th/9403040, *Phys. Lett.* **329B** (1994) 435.
7. M.B. Green, Dynamical pointlike structure and dual strings, *Phys. Lett.* **89B** (1977) 89.
8. A. Cohen, G. Moore, P. Nelson, J. Polchinski, Semi off-shell string amplitudes, *Nucl. Phys.* **B281** (1986) 127.
9. D.J. Gross and P.F. Mende, String theory beyond the Planck scale, *Nucl. Phys.* **B303** (1988) 407.
10. M.B. Green, The influence of worldsheet boundaries on critical closed string theory, *Phys. Lett.* **302B** (1993) 29;
M. Gutperle, Multiboundary effects in Dirichlet string theory, hep-th/9502106, *Nucl. Phys.* **B444** (1995) 487.
11. J. Polchinski, Combinatorics of boundaries in string theory, hep-th/9407031, *Phys. Rev.* **D50** (1994) 6041.
12. M.B. Green, G.W. Gibbons and M.J. Perry, The instanton and seven-brane of type IIB supergravity, hep-th/9511170.
13. M.B. Green and M. Gutperle, Effects of D-instantons, hep-th/9701093, *Nucl. Phys.* **B498** (1997) 195.
14. M.B. Green and P. Vanhove, D-instantons, strings and M-theory, hep-th/9704145, *Phys. Lett.* **408B** (1997) 122.
15. M.B. Green, M. Gutperle and P. Vanhove, One loop in eleven dimensions, hep-th/9706175, *Phys. Lett.* **409B** (1997) 177.
16. M.B. Green, M. Gutperle and H. Kwon, Sixteen fermion and related terms in M theory on T^2, hep-th/9710151.
17. J.H. Schwarz, Covariant field equations of chiral N=2 D=10 supergravity, *Nucl. Phys.* **B226** (1983) 269.
18. P.S. Howe and P.C. West, The complete N=2 D=10 supergravity, *Nucl. Phys.* **B238** (1984) 181.
19. D.J. Gross and E. Witten, Superstring modifications of Einstein's equations *Nucl. Phys.* **B277** (1986) 1.
20. M.B. Green and J.H. Schwarz, Supersymmetric dual string theory III, *Nucl. Phys.* **B198** (1982) 441.
21. I. Antoniadis, B. Pioline and T.R. Taylor, Calculable $e^{1/\lambda}$ effects, hep-th/9707222.
22. N. Berkovits, Construction of R^4 terms in N=2 D=8 superspace, hep-th/9709116.
23. S. Sethi and M. Stern, D-brane bound states redux, hep-th/9705046;
P. Yi, Witten index and threshold bound states of D-branes, *Nucl. Phys.* **B505** (1997) 307, hep-th/9704098.
24. M.B. Green and M. Gutperle, D-particle bound states and the D-instanton measure, hep-th/9711107.
25. A. Kehagias and H. Partouche, The exact quartic effective action for the type IIB superstring, hep-th/9710023.

SOME CLASSICAL SOLUTIONS OF MEMBRANE MATRIX MODEL EQUATIONS

JENS HOPPE
Institute for Theoretical Physics, ETH Hönggerberg, CH-8093 Zürich, Switzerland

Let $X_i(t)$, $i = 1, ..., d$, be time-dependent elements of some (Lie-) algebra $\mathcal{A}$, satisfying (cp. [1, 2, 3, 4])

$$\ddot{X}_i = -\sum_{j=1}^{d} [[X_i, X_j], X_j] \tag{0.1}$$

$$\sum_{i=1}^{d} [X_i, \dot{X}_i] = 0.$$

The Ansatz

$$X_i(t) = x(t) r_{ij}(t) M_j$$

(with $(r_{ij})_{ij=1...d} = e^{\mathbf{A}(t)} \in SO(d)$ a time dependent rotation, $x(t)$ an overall pulsation, $\mathbf{A}(t) = \varphi(t)\mathbf{A}$, $\mathbf{A}^2\vec{M} = -\mu\vec{M}$, $x^2(t)\dot{\varphi}(t) = L = const.$) reduces (0.1) to the equation(s)

$$\sum_{j=1}^{d} [[M_i, M_j], M_j] = \lambda M_i, \tag{0.2}$$

while $x(t)$ is then given as a solution of $\ddot{x} + \lambda x^3 - \frac{\mu L^2}{x^2} = 0$, i.e.

$$\frac{1}{2}\dot{x}^2 + \frac{\lambda}{4}x^4 + \frac{\mu L^2}{2}\frac{1}{x^2} = E \quad ;$$

for $L \neq 0$ (when $L = 0$, (2) automatically holds), (2) may e.g. be satisfied by choosing half (or more) of the components of $\vec{M} = (N_1, N_2, ... N_{\underset{\sim}{d}}, 0, ..., 0)$ to vanish, and $\mathbf{A}_{ij}$ to vanish whenever M_i and M_j don't, such as in

$$X_i = x(t)(cos\varphi(t)\vec{N}, sin\varphi(t)\vec{N}, 0...0). \tag{0.3}$$

L. Baulieu et al. (eds.), Strings, Branes and Dualities, 423–427.

Equation (0.2), which (for $\vec{M}^2 = \mathbf{1}$) may be considered as a discrete, or "quantized", analogue of minimally embedding a compact (2 dimensional) surface into a higher dimensional unit sphere (cp. [2], eq. (30)), has a very rich spectrum of solutions-some of which will now be given:

I. Let $\vec{N}$ be a basis of a representation of some Lie-algebra $\mathcal{G}$, satisfying

$$[N_a, N_b] = if_{abc}N_c, \quad a,b,c = 1,\ldots \underset{\sim}{d}, \qquad f_{abc}f_{cba'} = -\lambda\delta_{aa'} \tag{0.4}$$

For $\mathcal{G}$ = SO(3), X_i (given by (0.3)) may be thought of as a discrete version of a spherical membrane. For higher rank groups, the corresponding solutions X_i will be discrete analogues of higher(!)-dimensional extended objects, rotating and pulsating in space. This has to do with the (not so much noticed) fact, that just as the Poisson-algebra of functions on a two-dimensional surface are related to certain limits of matrix-algebras, there also exist $N \to \infty$ limits of su(N) yielding Poisson-algebras of functions on higher dimensional coadjoint orbits [7], or more generally, higher dimensional Kähler manifold [8].

I. Note that the Ansatz

$$X_a(t) = X(t) \otimes N_a$$

would reduce (0.1) to the (solvable) matrix model equations

$$\ddot{X} + X^3 = 0$$

with (2) being satisfied provided $[X(0), \dot{X}(0)] = 0$.

II. Suppose $U, V \in \mathcal{A}$ satisfy

$$VU = \omega UV$$

with $\omega = e^{4\pi i\Lambda}$, $\Lambda \in \mathbf{R}$;

$$\vec{M} = \frac{1}{\sqrt{2}}\left(\frac{U+U^{-1}}{2}, \frac{U-U^{-1}}{2i}, \frac{V+V^{-1}}{2}, \frac{V-V^{-1}}{2i}, 0\ldots 0\right) \tag{0.5}$$

will then satisfy (0.2) with

$$\lambda = 2sin^2(2\pi\Lambda)$$

-as for the above choice of $\vec{M}$ one has

$$\sum_j [[\cdot, M_j], M_j] = \frac{1}{2}([[\cdot, U], U^{-1}] + [[\cdot, V], V^{-1}]),$$

which is proportional to the "quantum torus Laplace operator"

$$\Delta_\Lambda := \frac{-1}{16\pi^2\Lambda^2}([[\cdot, U], U^{-1}] + [[\cdot, V], V^{-1}]),$$

$$\Delta_\Lambda(U^{m_1}V^{m_2}) = -\frac{sin^2(2\pi\Lambda m_1) + sin^2(2\pi\Lambda m_2)}{4\pi^2\Lambda^2}(U^{m_1}V^{m_2}),$$

whose eigenvalues for the 4 different components of $\vec{M}$ are identical. For rational Λ, $\Lambda = \frac{M}{N}$, U and V may be taken to be finite-dimensional matrices. The $N \to \infty$ limit of the solutions to the corresponding matrix equations (0.1) then either leads to solutions of the continuous membrane equations (cp. [1], [2]), (or choosing $M = M(N)$ such that $\frac{M}{N} \to \Lambda_\infty \neq 0$, see e.g. p. 53 of [5]) to solutions of the "star-product membrane equations" [6].

In the former case, it is known that the analogue of (0.2) has solutions describing minimal embeddings into S^3 of 2-dimensional surfaces of arbitrary genus (cp. [2]). The continuous limit of (0.5) and $(7)_{SO(3)}$, e.g.,

$$\vec{m} = \frac{1}{\sqrt{2}}(cos\varphi_1, sin\varphi_1, cos\varphi_2, sin\varphi_2, 0...0),$$

$$\vec{m} = (sin\theta cos\varphi, sin\theta sin\varphi, cos\theta, 0...0),$$

(note that just as $\vec{m}^2 = 1$, (0.5) also implies $\vec{M}^2 = \mathbf{1}$, and in the case of (0.4) irreducible representations could be normalized to have $\vec{M}^2 = 1$) satisfy

$$-\{\{m_i, m_j\} m_j\} = 2m_i,$$

where $\{\ ,\ \}$ denotes Poisson brackets for functions on T^2, resp. S^2.

III. Let

$$\begin{array}{rcl} M_1 &=& \frac{1}{4}(S_{kl} + S_{-k-l} + S_{-kl} + S_{k-l}), \\ M_2 &=& \frac{-i}{4}(S_{kl} - S_{-k-l} - S_{-kl} + S_{k-l}), \\ M_3 &=& \frac{-i}{4}(S_{kl} - S_{-k-l} + S_{-kl} - S_{k-l}), \\ M_4 &=& \frac{-1}{4}(S_{kl} + S_{-k-l} - S_{-kl} - S_{k-l}), \\ M_{i>4} &=& 0, \end{array} \tag{0.6}$$

, where

$$S_{\vec{m}} := \omega^{\frac{1}{2}m_1 m_2} U^{m_1} V^{m_2},$$

$\vec{m} = (m_1, m_2) \in \mathbf{Z}^2$, implying

$$S_{\vec{m}} S_{\vec{n}} = \omega^{-\frac{1}{2}(m_1 m_2 - m_2 m_1)} S_{\vec{m}+\vec{n}},$$

$$S_{\vec{n}} S_{\vec{m}} S_{-\vec{n}} = \omega^{(\vec{m} \times \vec{n})} S_{\vec{m}}.$$

Using

$$\sum_{i=1}^{4} [[X, M_j], M_j] = \frac{1}{2}\big([[X, S_{kl}], S_{-k-l}] + [[X, S_{-kl}], S_{k-l}]\big).$$

and

$$[[S_{\vec{m}}, S_{\vec{n}}], S_{-\vec{n}}] = 4sin^2\big(2\pi\Lambda(\vec{m} \times \vec{n})\big) S_{\vec{m}}$$

one finds that (0.6) satisfies (0.2), with

$$\lambda = 2sin^2\big(2\pi\Lambda(2kl)\big)$$

and $\vec{M}^2 = S_{00} \equiv \mathbf{1}$. (Note that, as long as Λ is left arbitrary, the kl degree of freedom is somewhat redundant, and without loss of generality one could have restricted to $k = 1 = l$, as was done in II). Also note that the continuous ($\Lambda \to 0$) limit of (0.6) gives minimal embedding into S^3 of two-dimensional surfaces with self-intersections, described by

$$\vec{m} = (cosk\varphi_1 cosl\varphi_2, sink\varphi_1 cosl\varphi_2, cosk\varphi_1 sinl\varphi_2, sink\varphi_1 sinl\varphi_2, 0, ...0).$$

Finally, I would like to add a class of solutions of slightly different type:

$$\begin{aligned} \vec{X} &= \textstyle\sum_\alpha r_\alpha(t)\vec{E}_\alpha, \\ \vec{E}_1 &= \tfrac{1}{2}(S_{\vec{m}_1} + S_{-\vec{m}_1}, -i(S_{\vec{m}_1} - S_{-\vec{m}_1}), 0, ..., 0), \\ \vec{E}_2 &= \tfrac{1}{2}(0, 0, S_{\vec{m}_2} + S_{-\vec{m}_2}, -i(S_{\vec{m}_2} - S_{-\vec{m}_2}), 0, ..., 0), ... \end{aligned}$$

$$\ddot{r}_\alpha = -4r_\alpha \sum_\beta sin^2(2\pi\Lambda(\vec{m}_\alpha \times \vec{m}_\beta))r_\beta^2;$$

note that $\vec{E}_\alpha^2 = \mathbf{1}$, and

$$[[Z, \vec{E}_\alpha], \vec{E}_\alpha] = [[Z, S_{\vec{m}_\alpha}], S_{-\vec{m}_\alpha}],$$

which equals $4sin^2(2\pi\Lambda(\vec{m}_\alpha \times \vec{m}_\beta))Z$ for Z being any component of $\vec{E}_\beta$. Their continuum analogue,

$$\begin{aligned} \vec{x}(t, \varphi^1, \varphi^2) &= \textstyle\sum_\alpha r_\alpha(t)\vec{e}_\alpha, \\ \vec{e}_1 &= (cos(\vec{m}_1\vec{\varphi}), sin(\vec{m}_1\vec{\varphi}), 0, ..., 0), \\ \vec{e}_2 &= (0, 0, cos(\vec{m}_2\vec{\varphi}), sin(\vec{m}_2\vec{\varphi}), 0, ..., 0), ... \end{aligned}$$

$$\ddot{r}_\alpha = -r_\alpha \sum_\beta \frac{(\vec{m}_\alpha \times \vec{m}_\beta)^2}{\rho^2} r_\beta^2, \quad \rho = const.,$$

describes a membrane embedded into a sphere of timedependent radius, $\sum_\beta r_\beta^2$, in such a way that the 3-dimensional manifold swept out in space-time is extremal;

$$\dot{\vec{x}}\partial_1\vec{x} = 0 = \dot{\vec{x}}\partial_2\vec{x}$$

,

$$g = det(\partial_r\vec{x}\partial_s\vec{x}) = \sum_{\alpha<\beta} r_\alpha^2 r_\beta^2 (\vec{m}_\alpha \times \vec{m}_\beta)^2,$$

$$\dot{\vec{x}}^2 + g/\rho^2 = 1,$$

$$\ddot{\vec{x}} = \{\{\vec{x}, x_j\}x_j\} = \frac{1}{\rho}\partial_r(\frac{gg^{rs}}{\rho}\partial_s\vec{x}) = g/\rho^2 \frac{1}{\sqrt{g}}\partial_r(\sqrt{g}g^{rs}\partial_s\vec{x}).$$

References

1. J. Hoppe; "Quantum Theory of a Relativistic Surface", in: "Constraint's Theory and Relativistic Dynamics" (eds. G. Longhi, L. Lusanna) Arcetri, Florence 1986, World Scientific.
2. J. Hoppe, H. Nicolai; Phys. Lett. B 196(1987)451.
3. B. de Wit, J. Hoppe, H. Nicolai; Nucl. Phys. B305 [FS23](1988)545.
4. T. Banks, W. Fischler, S.H. Shenker, L. Susskind; "M Theory as a Matrix Model: a Conjecture", hep-th/9610043.
5. J. Hoppe, "Lectures On Integrable Systems", Springer Lecture Notes on Physics, 1992.
6. J. Hoppe, Phys. Lett. B250(1990)44.
7. J. Hoppe, Int. J. Mod. Phys. A4(1989)5235; hep-th/9612024.
8. M. Bordemann, J. Hoppe, P. Schaller, and M. Schlichenmaier, Comm. Math. Phys. 138(1991)209; M. Bordemann, E. Meinrenken, and M. Schlichenmaier, Comm. Math. Phys. 165(1994)281.

Part III

Gong Show

D-BRANE ACTIONS, INTRINSIC GEOMETRY AND DUALITY

M. ABOU ZEID
Physics Department, Queen Mary and Westfield College,
Mile End Road, London E1 4NS, U. K.

1. Supersymmetric D-p-Brane Actions

The effective action for superstring theory in the presence of a D-p-brane is the sum $S = S_D + S_{(p)}$ of the bulk supergravity action S_D and the effective world-volume action $S_{(p)}$. The latter is the sum of a Dirac-Born-Infeld type action [1, 2] and a Wess-Zumino term [3, 4], and it was shown in refs. [5, 6, 7, 8], to be invariant under a local fermionic symmetry which reflects the fact that the D-brane is BPS-satured and breaks half the space-time supersymmetries [9].

We begin with a brief review of the usual form of $S_{(p)}$. The (flat) superspace coordinates are the $D = 10$ space-time coordinates X^i and the Grassmann coordinates θ, which are space-time spinors and world-volume scalars. For the IIa superstring (even p), θ is Majorana but not Weyl while in the IIb superstring there are two Majorana-Weyl spinors θ_α $(\alpha = 1, 2)$ of the same chirality. The superspace (global) supersymmetry transformations are

$$\delta_\epsilon \theta = \epsilon \quad , \quad \delta_\epsilon X^i = \bar{\epsilon}\Gamma^i\theta. \tag{1.1}$$

The world-volume theory has global type IIa or type IIb super-Poincaré symmetry and is constructed using the supersymmetric one-forms $\partial_\mu\theta$ and

$$\Pi^i_\mu = \partial_\mu X^i - \bar{\theta}\Gamma^i\partial_\mu\theta. \tag{1.2}$$

The induced world-volume metric is

$$G_{\mu\nu} = G_{ij}\Pi^i_\mu\Pi^j_\nu. \tag{1.3}$$

The supersymmetric world-volume gauge field-strength $\mathcal{F}$ is defined by $\mathcal{F} = F - B$, and a convenient choice for the two-form B is [10]

$$B = -\bar{\theta}\Gamma_{11}\Gamma_i d\theta(dX^i + \frac{1}{2}\bar{\theta}\Gamma^i d\theta) \tag{1.4}$$

when p is even or the same formula with Γ_{11} replaced with the Pauli matrix τ_3 when p is odd. With the choice (1.4), $\delta_\epsilon B$ is an exact two-form and $\mathcal{F}$ is supersymmetric for an appropriate choice of $\delta_\epsilon A$ [10, 6].

L. Baulieu et al. (eds.), Strings, Branes and Dualities, 431–436.

The leading contribution to the effective supersymmetric D-p-brane action with constant dilaton and zero cosmological constant then takes the form

$$S_{(p)} = -T_p \int d^n\sigma e^{-\phi}\sqrt{-\det(G_{\mu\nu} + \mathcal{F}_{\mu\nu})} + T_p \int_{W_{p+1}} Ce^{\mathcal{F}} \tag{1.5}$$

The constant T_p is the p-volume tension with mass dimension $p+1$, and W_{p+1} denotes the $(p+1)$-dimensional worldvolume of the brane. Here C represents the complex of differential forms $C = \sum_{r=0}^{9} C^{(r)}$ where the $C^{(r)}$ are the pull-backs of superspace forms $C^{(r)} = d\bar{\theta}T^{(r-2)}d\theta$ for certain $r-2$ forms $T^{(r-2)}$ given explicitly in [7, 8, 5, 6], and it is understood that the $p+1$ form part of $Ce^{\mathcal{F}}$, which is $C^{(p+1)} + C^{(n-2)}\mathcal{F} + \frac{1}{2}C^{(n-4)}\mathcal{F}^2 + \ldots$, is selected. The bosonic part of the Wess-Zumino term gives the coupling of the brane to the background Ramond-Ramond r-form gauge fields (where r is odd for type IIa and even for type IIb). The case of the 9-form potential is special because its field equation forces the dual of its field strength to be a constant m. The backgrounds with $m = 0$ are the familiar type IIa ones, while backgrounds with $m \neq 0$ are those of massive type IIa supergravity [11]. In the $m \neq 0$ case the Wess-Zumino term given in (1.5) requires m-dependent Chern-Simons modifications, and the bosonic part of these were derived in [12]. The constant m will be taken to be zero here, so that the 9-form potential vanishes; the generalisation of the discussion below to the more general situation will be given elsewhere.

The action (1.5) is supersymmetric and invariant under local kappa symmetry [7, 8, 5, 6]. However, just as in the case of the usual Nambu-Goto action for a p-brane, the non-linearity of (1.5) is inconvenient for many purposes. In particular, dualising the action has proved difficult in this approach, and has only been achieved for $p < 5$ [10, 13, 14, 15, 16, 17]. It is therefore useful to generalise the simplification of the Nambu-Goto action using an auxiliary world-volume metric $g_{\mu\nu}$ to the case of actions involving world-volume gauge fields in addition to the induced metric $G_{\mu\nu}$. This was achieved in ref. [18] by introducing an auxiliary tensor field

$$k_{\mu\nu} \equiv g_{\mu\nu} + b_{\mu\nu} \tag{1.6}$$

with both a symmetric part $g_{\mu\nu}$ and an antisymmetric part $b_{\mu\nu}$. The supersymmetric D-brane action which is classically equivalent to (1.5) is

$$S = -\frac{1}{2}T'_p \int d^n\sigma e^{-\phi}\sqrt{-k}\,[k^{\mu\nu}(G_{\mu\nu} + \mathcal{F}_{\mu\nu}) - (p-1)\Lambda] + T_p \int_{W_n} Ce^{\mathcal{F}}, \tag{1.7}$$

where $k \equiv \det k_{\mu\nu}$ and Λ is a constant. The inverse tensor $k^{\mu\nu}$ satisfies

$$k^{\mu\nu}k_{\nu\rho} = \delta^{\mu}_{\rho}. \tag{1.8}$$

For $p \neq 1$, the $k_{\mu\nu}$ field equation implies

$$G_{\mu\nu} + \mathcal{F}_{\mu\nu} = \Lambda k_{\nu\mu} \tag{1.9}$$

and substituting back into (1.7) yields the supersymmetric D-brane action in the DBI form (1.5) , with the constants T_p ,T'_p related by

$$T'_p = \Lambda^{\frac{1}{2}(p-1)}T_p. \tag{1.10}$$

For $p = 1$, the action(1.7) is invariant under the generalised Weyl transformation

$$k_{\mu\nu} \to \omega(\sigma)k_{\mu\nu} \tag{1.11}$$

and the $k_{\mu\nu}$ field equation implies

$$G_{\mu\nu} + \mathcal{F}_{\mu\nu} = \frac{1}{2}k^{\rho\sigma}(G_{\rho\sigma} + \mathcal{F}_{\rho\sigma})k_{\nu\mu}. \tag{1.12}$$

The first term in the action (1.7) is *linear* in $(G_{\mu\nu}+\mathcal{F}_{\mu\nu})$ and *quadratic* in ∂X, so it is easier to analyse than (1.5). Moreover, being classically equivalent to (1.5), the action (1.7) is supersymmetric and invariant under local kappa symmetry; this can also be checked explicitly.

2. Dual Super D-p-Brane Actions

The dualisation of the action (1.7) can be achieved by adding a Lagrange term $\frac{1}{2}\tilde{H}^{\mu\nu}(F_{\mu\nu} - 2\partial_{[\mu}A_{\nu]})$ imposing the constraint $F = dA$. Here the anti-symmetric tensor density $\tilde{H}^{\mu\nu}$ is a Lagrange multiplier which can be integrated out to regain the original action (1.7). Alternatively, integrating out over A_μ imposes the constraint

$$\partial_\mu \tilde{H}^{\mu\nu} = 0, \tag{2.1}$$

which can be solved in terms of an $p-2$ form $\tilde{A}$:

$$\tilde{H}^{\mu\nu} = \frac{1}{(p-1)!}\epsilon^{\mu\nu\rho\gamma_1\dots\gamma_{p-2}}\partial_{[\rho}\tilde{A}_{\gamma_1\dots\gamma_{p-2}]}, \tag{2.2}$$

where $\epsilon^{\mu\nu\rho\dots}$ is the alternating tensor density. Now F is an auxiliary 2-form occuring algebraically; we emphasize this by rewriting $F \to L$, so that the action is

$$-\frac{1}{2}T'_p \int d^{p+1}\sigma \left\{ \sqrt{-k}[k^{\mu\nu}(N_{\mu\nu} + L_{\mu\nu}) - (p-1)\Lambda] + \frac{1}{2}\tilde{H}^{\mu\nu}L_{\mu\nu} \right\}$$
$$+T_p \int_{W_{p+1}} Ce^{L-B}. \tag{2.3}$$

The field equation for L is then

$$\sqrt{-k}k^{[\mu\nu]} + \frac{1}{2}\tilde{H}^{\mu\nu} + \frac{\delta f(L)}{\delta L_{\mu\nu}} = 0, \tag{2.4}$$

where the potential $f(L) = Ce^{L-B}$ is a polynomial of order $[\frac{1}{2}(p+1)]$ in L (i. e. the integer part of $\frac{1}{2}(p+1)$), so that the field equation (2.4) is of order $[\frac{1}{2}(p+1)-1]$, which is at most quartic for $p < 9$. In particular, it is quadratic for $p < 5$, so that the dual supersymmetric actions for the corresponding D-branes should be obtainable straigthforwardly; this will be illustrated in an example below, and will be discussed further elsewhere. Here we will consider only the cases in which the action (1.7) is linear in F.

3. The D-Membrane

The dualisation of (the bosonic part of) the D-string action starting from the form (1.7) was discussed in ref. [18]; here we consider the case of the D-membrane of type IIa superstring theory. For $p = 2$, (1.7) reduces to the following form of the D-2-brane action in the string metric,

$$S = -\frac{1}{2}T_2' \int d^3\sigma e^{-\phi}\sqrt{-k}\left[k^{\mu\nu}(G_{\mu\nu} + \mathcal{F}_{\mu\nu}) - \Lambda\right] + T_2 \int_{W_3} C^{(3)} + C^{(1)}\mathcal{F} \quad (3.1)$$

with $T_2' = \sqrt{\Lambda}T_2$. Adding a Lagrange multiplier term $\frac{1}{2}\tilde{H}^{\mu\nu}(F_{\mu\nu} - 2\partial_{[\mu}A_{\nu]})$ to this action and integrating out the gauge field A gives the constraint (2.1), which is solved in $p+1 = 3$ dimensions by

$$\tilde{H}^{\mu\nu} = \epsilon^{\mu\nu\rho}\partial_\rho S \quad (3.2)$$

for some scalar field S. Substituting this back in the action and using a density $\tilde{k}^{\mu\nu} = \sqrt{-k}k^{\mu\nu}$ with $\tilde{k} \equiv \det\tilde{k}^{\mu\nu} = -\sqrt{-k}$ gives

$$\begin{aligned} S &= -\frac{\sqrt{\Lambda}}{2}T_2 \int d^3\sigma \left\{e^{-\phi}\left[\tilde{k}^{\mu\nu}(N_{\mu\nu} + L_{\mu\nu}) + \Lambda\tilde{k}\right] + \epsilon^{\mu\nu\rho}K_\rho L_{\mu\nu}\right\} \\ &\quad + T_2 \int_{W_3} C^{(3)} - C^{(1)}B, \end{aligned} \quad (3.3)$$

where L is an auxiliary two-form and we have defined

$$K_\rho \equiv \frac{1}{2}\partial_\rho S - \frac{1}{\sqrt{\Lambda}}C^{(1)}_\rho. \quad (3.4)$$

Integrating out L yields the constraint

$$e^{-\phi}\tilde{k}^{[\mu\nu]} + \epsilon^{\mu\nu\rho}K_\rho = 0, \quad (3.5)$$

so that

$$\tilde{k}^{\mu\nu} = \tilde{g}^{\mu\nu} + \tilde{\mathcal{H}}^{\mu\nu} \quad (3.6)$$

where $\tilde{\mathcal{H}}^{\mu\nu} = -\epsilon^{\mu\nu\rho}e^{\phi}K_\rho$, and $\tilde{g}^{\mu\nu}$ is the symmetric tensor density defined by $\tilde{g}^{\mu\nu} = \tilde{k}^{(\mu\nu)}$. Then $\det(\tilde{k}^{\mu\nu}) = \det(\tilde{g}^{\mu\nu})\Omega$, where $\Omega = \det(\delta^\nu{}_\rho + \tilde{\mathcal{H}}^\nu{}_\mu)$. Using the identity $\det(1+X) = 1 - \frac{1}{2}trX^2$ valid for $n = 3$ and X antisymmetric, we find

$$\begin{aligned} \Omega &= 1 - \frac{1}{2}\tilde{\mathcal{H}}^\nu{}_\mu\tilde{\mathcal{H}}_\nu{}^\mu = 1 - \frac{1}{2}e^{2\phi}\tilde{g}_{\mu\rho}\tilde{g}_{\nu\sigma}\epsilon^{\rho\nu\kappa}\epsilon^{\sigma\mu\delta}K_\kappa K_\delta \\ &= 1 + \frac{1}{2}e^{2\phi}g^{\mu\nu}K_\mu K_\nu, \end{aligned} \quad (3.7)$$

where $g^{\mu\nu} = \frac{1}{\sqrt{-g}}\tilde{g}^{\mu\nu}$ is a symmetric tensor with inverse $g_{\mu\nu}$, and $g \equiv \det(g_{\mu\nu})$. Using these results, the first term in (3.3) becomes

$$S = -\frac{\sqrt{\Lambda}}{2}T_2 \int d^3\sigma e^{-\phi}\sqrt{-g}\left[g^{\mu\nu}G_{\mu\nu} - \Lambda\Omega\right] + \frac{\sqrt{\Lambda}}{2}T_2 \int_{W_3} B \wedge K. \quad (3.8)$$

The field equation for the metric $g_{\mu\nu}$ implies

$$g_{\mu\nu} = \frac{1}{\Lambda}\left(G_{\mu\nu} - \frac{\Lambda}{2}e^{2\phi}K_\mu K_\nu\right). \tag{3.9}$$

Substituting this back in (3.8) and adding the second term in (3.3) yields the dual action

$$\begin{aligned} S &= -T_2\int d^3\sigma e^{-\phi}\sqrt{-\det\left[G_{\mu\nu} - \frac{\Lambda}{8}e^{2\phi}(\partial_\mu S - \frac{2}{\sqrt{\Lambda}}C^{(1)}_\mu)(\partial_\nu S - \frac{2}{\sqrt{\Lambda}}C^{(1)}_\nu)\right]} \\ &\quad +T_2\int_{W_3}\left(-C^{(3)} + \frac{\sqrt{\Lambda}}{2}BdS\right) \end{aligned} \tag{3.10}$$

which agrees with that of [10, 13, 15, 16, 17] if we set $\Lambda = 4$. The dilaton dependence can be removed by a rescaling of the superspace coordinates, and this yields the standard M theory membrane action with a circular dimension, as expected on the basis of the relationship between type IIa superstring theory and M theory.

Acknowledgements

I would like to thank Christopher Hull for collaboration and suggestions, and Vijay Balasubramanian, Emil Martinec, Dimitri Sorokin and Paul Townsend for helpful discussions. I am also grateful to the organisers of the school for their support.

References

1. Leigh, R. G. (1989) Dirac-Born-Infeld Action from Dirichlet Sigma Model, *Mod. Phys. Lett.* **A4**, p. 2767.
2. Fradkin, E. S. and Tseytlin, A. A. (1985), *Phys. Lett.* **B163**, p. 123.
3. Li, M. (1996), *Nucl. Phys.*, **B460**, p. 351, [hep-th/9510161].
4. Douglas, M. (1995), Branes Within Branes, [hep-th/9512077].
5. Aganagic, M., Popescu, C. and Schwarz, J. H. (1997) D-Brane Actions with Local Kappa Symmetry, *Phys. Lett.*, **B393**, p. 311, [hep-th/9610249].
6. Aganagic, M., Popescu, C. and Schwarz, J. H. (1997) Gauge-Invariant and Gauge-Fixed D-Brane Actions, *Nucl. Phys.*, **B495**, p. 99, [hep-th/9612080].
7. Cederwall, M., von Gussich, A., Nilsson, B. E. W. and Westerberg, A. (1997), The Dirichlet Super-Three-Brane in Ten-Dimensional Type IIB Supergravity, *Nucl. Phys.*, **B490**, p. 163, [hep-th/9610148]; Cederwall, M., von Gussich, A., Nilsson, B. E. W., Sundell, P. and Westerberg, A. (1997), The Dirichlet Super-p-Branes in Ten-Dimensional Type IIa and IIb Supergravity, *Nucl. Phys.*, **B490**, p. 197, [hep-th/9611159]; Cederwall, M . (1996), Aspects of D-Brane Actions, [hep-th/9612153].
8. Bergshoeff, E. and Townsend, P. K. (1997) Super D-Branes, *Nucl. Phys.*, **B490**, p. 145, [hep-th/9611173].
9. Polchinski, J. (1995) Dirichlet-Branes and Ramond-Ramond Charges, *Phys. Rev. Lett.*, **75**, p. 4724, [hep-th/9510017].
10. Townsend, P. K. (1996), D-Branes from M-Branes, *Phys. Lett.*, **B373**, p. 68, [hep-th/9512062].
11. Romans, L. J. (1986) Massive $N = 2A$ Supergravity in Ten Dimensions, *Phys. Lett.*, **B169**, p. 374.

12. Green, M. B. Hull, C. M. and Townsend, P. K. (1996) D-Brane Wess-Zumino Actions, T-Duality and the Cosmological Constant, *Phys. Lett.*, **B382**, p. 65, [hep-th/9604119].
13. Schmidhuber, C. (1996), D-Brane Actions, *Nucl. Phys.*, **B467**, p. 146, [hep-th/9601003].
14. de Alwis, S. P. and Sato, K. (1996), D-Strings and F-Strings from String Loops, *Phys. Rev.*, **D53**, p. 7187, [hep-th/9601167].
15. Tseytlin, A. A. (1996), Self-Duality of Born-Infeld Action and Dirichlet 3-Brane of Type IIB Superstring Theory, *Nucl. Phys.*, **B469**, p. 51, [hep-th/9602064].
16. Lozano, Y. (1997) D-Brane Dualities as Canonical Transformations, *Phys. Lett.*, **B399**, p. 233.
17. Aganagic, M., Park, J., Popescu, C. and Schwarz, J. H. (1997) Dual D-Brane Actions, *Nucl. Phys.*, **B496**, p. 191, [hep-th/9702133].
18. Abou Zeid, M. and Hull, C. M. (1997) Intrinsic Geometry of D-Branes, *Phys. Lett.*, **B404**, p. 264, [hep-th/9704021].

BRANES AND CHIRAL SYMMETRY

J.H. BRODIE
Princeton University
Princeton, NJ 08540, USA

1. Introduction

D-branes of type II string theory have provided a very picturesque and intuitive way of studying supersymmetric gauge theories. In a paper by Hanany and Witten [1], it was discovered that when a D5-brane passes through an orthogonally oriented NS 5-brane, a D3-brane is created. This transition was subsequently verified in a number of papers [2]. In [3], Seiberg's duality in $N = 1$ supersymmetric QCD [4] was found to arise naturally in the framework of D-branes using the Hanany-Witten transition. A prominent feature of $N = 1$ supersymmetric QCD and non-supersymmetric QCD is chiral symmetry. Here we review how chiral symmetry can be realized in the D-brane construction. This review is based on the papers [5], [7].

2. The brane construction

In the type IIA string theory brane construction we will be considering, we have a NS 5-brane extending in the directions $(x^0, x^1, x^2, x^3, x^4, x^5)$, another NS' 5-brane extending in the directions $(x^0, x^1, x^2, x^3, x^8, x^9)$ N_f D6-branes extending in the directions $(x^0, x^1, x^2, x^3, x^7, x^8, x^9)$, and N_c D4-branes extending in the directions $(x^0, x^1, x^2, x^3, x^6)$. The D4-branes stretch between the NS and the NS' 5-brane in the x^6 direction. The four-dimensional gauge theory lives on the intersection of the branes in the directions (x^0, x^1, x^2, x^3). One can check that such a brane configuration preserves 1/8 of the supersymmetry of type IIA string theory. A $U(N_c)$ non-Abelian gauge symmetry results from the relatively light D4-branes which are forced to be on top of each other by the 5-branes. In [8], it was shown that when one takes into account the bending of the 5-branes the $U(N_c)$ is actually reduced to $SU(N_c)$ in order to maintain finite energy solutions of the 5-branes. The gauge coupling of the non-Abelian gauge fields is related to the distance between the NS and NS' 5-brane in the x^6 direction, $\frac{1}{g^2} = \frac{x^6 M_s}{g_s}$ where M_s is the string scale and g_s is the string coupling. The N_f D6-brane when all on top of each other in the (x^4, x^5, x^6) directions give a global $SU(N_f)$ flavor symmetry group. Notice

L. Baulieu et al. (eds.), Strings, Branes and Dualities, 437–440.

that this is a problem. $N = 1$ supersymmetric QCD should have chiral symmetry $SU(N_f)_R \times SU(N_f)_L$ where the right-handed quarks can be rotated independently of the left-handed quarks. An $SU(N_f)$ symmetry indicates that the chiral symmetry has been broken to the diagonal subgroup. It would be unfortunate if our $N = 1$ SQCD did not have this important feature of real QCD. The solution, as was first suggested in [5], is that the D6-branes can intersect the NS' 5-brane in the x^7 direction. When this happens, the D6-branes split into two halves: one halve ending on the right side of the NS' 5-brane and the other halve ending on the left side of the NS' 5-brane in the x^7 direction. Strings extending from the right D6-branes to the D4-branes contribute N_f "right handed" *chiral* multiplets, Q_i while the strings extending from the left set of D6-branes contribute N_f "left handed" chiral multiplets, $\tilde{Q}_i$. When the D6-branes do not intersect the NS' 5-brane, then each string contributes a *hyper-multiplet*, which consists of two $N = 1$ chiral multiplets, to the low energy spectrum. When the D6-branes are away from the NS' 5-brane, it looks locally in the string theory like 1/4 supersymmetry is preserved at the point where the D6-branes intersect the D4-branes. In contrast, at the point where the D4-branes, D6-branes and NS' 5-brane all intersect, 1/8 supersymmetry is preserved locally. Therefore, each string state contributes a chiral multiplet rather than a hyper multiplet.

Let's see if these ideas match with known mechanisms of breaking chrial symmetry. In $N = 2$ SQCD the coupling in the superpotential $W = \lambda \tilde{Q} \phi Q$ breaks chiral symmetry where ϕ is the adjoint field of the $N = 2$ vector multiplet. In order to construct $N = 2$ SQCD using branes we must rotate the NS' 5- brane from the x^8, x^9 direction to the x^4, x^5 direction. Now we notice that it is impossible to have chiral symmetry in the branes by the mechanism discussed above because there is no longer an NS' 5-brane on which the D6-brane can break. This is indeed consistent with field theory. In fact, the parameter λ in the $N = 2$ superpotential we can identify with the direction of the D6-branes in the x^4, x^5, x^8, x^9 plane. $\lambda = \sqrt{2}$ when the D6-branes point purely in the x^8, x^9 direction and $\lambda = 0$ when they point in the x^4, x^5 direction. There should be a restoration of chiral symmetry when the D6-branes extend in the $x^0, x^1, x^2, x^3, x^4, x^5, x^7$ direction, and this agrees with the chiral symmetry mechanism proposed above: When the D6-branes point in the x^4, x^5 direction with two NS 5-branes, then 1/8 supersymmetry is broken. Furthermore, it is now possible for the D6-branes to break on the NS 5-branes in the x^7 direction producing chiral symmetry. Although this theory, $SU(N_c)$ gauge theory with N_f fundamentals and anti-fundamentals and an adjoint chiral superfield, is not $N = 1$ SQCD, it should still have chiral symmetry. This is in agreement with our proposal for chiral symmetry in the brane constructions.

Another limit we can check is perturbing the $N = 1$ SQCD by a mass term $W = mQ\tilde{Q}$. Such a term breaks chiral symmetry. We can understand this in the brane construction because motion of the D6-branes in the x^4, x^5 direction corresponds to giving the quark fields a mass. This is clear since strings joining the D6-branes to the D4-branes become very tense. Since the NS' 5- brane is a point in the directions x^4, x^5, once the D6-branes move in those directions they must move off the NS' 5-brane, restoring chiral symmetry.

One should take note that a single D6-branes ending on a NS' 5-brane is not a consistent configurations: the flux has no where to go. There are no fields on the worldvolume of the NS' 5-brane which can couple to the charge carried by the D6-brane. However, it is possible to have an D6-brane end on a NS' 5-brane provided that there is another D6-brane on the other side to take away the flux. This is in some sense similar to the situation with D0-branes. D0-branes must have two strings ending on them for consistency [10]. However, it was shown in [7] that by including D8 branes, one can have a single D6 brane ending on a NS 5-brane.

3. Two-dimensions and the $\mathbf{CP^{n-1}}$ model.

The mechanism for generating chiral symmetry in four dimensional gauge theories from brane constructions can be used to make chiral theories in two-dimensions. Two get to two-dimensions form four, we simply T-dualize along the directions x^2 and x^3. Here we can have D4-branes ending on the NS' 5-brane. There is no problem with charge conservation since from the M-theory prospective the D4-brane is simply part of the NS' 5-brane. We can therefore construct chiral models in two-dimensions: Strings connecting a semi-infinite D4-branes to a D2-brane contributes a chiral multiplet. Therefore, we can construct the $\mathbf{CP^{n-1}}$ model, $U(1)$ with n right-handed chiral multiplets, Q_i and no left handed chiral multiplets. Notice that the global $SU(n)$ symmetry can alternatively be viewed as $N = 1$ super Yang-Mills. In fact, such a correspondence between the $\mathbf{CP^{n-1}}$ model and N=1 SYM was noticed long ago (see for example [11]). Here we see in the branes a manifest connection between the $\mathbf{CP^{n-1}}$ model and $SU(n)$ $N = 1$ SYM.

Acknowledgements

I would like to thank A. Hanany for collaboration and A. Hashimoto, K. Intiligator, R. Plesser, S. Ramgoolam, and N. Seiberg for valuable discussions. This work was supported by an NSF-ASI grant number 961017 and DOE grant number DE-FG02-91ER-40671.

References

1. Hanany, A. and Witten, E. (1996) Type IIB Superstrings, BPS Monopoles, and Three-Dimensional Gauge Dynamics, *Nuclear Physics*, **B492** pp.152-190.
2. Danielsson, U.H., Ferretti, G., and Klebanov, I.R. (1997) Creation of Fundamental Strings by Crossing D-branes, *Phys.Rev.Lett.* **79** pp. 1984-1987. Bachas, C., Douglas, M.R., and Green, M.B. (1997) Anomalous Creation of Branes, hep-th/9705074.
3. Elitzur, S., Giveon, A., and Kutasov, D. (1997) Branes and N=1 Duality in String Theory, *Physics Letters* **B400** pp. 269-274.
4. Seiberg, N. (1994) Electric-Magnetic Duality in Supersymmetric Non-Abelian Gauge Theories, *Nuclear Physics*, **B435** pp.129-146.
5. Brodie, J. and Hanany, A. (1997) Type IIA Superstrings, Chiral Symmetry, and N=1 4D Gauge Theory Dualities, *Nuclear Physics*, **B**, in press, hep-th/9704043.
6. Hanany, A. and Hori, K. (1997) Branes and N=2 Theories in Two Dimensions, hep-th/9707192.

7. Hanany, A. and Zaffaroni, A. (1997) Chiral Symmetry from Type IIA Branes, hep-th/9706047.
8. Witten, E. (1997) Solutions Of Four-Dimensional Field Theories Via M Theory, *Nuclear Physics*, **B500** pp.3-42.
9. Aharony, O. and Hanany, A. (1997) Branes, Superpotentials and Superconformal Fixed Points, *Nuclear Physics*, **B504** pp.239-271.
10. Strominger, A. (1995) Open P-Branes, *Physics Letters* **B383** pp. 44-47.
11. Coleman, S. (1993) 1/N , *Aspects of Symmetry*, Cambridge University Press, Newcastle upon Tyne.

REGULARISATION OF CLASSICAL SELF INTERACTION IN STRINGS

B. CARTER
Département d'Astrophysique Relativiste et de Cosmologie, C.N.R.S., Observatoire de Paris, 92195 Meudon, France.

Concentration on supersymmetric quantum string theory has diverted attention from basic problems that remain to be dealt with in classical string theory. However even while laying the foundations of quantum electrodynamics, Dirac found time to obtain a regularised Lorentz covariant treatment of a classical electromagnetically self interacting point particle. Following this example, my purpose here is to treat the analogous problem of self interaction in classical relativistic string models.

I shall consider only the kinds of long range interaction that are experimentally familiar, namely electromagnetic interactions as governed by Maxwell's equations for A_μ, and weak gravitational interactions as governed by the linearised Einstein equations for a small amplitude but perhaps rapidly varying perturbation $h_{\mu\nu} = \delta g_{\mu\nu}$ of a slowly varing spacetime metric $g_{\mu\nu}$ characterising a 4-dimensional background with local coordinates x^μ. In a suitable gauge, the field equations reduce to the standard forms

$$\nabla_\sigma \nabla^\sigma A^\mu = -4\pi \hat{j}^\mu , \tag{0.1}$$

and

$$\nabla_\sigma \nabla^\sigma h^{\mu\nu} = -8\pi \mathrm{G} \big(2\hat{T}^{\mu\nu} - \hat{T}_\sigma^{\ \sigma} g^{\mu\nu} \big) , \tag{0.2}$$

in which $\hat{j}^\mu$ is the electromagnetic current density, $\hat{T}^{\mu\nu}$ is the stress momentum energy density tensor, and G is Newton's constant.

The problem of ultraviolet divergences for point particle or string models arises because, in these cases, the relevant source densities $\hat{j}^\mu$ and $\hat{T}^{\mu\nu}$ are not regular functions: they will be Dirac type distributions that vanish outside the relevant one or two dimensional worldsheets. In the case of a string with local worldsheet coordinates σ^a $(a = 0, 1)$ and induced metric $\gamma_{ab} = g_{\mu\nu} x^\mu{}_{,a} x^\mu{}_{,b}$, the latter will be expressible using the terminology of Dirac delta "functions" in the form

$$\hat{T}^{\mu\nu} = \|g\|^{-1/2} \int \overline{T}^{\mu\nu} \, \delta^4[x - x\{\sigma\}] \, \|\gamma\|^{1/2} \, d^2\sigma , \tag{0.3}$$

and there will be a similar relation between $\hat{j}^\mu$ and $\overline{j}^\mu$, where the surface stress momentum energy density $\overline{T}^{\mu\nu}$, and the surface current $\overline{j}^\mu$ are *regular* tensorial

L. Baulieu et al. (eds.), Strings, Branes and Dualities, 441–444.

functions on the worldsheet (but undefined off it). The string case is more awkward than that of a point particle, since is beset by infrared as well as ultraviolet divergences. This complication has so far prevented the construction of a satisfactorily Lorentz covariant string analogue of Dirac's (finite) radiation reaction formula for the point particle case.

What *can* be done for the string case is the analogue of the familiar treatment of the dominant (lowest differential order) contribution, which is divergent, and needs to be regularised by a cut-off procedure, following which – in the point particle case – it turns out to be absorbable by a simple mass renormalisation. It will be seen that an analogous, but not so simple, renormalisation is also possible in the string case. The basic force balance equation will be expressible in the form

$$\overline{\nabla}_\nu \overline{T}^{\mu\nu} = f_{\rm e}{}^{\mu} + f_{\rm g}{}^{\mu} , \tag{0.4}$$

in which the relevant tangentially projected differentiation operator is defined by $\overline{\nabla}_\nu = \eta_\nu{}^\mu \nabla_\mu$ where $\eta^{\mu\nu} = \gamma^{ab} x^\mu{}_{,a} x^\nu{}_{,b}$ is the fundamental tensor of the worldsheet, while the electromagnetic force density contribution will be given by the familiar formula

$$f_{\rm e}{}^{\mu} = F^\mu{}_\nu \overline{j}^\nu , \qquad F_{\mu\nu} = \nabla_{[\mu} A_{\nu]} , \tag{0.5}$$

and, by the results of a recent gravitational perturbation analysis[1], the gravitational force contribution will be given in terms of the relevant worldsheet supported hyper-Cauchy tensor $\overline{C}^{\mu\nu\rho\sigma}$ by an expression of the form

$$f_{\rm g}{}^{\mu} = \tfrac{1}{2} \overline{T}^{\nu\sigma} \nabla^\mu h_{\nu\sigma} - \overline{\nabla}_\nu \big(\overline{T}^{\nu\sigma} h_\sigma{}^\mu + \overline{C}^{\mu\nu\rho\sigma} h_{\rho\sigma} \big) . \tag{0.6}$$

These force densities would evidently be well behaved if the fields A_μ and $h_{\mu\nu}$ were due just to passing radiation. However we are concerned with the case in which they are obtained as the appropriate Dalembertian Green function solutions of the source equations (0.1), (0.2) , which give values that (while finite outside) are divergent on the worldsheet where they are needed.

As in the point particle case, one can obtain realistically regularised values, $\widehat{A}_\mu$ and $\widehat{h}_{\mu\nu}$ say, on the string worldsheet itself, by taking account of the fact that the physical system one wishes to describe (a vacuum vortex defect in the cosmic string case) will not really be infinitely thin but will have a finite thickness that provides an appropriately microscopic "ultraviolet" cut-off length, δ_* say. In the string case it is also be necessary to introduce a long range "infrared" cut-off length, Δ say, that might represent the macroscopic mean distance between neighbouring strings. The relevant Green function integrals will then be proportional to a logarithmic regularisation factor of the form $\widehat{l} = \ln\{\Delta^2/\delta_*^2\}$. More specifically (as pointed out in his original discussion of "superconducting" cosmic strings by Witten) the dominant contribution to the regularised electromagnetic self field $\widehat{A}_\mu$ on the string will be obtained [2] in the simple form

$$\widehat{A}_\mu = \widehat{l}\, \overline{j}_\mu . \tag{0.7}$$

By comparing (0.1) and (0.2), it can thus be seen that the corresponding expression for the regularised gravitational self field, $\widehat{h}_{\mu\nu}$ say, will have the form

$$\widehat{h}_{\mu\nu} = 2\mathrm{G}\, \widehat{l} \big(2\overline{T}_{\mu\nu} - \overline{T}_\sigma{}^\sigma g_{\mu\nu} \big) . \tag{0.8}$$

(If the microscopic electromagnetic source distribution were very different from that of the gravitational source distribution, the natural cut-off δ_* that would be most appropriate for the former might be somewhat different from what would be most appropriate for the latter, but since the dependence on the cut off is only logarithmic there will not usually be any significant loss of accuracy in using the same regularisation factor $\widehat{l}$ for both cases.)

For substitution in (0.5) and (0.6) – to get correspondingly regularised self force contributions $\widehat{f}_{\rm e}^{\,\mu}$ and $\widehat{f}_{\rm g}^{\,\mu}$ – knowledge of the simple regularised self fields $\widehat{A}_\mu$ and $\widehat{h}_{\mu\nu}$ is not sufficient. These regularised values are well defined only on the worldsheet and so do not provide what is needed for a direct evaluation of the gradients that are required: there is no difficulty for the terms involving just the tangentially projected gradient operator $\overline{\nabla}_\nu$, but there are also contributions from the unprojected gradient operator ∇_ν, which is directly meaningfull only when acting on fields whose support extends off the worldsheet.

It fortunately turns out that that this problem has a very simple and elegant solution [2]. The appropriate regularisation of the gradients on the string worldsheet turns out to be obtainable simply by replacing the ill defined operator ∇_ν by a corresponding regularised gradient operator that is given in terms of the worldsheet curvature vector, $K^\mu = \overline{\nabla}_\nu \eta^{\mu\nu}$, by the formula

$$\widehat{\nabla}_\nu = \overline{\nabla}_\nu + \tfrac{1}{2} K_\nu \, . \tag{0.9}$$

For the appropriately regularised electromagnetic field tensor on the string worldsheet this gives

$$\widehat{F}_{\mu\nu} = 2\widehat{\nabla}_{[\mu}\widehat{A}_{\nu]} = \widehat{l}\left(2\overline{\nabla}_{[\mu}\overline{j}_{\nu]} + K_{[\mu}\overline{j}_{\nu]}\right) , \tag{0.10}$$

which, by the surface current conservation condition $\overline{\nabla}_\nu \overline{j}^\nu = 0$, implies that the corresponding electromagnetic self force contribution in (0.5) will be expressible in the form

$$\widehat{f}_{\rm e}^{\,\mu} = -\overline{\nabla}_\nu \widehat{T}_{\rm e}^{\,\mu\nu} \, , \tag{0.11}$$

where the relevant stress momentum energy density contribution from the electromagnetic self interaction is

$$\widehat{T}_{\rm e}^{\,\mu\nu} = \widehat{A}^\mu \overline{j}^\mu - \tfrac{1}{2}\widehat{A}_\rho \overline{j}^\rho \eta^{\mu\nu} \, . \tag{0.12}$$

It can be seen from (0.6) that the regularised gravitational self force contribution will be similarly expressible in the form

$$\widehat{f}_{\rm g}^{\,\mu} = -\overline{\nabla}_\nu \widehat{T}_{\rm g}^{\,\mu\nu} \, , \tag{0.13}$$

in which the relevant self gravitational stress momentum energy density contribution works out as

$$\widehat{T}_{\rm g}^{\,\mu\nu} = \widehat{h}_\sigma{}^\mu \overline{T}^{\nu\sigma} - \tfrac{1}{4}\widehat{h}_{\rho\sigma}\overline{T}^{\rho\sigma}\eta^{\mu\nu} + \widehat{h}_{\rho\sigma}\overline{C}^{\rho\sigma\mu\nu} \, . \tag{0.14}$$

It evidently follows that the self force contributions will be absorbable by a renormalisation whereby the original "bare" stress momentum energy density tensor $\overline{T}^{\mu\nu}$ is replaced by the "dressed" stress momentum energy density tensor

$$\widetilde{T}^{\mu\nu} = \overline{T}^{\mu\nu} + \widehat{T}_{\rm e}{}^{\mu\nu} + \widehat{T}_{\rm g}{}^{\mu\nu} \, , \tag{0.15}$$

so that – in the absence of radiation from outside – the basic force balance equation (0.4) will reduce to a surface energy momentum conservation law of the simple form $\overline{\nabla}_\nu \widetilde{T}^{\mu\nu} = 0$.

References

1. Battye, R.A., Carter, B. (1995) Gravitational perturbations of relativistic membranes and strings, *Phys. Letters* **B357**, pp 29-35 [hep-th/9507059]
2. Carter, B. (1997) Electromagnetic self interaction in strings, *Phys. Letters* **B404**, pp 246-252 [hep-th/9704210]

D-BRANE BOUND STATES

MIGUEL S. COSTA
DAMTP, University of Cambridge
Silver Street, Cambridge CB3 9EW, U.K.

1. Introduction

D-branes have played a crucial role in understanding many aspects of string theory ([1] and references there in). At the heart of Polchinski's proposal is the fact that the excitations of these solitons have a well defined quantum mechanical description: *D-branes are extended objects with the property that open strings can end on them.* The D-brane 'revolution' has endowed string theory with a vast calculation power. In particular, it has made possible to describe quantum-mechanically certain extreme and near-extreme black holes carrying charge in the Ramond-Ramond sector of the theory successfully reproducing the Bekenstein-Hawking entropy formula [2] (see [3] for a complete list of references).

In this note we shall describe how to construct supersymmetric configurations of intersecting D-branes with non-vanishing worldvolume gauge fields. We start by describing how to generated non-threshold D-brane bound states by performing T-duality at an angle. The resulting D-p-brane also carries D-$(p-2)$-brane charge due to a magnetic condensate on the brane worldvolume. We shall then consider two examples of supersymmetric intersections of D-4-branes also carrying D-2-brane charge. One of these configurations may be used to construct a 5-dimensional black hole with finite entropy. The corresponding entropy formula is then explained in terms of the Landau levels for strings stretching between different branes.

2. T-duality at an angle

The worldvolume bosonic fields that describe the low energy excitations of a D-p-brane are $9-p$ scalar fields ϕ^m $(m = p+1, ..., 9)$ and a 1-form gauge potential A_α $(\alpha = 0, ..., p)$. These fields are interchanged by T-duality transformations. The scalar fields are associated to massless excitations of the brane that describe transverse displacements, and the gauge field is associated to worldvolume massless excitations on the brane.

Let us start by considering a D-$(p-1)$-brane parallel to the $x^1, ..., x^{p-2}$-directions and wrapping the x^{p-1}, x^p 2-torus along a (m,n)-cycle. A bound state of a D-p-brane with a D-$(p-2)$-brane, i.e. a $(p|p-2)$ D-brane bound state, may be obtained by performing T-duality along the x^p-direction on the previously de-

L. Baulieu et al. (eds.), Strings, Branes and Dualities, 445–448.

scribed D-$(p-1)$-brane[4]

$$\begin{array}{llcll} p-1: & X^1,...,X^{p-1} & \longrightarrow & p: & X^1,...,X^p \\ & \phi^p = \tan\zeta\ X^{p-1} & & & A_p = \tan\zeta\ X^{p-1} \end{array} \tag{2.1}$$

where $\tan\zeta = (mR_{p-1})/(nR_p)$ with R_i the radius of the circle in the i-direction. The resulting D-brane bound state wraps the x^{p-1}-direction with winding number m. Therefore, in the linear approximation the corresponding worldvolume theory is the $U(m)$ super Yang Mills theory reduced to $(p+1)$-dimensions [5]. The fields in (2.1) represent the $U(1)$ center of the group. This D-p-brane carries n units of flux in the $x^{p-1}x^p$ 2-torus. This induces n units of D-$(p-2)$-brane charge. This fact may be understood if we consider a D-$(p-2)$-brane probe in a D-p-brane background. Since the corresponding potential is attractive the $(p-2)$-brane is attracted to the p-brane and it gets dissolved on the p-brane worldvolume leaving flux.

The same ideas may be implemented at the level of the supergravity solutions. The metric and dilaton field for the $(p|p-2)$ D-brane bound state are

$$\begin{aligned} ds^2 &= H^{\frac{1}{2}}\left[H^{-1}ds^2(\mathbf{M}^{p-1}) + \tilde{H}^{-1}ds^2(\mathbf{E}^2) + ds^2(\mathbf{E}^{9-p})\right] , \\ e^{2(\phi-\phi_\infty)} &= H^{\frac{1-p}{2}}\tilde{H} , \end{aligned} \tag{2.2}$$

where H is a harmonic function on the Euclidean space $\mathbf{E}^{9-p}$, $\tilde{H} = \sin^2\zeta + \cos^2\zeta\ H$ and $\mathbf{M}^{p-1}$ is the $(p-1)$-dimensional Minkowski space.

3. Intersecting D-brane bound states

D-branes may intersect while preserving some spacetime supersymmetry. The rule for a D-p-brane and a D-q-brane intersecting orthogonally over a r-brane is that $p+q = 2r$ mod 4. This means that the number of worldsheet fields satisfying ND boundary conditions is either $0, 4$ or 8. For example, two D-p-branes intersect orthogonally over a $(p-2)$-brane. The intersection rule for the $(p|p-2)$ D-brane bound states is that all the constituent D-branes (including the $(p-2)$-branes that are dissolved in the p-branes) should intersect in a manner compactible with the previous rule. For example, two $(4|2)$ D-brane bound states intersect in the following way

$$\begin{array}{llcll} 4_1: & X^1,X^2,X^3,X^4 & ; & 4_2: & X^2,X^4,X^5,X^6 \\ & F_{34} = \tan\zeta_1 & & & F_{26} = \tan\zeta_2 \end{array} \tag{3.1}$$

where $F = dA$. The corresponding supergravity solution may be obtained by considering the solution (2.2) and by using the harmonic function rule.

We have seen how to use T-duality at an angle to generate the $(p|p-2)$ D-brane bound state and how to intersect such states. However, there are other possibilities to construct intersecting D-p-branes with magnetic condensates on the corresponding worldvolume theories . To be definite we shall consider an example of two intersecting D-4-branes contrasting with the configuration (3.1). Start with two orthogonally intersecting D-3-branes described by

$$\begin{array}{llcll} 3_1: & X^1, X^2, X^5 & ; & 3_2: & X^1, X^3, X^5 \\ & \phi^4 = -\tan\zeta\; X^5 & & & \phi^4 = \cot\zeta\; X^5 \end{array} \tag{3.2}$$

The i-th D-3-brane wraps the x^4, x^5 2-torus on a (q_i, p_i)-cycle with

$$\tan\zeta = (q_1 R_4)/(p_1 R_5) \text{and} \cot\zeta = (q_2 R_4)/(p_2 R_5)$$

. Thus, the moduli R_4 and R_5 are related by $(R_4/R_5)^2 = (p_1 p_2)/(q_1 q_2)$. A T-duality transformation along the x^4-direction will result in the bound state

$$\begin{array}{llcll} 4_1: & X^1, X^2, X^4, X^5 & ; & 4_2: & X^1, X^3, X^4, X^5 \\ & F_{45} = \tan\zeta & & & F_{45} = -\cot\zeta \end{array} \tag{3.3}$$

Using these ideas at the level of the supergravity we obtain the following metric and dilaton field describing this configuration

$$\begin{aligned} ds^2 &= (H_1 H_2)^{\frac{1}{2}} \Big[(H_1 H_2)^{-1} ds^2(\mathbf{M}^2) + H_1^{-1} dx_2^2 \\ &\quad + H_2^{-1} dx_3^2 + \tilde{H}_{12}^{-1}(dx_4^2 + dx_5^2) + ds^2(\mathbf{E}^4) \Big] , \\ e^{2(\phi - \phi_\infty)} &= \tilde{H}_{12}^{-1} (H_1 H_2)^{\frac{1}{2}} , \end{aligned} \tag{3.4}$$

where H_i are harmonic functions on the Euclidean space $\mathbf{E}^4$ and $\tilde{H}_{12} = \cos^2\zeta\; H_1 + \sin^2\zeta\; H_2$. The same solution may be obtained by using the $SL(2,\mathbf{R})$ duality symmetry of $D = 8$ $N = 2$ supergravity (broken to $SL(2,\mathbf{Z})$ in the quantum theory). Comparing the configurations (3.1) and (3.3) we see that the worldvolume fields play an essential role in ensuring supersymmetry of this D-brane system.

The D-brane system (3.3) may be excited by allowing the open strings ending on these branes to carry some (left)-moving momentum along the x^1-direction while preserving 1/8 of the spacetime supersymmetry. The corresponding supergravity solution is obtained by replacing the 2-dimensional Minkowski metric in (3.4) by a plane wave metric. The resulting 5-dimensional compactified solution describes a black hole with a regular horizon. The corresponding Bekenstein-Hawking entropy may be calculated in terms of the string theory quantities defining our D-brane system (3.3). The result is

$$S = 2\pi\sqrt{N N_1 N_2 n_L} , \qquad n_L = p_1 q_2 + p_2 q_1 , \tag{3.5}$$

where N is the number of momentum quanta in the x^1-direction and N_i the number of D-4-branes described by (3.3).

4. Conclusion

The entropy formula (3.5) ought to have a microscopic explanation within the string theory description of D-branes. We have to study open strings ending on the constituent D-branes. The open strings that are relevant for the entropy counting are those ending on different branes. The corresponding worldsheet fields X^4 and X^5 are coupled to the magnetic condensates on the D-branes due to the boundary term in the open string action. After finding the mode expansion for these fields and quantising them it is seen that the corresponding zero modes are non-commuting variables. This results in the existence of n_L Landau levels for these open strings. The resulting massless spectrum is described by $4N_1N_2n_L$ bosonic and fermionic excitations. To carry the momentum N/R_1 the system is at level N resulting in the entropy formula (3.5) for $N \gg N_1N_2n_L$. If the branes are wrapped along the x^1-direction we have $4n_L$ bosonic and fermionic excitations but the momentum is quantised in units of $1/(R_1N_1N_2)$. The system is at level $N' = NN_1N_2$ and we again obtain the entropy formula (3.5) but this time for $NN_1N_2 \gg n_L$.

Finally, the Landau levels have also to appear in the gauge theory description of the low-lying states on the D-brane system (3.3). The $U(1)$ flux on the branes induces a t'Hooft twist and the fields obey twisted boundary conditions on the x^4x^5 2-torus. The spectrum of this theory contains several massless torons that exactly match the entropy formula (3.5).

Acknowledgements

Most of the work here presented was done in collaboration with M. Cvetič, G. Papadopoulos and M. Perry. The financial support of JNICT (Portugal) and NATO is gratefully acknowledge. It is a pleasure to thank the organisers for an excellent summer school and a very pleasant time.

References

1. Polchinski, J. (1996) TASI Lectures on D-branes, hep-th/9611050.
2. Strominger, A. and Vafa, C. (1996) Microscopic Origin of the Bekenstein-Hawking Entropy, *Phys. Lett.*, **B379**, 99-104; Callan, C. and Maldacena, J. (1996) D-brane Approach to Black Hole Quantum Mechanics, *Nucl. Phys.*, **B472**, 591-610.
3. Maldacena, J. (1996) Black Holes in String Theory, Ph.D. thesis, hep-th/9611050.
4. Bachas, C. (1996) D-brane dynamics, *Phys. Lett.*, **B374**, 37-42; Douglas, M. (1995) Branes within Branes, hep-th/9512077.
5. Witten, E. (1996) Bound States Of Strings And *p*-Branes, *Nucl. Phys.*, **B460**, 335-350.

SPECIAL KÄHLER GEOMETRY

Does there Exist a Prepotential?

B. CRAPS, F. ROOSE, W. TROOST AND A. VAN PROEYEN
Instituut voor Theoretische Fysica
Katholieke Universiteit Leuven, B-3001 Leuven, Belgium

1. Introduction

Special (Kähler) geometry [2] is, by definition, the geometry of vectormultiplet scalars in $N = 2$ supergravity. However, one would like to define this geometry only referring to these scalars, not to any other fields. Therefore one needs to know the most general way of coupling vectormultiplets to supergravity. Originally, supergravity actions for vectormultiplets were constructed using a holomorphic 'prepotential'. As turned out later, duality transformations can lead to actions for which a prepotential does not exist [4]. In [1] a formulation ('definition') of special geometry was given which is manifestly invariant under duality transformations. It was proved that this formulation is equivalent to the original one, in the sense that it is always possible to perform a duality transformation such that a prepotential exists in the dual formulation of the theory. Moreover, it describes all presently known examples of special geometry. All constraints imposed in this definition have a nice physical interpretation (related to duality invariance and positivity of the kinetic energy), except for one constraint in the special case of only one vectormultiplet. This exception suggests that one could try to construct a more general supergravity theory for one vectormultiplet, one that could not be encoded in a holomorphic prepotential.

The aim of this contribution is to review this evolution, simplifying the discussion by omitting a lot of details and concentrating on the points essential for the interpretation of the constraints imposed in [1].

2. $N = 2$ supergravity coupled to vectormultiplets

The (abelian) theories we are going to consider describe a SUGRA multiplet (whose bosonic dynamical degrees of freedom are the graviton and a vector, the graviphoton) and n vectormultiplets (each a vector and a complex scalar). Thus we have $n + 1$ vectors (described by $\mathcal{F}^I_{\mu\nu}$, $I = 0, \ldots, n$) and n complex scalars z^α, $\alpha = 1, \ldots, n$.

L. Baulieu et al. (eds.), Strings, Branes and Dualities, 449–454.

The scalars are interpreted as coordinates on a manifold $\mathcal{M}$. This manifold allows projective coordinates

$$Z^I(z^\alpha) \sim e^{f(z^\alpha)}\, Z^I(z^\alpha)\ .$$

There exists a holomorphic function $F(Z^I)$, homogeneous of second degree, named the 'prepotential', from which the following $2n+2$ component vector can be defined:

$$v \equiv \begin{pmatrix} Z^I \\ \partial F/\partial Z^I \end{pmatrix} \equiv \begin{pmatrix} Z^I \\ F_I \end{pmatrix}\ . \tag{2.1}$$

The kinetic term $\mathcal{L}_0$ of the scalars is given by

$$\mathcal{L}_0 = -e\, g_{\alpha\bar\beta}\, \partial_\mu z^\alpha\, \partial^\mu \bar z^\beta\ , \tag{2.2}$$

where e is the determinant of the vierbein and $g_{\alpha\bar\beta}$ is a Kähler metric with Kähler potential

$$K = -\ln[i < \bar v, v >]\ . \tag{2.3}$$

Here the bracket denotes the symplectic inproduct

$$\langle V, W\rangle \equiv V^T \Omega W \qquad \text{with} \qquad \Omega = \begin{pmatrix} 0 & \mathbb{1} \\ -\mathbb{1} & 0 \end{pmatrix}\ .$$

The metric $g_{\alpha\bar\beta}$, describing the geometry of the scalar manifold $\mathcal{M}$, is called *special Kähler.*

The vector kinetic term is

$$\mathcal{L}_1 = e\left(\tfrac{1}{4}(\mathrm{Im}\,\mathcal{N}_{IJ})\mathcal{F}^I_{\mu\nu}\mathcal{F}^{\mu\nu J} - \tfrac{i}{8}(\mathrm{Re}\,\mathcal{N}_{IJ})\varepsilon^{\mu\nu\rho\sigma}\mathcal{F}^I_{\mu\nu}\mathcal{F}^J_{\rho\sigma}\right)\ , \tag{2.4}$$

where

$$\mathcal{N}_{IJ}(Z) = \bar F_{IJ}(\bar Z) + 2i\,\frac{\mathrm{Im}\,F_{IK}(Z)\,\mathrm{Im}\,F_{JL}(Z)\,Z^K Z^L}{\mathrm{Im}\,F_{KL}(Z)\,Z^K Z^L}\ .$$

The combination of equations of motion and Bianchi identities is invariant under symplectic *duality transformations*

$$v \;\to\; \begin{pmatrix} A & B \\ C & D \end{pmatrix} v \tag{2.5}$$

$$\mathcal{N} \;\to\; (C + D\mathcal{N})(A + B\mathcal{N})^{-1}\ , \tag{2.6}$$

with $\begin{pmatrix} A & B \\ C & D \end{pmatrix} \in Sp(2n+2, \mathbb{R})$. These transformations leave $g_{\alpha\bar\beta}$ invariant.

As an *example*, consider $n = 1$ and $F(Z^0, Z^1) = -iZ^0 Z^1$, with $Z^0 = 1$ and $Z^1 = z$. Thus

$$v = \begin{pmatrix} Z^I \\ F_I \end{pmatrix} = \begin{pmatrix} 1 \\ z \\ -iz \\ -i \end{pmatrix}\ ,$$

$e^{-K} = 2(z+\bar{z})$ and $g_{z\bar{z}} = (z+\bar{z})^{-2}$. Now consider the following symplectic transformation:

$$v \to \tilde{v} = \begin{pmatrix} 1 & 0 & 0 & 0 \\ 0 & 0 & 0 & -1 \\ 0 & 0 & 1 & 0 \\ 0 & 1 & 0 & 0 \end{pmatrix} v = \begin{pmatrix} 1 \\ i \\ -iz \\ z \end{pmatrix}$$

It is clear that no prepotential exists for $\tilde{v}$, since $\tilde{v}$ is not of the form (2.1) (its lower part is independent of its upper one).

3. Symplectically invariant definition

We define a special Kähler metric as a Kähler metric[1] for which there exists a $2n+2$ component vector $v(z^\alpha)$ such that (2.3) is satisfied and

$$< v, \mathcal{D}_\alpha v >= 0\ , \tag{3.1}$$

where $\mathcal{D}_\alpha \equiv \partial_\alpha + (\partial_\alpha K)$ is covariant under $v \to e^{f(z)}\, v$.

Using the notation $v = \begin{pmatrix} Z^I \\ F_I \end{pmatrix}$, where F_I need not be the derivative of a prepotential, we can write down the following manifestly symplectically covariant form of $\mathcal{N}$:

$$\mathcal{N}_{IJ} = (\ \mathcal{D}_{\bar{\alpha}}\bar{F}_I \quad F_I\ )(\ \mathcal{D}_{\bar{\alpha}}\bar{Z}^J \quad Z^J\ )^{-1}\ . \tag{3.2}$$

It was proved in [1] that this definition is equivalent to the prepotential–definition and that it includes all (presently) known supergravity actions.

Let us remark that the matrix $(\ \mathcal{D}_{\bar{\alpha}}\bar{Z}^J \quad Z^J\ )$ is always invertible, whereas the invertibility of $(\ \mathcal{D}_\alpha Z^J \quad Z^J\ )$ is equivalent to the existence of a prepotential for v.

4. Interpretation of the constraint $< v, \mathcal{D}_\alpha v >= 0$

The constraint (3.1) implies the positivity of the following matrix:

$$\begin{pmatrix} \mathcal{D}_\alpha Z^I \\ \bar{Z}^I \end{pmatrix} i\, e^K (\mathcal{N} - \mathcal{N}^\dagger)_{IJ} \left(\mathcal{D}_{\bar{\beta}}\bar{Z}^J \quad Z^J\right) = \begin{pmatrix} g_{\alpha\bar{\beta}} & i < \mathcal{D}_\alpha v, v > \\ i < \bar{v}, \mathcal{D}_{\bar{\beta}}\bar{v} > & e^{-K} \end{pmatrix}, \tag{4.1}$$

which in turn implies both that $(\ \mathcal{D}_{\bar{\alpha}}\bar{Z}^J \quad Z^J\ )$ is invertible (guaranteeing that $\mathcal{N}$ is well-defined) and, using the symmetry of $\mathcal{N}$, that Im $\mathcal{N}$ is negative (ensuring the positivity of the vector kinetic energy).

On the other hand, the symmetry of $\mathcal{N}$ defined in (3.2) (needed for duality invariance) is equivalent to

$$< \mathcal{D}_\alpha v, \mathcal{D}_\beta v >= 0\ . \tag{4.2}$$

Whereas (3.1) implies (4.2), the reverse only holds for $n > 1$.

[1] We include positivity in the definition of a Kähler metric.

This raises the question whether the constraint (3.1) can be relaxed for $n=1$ to the combination of the positivity of (4.1), and (4.2). Solving this question means verifying whether it is possible to construct an $N=2$ supergravity theory for a single vectormultiplet with a vector v satisfying both the positivity of (4.1) and (4.2) but not (3.1). This problem has not been solved yet.

5. Special geometry and Calabi-Yau moduli spaces

The low-energy description of a type II string theory compactified on a Calabi-Yau threefold is a $d=4, N=2$ supergravity theory. As such the (vector multiplet) scalar sector involved in such compactifications is expected to yield a special Kähler manifold. In this section we will outline how the various constraints of special Kähler geometry are realised in this concrete setting.

The interpretation of the special geometry relations goes basically via the following isomorphism

$$\begin{array}{ccccc} H^3(X) & \to & & & \Sigma \\ \omega & \mapsto & v_\omega & = & \begin{pmatrix} \int_{A^I}\omega \\ \int_{B_I}\omega \end{pmatrix} , \end{array} \tag{5.1}$$

where X is some generic CY threefold and $\{A^I, B_I\}$ a canonical basis of 3-cycles. Σ is a $2h^{2,1}+2$ complex dimensional vector space. The threeform cohomology may be endowed with the sesquilinear form Q:

$$Q(\eta,\omega)=i\int_X \eta\wedge\bar{\omega} ,$$

which corresponds to a symplectic form $i\langle v_\eta, v_\omega\rangle$ on Σ. Note that a symplectic transformation on Σ correspond to a change of canonical homology basis. The form Q can be shown [5] to enjoy the following properties:

1. It is block-diagonal in an appropriate cohomology basis. More explicitly, when expressed with respect to any cohomology basis adapted to the Hodge decomposition $H^3 = H^{3,0}\oplus H^{2,1}\oplus H^{1,2}\oplus H^{0,3}$, Q takes the following form

$$\begin{pmatrix} H & 0 & 0 & 0 \\ 0 & -H^*_{\bar{\alpha}\beta} & 0 & 0 \\ 0 & 0 & H_{\alpha\bar{\beta}} & 0 \\ 0 & 0 & 0 & -H \end{pmatrix} .$$

2. Q is positive-definite on $H^{3,0}\oplus H^{1,2}$.

Now consider a family of CY threefolds with varying complex structure and fixed Kähler class. In the generic case $h^{2,1}$ complex parameters z^α specify one particular member in this family. Furthermore fix a complex structure on this space of moduli such that the unique holomorphic threeform Ω on X depends only holomorphically on these moduli z^α. As small variations of a (p,q) form give at most $(p\pm 1, q\mp 1)$ forms, we find

$$\begin{array}{ccccc} \partial_\alpha\Omega & = & -\kappa_\alpha\Omega & \oplus & \Omega_\alpha , \\ & \in & H^{3,0} & \oplus & H^{2,1} . \end{array}$$

Thus we may define $\mathcal{D}_\alpha\Omega \equiv [\partial_\alpha + \kappa_\alpha]\Omega$. Moreover this set can be shown to constitute a basis for $H^{2,1}$. We define further $\mathcal{D}_{\bar\alpha}v \equiv \partial_{\bar\alpha}v = 0$. Under the mapping (5.1) the threeforms $\{\Omega, \Omega_\alpha\}$ are mapped to a corresponding set $\{v, \mathcal{D}_\alpha v\}$ of symplectic vectors in Σ, with the mentioned properties induced.

Putting things together we are now in a position to state the main results of this whole setup. After specifying some arbitrary canonical homology basis we define v to be the image of the unique holomorphic threeform Ω under the mapping (5.1). This object v is the one central object in section 3. Furthermore $\mathcal{D}$ is defined as in the above paragraph.

The various properties of Q may now be properly translated into relations satisfied by v and $\mathcal{D}_\alpha v$. The quantity $-i\langle v, \bar v\rangle$ is strictly positive as is correspondingly the matrix entry H in the bilinear form Q on cohomology. As a result we may define a real Kähler potential through $-i\langle v, \bar v\rangle = e^{-K}$. It is an easy exercise to check that the Kähler potential thus defined appears in the derivative $\mathcal{D}_\alpha = \partial_\alpha + (\partial_\alpha K)$, precisely as in section 3. The positivity of Q on $H^{1,2}$ is equivalent to the positivity of the metric which we define to be $g_{\alpha\bar\beta} = -i\langle \mathcal{D}_\alpha v, \mathcal{D}_{\bar\beta}\bar v\rangle$. This metric coincides with the one obtained from the Kähler potential, $g_{\alpha\bar\beta} = \partial_\alpha\partial_{\bar\beta}K$. The derivative $\mathcal{D}$ is covariant under holomorphic rescalings $v \to e^{f(z)}v$, which in turn correspond to Kähler transformations on the moduli space. From the block-diagonal form of Q in the specified basis the following relations are immediate consequences:

$$\langle v, \mathcal{D}_\alpha v\rangle = 0\,, \tag{5.2}$$

$$\langle v, \mathcal{D}_{\bar\alpha}\bar v\rangle = 0\,. \tag{5.3}$$

As to the invertibility of the matrix $\begin{pmatrix} \mathcal{D}_{\bar\alpha}\bar Z^J & Z^J \end{pmatrix}$ in section 3, first notice that it corresponds to $\begin{pmatrix} \int_{A^J}\bar\Omega_{\bar\alpha} & \int_{A^J}\Omega \end{pmatrix}$ in the present context. The positivity of Q on $H^{3,0}\oplus H^{1,2}$ guarantees its invertibility [1]. Replacing $H^{1,2}$ with $H^{2,1}$, no similar statement can be made, as a result of which $\begin{pmatrix} \mathcal{D}_\alpha Z^J & Z^J \end{pmatrix}$, or equivalently $\begin{pmatrix} \int_{A^J}\Omega_\alpha & \int_{A^J}\Omega \end{pmatrix}$, may not be invertible. Via an appropriate symplectic rotation on Σ it is always possible to make it invertible, though [1]. A prepotential formulation of the special geometry of the CY moduli space is absent if the above matrix is non-invertible. This non-invertibility is due to a particular choice of 3-cycles.

Concerning section 4 it is obvious that in the present case the stronger constraint $\langle v, \mathcal{D}_\alpha v\rangle$ is always obeyed, as is implied by Q and its properties once more.

Acknowledgements

B.C. and F.R. would like to thank the organizers for the opportunity to give these talks. B.C. (Aspirant FWO), W.T. (Onderzoeksleider FWO) and A.V.P. (Onderzoeksdirecteur FWO) thank the Fund for Scientific Research–Flanders for financial support. This work was supported by

the European Commission TMR programme ERBFMRX-CT96-0045.

References

1. Craps, B., Roose, F., Troost, W. and Van Proeyen, A. (1997) What is Special Kähler Geometry? *Nucl. Phys.* **B 503**, pp. 565–613; hep-th/9703082
2. De Wit, B., Lauwers, P.G., Philippe, R., Su, S.-Q. and Van Proeyen, A. (1984) *Phys. Lett.* **B 134**, 37;
   De Wit, B. and Van Proeyen, A. (1984) *Nucl. Phys.* **B 245**, 89
3. Strominger, A. (1990) *Comm. Math. Phys.* **133**, 163
4. Ceresole, A., D'Auria, R., Ferrara, S. and Van Proeyen, A. (1995) *Nucl. Phys.* **B 444**, 92, hep-th/9502072
5. Griffiths, P. and Harris, J. (1951), *Principles of algebraic geometry.* Wiley, New York.

COMPACTIFICATION OF M-THEORY AND SUSY BREAKING

C. GROJEAN
Service de Physique Théorique
CEA-SACLAY, F-91191 Gif-sur-Yvette, France

To Véronique, with love ...

1. Introduction

This work [1] fits into the old context of supersymmetric extension of the standard model of electroweak and strong interactions. A decade ago, a huge revival had followed the construction of the $E_8 \times E_8$ heterotic string. Nevertheless, it has soon appeared [2] that the string weakly coupled regime can hardly be phenomenologically acceptable leading to far too large a Newton constant. So phenomenological attention must be paid to strongly coupled strings. Unfortunately, non-perturbative studies were out of reach until only recent works [3] where a lot of progress has been made in understanding this new physics: the strongly coupled regime is now viewed as the low energy limit of M-theory and, in particular, the strongly coupled $E_8 \times E_8$ heterotic string can be described, in the low energy limit, by the eleven-dimensional supergravity with the two E_8 gauge factors living each on a 10-d boundary. The radius of the eleventh dimension is related to the 10-d string coupling by $R_{11} \sim \lambda_{st}^{2/3}$.

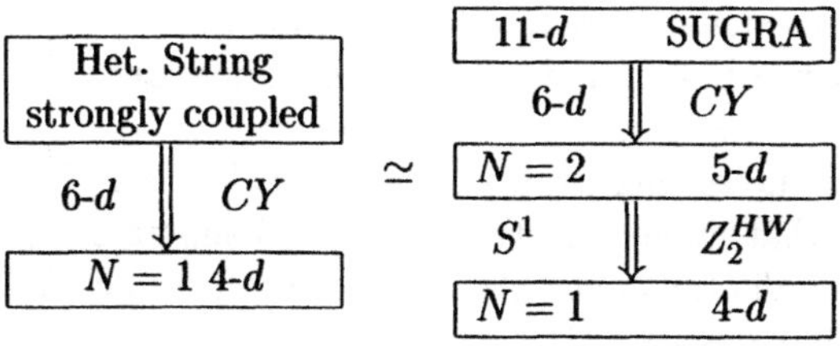

2. Ordinary dimensional reduction

In 11-d the spectrum consists in the metric tensor g and a three-form C plus gauge fields A^a living on the 10-d boundaries. Symbolically, the 11-d action reads

$$S^{(11)} = \int_{\mathcal{M}^{11}} \left(\mathcal{R}^{(11)} + (dC)^2 + C \wedge dC \wedge dC\right) + \int_{\mathcal{M}^{10}} F_{ab}F^{ab} . \quad (2.1)$$

L. Baulieu et al. (eds.), Strings, Branes and Dualities, 455–458.

To reproduce the spectrum of the 10-*d* string, we have to supplement the above Lagrangian by the Hořava-Witten Z_2^{HW} projection, which projects out one gravitino and part of the other fields on the boundary.

Our goal is to compactify the Lagrangian to 4-*d* in a way compatible with $N = 1$ 4-*d* supersymmetry. However the M-theory is just known perturbatively and, in particular, there is no full quantum description. Therefore an explicit compactification is still out of reach. So we adopt an alternative way that was proposed by Witten [4] to mimic compactification of weakly coupled strings: a truncation of the spectrum, by imposing a symmetry on the compact space, such as to maintain $N = 1$ supersymmetry in 4-*d*. Of course, in this case we can describe only the analogue of untwisted fields of string theories, originating from 11-*d* and 10-*d* fields.

The consistency with a true compactification on a CY space is checked by identifying the Kähler structure for the moduli fields describing the shape of the internal manifold.

3. Supersymmetry breaking by compactification *à la Scherk-Schwarz*

The ordinary dimensional reduction as presented in the last section, *i.e.* without any dependence on the internal coordinates for the fields, suffers from three main difficulties to be 4-*d* phenomenologically acceptable:

1. it usually does not generate non-abelian gauge-group;
2. it does not generate masses: starting with a massless theory in high dimension, the reduced theory still remains massless;
3. as explained, it leads to a 4-*d* theory with an exact $N = 1$ supersymmetry whereas we would prefer a spontaneously broken $N = 1$ supersymmetry.

These three difficulties are solved by considering a more elaborated dimensional reduction proposed originally by Scherk and Schwarz in a supergravity context [5] and then generalized to superstrings [6].

The Scherk–Schwarz mechanism is a generalized dimensional reduction that allows for the fields a dependence on the compact coordinates. Typically, the fields in the higher-dimensional theory take the form: $\hat{T}(x,y) = \mathcal{U}(y)\,T(x)$, where y denotes compact coordinates and x non-compact ones and $\mathcal{U}$ is a symmetry of the 5-*d* theory.

A particularly interesting case is the use, as $\mathcal{U}$, of a $SU(2)_R$ subgroup of the $Sp(8)$ symmetry that appears in the the $N = 8$ supergravity in 5-*d*, generating a potential for the dilaton superfield S. After compactification, the resulting 4-*d* theory is completely described by

$$\mathcal{K} = -\ln(S + S^\dagger) - 3\ln(T + T^\dagger)\ , \quad W = 2m(1 + S)\ . \tag{3.1}$$

Notice that this superpotential for S should correspond to a non-perturbative effect from the heterotic string point of view. The minimum of the scalar potential corresponds to $S = 1$ and a spontaneously broken supergravity model with a zero cosmological constant. The order parameter for supersymmetry breaking is the gravitino mass $m_{3/2}^2 = e^{\mathcal{K}}|W|^2 = 8m^2/(T + T^\dagger)^3$.

4. Phenomenological study

In appropriate units, the dimensional-reduced action takes the symbolic form:

$$S^{(4)} = \int_{\mathcal{M}^4} tM_{11}^2 \left(\mathcal{R}^{(4)} + \text{kin.}\right) + sF_{\mu\nu}F^{\mu\nu} . \tag{4.1}$$

This allows us to identify the gauge coupling and the 4-d reduced Planck mass to be ${\alpha_{GUT}}^{-1} \sim s/(4\pi)^{5/3}$ and ${M_{Pl}}^2 = tM_{11}^2$. Combining these equalities, we obtain the an expression for the Newton constant expressed in 4-d Planck scale, defining $t = M_{Pl}/R_5^{-1}$:

$$G_{\mathcal{N}} \sim \frac{R_5^{-1}\alpha_{GUT}^{1/3}}{M_{Pl}M_{GUT}^2} . \tag{4.2}$$

So, in order to be compatible with experimental constraints, the "radius" of the fifth dimension has an intermediate scale value, as already suggested by [7],[8] $R_5^{-1} \sim 10^{12-13}$GeV.

In that way, the problem of the Newton constant appearing in the weakly coupled string compactifications can be cured. Nevertheless, we know that the Scherk–Schwarz mechanism generates a mass for the gravitino which is related to R_5 and thus it is important to wonder whether this value is phenomenologically acceptable. In the expression for gravitino mass obtained from (3.1), the mass parameter m that appears in the Scherk–Schwarz mechanism is naturally of the order of the M-theory scale, so we obtain $m_{3/2} \sim M_{11}/\sqrt{t^3}$, which in term of the radius R_5 becomes $m_{3/2} \sim R{c_5}^{-2}/M_{Pl}$. This expression is similar to a usual supergravity form in which an auxiliary field would have developed a *vev* of the order of R_5^{-2}. Using the value of R_5 obtained from the Newton constant, it gives $m_{3/2} \sim 10^6$ GeV. We won about ten orders of magnitude comparing to the usual expression encountered in string compactifications ($m_{3/2} \sim 1/R_{CY} \sim M_{GUT}$). Nevertheless, this value is still a bit too high to be transmitted directly to the observable sector. Actually, if all the usual matter fields come from 10-d gauge fields on the boundary, they do not feel the eleventh dimension and consequently they do not get masses by the Scherk-Schwarz mechanism. In the class of models that we discussed in the previous sections, this can be checked explicitly at tree-level of the supergravity models. So the soft terms in the observable sector must be the result of additional loop effects. One possibility would be the well-known threshold corrections to the gauge and the Yukawa couplings. Another possibility, proposed in [9] , is to compute supergravity one-loop corrections to scalar masses by including the whole tower of Kaluza-Klein excitations running in the loop. In this case, the result, which is finite in the ultraviolet and the infrared, is qualitatively [9] $m_{soft} \sim m_{3/2}^2/M_{Pl}$. A new gravitational suppression appear which is welcome and feed the supersymmetry breaking in the observable sector at moderate values.

An interesting question which is not answered in the present context is a dynamical reason which fixes the large value of the fifth dimensional radius. From the point of view of the Scherk-Schwarz mechanism, it is a completely flat direction. If lifted at low energy only and if the soft terms in the observable sector are gravitationally suppressed compared to the gravitino mass as discussed above, then we could dynamically get the required value for the radius. If this scenario is valid remains to be seen.

Acknowledgements

It is a great pleasure to thank E. Dudas for his scientific skills and the Cargèse ASI'97 organization committee for their financial support and their warm hospitality.

References

1. E. Dudas and C. Grojean, *hep-th/***9704177**, to appear in *Nucl. Phys.* **B**.
2. M. Dine and N. Seiberg, *Phys. Rev. Lett.* **55** (1985) 366;
   V.S. Kaplunovsky, *Phys. Rev. Lett.* **55** (1985) 1036;
   M. Dine and N. Seiberg, *Phys. Lett.* **162B** (1985) 299.
3. P. Hořava and E. Witten, *Nucl. Phys.* **B460** (1996) 506 and **B475** (1996) 94.
4. E. Witten, *Phys. Lett.* **155B** (1985) 151.
5. J. Scherk and J.H. Schwarz, *Nucl. Phys.* **B153** (1979) 61, and *Phys. Lett.* **B82** (1979) 60.
6. C. Kounnas and M. Porrati, *Nucl. Phys.* **B310** (1988) 355;
   S. Ferrara, C. Kounnas, M. Porrati and F. Zwirner, *Nucl. Phys.* **B318** (1989) 75.
7. T. Banks and M. Dine, *Nucl. Phys.* **B479** (1996) 173, *hep-th/***9609046**;
   E. Caceres, V.S. Kaplunovsky and I.M. Mandelberg, *Nucl. Phys.* **B493** (1997) 73;
   T. Li, J.L. Lopez and D.V. Nanopoulos, *hep-ph/***9702237**.
8. I. Antoniadis and M. Quiros, *Phys. Lett.* **B392** (1997) 61.
9. I. Antoniadis and M. Quiros, *hep-th/***9705037**.

THREEBREANES IN TWELVE DIMENSIONS

S. F. HEWSON
Cambridge University
DAMTP, Silver Street, CB3 9EW, England

1. Introduction

Recently we have seen great increases in the understanding of non-perturbative string theory in terms of the eleven dimensional M-theory structure [1, 2]. This understanding has been greatly enhanced by the realisation that many problems may be solved by the introduction of a signature (10,2) 'F-theory' [3]. To investigate this proposal further we discuss the formulation of a Green-Schwarz threebrane in twelve dimensions [4].

2. Supersymmetric threebrane in 12D

Traditional supersymmetry p-branes propagate in a superspace which is invariant under the action of some supersymmetric extension of the Poincaré group. In 11D this essentially involves the addition of the term

$$\{Q_{11}^{\alpha}, Q_{11}^{\beta}\} = (\gamma^{p})^{\alpha\beta} P_{p} + (\gamma^{p_1 p_2})^{\alpha\beta} Z_{p_1 p_2} + (\gamma^{p_1 \cdots p_5})^{\alpha\beta} Z_{p_1 \cdots p_5}, \tag{2.1}$$

where γ^{p} satisfy the 11D Clifford algebra and Q_{11} is a Majorana spinor. This algebra is maximal, possessing (528+528) degrees of freedom, and also reproduces the type IIA and IIB algebras with the help of dimensional reduction and T-duality [5]. However, all of these algebras may be obtained via a reduction of another maximal system: the 'F-algebra', given by

$$\begin{aligned} \{Q^{\alpha}, Q^{\beta}\}_{(10,2)} &= \frac{1}{2!}\left(C\Gamma^{\mu\nu}\right)^{\alpha\beta} Z_{\mu\nu} + \frac{1}{6!}\left(C\Gamma^{\mu\nu\rho\sigma\lambda\delta}\right)^{\alpha\beta} Z^{+}_{\mu\nu\rho\sigma\lambda\delta} \\ [M_{\mu\nu}, M_{\rho\sigma}] &= M_{\nu\sigma}\eta_{\mu\rho} + M_{\mu\rho}\eta_{\nu\sigma} - M_{\nu\rho}\eta_{\sigma\mu} - M_{\sigma\mu}\eta_{\nu\rho}, \end{aligned} \tag{2.2}$$

where $Z_{\mu\nu}$ and $Z^{+}_{\mu\mu\rho\sigma\lambda\delta}$ are central charges, $Z^{+}_{\mu\mu\rho\sigma\lambda\delta}$ being self-dual, and Q^{α} is Majorana-Weyl. This seems to be a natural algebra to study in twelve dimensions [6], and leads to the existence of a threebrane solution. We initially formulate

L. Baulieu et al. (eds.), Strings, Branes and Dualities, 459–462.

the superspace underlying the F-algebra and then discuss the formulation of the threebrane[1].

2.1. THE SUPERSPACE

Given a super Lie group G and some subgroup H we may define a coset supermanifold $\mathcal{M} = G/H$. Ordinary superspace is formed by considering the coset of the super Poincaré group by the Lorentz rotations. In 12D we perform an analogous procedure for the F-algebra (2.2). We define

$$G(X,\theta,\omega) = \exp(\theta_\alpha Q^\alpha + X^{\mu\nu}Z_{\mu\nu})\exp(\frac{1}{2}\omega^{\mu\nu}M_{\mu\nu}). \qquad (2.3)$$

To generate an infinitesimal change in the coordinates we act on the left with an infinitesimal group element $G(\delta X, \delta\theta, \delta\omega)$. From these changes we may generate invariant forms on the superspace

$$\Pi^\alpha = d\theta^\alpha\,, \quad \Pi^{\mu\nu} = dX^{\mu\nu} - \frac{1}{2}\theta_\alpha(\Gamma^{\mu\nu})^{\alpha\beta}d\theta_\beta\,. \qquad (2.4)$$

It is natural to consider the F-algebra as descending from a curved space de Sitter algebra [7], in which the 2-form $Z_{\mu\nu}$ obeys the same algebraic relationships as the Lorentz generator $M_{\mu\nu}$ with itself and Q^α. Due to the non-commuting nature of the $Z_{\mu\nu}$ generator in this case, the action of the supergroup on the manifold is highly non-linearly realised. However, a careful analysis of the expansion of the transformed group elements shows that we may obtain a *linear* realisation of the supersymmetry if one of the following conditions is met:

$$\text{(i)}\ \ (\Gamma_{\mu\nu})^{\alpha[\beta}(\Gamma^{\mu\nu})^{\gamma]\delta} = 0 \qquad\qquad \text{(ii)}\ \ \widetilde{\mathcal{P}}C\Gamma^{\mu\nu}\mathcal{P} = 0\,. \qquad (2.5)$$

In case (i) the invariant forms are still written as in (2.4), whereas if the so-called 'purity' condition (ii) is met then the spin space becomes trivial:

$$\Pi^\alpha \to d\theta^\alpha\,, \quad \Pi^{\mu\nu} \to dX^{\mu\nu}\,. \qquad (2.6)$$

2.2. THE THREEBRANE

The basic threebrane action in the new superspace is defined in exact analogy with the traditional Green-Schwarz string and membrane:

$$S = \int d^4\xi\sqrt{\det(\Pi^{\mu\nu}{}_i\Pi_{\mu\nu i})}\,, \qquad (2.7)$$

where the $\Pi^{\mu\nu}{}_i$ are the pullback of the forms (2.4) to the (2+2)-dimensional worldvolume. By construction, this algebra is manifestly *spacetime* supersymmetric, whereas in order for *worldvolume* supersymmetry to occur we must obtain a

[1] In the following discussion of the properties of the threebrane we set the central 6-form charge to zero for simplicity.

matching of bosonic and fermionic excitations on the brane. To see that this is the case even for the F-algebra, note that a superspace consists of the disjoint union of a bosonic piece $\mathcal{B}$ and a fermionic piece $\mathcal{F}$. The supersymmetry generator Q maps $\mathcal{B}$ onto a proper subspace of $\mathcal{F}$ and vice-versa. Due to the group structure of the superspace this implies that the mapping is a bijection, and therefore the bosonic and fermionic degrees of freedom must match up. The (2+2)(=3)-brane has eight transverse scalar fields propagating on the worldsheet, whereas the on-shell Majorana-Weyl spinors have sixteen real degrees of freedom. In order to obtain a match we must project out half of the remaining spin degrees of freedom. For super-Poincaré brane theories, the additional fermionic excitations are lost due to a κ-symmetry of the full action consisting of a Dirac term of type (2.7) plus a Wess-Zumion integral [5]. No such symmetry may occur in the 12D theory [4]. However, to obtain a linear realisation of the superspace theory of the underlying *de Sitter* algebra we need to make a projection of the spinor operators such that (2.5) holds. This has the same effect as a κ-symmetry and gives us the required (8+8) worldvolume degrees of freedom. Note that since branes are microscopic objects, and therefore insensitive to large scale structure, we choose to keep this restriction in the flat space limit.

2.2.1. *Reduction to lower dimensions*

We now consider the reduction of the 12D threebrane. We reduce on the timelike direction 12 to 11D and on a null torus to 10D. Reduction to 10D and 11D requires us to act on the spinor indices with the projectors $\mathcal{P}_{10} = \frac{1}{2}(1+\Gamma_0\ldots\Gamma_9)$ and $\mathcal{P}_{11} = \frac{1}{2}(1+\Gamma_0\ldots\Gamma_9\Gamma_{11})$ respectively. It is a simple matter to show that one obtains two 10D spinors of the *same* chirality and one 11D spinor after these reductions. Hence the theories are projected down onto the IIB and 11D sectors of M-theory. We now consider the reduction of the action (2.7), concentrating on the circle compactification to 11D. Upon reduction we find that the $Z^{p12} \to P^p_{11}$ and $Z^{pq} \to Z^{pq}_{11}$, in an obvious notation. There are two ways in which the superspace forms will reduce. In the case for which the first equation of (2.5) holds then the forms possess a torsion term as in (2.4) and reduce as follows

$$\Pi^{p12} = dX^{p12} - \frac{1}{2}\theta_\alpha(\Gamma^{p12})^{\alpha\beta}d\theta_\beta \longrightarrow \Pi^p = dX^p - \frac{1}{2}\theta_\alpha(\gamma^p)^{\alpha\beta}d\theta_\beta\,, \tag{2.8}$$

where we set the superspace terms Z^{pq}_{11} equal to zero as usual. Thus we obtain the usual Green-Schwarz actions in lower dimensions. In the case for which the spinors are restricted so as to satisfy the equation $\widetilde{\mathcal{P}}C\Gamma^{\mu\nu}\mathcal{P} = 0$ the superspace is trivially realised; there then exists a very simple (2+2)-brane action which possesses a manifest supersymmetry between the bosons and fermions:

$$S = \int d^4\xi\sqrt{\det(\Pi_i^A\Pi_j^B\mathcal{G}_{AB})}\,, \tag{2.9}$$

in which the superspace 'metric' is defined by the union of the flat spacetime metric η and the charge conjugation matrix C, as $\mathcal{G} = \mathrm{diag}(\eta, C)$. The forms in the action are now trivial: $\Pi^A = (dX^\mu, d\theta^\alpha)$. In this situation it is less clear that the action will reduce to the correct *super-Poincaré* invariant action under dimensional reduction. However, when one performs a superspace compactification then it becomes clear that a natural torsion term of the form $\theta_\alpha(\gamma^\mu)^{\alpha\beta}d\theta_\beta$ may be introduced into the expression for Π^μ [4]. Thus we do still recover the traditional Green-Schwarz actions in lower dimension. However, the addition of this torsion term is not compulsory and thus reduction may also lead to a torsion-free superspace sector of M-theory. A discussion of such supersymmetries is presented elsewhere [8].

3. Conclusion

We have outlined the formulation of the three-brane in twelve dimensions. Upon reduction we find that the superspaces associated with the F-algebra reduce to those for the Poincaré algebra, and that the threebrane action reduces to that for a usual Green-Schwarz string or membrane. In addition, reduction of the 12D spinors leads to one spinor in 11D and two spinors of the *same* chirality in 10D. Thus we obtain the classical representations of the type IIB string and the M-2-brane, directly from twelve dimensions.

Acknowledgements

I would like to acknowledge financial support from NATO and Queens' College, Cambridge. I also extend my sincere thanks to the organisers of Cargese 97, where this work was presented.

References

1. Duff, M. J. (1996) M-theory (the Theory formerly known as strings), *Int. J. Mod. Phys.* **A11**, pp. 5623–5642.
2. Schwarz, J. H. (1996) Lectures on superstring and M-theory dualities, *Nucl. Phys. Proc. Suppl.* **55B**: pp. 1, hep-th/9607201.
3. Vafa, C. (1996) Evidence for F-theory, *Nucl. Phys.* **B469**, pp. 403–418.
4. Hewson S. F. and Perry M. J. (1997) The twelve-dimensional super (2+2)-brane, *Nucl. Phys.* **B492**, pp. 249–277.
5. Townsend, P. K. Four lectures on M-theory, hep-th/9612121.
6. Hewson, S. F. An approach to F-theory, hep-th/9712017.
7. van Holten, J. W. and Van Proeyen, A (1982) N=1 Supersymmetry algebras in D=2,3,4 mod8, *J. Phys.* **A15**, pp. 3763.
8. Hewson, S. F. (1997) Generalised Poincaré supersymmetry and *p*-brane actions, *Nucl. Phys.* **B501**, pp. 445–468.

STRING SOLITONS AND DIMENSIONAL REDUCTION

GERMAR SCHRÖDER
Max-Planck-Institut für Gravitationsphysik
Albert-Einstein-Institut, Schlaatzweg 1, D-14473 Potsdam
Germany

1. Introduction

Classical black hole solutions in supergravity theories that preserve a certain number of supersymmetries saturate the gravitational version of the BPS-bound. Such solutions can be identified with massive solitonic string states [1] for an investigation of the conjectured non-perturbative duality symmetries of string theory, and they can be used to study the microscopic origin of black hole entropy [2]. In section 2 and 3, I will review aspects of BPS-solitons for the type II string compactified on T^6.
Usually black hole solutions in this context are spherically symmetric or axisymmetric and therefore can be studied in a dimensionally reduced theory which will be (a truncation of) integrable 2d N=16 supergravity [3]. In section 4, I will outline this reduction for four-dimensional type II BPS solitons.

2. Maximal supergravity in four dimensions

The low energy effective theory of the type II string compactified on T^6 is $N = 8$ supergravity in four dimensions [4]. The bosonic action is

$$\mathcal{L} = -\frac{1}{4}e\mathcal{R} + \frac{1}{8}eF^{ij}_{\mu\nu}\tilde{H}^{\mu\nu}_{ij} + \frac{1}{4}e(P^7_\mu, P^{7\mu}) \tag{2.1}$$

containing the vierbein $e_\mu{}^\alpha$, 28 $U(1)$ gauge fields $F^{ij}_{\mu\nu}$ with $\{ij\}$ antisymmetric and indices i, j running from 1 to 8, and 70 scalars combined in a field $\mathcal{V}^7 \in E_{(7)(+7)}/SU(8)$ in the fundamental representation with

$$\partial_\mu \mathcal{V}^7(\mathcal{V}^7)^{-1} = Q^7_\mu + P^7_\mu, \quad Q^7_\mu \in su(8), \quad P^7_\mu \in e_{7(+7)} - su(8). \tag{2.2}$$

$(\cdot,\cdot)$ is the Killing form. $\tilde{H}^{\mu\nu}_{ij}$, the space-time dual of $H^{\mu\nu}_{ij}$, is defined by

$$\mathcal{F} = \begin{pmatrix} F^{ij}_{\mu\nu} \\ H_{\mu\nu ij} \end{pmatrix}, \quad \mathcal{V}\mathcal{F} = \Omega\mathcal{V}\tilde{\mathcal{F}} \text{ with } \Omega = \begin{pmatrix} 0 & -I_{28} \\ I_{28} & 0 \end{pmatrix}.$$

L. Baulieu et al. (eds.), Strings, Branes and Dualities, 463–466.

The theory has a classical $E_{7(+7)}$ symmetry of the equations of motion:

$$\mathcal{F} \to \Lambda\mathcal{F}, \quad \mathcal{V} \to \mathcal{V}\Lambda^{-1}, \quad \Lambda \in E_{7(+7)}. \tag{2.3}$$

In a quantum theory, charge quantization implies that $E_{7(+7)}$ is broken to $E_{7(+7)}(\mathbb{Z})$. This is the proposed U-duality group of the type II string on T^6 as defined in [1].

3. BPS Solitons

Purely bosonic BPS solutions may be obtained by setting all fermions to zero and demanding that supersymmetry transformations of the gravitini and spin-1/2 fermions vanish.

The simplest example is obtained if we set all scalars to zero, leading to (multi-)Reissner-Nordstrøm black holes that preserve one supersymmetry.

If the solution preserves higher degrees of supersymmetry, the scalars are nonzero. As an example, such a solution shall be given here, corresponding to the "three form black hole solutions" of [5].

The following Ansätze are implemented: The scalar sector is

$$\mathcal{V} = \begin{pmatrix} v^{[a}_{[i} v^{b]}_{j]} & 0 \\ 0 & v_{[a}^{[i} v_{b]}^{j]} \end{pmatrix}, \quad v_i{}^a = \Delta^{-\frac{1}{8}} \begin{pmatrix} E_{\hat{i}}{}^{\hat{a}} & 0 \\ 0 & \sqrt{\Delta} \end{pmatrix}, \quad v_i{}^b v_a{}^i = \delta_a{}^b,$$

with indices $\hat{\imath}, \hat{a} = 1...7$. $G_{\hat{\imath}\hat{\jmath}} = -E_{\hat{\imath}}{}^{\hat{a}} E_{\hat{\jmath}}{}^{\hat{a}}$ can be identified with the "internal metric" of [5], $\sqrt{\Delta} = \det E$. $G_{\hat{\imath}\hat{\jmath}}$ is chosen to be diagonal.

Two gauge fields strengths are nonzero, the first electrically, the second magnetically charged:

$$F^{12}_{tr} = G_{11}^{-1} G_{22}^{-1} \sqrt{\Delta}\, \frac{\tilde{Q}}{R}, \qquad F^{13}_{\theta\phi} = \tilde{P} \sin\theta. \tag{3.1}$$

The space-time metric is spherically symmetric,

$$ds^2 = \lambda(r) dt^2 - \lambda(r)^{-1} dr^2 - R(r) d\Omega^2.$$

The solution is then given by

$$\lambda = \frac{r - 2P - 2Q}{(r-2P)^{\frac{1}{2}}(r-2Q)^{\frac{1}{2}}}, \quad R = (r - 2P - 2Q)(r-2P)^{\frac{1}{2}}(r-2Q)^{\frac{1}{2}},$$

$$\Delta = \left(\frac{r-2Q}{r-2P}\right)^{\frac{1}{3}}, G_{11} = \Delta^{-2}, \quad G_{22} = \Delta^{-\frac{1}{2}}\lambda^{-1}, \quad G_{33} = \Delta^{-\frac{1}{2}}\lambda, \quad G_{\hat{l}\hat{l}} = \Delta \tag{3.2}$$

for $l \neq 1,2,3$ with $Q \equiv \Delta_\infty^{1/4} G_{11\infty}^{-\frac{1}{2}} G_{22\infty}^{-\frac{1}{2}} \tilde{Q}$, $P \equiv \Delta_\infty^{-1/4} G_{11\infty}^{\frac{1}{2}} G_{22\infty}^{\frac{1}{2}} \tilde{P}$, where the subscript ∞ indicates asymptotic values at $r \to \infty$.

The action of $E_{7(+7)}$, that can be seen explicitly as in (2.3) in this notation, can now be used to generate the full $E_{7(+7)}$-multiplet. These solutions preserve 1/4 of supersymmetry and correspond to black holes with null singularity. If one charge is zero, 1/2 of supersymmetry is preserved and the solution corresponds to a naked singularity [5].

4. Reduction from four to two dimensions

Black hole solutions discussed in the context of string theory usually admit enough Killing symmetries for a reduction to two dimensions. It is tempting to speculate that BPS-conditions, duality symmetries and solitons of higher dimensional string theories can be discussed in a unified way in the context of the integrable structure of this theory with affine (e.g. E_9) symmetry.

The field redefinitions needed for this reduction shall be described in the following.[1] For a reduction to three dimensions, the vierbein is taken to be

$$e_{\bar{\mu}}^{(4)\,\bar{\alpha}} = \begin{pmatrix} e^{\frac{\varphi}{2}} e_{\mu}^{(3)\alpha} & e^{-\frac{\varphi}{2}} A_{\mu} \\ 0 & e^{-\frac{\varphi}{2}} \end{pmatrix}.$$

where indices μ, α now run from 1 to 3.

Defining $F_{\mu\nu} = 2\partial_{[\mu} A_{\nu]}$, all field strengths are dualized to scalars $\bar{\eta}^{ij}$ and f in three dimensions. The gauge field components along the reduced dimension are taken to be pseudoscalars η^{ij}. This yields

$$\begin{aligned} \mathcal{L}^{(3)} &= -\frac{1}{4} e^{(3)} \mathcal{R} + \frac{1}{8} e^{(3)} \partial_{\mu}\varphi \partial^{\mu}\varphi + \frac{1}{4} e^{(3)} (P_{\mu}^{7}, P^{7\mu}) \\ &\quad + \frac{1}{8} e^{(3)} e^{2\varphi} (\partial_{\rho} f + Y^{t} \Omega \partial_{\rho} Y)(\partial^{\rho} f + Y^{t} \Omega \partial^{\rho} Y) \\ &\quad + \frac{1}{4} e^{(3)} e^{\varphi} (\partial_{\rho} Y^{t}\, \mathcal{V}^{7t} \mathcal{V}^{7}\, \partial^{\rho} Y) \end{aligned} \tag{4.1}$$

with $Y^{t} = (\eta, \bar{\eta})$. This can be simplified to

$$\mathcal{L}^{(3)} = -\frac{1}{4} e^{(3)} \mathcal{R} + \frac{1}{4} e^{(3)} (P_{\mu}^{8}, P^{\mu 8}), \tag{4.2}$$

with $\mathcal{V}^{8} \in E_{(8)(+8)}/SO(16)$ and P_{μ}^{8} in analogy to (2.2) by considering the following: For an Iwasawa decomposition of $\mathcal{V}^{8}$, the generators of $e_{8(+8)}$ corresponding to positive roots can be grouped as τ^{+}, the generator corresponding to the highest positive root, generators e_{α_i} that correspond to positive roots of the $e_{7(+7)}$ subalgebra, and S_i^{+} generators that carry the fundamental representation of the $e_{7(+7)}$ subalgebra. The Cartan subalgebra can be taken to be h_i^{7}, h_{τ}. Iwasawa decomposition of $\mathcal{V}^{8}$ then yields

$$\begin{aligned} \mathcal{V}^{8} &= \exp(\Delta_i h_i^{7}) \exp(\mu_i e_{\alpha_i}) \exp(\Delta h_{\tau}) \exp(\lambda_i S_i^{+} + B\tau^{+}) \\ &= \mathcal{V}^{7} \exp(\Delta h_{\tau}) \exp(\lambda_i S_i^{+} + B\tau^{+}). \end{aligned}$$

The dimensionally reduced Lagrangian takes the form (4.2) provided

$$2\Delta = \varphi, \quad \sqrt{2}^{-1} \Omega \lambda = Y, \quad B = f.$$

[1] This work was done in collaboration with Jörg Teschner (U. Montpellier).

For the explicit BPS solution (3.2), one gets

$$f = 0,\ \varphi = -\ln(R\sin^2\theta),\ \eta^{13} = \tilde{P}(1-\cos\theta),$$
$$\bar{\eta}_{12} = \tilde{Q}R^2(-\tfrac{1}{5}\sin^4\theta\cos\theta - \tfrac{4}{15}\sin^2\theta\cos\theta - \tfrac{8}{15}\cos\theta),$$

all other $\eta, \bar{\eta}$ zero.

For a reduction to two dimensions, the dreibein takes the form

$$e_{\bar{\mu}}^{(3)\,\bar{\alpha}} = \begin{pmatrix} \sigma\delta_{\mu}{}^{\alpha} & 0 \\ 0 & \rho \end{pmatrix},$$

where μ, α now run from 1 to 2.

For (3.2), $\rho = \sqrt{\lambda R}\sin\theta$ and $\sigma = R^2\sin^2\theta$ with the coordinate change $r, \theta \to y, \theta$ and $dy/dr = (R\lambda)^{-\frac{1}{2}}$.

With these definitions, classes of BPS solutions can now be studied in connection to the integrable system as given in [3].

References

1. C.M. Hull and P.K. Townsend, "Unity of Superstring Dualities", Nucl. Phys. **B438**, 109 (1995)
2. J.M. Maldacena, "Black Holes in String Theory", Princeton U. preprint UMI-96-27605-mc, hep-th/9607235
3. H. Nicolai and N.P. Warner, "The Strucure of $N = 16$ Supergravity in Two Dimensions", Commun. Math. Phys. **125**, 369 (1989)
4. E. Cremmer and B. Julia, "The $SO(8)$ Supergravity", Nucl. Phys. **B159**, 141 (1979)
5. M. Cvetic and D. Youm: "Four-dimensional Supersymmetric Dyonic Black Holes in Eleven-dimensional Supergravity", Nucl. Phys. **B453**, 259 (1995)

BPS SATURATED AMPLITUDES AND NON-PERTURBATIVE STRING THEORY

PIERRE VANHOVE
Centre de Physique Théorique
École Polytechnique
91128 Palaiseau, France

Computations of non-perturbative contributions is still a difficult task. In (supersymmetric) Field theory a direct computation requires the construction of the measure of integration over vacua containing n-instantons, which is not yet available for every n. The use of dualities is still the main tool for getting the answer. In String theory the D-instantons contributions for the R^4 were conjectured thanks to the $Sl(2,\mathbf{Z})$ invariance of the type IIb theory [1], and for the Spin(32)/$\mathbf{Z}_2$ some *ad hoc* rules for the contributions for the R^4 and F^4 were derived from the S-duality relation between the Heterotic and Type I strings [2]. In this proceeding I will explain how these non-perturbative contributions can be computed exactly, concentrating on the eight-dimensional case. Details and generalizations will be found in [3].

On the heterotic side the F^4 and R^4 terms are extracted from the fourth point amplitude on the torus involving four gauge bosons. In order to get a non zero contribution, eight fermionic zero modes have to be soaked up. So all the fermionic zero modes from the supersymmetric side of the vertex operators have to be saturated, leaving a complete freedom on the non-supersymmetrical current algebra side. These contributions involve only states belonging to the short multiplets of the $N_{4d} = 4$ supersymmetry algebra [4]. The non-perturbative contributions in eight-dimensions reads

$$\mathcal{L}_{\text{non-pert}} = -\frac{2V^{(8)}}{2^{10}\pi^6}\sum_{N>0} e^{-2i\pi N T_s \mathcal{T}_2}\frac{1}{N}\sum_{\substack{N=mn\\ 0\leq j<n}}[\mathcal{D}\mathcal{A}]\left(\frac{m\mathcal{U}+j}{n}\right). \tag{1}$$

In this expression $\mathcal{T}$ and $\mathcal{U}$ are the moduli of the two-torus of compactification, T_s is the tension of the soliton and the $\mathcal{A}(q,F,R)$ is a modular invariant partition function computed in the Ramond sector expanded at the eight order with respect to the space-time fermionic zero modes [5]. For type II string for the R^4 terms we don't have anymore freedom for a current algebra, thus $\mathcal{A}=1$. $\mathcal{D}$ is a differential operator whose appearance is related to non-holomorphicities of the elliptic genus [2]. Hereafter only the leading part is needed $\mathcal{D} = 1 + \ldots$. These fluctuations around

L. Baulieu et al. (eds.), Strings, Branes and Dualities, 467–470.

the wrapped D1-brane are described by a SO($2N$) Super-Yang-Mills (SYM) theory. Thanks to the holomorphicity of $\mathcal{A}$, (1) can be computed at the infra-red (IR) fixed point of this SYM theory: $g_{\rm SYM} \to \infty$ with $\alpha' g_{\rm SYM} g_s = 1$ [6]. This limit is a (8,0) supersymmetrical σ-model with target space the symmetric orbifold

$$\mathcal{M} = (\mathbf{R}^8)^N / \mathcal{S}_N \ . \tag{2}$$

The Euclidean action in the light-cone gauge

$$\begin{aligned} S_{\rm E} = \int dt \int_0^{2\pi} \frac{d\sigma}{2\pi} \mathrm{Tr}\, \big[& \partial \mathbf{X}^i \left(\bar\partial \delta_{ij} + i \mathcal{R}_{ij}/2\pi \right) \mathbf{X}^j + \mathbf{S}_a \partial \mathbf{S}_a + \\ & + {}^{\rm T}\lambda_A^{(P)} \left(\bar\partial \delta^{AB} + i \mathcal{F}^{AB}/2\pi \right) \lambda_B^{(P)} + {}^{\rm T}\lambda_A^{(A)} \left(\bar\partial \delta^{AB} + i \mathcal{F}^{AB}/2\pi \right) \lambda_B^{(A)} \big] \end{aligned} \tag{3}$$

describes the coupling between the fluctuations, $\mathbf{X}^i$, $\mathbf{S}_a$ living in the adjoint representation of SO($2N$) and λ_A in vectorial representation of SO($2N$), with the background fields

$$\begin{aligned} (\mathcal{R}_{ij})_{MN} &= -\frac{1}{8} \mathcal{R}_{ijkl} (\mathbf{S}_0^a)_{MP} \gamma_{ab}^{kl} (\mathbf{S}_0^b)_{PN} \\ (\mathcal{F}^{AB})_{MN} &= -\frac{1}{8} \mathcal{F}_{kl}^{AB} (\mathbf{S}_0^a)_{MP} \gamma_{ab}^{kl} (\mathbf{S}_0^b)_{PN} \end{aligned} \qquad M, N, P \in \{1, \cdots, 2N\} \ . \tag{4}$$

It is important to remark that the space-time fermionics zero modes $\mathbf{S}_0^a$ enter only through these tensors. The various twisted sectors of the Hilbert space constructed on $\mathcal{M}$ are labeled by the action of the permutations $\sigma \in \mathcal{S}_N$ decomposed in disjoint cycles as $\sigma = [1^{N_1}, \ldots, N^{N_N}]$ with $N = \sum_i i N_i$ [6]. A sector with boundary conditions twisted by σ will contain $8 \times \sum_i N_i$ fermionic zero modes. As only terms with eight fermionic zero modes are needed, we must have $\sigma = [N^1]$. According to the rules of orbifold computation the determinant is given by the partition function on a torus computed over the invariant states [7]:

$$Z = \frac{1}{N!} \sum_{\omega = [N^1]} \sum_{\substack{\eta \\ \eta\omega = \omega\eta}} {}_{\eta}\square_{\omega} \ , \tag{5}$$

${}_{\eta}\square_{\omega}$ represents the functional integral

$${}_{\eta}\square_{\omega} = \int_{\rm Torus} \mathcal{D}X \mathcal{D}S^a \mathcal{D}\lambda \; e^{-S_{\rm E}}$$

calculated for the boundary conditions specified by η and ω. Yielding exactly the determinant of (1)

$$Z = \frac{1}{N} \left[\mathcal{A}(N\mathcal{U}) + \sum_{\substack{n_1 n_2 = N \\ n_2 \neq 1}} \sum_{j=0}^{n_2 - 1} \mathcal{A}\left(\frac{n_1 \mathcal{U} + j}{n_2} \right) \right] \ . \tag{6}$$

The Type II case, is very similar in spirit, the fluctuations around a vacua containing wrapped D-branes are described by a U(N) SYM model with an IR limit described by a (8,8) supersymmetrical σ-model (coupled to background for terms involving less zero modes). The big difference here is the existence of a U(1) gauge field freedom allowing 't Hooft fluxes [8]. Theses fluxes constraint the twisted states to appear as bound states of (p,q)-strings, $T_s = |p + q\tau|$ in (1). This gauge field is a remnant of the invariance under volume preserving diffeomorphism of the D-brane's world-volume theory. For the Spin(32)/$\mathbf{Z}_2$ case it was projected out by the Ω parity projection. As briefly explained in [9], seeing the global topology of the solitonic configuration, the gauge configuration contributions to the determinant of fluctuations are responsible for the appearance of the Bessel function $K_1(\cdots)$. In a T-dual language, this is the origin of the fractionalization of the D-instanton charge $N = nm$ [1, 10]. The D-instanton case is more tricky because the associated topological SYM model does not have a well defined IR limit. Nevertheless, [11] shown that these contributions stem from a one-loop computation involving the super-graviton compactified on a two-torus with complex structure $\tau = C^{(0)} + ie^{-\phi}$

$$\begin{aligned}\mathcal{L}^{FT} &= \int_0^\infty \frac{dt}{t}\,\mathrm{Str}\left((\mathcal{R}_{ijkl}M^{ij}\tilde{M}^{kl})^4 e^{-tS}\right)\\ &= t_{(8)}t_{(8)}\mathcal{R}^4\frac{\pi^{3/2}}{\Gamma(3/2)}\int_0^\infty \frac{dt}{t}\, t^{3/2}\sum_{(l_1,l_2)\neq(0,0)} e^{-\pi t\frac{|l_1+l_2\tau|^2}{\tau_2}}\ .\end{aligned} \tag{7}$$

M^{ij} ($\tilde{M}^{kl}$) is a Lorentz generator in the adjoint representation of SO(8) containing two left (right) fermionic zero modes [12], $\mathcal{R}_{ijkl}$ is the curvature. The super-trace being computed over 1/2-BPS multiplet of the IIb supergravity algebra. In lower dimensions the super-trace computed over the 1/2-BPS momentum multiplets of the U-duality group [9], yields the weight-3/2 Poincaré serie, of course 1/2-BPS states from fluxes multiplets can contribute too [13]. In ten dimension this expression can be rewritten as

$$\begin{aligned}\mathcal{L}^{FT} &= 2\zeta(3)\tau_2^{3/2} + \frac{2\pi^2}{3}\tau_2^{-1/2} + 8\pi\sum_{N\geq 1}\left(e^{2i\pi N\tau} + c.c.\right)\times\\ &\times\left[\sum_{N=mn}\frac{1}{m^2}\right]\left\{\sqrt{N}\left(1+\sum_{k=1}^{\infty}\frac{(-1)^k}{(4\pi N\tau_2)^k}\frac{\Gamma\left(k-\frac{1}{2}\right)\Gamma\left(k+\frac{3}{2}\right)}{\Gamma\left(\frac{1}{2}\right)\Gamma\left(\frac{3}{2}\right)k!}\right)\right\}\ .\end{aligned} \tag{8}$$

The contribution in square bracket $[\cdots]$ should come from the same counting of zero modes as before [10]. The contribution in braces $\{\cdots\}$ is the signature of higher loops corrections around the D-instantons.

Acknowledgements

I want to thank the (*gentils*) organizers of the Cargèse summer school for financial support, and allowing me to give a talk. These works were done in collaboration with C. Bachas, C. Fabre, M.B. Green, M. Gutperle, E. Kiritsis, N.A. Obers.

Moreover it is a pleasure to thanks K. Förger, P. Windey, F. Laudenbach and N. Berline for interesting discussions.

References

1. M.B. Green and M. Gutperle, Nucl. Phys. **B498** (1997) 195.
2. C. Bachas, C. Fabre, E. Kiritsis, N.A. Obers and P. Vanhove Nucl. Phys. **B509** (1998) 33.
3. C. Bachas, *Heterotic versus Type I*, hep-th/9710102;
   P. Vanhove, PhD dissertation École polytechnique France, CPTH-T602.0498.
4. C. Bachas and E. Kiritsis, Nucl. Phys. Proc. Suppl. **55B** (1997) 194.
5. W. Lerche, B. Nilsson and A. Schellekens, Nucl. Phys. **B289** (1987) 609;
   W. Lerche, B. Nilson, A. Schellekens and N. Warner, Nucl. Phys. **B299** (1988) 91;
   W. Lerche, Nucl. Phys. **B308** (1988) 102
6. W. Taylor, Phys. Rev. Lett. **177** (1996) 394;
   R. Dijkgraaf, E. Verlinde and H. Verlinde, Nucl. Phys. **B500** (1997) 43 and *Notes on Matrix and Micro Strings*, hep-th/9709107;
   S.J. Rey, Nucl. Phys. B502 (1997) 170;
   D.A. Lowe, Phys.Lett. B403 (1997) 243.
7. P. Ginsparg, *Applied Conformal Field Theory* Les Houches session XLIX, ed. E. Brézin et J. Zinn-Justin Elsevier Science Publisher B.V. 1989.
8. G. 't Hooft, Nucl. Phys. **B153** (1979) 141;
   O.J. Ganor, S. Ramgoolam and W. Taylor IV, Nucl. Phys. **B492** (1997) 191.
9. M.B. Green and P. Vanhove, Phys. Lett. **408B** (1997) 122.
10. M.B. Green and M. Gutperle, *D-particle bound states and the D-instanton measure*, hep-th/9711107.
11. M.B. Green, M. Gutperle and P. Vanhove, Phys. Lett. **409B** (1997) 177.
12. M.B. Green, J.H. Schwarz and E. Witten *Superstring theory*, Appendix 9.A, Cambridge Univ. Press. (1987).
13. S. Elitzur, A. Giveon, D. Kutasov and E. Rabinovici, *Algebraic Aspects of Matrix Theory on T^d*, hep-th/9707217.

ON DOMAIN WALL IN MQCD

ANASTASIA VOLOVICH[†]
LPTHE, Universités Paris VI - Paris VII
4 place Jussieu, F-75252 Paris Cedex 05, France

Recently, many interesting results about supersymmetric gauge theories in different dimensions have been obtained by embedding gauge theories in string/M-theory and using properties of the latter to study the former [1]. Witten used this method to explore the minimal $N = 1$ model with an $SU(n)$ vector multiplet in four dimensions [2]. He showed how for this model some of the outstanding properties of the ordinary QCD such as confinement, a mass gap and spontaneous breaking of a discrete chiral symmetry can be approached from M-theory point of view.

To describe the M5-brane version of QCD (MQCD) [2], let us start from the brane configuration in type IIA superstring theory with space-time coordinates $(x^0, x^1, ..., x^9)$ and study a configuration arising from n D4-branes suspended between two NS5-branes located at $x^6 = 0$ and $x^6 = S_0$ [2]–[3]. D4-branes world-volumes occupy $(x^0, x^1, x^2, x^3, x^6)$, with $0 \leq x_6 \leq S_0$, NS5-brane's world-volume is spanned by $(x^0, x^1, x^2, x^3, x^4, x^5)$ and another NS'5-brane's by $(x^0, x^1, x^2, x^3, x^7, x^8)$, where S_0 is an arbitrary length. Then the world-volume theory on D4s is four dimensional Super Yang Mills with gauge group $SU(n)$ and $N = 1$ supersymmetry. Elevating to the M-theory picture by adding the coordinate x^{10} makes possible a solution of the theory as follows [2]. Reinterpreted as a brane configuration embedded in 11d, the entire brane configuration corresponds to a single smooth M5-brane with world-volume $R^{1,3} \times \Sigma$, where Σ is a Riemann surface, embedded in three-dimensional space Y with coordinates $v, w, t = e^{-s}$ as $vw = \zeta,\ v^n = t$, where $v = x^4 + ix^5,\ w = x^7 + ix^8,\ s = x^6 + ix^{10}, 0 \leq x^{10} \leq 2\pi$ and ζ is a complex constant. Analyzing the symmetries, one can notice [2] that Z_n symmetry: $t \to t,\ v \to v, w \to e^{2\pi i/n} w$ is only symmetry at infinity, which doesn't leave the first equation defining Σ invariant. Thus [2], this symmetry is spontaneously broken and the theory has n distinct vacua, specified by the curves $w = exp(2\pi i/n)\zeta v^{-1}, t = v^n$.

A consequence of the spontaneously broken chiral symmetry is that there can be a domain wall separating different vacua. BPS–saturated domain walls in four dimensional supersymmetric gauge theories have been considered in [6]. Witten

[†]On leave from Moscow State University and L. D. Landau Institute for Theoretical Physics, Moscow, Russia.

L. Baulieu et al. (eds.), Strings, Branes and Dualities, 471–474.

has suggested that a BPS- saturated domain wall exists in MQCD and that it can be represented as a supersymmetric three-cycle in the sense of Becker *et al* [4]-[5] with a prescribed asymptotic behavior. The aim of this note is to study the Witten's conjecture. First, we derive a differential equation for the supersymmetric three-cycle, then we investigate solutions of this equation. Details can be found in [7].

The domain wall is described [2] as an M-theory fivebrane of the form $\mathbf{R}^3 \times S$, where $\mathbf{R}^3$ is parameterized by x^0, x^1, x^2 and S is a three-surface in the seven manifold $\tilde{Y} = \mathbf{R} \times Y$, here $\mathbf{R}$ is the copy of x^3 direction. Near $x^3 = -\infty$, S should look like $\mathbf{R} \times \Sigma$, where Σ is the Riemann surface defined by $w = \zeta v^{-1}$, $t = v^n$. Near $x^3 = +\infty$, S should look like $\mathbf{R} \times \Sigma'$, where Σ' is the Riemann surface of an "adjacent" vacuum, defined by $w = exp(2\pi i/n)\zeta v^{-1}, t = v^n$. We have to find such S which is a supersymmetric three-cycle in $\tilde{Y}$ with the described asymptotic behavior.

A supersymmetric cycle is defined by the property that the world-volume theory of a brane wrapping around it is supersymmetric. To study supersymmetric cycles one uses the concept of calibration [5]. A calibration is a closed p-form Φ on a Riemannian manifold of dimension n such that Φ has comass 1. Submanifolds for which there is equality are said to be calibrated by Φ. The property of calibrated submanifold to have the least volume in its homology class provides a natural geometrical interpretation of the BPS bound for D-branes wrapped around such submanifolds, with the calibrated submanifolds corresponding to BPS-states, which saturate the bound.

In the case of domain walls in MQCD [2] one deals with a seven dimensional flat manifold of G_2 holonomy and with the associative calibration. Just as in Calabi-Yau case in the case of G_2 holonomy manifold there is a canonical 3 form Φ and its Hodge dual $*\Phi$ which are covariantly constant [9]. If we choose the local veilbein so that the metric on $\tilde{Y}$ is $\sum_{i=1}^{n} e_i \otimes e_i$, locally the G_2 invariant 3-form Φ can be written as [10]

$$\Phi = e_1 \wedge e_2 \wedge e_7 + e_1 \wedge e_3 \wedge e_6 + e_1 \wedge e_4 \wedge e_5 + e_2 \wedge e_3 \wedge e_5 -$$

$$e_2 \wedge e_4 \wedge e_6 + e_3 \wedge e_4 \wedge e_7 + e_5 \wedge e_6 \wedge e_7.$$

A supersymmetric three-cycle S in $\tilde{Y}$ is one for which the pullback of this three-form is a constant multiple of the volume element [5]. This condition is written below as a partial differential equation in terms of the Poisson brackets.

Let us choose $v, \bar{v}, x^3$ as coordinates on S. The 3-cycle S is embedded in $\tilde{Y}$ as

$$w = w(x^3, v, \bar{v}), \qquad s = s(x^3, v, \bar{v}). \tag{1}$$

Here functions s and w have asymptotic behavior of the form

$$w(+\infty, v, \bar{v}) = e^{\frac{2\pi i}{n}} \zeta v^{-1} \equiv e^{\frac{2\pi i}{n}} (c(x^4, x^5) + id(x^4, x^5)), \tag{2}$$

$$w(-\infty, v, \bar{v}) = \zeta v^{-1} \equiv c(x^4, x^5) + id(x^4, x^5), \tag{3}$$

$$s(\pm\infty, v, \bar{v}) = -n \ln v \equiv a(x^4, x^5) + ib(x^4, x^5). \tag{4}$$

We denote $x^6 = A$, $x^7 = C$, $x^8 = D$, $x^{10} = B$ with A, B, C and D being functions of x^3, x^4, x^5 and set

$$e_1 = dx^{10}, e_2 = dx^5, e_3 = dx^3, e_4 = dx^7, e_5 = dx^4, e_6 = dx^6, e_7 = dx^8, \tag{5}$$

then we can prove the following

Lemma.[7] The surface S embedded in $\tilde{Y}$ as (1) is a supersymmetric three-cycle iff A, B, C, D as functions of x^3, x^4, x^5 satisfy the following non-linear partial differential equation

$$1 + \{A, B\}_{(1,2)} + \{C, D\}_{(1,2)} + \{B, D\}_{(1,3)} +$$

$$\{B, C\}_{(2,3)} + \{C, A\}_{(1,3)} + \{A, D\}_{(2,3)} = \sqrt{g}, \tag{6}$$

where we denoted $x^4 = y_1$, $x^5 = y_2$, $x^3 = y_3$ and defined the Poisson brackets as

$$\{f, g\}_{(i,j)} = \frac{\partial f}{\partial y_i}\frac{\partial g}{\partial y_j} - \frac{\partial g}{\partial y_i}\frac{\partial f}{\partial y_j}.$$

Here $f = f(y_1, y_2, y_3)$, $g = g(y_1, y_2, y_3)$, $i, j = 1, 2, 3$ and g is the determinant of the induced metric

$$g_{ij} = \delta_{ij} + \partial_i A \partial_j A + \partial_i B \partial_j B + \partial_i C \partial_j C + \partial_i D \partial_j D. \tag{7}$$

We will consider here the case $n = 2$ and ζ being a real number. To solve equation (6) with boundary conditions (2)–(4) the following ansatz [7] can be used

$$A = a(y_1, y_2) + \sum_{k=1}^{\infty} \gamma^{2k}(y_3) a_{2k}(y_1, y_2), \tag{8}$$

$$B = b(y_1, y_2) + \sum_{k=1}^{\infty} \gamma^{2k}(y_3) b_{2k}(y_1, y_2), \tag{9}$$

$$C = \beta(y_3) c(y_1, y_2) + \sum_{k=1}^{\infty} \gamma^{2k}(y_3) c_{2k}(y_1, y_2), \tag{10}$$

$$D = \beta(y_3) d(y_1, y_2) + \sum_{k=1}^{\infty} \gamma^{2k}(y_3) d_{2k}(y_1, y_2), \tag{11}$$

here

$$\beta(y_3) = \frac{e^{y_3} - e^{-y_3}}{e^{y_3} + e^{-y_3}}, \quad \gamma(y_3) = \frac{1}{e^{y_3} + e^{-y_3}}.$$

The boundary conditions (2-4) are satisfied for the above ansatz. In order to solve equations (6) we have to find functions $a_{2k}(y_1, y_2)$, $b_{2k}(y_1, y_2)$, $c_{2k}(y_1, y_2)$, $d_{2k}(y_1, y_2)$. Let us require vanishing of the coefficients in front of γ^k for all k. This will lead us to the system of recursive differential equations for a_{2k}, b_{2k}, c_{2k}, d_{2k}, moreover we will have only one equation for a_{2k}, b_{2k}, c_{2k}, d_{2k} for each k, thus we can assume some relations between these coefficients, for example in [7] we set

$c_{2k} = d_{2k} = 0$. For the domain wall to be in the universality class of SQCD, S must be invariant under $U(1)$ symmetry [2]: $t \to e^{in\delta}t$, $v \to e^{i\delta}v$, $w \to e^{-i\delta}w$, so we have to assume $a_{2k} = a_{2k}(y_1^2 + y_2^2)$, $b_{2k} = b_{2k}(y_1^2 + y_2^2)$, $c_{2k} = c_{2k}(y_1^2 + y_2^2)$, $d_{2k} = d_{2k}(y_1^2 + y_2^2)$ this is under current investigation [8].

In this note we have reviewed the derivation of the equation for the domain wall in MQCD and discussed some properties of the solution of this equation. The problem of the rigorous proof of the existence of the domain wall solution of the equation requires a further consideration [8]. Recently a different approach was suggested in [11] where an explicit intersecting fivebrane configuration was found and it was interpreted as a domain wall in MQCD. Another very interesting approach to the problem of domain walls in MQCD has been explored in [12].

Acknowledgements

I would like to express my gratitude to the organizers of this summer institute for invitation, financial support and their intense effort to organize an extremely interesting school. I would like to thank Laboratoire de Physique Theorique et Hautes Energies and especially L. Baulieu for invitation and kind hospitality and acknowledge the support of the CNRS grant.

References

1. A. Hanany and E. Witten. (1997) Type IIB Superstrings, BPS Monopoles, and Three-Dimensional Gauge Dynamics, *Nucl. Phys.*, **B492** p. 152.
2. E. Witten, (1997) Branes and the dynamics of QCD, hep-th/9706109.
3. S. Elitzur, A. Giveon, D. Kutasov. (1997) Branes and $N = 1$ Duality in String Theory, *Phys. Lett.* **B400** p. 269.
4. K. Becker, M. Becker and A. Strominger. (1995) Fivebranes, membranes and non-perturpative string theory, *Nucl. Phys.* **B456** p. 130.
5. K. Becker, M. Becker, D. R. Morrison, H. Ooguri, Y. Oz, Z. Yin. (1996) Supersymmetric Cycles in Exceptional Holonomy Manifolds and Calabi– Yau 4–folds, *Nucl. Phys.*, **B480** p. 225.
6. G. Dvali and M. Shifman. (1997) Domain Walls In Strongly Coupled Theories, hep-th/9706089; A. Kovner, M. Shifman, and A. Smilga, Domain Walls in Supersymmetric Yang-Mills Theories, hep-th/9706089.
7. A. Volovich. (1997) Domain Wall in MQCD and Supersymmetric Cycles in Exceptional Holonomy Manifolds, hep-th/9710120.
8. A. Volovich. (1998) to appear.
9. D. Joyce. (1996) Compact Riemannian 7-Manifolds with Holonomy G_2, I;II, *J. Diff. Geom.* **43** p. 291; 329.
10. S. L. Shatashvili and C. Vafa. (1994) Superstrings and Manifolds of Exceptional Holonomy, hep-th/9407025.
11. A. Fayyazuddin and M. Spalinski. (1997) Extended Objects in MQCD, hep-th/9711083.
12. J. Sonnenschein, S. Theisen, private communications.

Appendix

KIDS WALKING ON A D-BRANE

Interviews by Nina Ilieva-Douglas

Hot spring-summer-time in Corsica. Quiet blue sea, the smell of algae, distant jingle bells from goats grazing on the rocks, and two half-stray dogs shaking out their ticks and yawning in front of the conference hall.

Two kids walk by and look into the hall. There are bunches of scientists and their students, scribbling endless equations on the board, speaking a weird language. But there was just one thing those kids wanted to know, a very simple thing:

How can a kid walk on a D-brane?

This is how the scientists and their students try to answer.

Paul Townsend

First of all, the question suggests that the kids would feel different walking on a D-brane than walking on other kinds of branes. Let's find out how they feel walking on any brane, and then on a D-brane.

Perhaps we could describe how it is to walk on water. It is some kind of analogy, because it is a surface, and that's the first point. If you look at the sea, you have a three dimensional space and a two dimensional surface. Now, that's some kind of analogy, because you can have waves on them. The first thing, I suppose, once you start to walk on them, you would create waves around you. It would be like walking on a trampoline. This is the closest I can get.

Two D-branes can have strings stretched between them, like hammocks between two trees. They are supposed to represent fluctuations like waves on the water. The two trees are two one-branes, and they have a tension. Left to themselves, they would collapse, so you'll have to keep them stretched between something to keep them from collapsing. You can think of a piece of elastic string. If you stretch it, it elongates, and if you let it go, it becomes shorter. But there comes a point where it doesn't become any shorter, while with this kind of string, they would become shorter and shorter, and eventually disappear. They would perhaps explode and turn into something else, create some particles.

I suppose you can have a D-brane that stretches across the universe. It would be a kind of cosmic D-brane. It would have to be a one brane, or two brane, or

L. Baulieu et al. (eds.), Strings, Branes and Dualities, 477–483.

zero brane, because, after all, there are only so many dimensions available to us. If you live in a higher dimension, you see many more possibilities. So if the children get to a higher dimensional space, I think it is going to be difficult for us to follow their adventure.

Emil Martinec

They have to slide their feet along the surface, but they cannot lift their feet up, because of the suction. The surface of the D-brane seems to be covered with some sticky stuff like oil. Their arms are free to move around, but they feel like they have magnets on their feet. The D-brane is something like a gigantic octopus with suckers everywhere. If they find another D-brane which intersects the one they are on, they can slide their feet into the other one, but they cannot lift their feet to put them on the other D-brane. And, if another D-brane passes them in a hand's distance, and they feel tempted to touch it, their hands will be stuck to the other D-brane too.

Lenny Susskind and Willy Fischler

Susskind:
The kids don't walk. They hold on to open strings attached to the D-brane, and the open strings slide along the D-brane, tossing them from side to side. That is extremely dangerous, because the strings move almost with the speed of light. A string can chop off your leg in a single stroke. If the kids get crushed into another open string, they can be annihilated and turn into black body radiation – that is a form of radiation, which is extremely dangerous. Well, these strings are not shoe laces. They have enormously high tension. They are strong and they can support enormous weights. Watch out if they snap!

Fischler:
The strings are made of other strings. They are very thin, but they are made from lots of closed strings themselves, which could bounce off. Bounce, and bounce, and bounce. That's about it.

Susskind:
If I was one of those kids, I would try to follow one of those strings, to get from one D-brane into another. If they are lucky, they might be traveling on a string that collides with another string and forms a closed string, and then they can leap across another brane, bounce off, and get reconnected to another brane! This is exciting, but dangerous. They can fall into the vacuum surrounding the brane, and they can get lost in eleven dimensions. The more dimensions you have, the more possible directions you can move. Eleven dimensions are a lot of dimensions to find your way in. It's too much freedom, and you can easily get lost.

Juan Maldacena

If the kids jump onto a single D-brane, they might bounce back. But a big pile of D-branes looks from the outside like a black hole.

When someone falls into a pile of D-branes, he instantly divides into two images. From each person you get two copies, and each copy would not realize that there was another. Each copy of the person would feel like he is a unique personality with his own thoughts, but in fact, he would be a mirror image of the copy. The two copies of each person would start to move in different directions. They would have more and more different thoughts, and become more and more different, but eventually they would die. The only way for them to be saved is to be reunited at some point, but that's probably not possible.

Then, the only possibility is to reconstruct them from the ashes! If the black hole evaporates, and you manage to gather all the particles from the evaporation, you can reconstruct them, but that's a very hard thing to do.

Michael Green

A D-brane would be very oppressive — very hot and very sticky. The kids' feet would be covered by a gooey slime of vibrating strings, which move around with their end-points fixed on the surface. Some of these strings are very long and tangled, zooming past the kids and enveloping them in a ghastly ghoulish web. I imagine they would be very scared – they would hardly be able to see where they were going and there would be lots of flashing sparks as the strings emit photons. Although many of these strings have both ends fixed to the surface on which the kids are walking, some of the strings have one end free, wiggling around like gigantic worms rising out of the earth. Sometimes a loop of closed string will break away from one of the writhing open strings and hurtle off into space. As the kids look up they see lots of these closed strings moving in the distance. Unfortunately, there are also closed strings hurtling towards them from some other region of space and hitting the surface of their D-brane. These are very energetic, and being hit by them is not very pleasant. Even worse – so terrifying that it hardly bears thinking about — there is the possibility of a collision with an anti D-brane and consequent utter and complete annihilation.

Looking up into space, the kids might see another D-brane pass close by and witness a spectacular shower of closed strings travelling between it and the D-brane they are on. They could perhaps grab a coil of closed string just as it splits off near them and hitch a ride on it to land on the other D-brane. Of course, there is a very real danger that this closed string, together with the kids, would drift in an unexpected direction. In this case the kids would might end up traveling through space-time on a fundamental closed string at the mercy of the elements with the grim prospect of eventually colliding with the E8 wall at the boundary of the Universe!

Herman Verlinde

The kids cannot walk on the D-brane like we walk on the earth. They have to be careful not to push themselves up and to move as little as possible. Like being on the moon, where you are not so heavy. They are almost weightless on the D-brane, because there is no gravitational force there. The surface is smooth, it's almost like sliding on ice. You cannot see your feet. If the kids are very, very small, they could see the strings. The strings are pretty fast moving things which they suddenly see shooting by, or moving slow, among their feet. They have to be careful not to step on a string. They are something like elastic bandages, and they move very fast. They appear and disappear suddenly and they send out flashes of light. The two end points of the strings are charged, one positively and one negatively. So if they get caught in between the two end points of the string, they could get an electric shock!

There are strings which go from one D-brane to the other, so they can grab a closed string, wrap it around one of them, and use it to go over. But they have to do it fast, because the string gets longer while they travel, and eventually breaks. So they have to do it in one smooth motion, just in one jump.

Erik Verlinde

They have to keep walking. You shouldn't stand still. If you stand still, it gets hot. The D-brane is a hot object. How hot is it? Depends on what kind of D-brane it is, but it can be very hot if you stay in the same place. But if you keep walking, then it cools off. Therefore, you have to move very quickly on a D-brane. So, keep running. If you are barefoot, you have to slide your feet. Although, you have to be careful. Sometimes there are objects on these D-branes, you may trip on something. There are some kind of ropes attached to the D-brane – these are the strings. They pop up and they disappear again. So you really have to watch out. You have to look for strings, which go from one D-brane to the other. If you see one of those you should grab on it and climb to the other one.

The D-branes are reddish, because they are hot objects. But these are the least hot, and there are also orange D-branes, yellow D-branes and bluish D-branes, which are the most dangerous. And sometimes, while walking on a D-brane, holes may appear in it, and if you slide through them, you can end up on another planet, or in another universe, without even noticing.

Robert Dijkgraaf

The D-brane can be of various dimensions. The easiest to imagine is a D-brane which is three dimensional. At this moment the kids could be sitting on it. If they drop on a two dimensional D-brane, they would have to squeeze into a plane, and even to a line, if they are on a one dimensional D-brane. In higher dimensional D-branes the kids would feel like a piece of paper, moving in an ordinary three dimensional space. Someone could take you and turn you upside-down, or switch your left side with your right side. Or, the kids could encounter intersecting branes. Imagine a three dimensional world, intersecting with a big plane, like a big bowl. So you are moving in your world, hit the wall, and suddenly start moving into a different world. If you have the kids live on a piece of paper, and glue this piece of paper onto another piece of paper, and let them move across the joint, they can enter the other world, like switching from one drawing into another drawing.

If you live on a D-brane, you got caught on it. But, if two kids live on a D-brane, and they meet, at the moment they meet, something strange happens to them, because they no longer feel glued to the surface of the D-brane, but can move off to the rest of the world. Then they can actually see the D-brane, like being on an airplane and seing the earth beneath you. The forces on the D-brane try to catch them, but cannot reach them. The forces are invisible, but there are also visible tenticles, which pop up from the surface of the D-brane and try to glue to them. If a string glues to them, it would stretch and stretch like an octopus, pulling them back to the D-brane.

One of the bizarre things is, you can have D-branes in D-branes in D-branes. So, the kids are on a two-dimensional D-brane, they escape, and find themselves on a three-dimensional D-brane, They escape and find themselves on a four-dimensional D-brane ...

Michael Douglas

How to move on a D-brane? Well, it depends who you are. If you are another D-brane, then you have strings sticking out of you in all directions. You move as if you are jumping rope, and the strings make circles around you. The D-brane is actually transparent except for all these fuzzy strings on it, like a sheet of glass with hair sprouting on it. You can shoot right through, but then the strings pull you back.

There are little D-branes and big D-branes. A D-brane like a long pole can be used like a railway line, with strings stretching all along the direction you are going. A D-brane can have more dimensions, but you can't see more than three dimensions at a time, so it would be something like a kaleidoscope, as it spins around. Sometimes it would look like a pole, sometimes like a sheet, or like a ball, and it can be growing and shrinking, spinning around, not because it is changing, but because you are seeing different sections of it.

Of course, a D-brane is much smaller than an atom, and you can't literally be so small, but if you are made out of D-branes, the D-branes may be very small, but you'll be very similar to what you were, just made out of some other kind of matter. There could be other beings walking on the D-brane, and maybe by walking on the D-brane you'll turn into one of them. Maybe their atoms are made out of D-branes. Maybe this is a different phase of the universe. The universe can have different regions, in different phases. One phase is our phase, and another phase might be five dimensional, eight dimensional, or whatever. In different phases, the matter can have different properties. Maybe there is a phase like ours, but with D-branes flying around, and other objects. Maybe a person can get himself into the other phase by having his matter transmuted into the appropriate matter for that phase.

Vijay Balasubramanian

Suppose there is something on a D-brane which is very close to the horizon of a black hole, and the adventurers have to retrieve it. They can try to throw a fishing line, to hook the thing up and bring it back. But they are going to find out that they need a tremendous force, because there are lots of funny forces near a black hole. If you are in a stationary orbit, as you come close to the horizon, you'll need an infinite force to pull the thing back, even if it is only an inch from the black hole. If you lower one person with the fishing line, he would age less than the one holding the line. If he falls in, he'll be smashed into pieces. But he isn't lost in a sense, because the information about him is radiated and can be gathered back together. Only his original self is lost.

There is a story in India: there is a saint, whose wife comes to him and asks him: "I have heard that when you die, your original self is lost. Is that the truth?" "Yes", he said. "Oh, how terrible", she said, "how can you live with a thing like that?" He said, "Bring me a bowl of water and some salt." So she brought them and he mixed the salt with the water. Then he asked her to taste in the middle – it tasted salty; and on the edge – it tasted salty. He said, "Give me the salt back!" and she couldn't. "The salt exists", he said, "just it is not really accessible". So, if you fall into a black hole, which becomes a D-brane, you'll become a little bump on the D-brane, but the information about you will be around as radiation.

So the question is, what does it mean to be saved. By getting the radiation you can determine, molecule by molecule, what it was that hit the D-brane. Then, if you can reassemble all the molecules in all the right places, you can reconstruct the person. It would take a very long time, almost infinite. And, if they don't wait long enough, maybe they'll reconstruct his nose a little bit crooked, because they didn't pick up all the correlations.

******

Hearing all this, the kids said nothing. They looked at each other, walked past the sleeping dogs by the door, and started walking towards the sea.

–Did you get it?

–I think yes. But why were they so different?

–There must be a reason, I guess, the taller one said.

–But you know what?, the other one looked him in the eyes. Now I know something for sure.

–To try to walk on a D-brane now?

–Yes, let's just do it before we turn into physicists.

The two kids walked away, jumping from stone to stone. One of the half stray dogs followed them lazily.

Note by the interviewer.

I am working on a children's book set in the world of D-branes, black holes and strings. During and after the Cargese summer school, I talked to some of the participants who were willing to discuss with me some of the recent developments in string theory in a popular way. The interviews presented here are answers to one of my questions.

Thanks to everybody who combined imagination with knowledge which can be inspiring for everybody.

LIST OF SPEAKERS

Antoniadis Igniatos École Polytechnique, CPT, 91128 Palaiseau, France, antoniad@cpht.polytechnique.fr

De Wit Bernard Utrecht University, Institute for Theoretical Physics, Princetonplein 5, 3508 TA Utrecht, The Netherlands, B.deWit@fys.ruu.nl

Dijkgraaf Robbert University of Amsterdam, Faculty of Maths & Computer Science, Plantage Muidergracht 24, 1018 TV Amsterdam, The Netherlands, rhd@fwi.uva.nl

Eguchi Tohru University of Tokyo, Department of Physics, Bunkyo-ku, Tokyo 113, Japan, eguchi@danjuro.phys.s.u-tokyo.ac.jp

Green Michael Cambridge University, DAMTP, Cambridge CB3 9EW, United Kingdom, M.B.Green@damtp.cam.ac.uk

Harvey Jeff University of Chicago, Enrico Fermi Institute, 5640 Ellis Avenue, Chicago, IL 60637, U.S.A., harvey@poincare.uchicago.edu

Julia Bernard École Normale Supérieure, Laboratoire de Physique Théorique, 24, rue Lhomond, 75231 Paris Cedex 05, France, Bernard.Julia@peterpan.ens.fr

Kounnas Costas CERN, Theory Division, 1211 Geneva 23, Switzerland, kounnas@nxth04.cern.ch

Louis Jan Martin–Luther–Universität Halle–Wittenberg, FB Physik, D-06099 Halle, Germany, jlouis@hermes.physik.uni-halle.de

Maldacena Juan Rutgers University, Department of Physics and Astronomy, P.O. Box 849, Piscataway, NJ 08855-0849, U.S.A., malda@physics.rutgers.edu

Martinec Emil University of Chicago, Enrico Fermi Institute, 5640 Ellis Avenue, Chicago, IL 60637, U.S.A., ejm@theory.uchicago.edu

Olive David University of Wales Swansea, Department of Physics, Swansea, SA2 8PP, Wales, United Kingdom, D.I.Olive@swansea.ac.uk

Polyakov Alexander Princeton University, Physics Department, Jadwin Hall, Princeton, NJ 08544-1000, U.S.A., polyakov@puhep1.princeton.edu

Susskind Lenny Stanford University, Physics Department, Stanford, CA 94305, U.S.A., susskind@sewerrat.stanford.edu

Townsend Paul Cambridge University, DAMTP, Cambridge CB3 9EW, United Kingdom, P.K.Townsend@damtp.cam.ac.uk

Verlinde Erik CERN, Theory Division, 1211 Geneva 23, Switzerland, verlinde@nxth04.cern.ch

L. Baulieu et al. (eds.), Strings, Branes and Dualities, 485–486.

Verlinde Herman University of Amsterdam, Institute for Theoretical Physics, Valckenierstraat 65, 1018 XE Amsterdam, The Netherlands, verlinde@phys.uva.nl

LIST OF PARTICIPANTS

Abou Zeid Mohab Queen Mary & Westfield College, Physics Department, Mile End Road, London E1 4NS, United Kingdom, M.Abouzeid@qmw.ac.uk

Aganagic Mina California Institute of Technology, Pasadena, CA 91125, U.S.A., mina@theory.caltech.edu

Antonov Alexander Université P. et M. Curie, Laboratoire de Physique Théorique et Hautes Énergies, Boite 126, 4 place Jussieu, 75252 Paris Cedex 05, France, antonov@lpthe.jussieu.fr

Argurio Ricardo Université Libre de Bruxelles, Service de Physique Théorique, Campus Plaine, C.P.225, Boulevard du Triomphe, B-1050 Bruxelles, Belgium, rargurio@ulb.ac.be

Balasubramanian Vijay Princeton University, Physics Department, Jadwin Hall, Princeton, NJ 08544-1000, U.S.A., vijayb@phoenix.princeton.edu

Bars Itzhak University of Southern California, Department of Physics and Astronomy, Los Angeles, CA 90089-0484, U.S.A., bars@physics1.usc.edu

Belavin Alexander Landau Institute For Theoretical, GSP-1, 117940 Moscow, V-334, Russia, belavin@itp.ac.ru

Bershadski Michael Harvard University, Lyman Laboratory of Physics, Cambridge, MA 02138, U.S.A., bershad@string.harvard.edu

Brodie John Princeton University, Physics Department, Jadwin Hall, Princeton, NJ 08544-1000, U.S.A., jhbrodie@phoenix.princeton.edu

Carter Brandon D.A.R.C., Observatoire de Paris, 92 Meudon, France, carter@obspm.fr

Costa Miguel Cambridge University, DAMTP, Cambridge CB3 9EW, United Kingdom, M.S.Costa@damtp.cam.ac.uk

Craps Ben Katholieke Universiteit Leuven, Instituut voor Theoretische Fysica, Celestijnenlaan 200D, B-3001 Leuven, Belgium, Ben.Craps@fys.kuleuven.ac.be

De Wild Propitius Mark Université P. et M. Curie, Laboratoire de Physique Théorique et Hautes Énergies, Boite 126, 4 place Jussieu, 75252 Paris Cedex 05, France, mdwp@lpthe.jussieu.fr

Derix Martijn Katholieke Universiteit Leuven, Instituut voor Theoretische Fysica, Celestijnenlaan 200D, B-3001 Leuven, Belgium, Martijn.Derix@fys.kuleuven.ac.be

L. Baulieu et al. (eds.), Strings, Branes and Dualities, 487–489.

Distler Jacques University of Texas at Austin, Physics Department, Theory Group, Austin, TX 78712, U.S.A., distler@golem.ph.utexas.edu

Fischler Willy University of Texas at Austin, Physics Department, Theory Group, Austin, TX 78712, U.S.A., fischler@utaphy.ph.utexas.edu

Fraser Chrisophe University of Wales Swansea, Department of Physics, Swansea, SA2 8PP, Wales, United Kingdom, c.fraser@swansea.ac.uk

Gabriel Claude Université de Mons-Hainaut, Mécanique et Gravitation, 15 avenue Maistriau, B-7000 Mons, Belgium, gabriel@sun1.umh.ac.be

Garcia Raquel Imperial College, Theoretical Physics Group, Blackett Laboratory, London SW7 2BZ, United Kingdom, garciars@ic.ac.uk

Gregori Andrea CERN, Theory Division, 1211 Geneve 23, Switzerland, agregori@vxcern.cern.ch

Grojean Christophe CERN, Theory Division, 1211 Geneve 23, Switzerland, Christophe.Grojean@cern.ch

Gukov Sergei Landau Institute For Theoretical, GSP-1, 117940 Moscow, V-334 Russia, gukov@itp.ac.ru

Gutperle Micahel Cambridge University, DAMTP, Cambridge CB3 9EW, United Kingdom, m.gutperle@damtp.cam.ac.uk

Helling Robert University of Hamburg, DESY, Luruper Chaussee 149, D-22761 Hamburg, Germany, helling@x4u2.desy.de

Hewson Stephen Cambridge University, DAMTP, Cambridge CB3 9EW, United Kingdom, sfh10@damtp.cam.ac.uk

Hoppe Jens Institute for Theoretical Physics, ETH Hönggerberg, CH-8093 Zürich, Switzerland, hoppe@itp.phys.ethz.ch

Houart Laurent Université Libre de Bruxelles, Service de Physique Théorique, Campus Plaine, C.P.225, Boulevard du Triomphe, B-1050 Bruxelles, Belgium, lhouart@ulb.ac.be

Jinzenji Masao University of Tokyo, Department of Physics, Bunkyo-ku, Tokyo 113, Japan, jin@kanquro.phys.s.u-tokyo.ac.jp

Kapustin Anton California Institute of Technology, Pasadena, CA 91125, U.S.A., kapustin@theory.caltech.edu

Kitanine Nikolai ENSLAPP, 46, allée d'Italie, 69007 Lyon, France, nkitanin@enslapp.ens-lyon.fr

Kraus Per California Institute of Technology, Pasadena, CA 91125, U.S.A., perkraus@theory.caltech.edu

Laroche Celine Université P. et M. Curie, Laboratoire de Physique Théorique et Hautes Énergies, Boite 126, 4 place Jussieu, 75252 Paris Cedex 05, France, laroche@lpthe.jussieu.fr

Michelson Jeremy University of California, Physics Department, Santa Barbara, California 93106, U.S.A., jeremy@cosmic1.physics.ucsb.edu

Nekrassov Nikita Harvard University, Lyman Laboratory of Physics, Cambridge, MA 02138, U.S.A., nikita@string.harvard.edu

Polhemus Gavin University of Chicago, Enrico Fermi Institute, 5640 Ellis Avenue, Chicago, IL 60637, U.S.A., g-polhemus@uchicago.edu

Polyakov Dimitri Institut des Hautes Etudes Scientifiques, 35, Route de Chartres, F-91440 Bures-sur-Yvette, France, polyakov@ihes.fr

Pouliot Philippe University of California, Physics Department, Santa Barbara, CA 93106, U.S.A., pouliot@sarek.physics.ucsb.edu

Ragoucy Eric CERN, Theory Division, 1211 Geneve 23, Switzerland, eric.ragoucy@cern.ch

Roose Frederik Katholieke Universiteit Leuven, Instituut voor Theoretische Fysica, Celestijnenlaan 200D, B-3001 Leuven, Belgium, frederik.roose@fys.kuleuven.ac.be

Rozenberg Alexander New York University, Physics Department, 4 Washington Place, New York, N.Y. 10003, U.S.A., sasha.rozenberg@nyu.edu

Russo Jorge Imperial College, Theoretical Physics Group, Blackett Laboratory, London SW7 2BZ, United Kingdom, j.russo@ic.ac.uk

Rychenkova Paulina Cambridge University, DAMTP, Cambridge CB3 9EW, United Kingdom, pr201@damtp.cam.ac.uk

Sadov Vladimir Institute for Advanced Study, School of Natural Science, Princeton, NJ 08540, U.S.A., sadov@sns.ias.edu

Schroeder Germar Max-Planck-Institut fuer Gravitationsphysik, Albert-Einstein-Institut, Schlaatzweg 1, D-14473 Potsdam, Germany, germar@aei-potsdam.mpg.de

Schwager Stefan Universität München, Sektion Physik, Theresienstrasse 37, 80333 Muenchen, Germany, Stefan.Schwager@Physik.Uni-Muenchen.de

Silva Sebastian École Normale Supérieure, Laboratoire de Physique Théorique, 24, rue Lhomond, 75231 Paris Cedex 05, France, Sebastian.Silva@lpt.ens.fr

Sorokin Dimitri Kharkov Institute of Physics and Technology, Kharkov, 310108, Ukraine, dsorokin@linda.kipt.kharkov.ua

Strigazzi Piero Università degli Studi di Torino, Dipartimenti di Fisica, Via Giuria 1, 10125 Torino, Italy, strigazzi@to.infn.it

Tatar Radu University of Miami, Department of Physics, P.O. Box 248046, Coral Gables, FL 33124, U.S.A., tatar@phyvax.ir.miami.edu

Terras Veronique ENSLAPP, 46, allée d'Italie, 69007 Lyon, France, vterras@enslapp.ens-lyon.fr

Van Den Broeck Chris Katholieke Universiteit Leuven, Instituut voor Theoretische Fysica, Celestijnenlaan 200D, B-3001 Leuven, Belgium, Chris@tfdec1.fys.kuleuven.ac.be

Van Der Schaar Jan Institute for Theoretical Physics, Nijenborgh 4, 9747 AG Groningen, The Netherlands, schaar@th.rug.nl

Vanhove Pierre École Polytechnique, CPT, 91128 Palaiseau, France, vanhove@cpht.polytechnique.fr

Volovich Anastasia Moscow State University, Department for Theoretical Physics, 119899, Moscow, Russia, nastya@arevol.mian.su

Wiegmann Paul The University of Chicago, James Franck Institute, 5640 South Ellis Avenue, Chicago, IL 60637-1433, U.S.A., wiegmann@rainbow.uchicago.edu

Wynter Thomas Service de Physique Théorique, CEN-Saclay, 91191 Gif-sur-Yvette, France, wynter@spht.saclay.cea.fr

Yin Zheng University of California, Physics Department, Berkeley, CA 94720-7300, U.S.A., zyin@physics.berkeley.edu

INDEX

L. Baulieu et al. (eds.), Strings, Branes and Dualities, 491–493.

GPSR Compliance

The European Union's (EU) General Product Safety Regulation (GPSR) is a set of rules that requires consumer products to be safe and our obligations to ensure this.

If you have any concerns about our products, you can contact us on ProductSafety@springernature.com

In case Publisher is established outside the EU, the EU authorized representative is:

Springer Nature Customer Service Center GmbH
Europaplatz 3
69115 Heidelberg, Germany

Batch number: 10755284

Printed by Printforce, the Netherlands